U0315195

KINGS FOR BID WINNING —BID PROPOSALS

得标为王—方案篇

2013-2014

下册

龙志伟 编著
Edited by Long Zhiwei

| 办公建筑 | 医疗建筑 | 学校建筑 | 交通建筑 | 住宅建筑 |

| Office Building | Medical Building | School Building | Transportation Building | Residence |

广西师范大学出版社
· 桂林 ·

序 Preface

“物竞天择，适者生存”这句流传了几千年的警句在当今这个竞争激烈的时代已被奉为至理名言，“优胜劣汰”这一法则在建筑设计领域也同样发挥了指挥棒的作用。如何提高公司的业务水平及竞争力，使己方提出的设计方案赢得招标方及大众的青睐和认可进而得标，是每一个设计师和设计团队都必须深思的问题。一个能在众多方案中脱颖而出、独占鳌头的方案，不仅关系到提案的经济性，更与提案的原创性、创新性、合理性、实用性和完整性息息相关。一个富于想象又不脱离实际、富有创意又经济实用、彰显个性又贴近生活的独创性方案才是招标方心中的首选。

《得标为王——方案篇 2013-2014》是一本大型的设计方案集锦。该书收录了阿特金斯、斯蒂文·霍尔建筑师事务所、Jaspers-Eyers & Partners、3LHD、蓝天组、澳大利亚 SDG、C.F.Møller、UNStudio、深圳天方、北京殊舍等近百家国内外优秀建筑设计公司的知名设计方案。本书精选了度假休闲、商业建筑、酒店建筑、购物中心、文化艺术建筑、城市综合体、办公建筑、医疗建筑、学校建筑、交通建筑、住宅建筑等类别的方案近 200 个。

无论是理念的创新、思维方式及构思角度的转换，还是最新技术的运用、生态节能材料的使用，抑或是独特的造型与外观，它们既使该方案变得独特而唯一，也使之成为备受客户认可和推崇的新设计、新理念、新方案。本书指明了当今建筑设计领域智能、仿生、生态的设计新趋势，阐释了当今备受世人关注的绿色、低碳、以人为本的设计理念，反映了使用者生理上和心理上的需求。这些方案的实施，不仅将促使许多新形式、新类别的建筑的诞生，使人与社会、人与自然和谐发展，同时也将改善人们的生活方式。

“Survival of the Fittest in Natural Selection” has been claimed to be words of wisdom in this fiercely competitive age. “Survival of the Fittest” also plays a leading role in architectural design. How to boost competitiveness and improve professional skills, making the design scheme be favored and accepted by tenderers and the public so as to win the bid has become a thought-provoking issue to every designer and each design team as well. The design scheme, which stands out from numerous proposals, must be not only economical, but also originative, innovative, rational, practical and integrated. Only the most unique proposal that is imaginative, practical, creative, economical, characteristic and close to life can be the priority for tenderers.

Kings for Bid Winning – Bid Proposals 2013-2014 is a large collection of design schemes. Well-known design proposals from nearly 100 national and international renowned architectural design companies, such as Atkins, Steven Holl Architects, Jaspers-Eyers Architects, 3LHD, COOP HIMMELB(L)AU, Shine Design Group, C.F. Møller, UNStudio, Shenzhen TAF Architect and Beijing Shushe Architecture, are included. New projects of about 200 cases involve Resort, Commercial Building, Hotel, Shopping Center, Culture & Art Building, Urban Complex, Office Building, Medical Building, School Building, Transportation Building and Residence.

The innovative concept, thinking mode, design perspective, or the application of latest technology and ecological energy-saving materials, or distinct shape and appearance, make the proposal a special and unique new design, new concept, new scheme recognized and highly appraised by clients. The book implies the architectural design trend for intelligent, bionic and ecological design solutions, interprets a design concept of “Green, Low-carbon and People-oriented” and reflects physical and psychological demands of users. Implementing these proposals not only means the emerging of new forms and new types of buildings, a harmonious development between man and society, man and nature, but also improves people’s lifestyle.

Contents

目录

办公建筑 8 Office Building

波兰华沙尖顶 10 Warsaw Spire

法国让蒂伊ZAC de la Porte de Gentilly办公楼 18 ZAC de la Porte de Gentilly

斯洛文尼亚卢布尔雅那太阳能办公室 22 Solar Power Offices

奥地利格拉茨Crossroad办公楼 30 Crossroad Office

丹麦哥本哈根LM海港大门 36 LM Harbor Gateway

天津TEDA MSD H2低碳示范楼 48 TEDA MSD H2 Low Carbon Building

医疗建筑 56 Medical Building

挪威贝尔根Haraldsplass 医院扩展 58 Haraldsplass Hospital Extension

北京新安贞医院 66 Beijing New Anzhen Hospital

伊朗拉什特Pars综合医院 74 Pars General Hospital, Rasht

冰岛雷克雅未克Landspitali大学医院 78 Landspitali University Hospital of Iceland

丹麦新奥尔胡斯大学医院 86 New University Hospital in Aarhus

丹麦Viborg地区医院新急性治疗中心 96 New Acute Treatment Center of Viborg Regional Hospital

丹麦哥本哈根Villa Vita癌症中心 104 Villa Vita Cancer Center, Copenhagen, Denmark

克罗地亚斯普利特私人医疗中心 110 Polyclinic St

学校建筑 114 School Building

奥地利维也纳新应用艺术大学 116 The New University of Applied Arts

澳大利亚未来教室 122 Classroom of the Future

奥地利阿姆施泰滕校园 128 Amstetten School Campus

澳大利亚昆士兰大学高级工程楼 132 Advanced Engineering Building of the University of Queensland

澳大利亚悉尼科技大学主楼 136 Tower Skin

波哥大哥伦比亚国立大学新博士楼 144 New Doctorate's Building, National University of Colombia

交通建筑 146 Transportation Building

比利时布鲁塞尔机场连接大楼 148 Brussels Airport Connector, Brussels, Belgium

拉脱维亚里加国际机场 156 Riga Airport

江苏常州机场 164 Changzhou Airport

台湾高雄港游轮码头 174 Kaoshiung Port Cruise Terminal

台湾基隆游轮码头 182 Cruise Terminal, Keelung, Taiwan

丹麦奥尔胡斯DSB地区和公交总站总体规划 190 DSB-areas and Bus Terminal Masterplan in Aarhus

广东佛山东平新城交通枢纽中心 202 Foshan Dongping New City Mass Transit Center

台湾基隆港新客运大楼 208 New Keelung Harbor Terminal Building

挪威奥斯陆中心车站 214 Oslo Central Station

保加利亚索菲亚20地铁站 222 Metro Station 20

比利时沙勒罗伊消防站 226 Charleroi Fire Station

住宅建筑 236 Residence

阿联酋阿布扎比雪花大厦 238 Snowflake Tower

马来西亚布城滨水住宅楼 244 Putrajaya Precinct 4 Waterfront Development, Kuala Lumpur

上海黑森林高级住宅三期 252 Black Forest Residences Phase Ⅲ

法国格勒诺布尔138住宅 258 138 Dwellings

澳大利亚达尔文市Gateway广场大楼 266 Gateway Plaza Building

澳大利亚达尔文市海滨公寓 272 Oceanfront Apartments

澳大利亚达尔文市日出大厦 276 Sunrise Towers

北京温都水城国际公寓 280 Beijing Hot Spring Leisure City International Apartment

印度班加罗尔住宅 286 Houses at Bangalore

意大利卡蒂尼亚诺Della Volpe住宅 294 Della Volpe House

澳大利亚达尔文港Aspire公寓 302 Aspire Apartments

澳大利亚悉尼The Eliza豪华公寓 306 The Eliza

办公建筑 Office Building

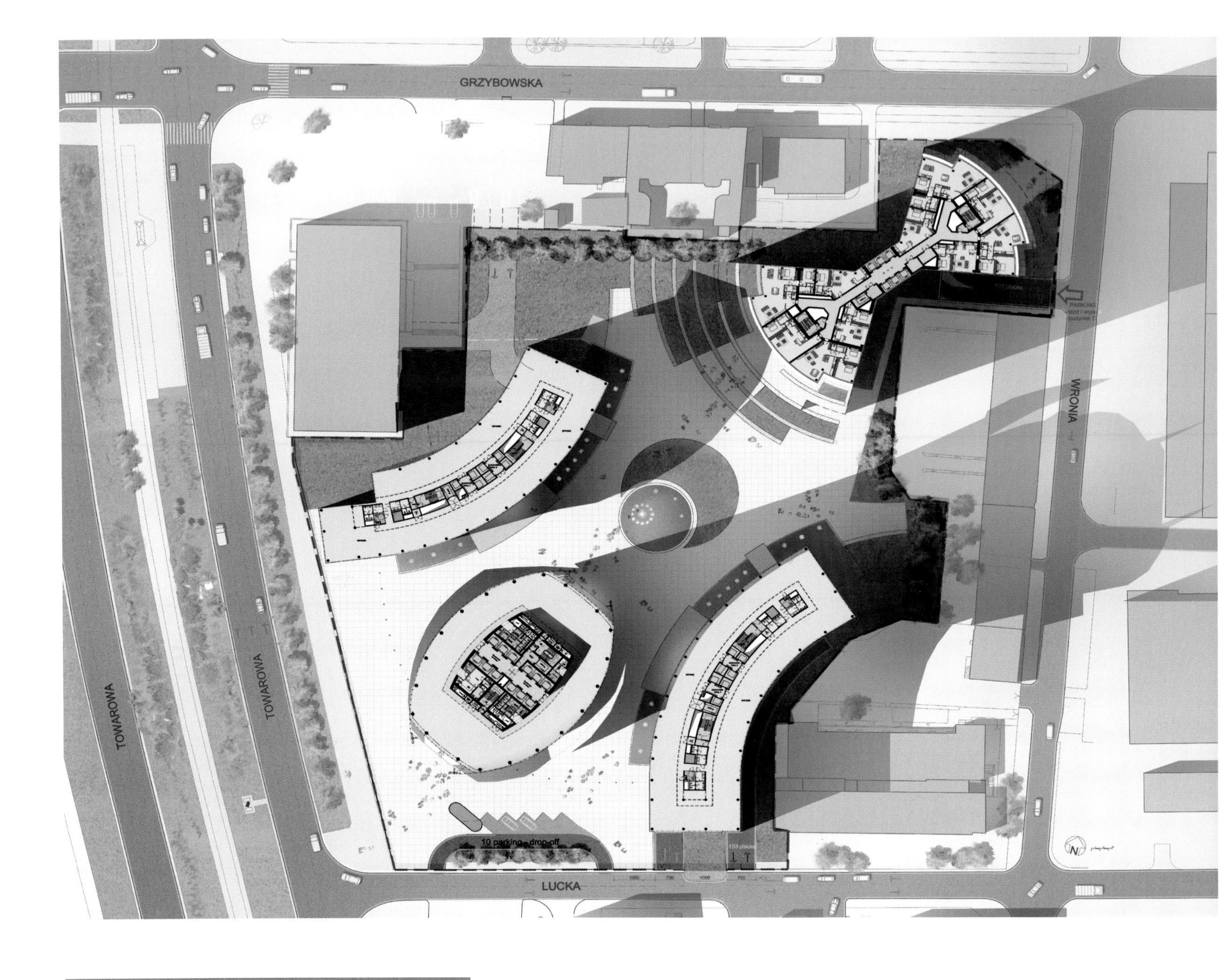

波兰华沙尖顶

Warsaw Spire

设计单位：Jaspers-Eyers Architects
开发商：Ghelamco
项目地址：波兰华沙

Designed by: Jaspers-Eyers Architects
Client: Ghelamco
Location: Warsaw, Poland

项目概况

项目靠近华沙的历史中心，位于著名的华沙文化科学宫和华沙起义博物馆之间。这座名为“华沙尖顶”的大厦，是现代办公大楼建筑愿景与城市空间布局新模式的独特组合，它既是华沙最高的办公楼，也将成为这个城市一个新的标志性建筑。

建筑设计

椭圆形的大厦曲线优美，玻璃立面好似披挂在楼身上的斗篷，盘旋上升的建筑线条和富于表现力的建筑形态，吸引了人们的眼球。建筑中心的楼层平面被两个玻璃外壳包裹着，延伸至顶部。螺旋状的形态强调了建筑的纤长，玻璃材质突出了建筑的轻盈，底座材质的变化则赋予了塔楼漂浮感。

华沙尖顶高达 220 米，其侧面与两栋高达 55 米的独立办公大楼相接。建筑的内部空间将高 160 米的住宅大楼肖邦塔与办公空间联系起来。除了办公空间，这栋现代的办公大楼还设置了地下停车场、地下档案馆、咖啡馆、餐厅等空间。

楼层空间围绕建筑核心来分布，以确保每个办公空间都能最大化地利用太阳能。2.7 米的层高以及全高的透明窗户，保证了建筑内部的自然光照，提高了工作环境的舒适度。

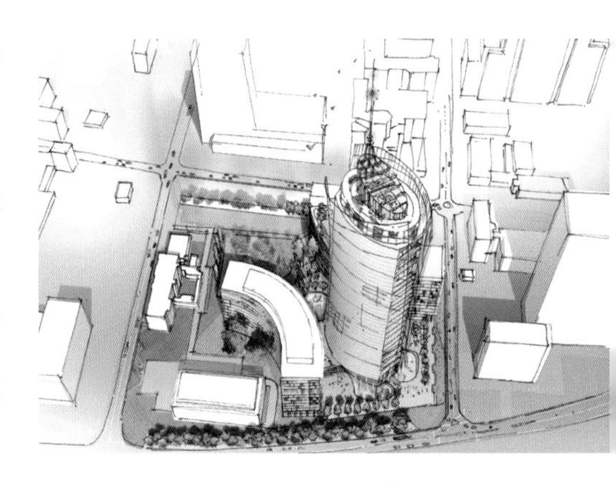

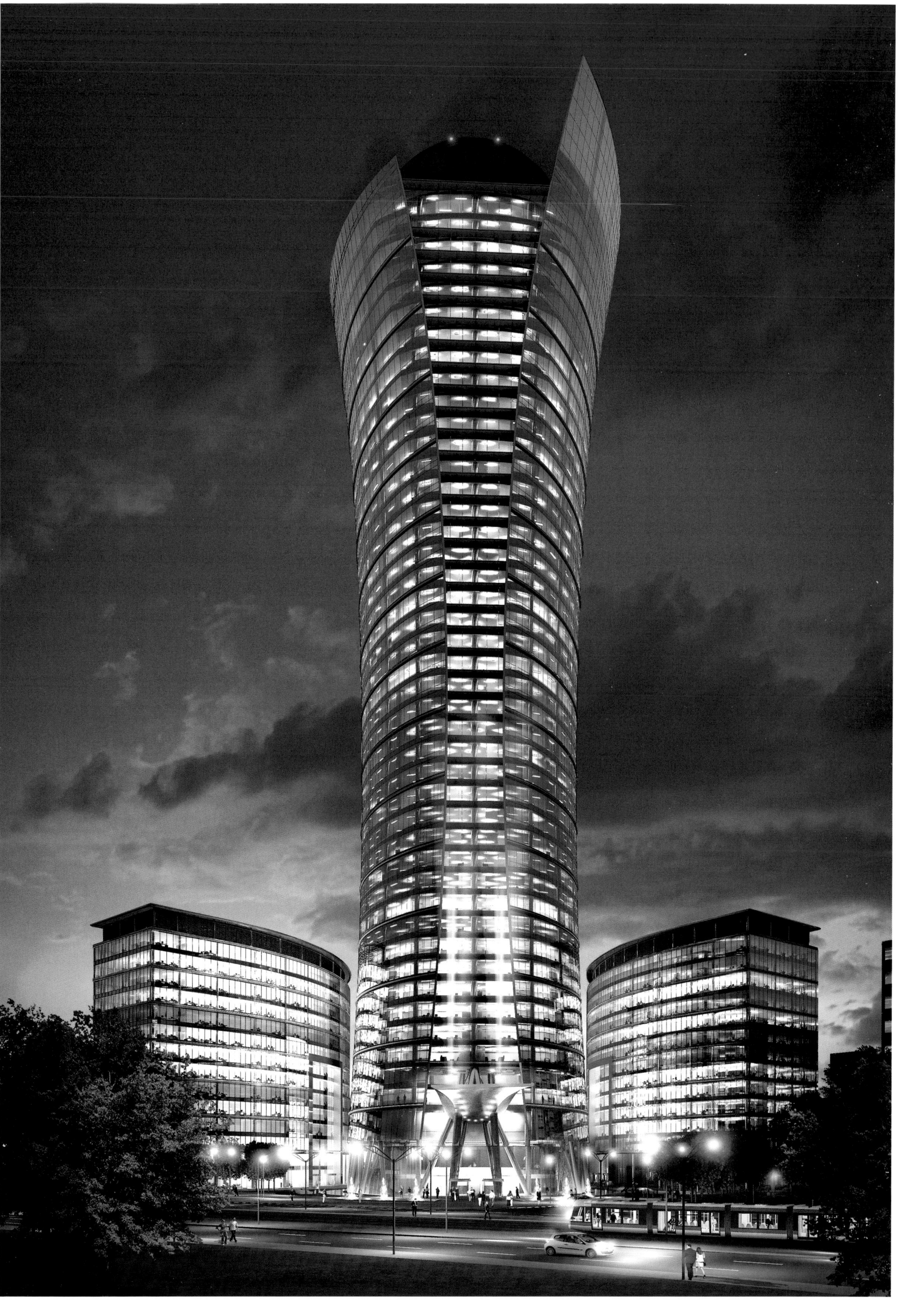

CHANEL
Dior
CAPTURE
TOTALE

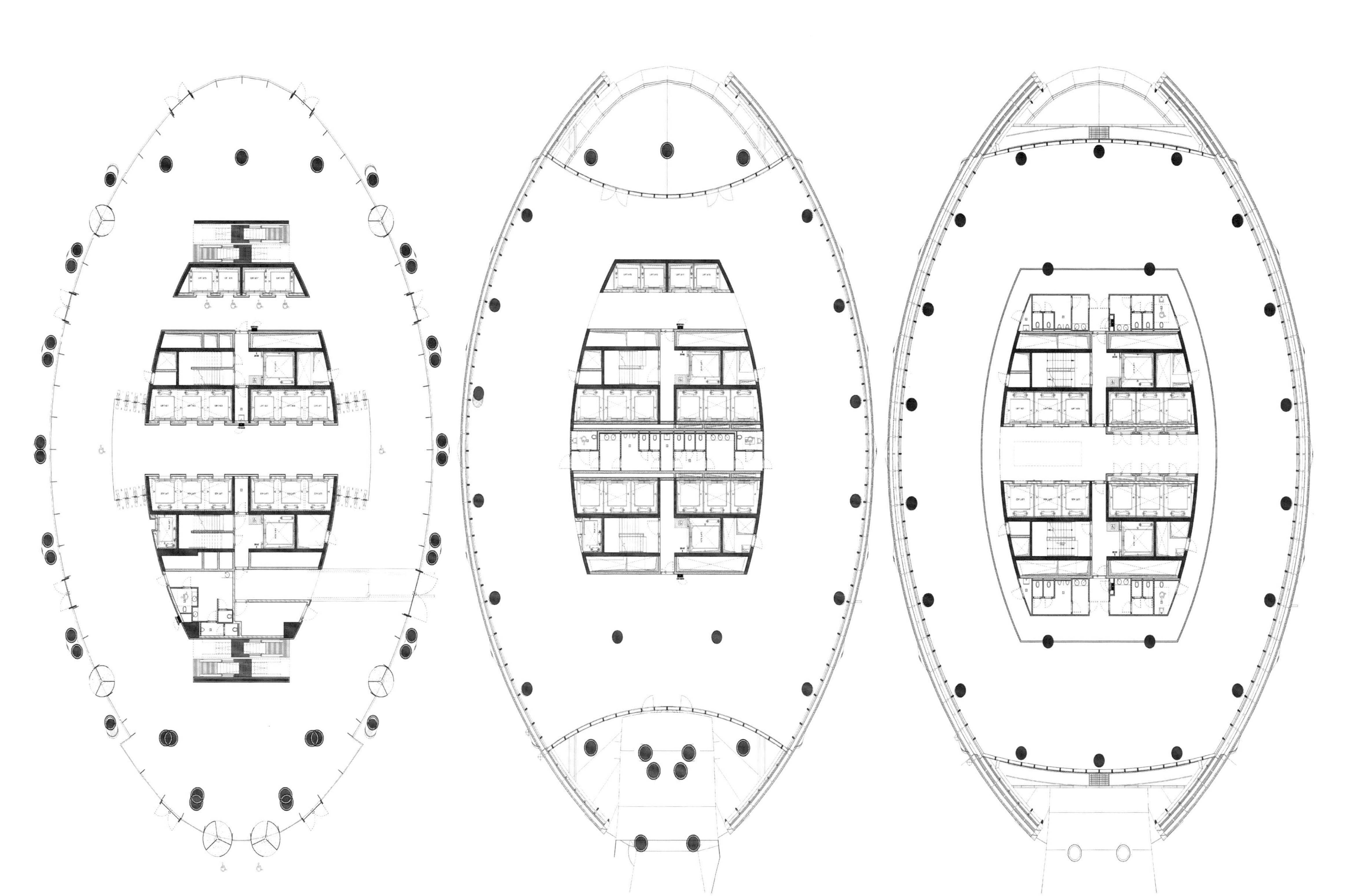

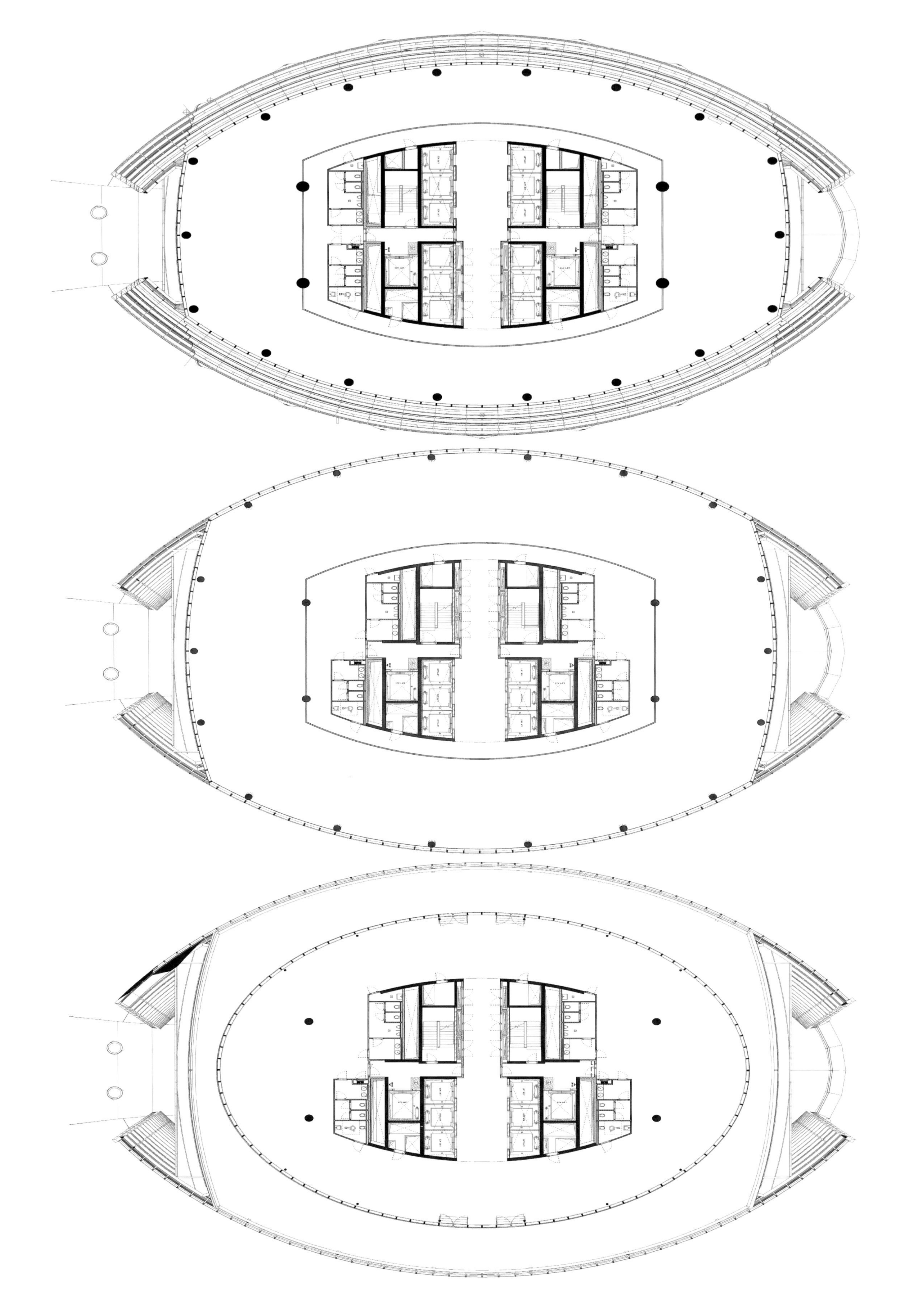

Profile

Next to the historic centre of Warsaw, nestled between the famous Warsaw Palace of Culture and Science and the Warsaw Rising Museum, stands the Warsaw Spire. Warsaw Spire is a unique combination of an architectural vision of a modern office building and a new approach to urban space arrangement. The Warsaw Spire is the highest office tower in Warsaw and will be erected as a new landmark for the city.

Architectural Design

The Warsaw Spire catches the eye with its attractive architecture, soaring lines and expressive form. The central floor platform is encircled by two glass "Shells" that extend at the top into a spiral form. The spiral shape emphasizes the slenderness of the tower. By only using glass on the outside of the building the lightness of the design is emphasized. At the same time, the glass "Shells" do not extend all the way to the ground, reinforcing the floating character of the tower.

The Warsaw Spire reaches a height of 220 meters including antennas and is flanked by two separate office buildings, each 55 meters high. A green inner area connects these office buildings with the 160 meters high Chopin Tower, a residential tower. The Warsaw Spire is a modern office building. In addition to the offices, the tower has space for underground parking, underground archive, a coffee bar and a restaurant.

The office floors are designed around a central core. In this way optimal use of daylight is made for every office space. Thanks to a clear floor height of 2.70 meters in combination with transparent windows that stretch from the floor to the ceiling, the level of lighting is maximized and work comfort increased.

JASPERS-EYERS

法国让蒂伊 ZAC de la Porte de Gentilly 办公楼

ZAC de la Porte de Gentilly

设计单位：ECDM Architectes
开发商：SODEARIF
项目地址：法国让蒂伊
项目面积：5 250 ㎡

Designed by: ECDM Architectes
Client: SODEARIF
Location: Gentilly, France
Area: 5,250 m²

项目概况

项目位于法国让蒂伊地区，设计旨在在视觉上联系周边的两大区域，同时将两大城市风景区自然地分隔开来。

设计特色

建筑由 3 个堆叠的体块组成，构成一个简约、清晰的几何图形。建筑就像是一把展开的扇子，逐步细化建筑的高度。整个建筑形成了 3 个开敞的区域：第一个区域位于底层，是接待区的延续；其他两个是位于第 4 层和第 7 层的露台。

办公楼的建筑轮廓使建筑因位移而形成独特的视角：从建筑较狭长的一面或是它的边角处都可以看到建筑正面。这种对复合建筑体的表现形式，避免了任何单边的视觉效果。

建筑外立面设置了一个 1.35 米的框架，这一框架的表现力通过先进的防晒外观得以加强，形成云纹效果，为整个项目增添了活力。连续、富有表现力的外立面裹覆了建筑外围，平衡了两个地块之间的视觉效果。

设计引入了标准化办公空间设计。每一个层次都呈现一种变化，前两个体量为 18 米，最后一个体量为 12 米。这一紧凑和理性的结构使建筑宛如一个楼梯式的香波城堡。

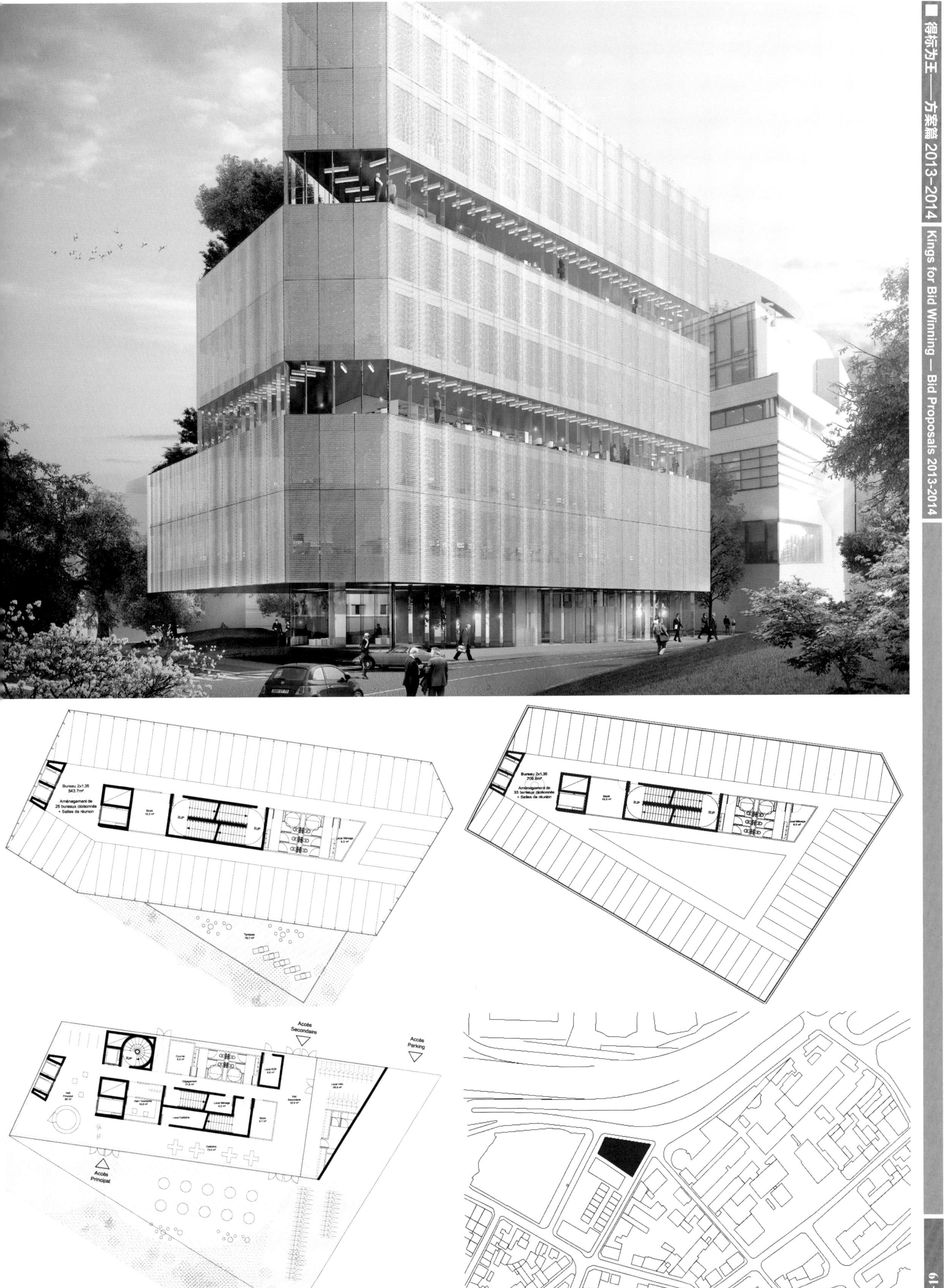
Accès
Secondaire
Accès
Parking
Accès
Principal

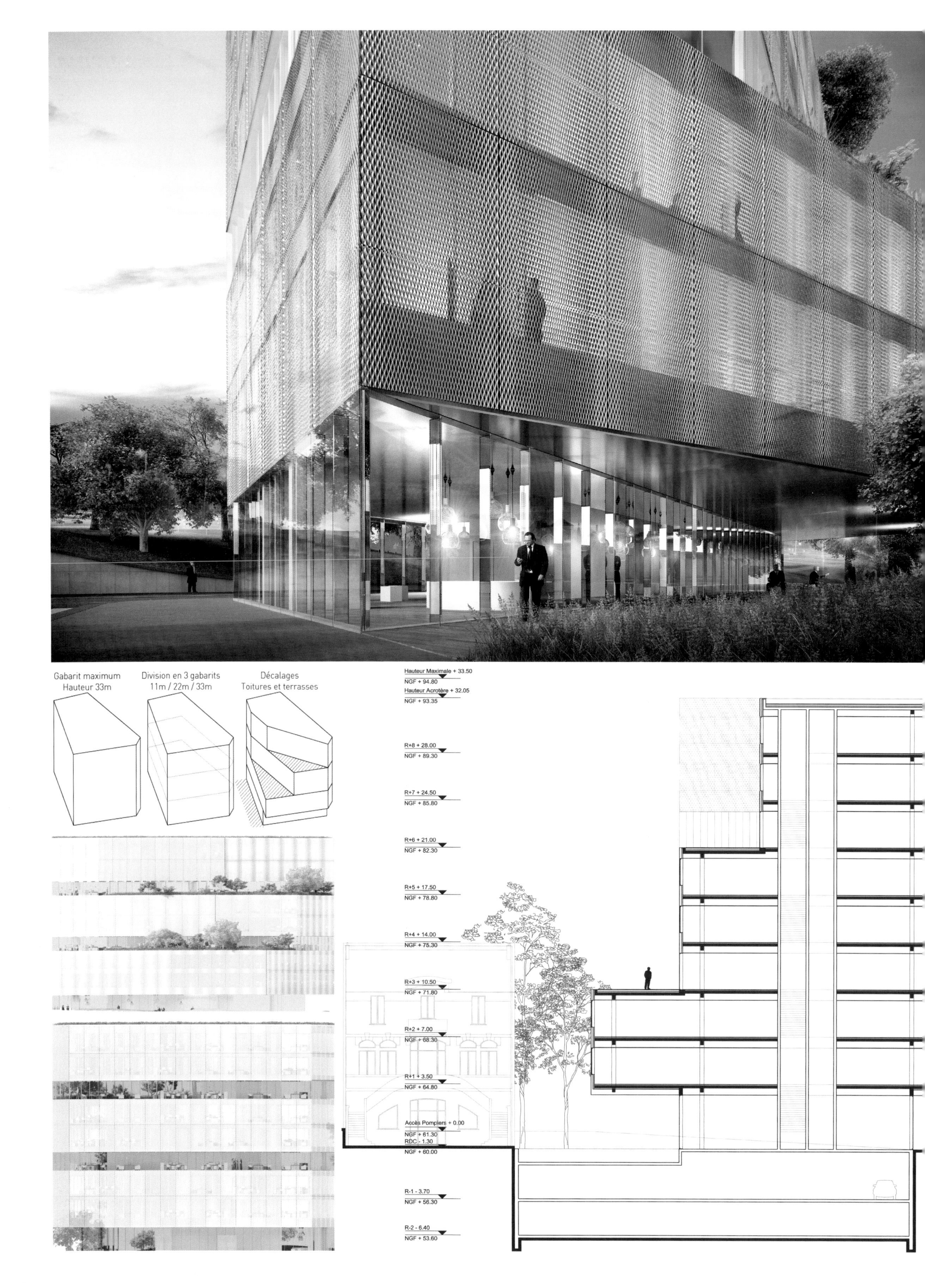
Gabarit maximum
Hauteur 33m
Division en 3 gabarits
11m / 22m / 33m
Décalages
Toitures et terrasses
Hauteur Maximale + 33.50
NGF + 94.80
Hauteur Acrotère + 32.05
NGF + 93.35
R+8 + 28.00
NGF + 89.30
R+7 + 24.50
NGF + 85.80
R+6 + 21.00
NGF + 82.30
R+5 + 17.50
NGF + 78.80
R+4 + 14.00
NGF + 75.30
R+3 + 10.50
NGF + 71.80
R+2 + 7.00
NGF + 68.30
R+1 + 3.50
NGF + 64.80
Accès Pompiers + 0.00
NGF + 61.30
RDC - 1.30
NGF + 60.00
R-1 - 3.70
NGF + 56.30
R-2 - 6.40
NGF + 53.60

Profile

The Lot 4 – ZAC de la Porte de Gentilly project is primarily a proposal to articulate two territories, two urban landscapes separated by the influence of the device.

Design Feature

The building consists of three superimposed volumes each with its own characteristics. The volumes constitute a simple clear geometric figure. The building opens like a fan to refine gradually gaining height. The whole building creates three open spaces. The first at ground level is a continuation of the reception area, then two successive terraces, one at the 4th level, the other at the 7th level.

The curvatures of the device and the notion of displacement generate a reading in perspective with always at least two fronts seen: the building will be seen from afar on the short sides, its angles. Designers work on the expression of a complex volume to avoid any single side effect.

This rationality is reflected in the writing of the facades that expresses strictly a frame of 1.35 meters of offices. The expression of the frame is enhanced by a working progressive external sun protection just muddy the reading by creating a moire effect, conferring a kinetic aspect to the whole. This pattern runs continuously to wrap the building in the manner of woven wallpaper and balance visual effect of these two areas.

The design introduces a standardized office space. Each level is a variation on archetypes of 18 meters for the first 2 volumes and 12 meters for the latter. This compactness and rationality allow designers to have an extremely compact core while having a staircase-like Chambord.

斯洛文尼亚卢布尔雅那太阳能办公室

Solar Power Offices

设计单位：OFIS 建筑事务所
开发商：ELES
项目地址：斯洛文尼亚卢布尔雅那
项目面积：35 500 ㎡
设计团队：Rok Oman　Spela Videcnik
Andrej Gregoric　Janez Martincic
Janja Del Linz　Katja Aljaz
Sergio Silva Santos　Marco Mazzotta
Grzegorz Ostrowski

Designed by: OFIS Arhitekti
Client: ELES
Location: Ljubljana, Slovenia
Area: 35,500 m^2
Project Team: Rok Oman, Spela Videcnik, Andrej Gregoric, Janez Martincic, Janja Del Linz, Katja Aljaz, Sergio Silva Santos, Marco Mazzotta, Grzegorz Ostrowski

项目概况

这是由 OFIS 建筑事务所设计的新商业办公建筑，其外观和布局都反映了斯洛文尼亚电力公司 ELES 的愿景和责任感。项目在建筑设计和设施装备上更加注重节能和新能源的利用，以实现真正意义上的碳中和。

设计特色

当地一年中大概有 187 天是晴天，日照充足，具备巨大的太阳能开发潜能。建筑外部采用的太阳能薄膜以及太阳能屋顶收集到的太阳能，可用来供暖和供热。另一方面，该城市的平均年降水量达到 1 402 毫米，设计师将雨水的利用纳入生态环保系统设计中，用于环保卫生、灌溉和洗车，实现对雨水的循环利用。

通风系统可根据空间的容量和空气质量监测的数据对室内空间进行自动调节和控制，且这些通风系统都与中央控制系统相连，使通风系统得以高效运作。

节能设计在建筑内部集中体现在综合型温室和天井上。天井的设置为室内空间提供了自然光照和自然通风，营造了纯粹、自然、舒适的办公环境。

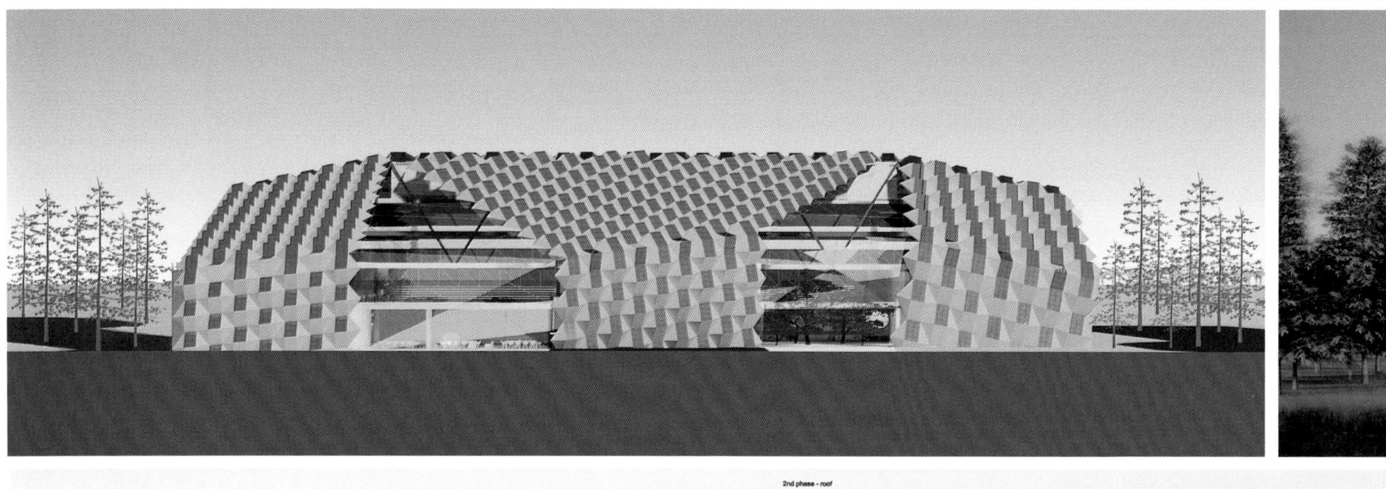

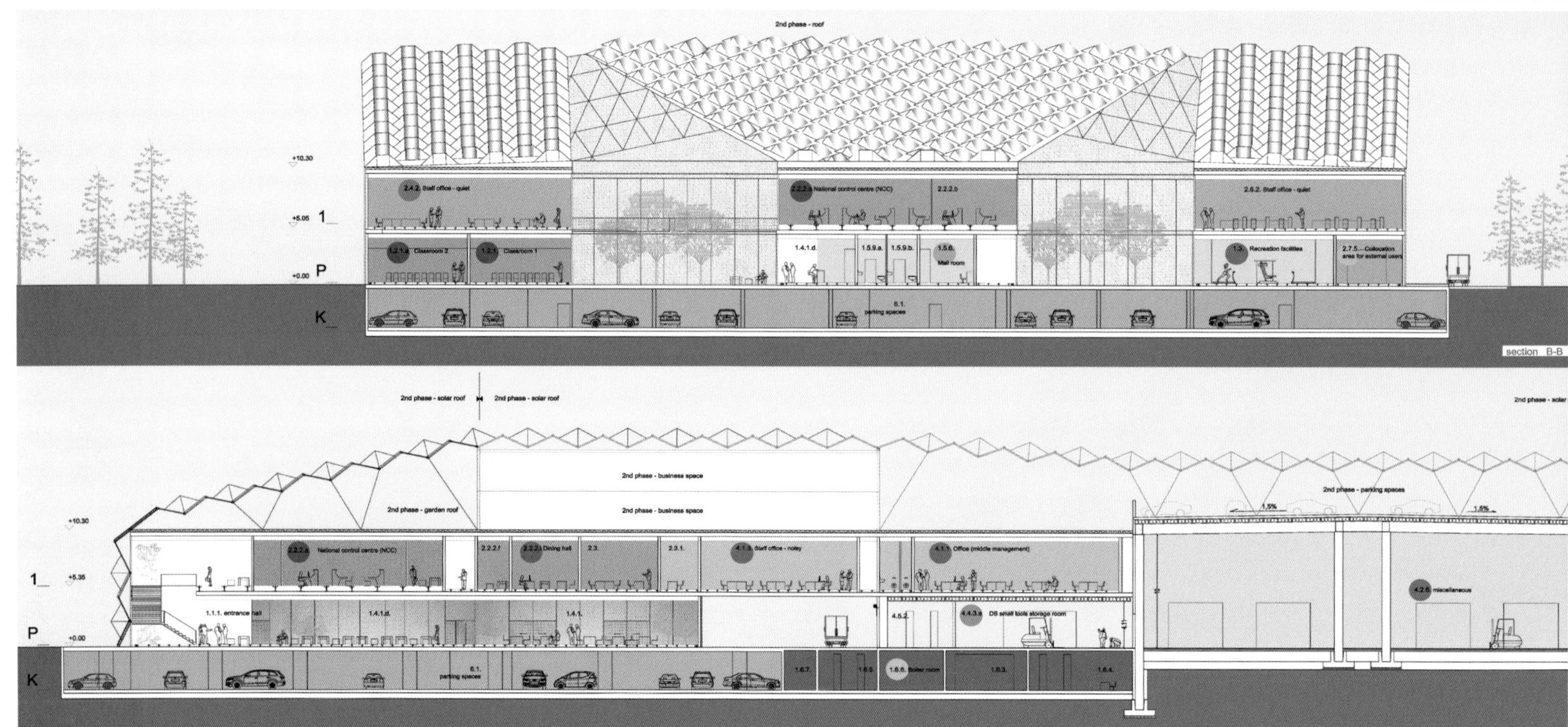
2nd phase - roof
+10.30
+5.05
+0.00
1
P
K
2.4.2. Staff office - quiet
2.2.2.a National control centre (NCC)
2.2.2.b
2.6.2. Staff office - quiet
1.2.1.a Classroom 2
1.2.1. Classroom 1
1.4.1.d.
1.5.9.a.
1.5.9.b.
1.5.6. Mail room
1.3. Recreation facilities
2.7.5. Collocation area for external users
6.1. parking spaces
section B-B
2nd phase - solar roof
2nd phase - solar roof
2nd phase - business space
2nd phase - business space
2nd phase - garden roof
2nd phase - parking spaces
1.5%
1.5%
+10.30
+5.35
+0.00
2.2.2.a National control centre (NCC)
2.2.2.f
2.2.2.i Dining hall
2.3.
2.3.1.
4.1.3. Staff office - noisy
4.1.1. Office (middle management)
4.2.6. miscellaneous
1.1.1. entrance hall
1.4.1.d.
1.4.1.
4.5.2.
4.4.3.a DS small tools storage room
6.1. parking spaces
1.6.7.
1.6.5.
1.6.6. Boiler room
1.6.3.
1.6.4.
section A-A

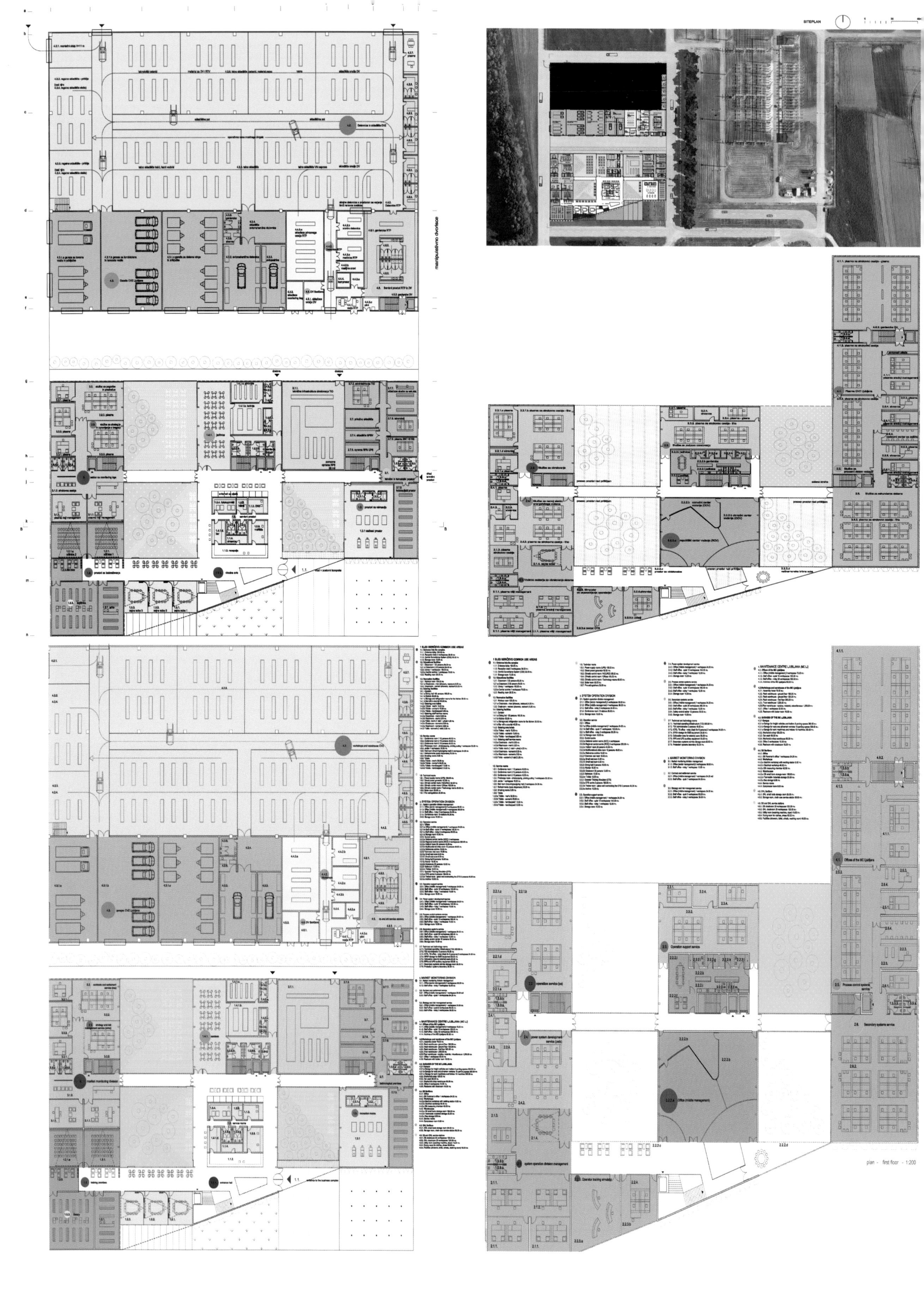
SITEPLAN
plan - first floor - 1:200

Profile

It is a new business facility for operational activities. The layout and appearance of the building reflect the vision and mission of Slovenian electricity company ELES. Based on fuel efficiency and alternative energy sources, the design of both the architecture and installations achieved complete carbon neutrality.

Design Feature

It was measured that on average, 187 days of the year are sunny, showing the potential for the exploitation of solar energy and the large scale harnessing of solar energy for heating and hot water. Depending on the design purpose and an average annual rainfall of 1,402 millimeters per year, it is reasonable to include the installation of rainwater for purposes such as sanitation, watering and car washing.

All ventilation systems are equipped with automatic regulation and control in the room depending on the room's occupancy and the monitored air quality. Ventilation systems are linked to a central control system which allows it to perform optimally.

The energy efficiency is also expressed in the interior with the use of integrated greenery and atria. The atria also allow natural light and air flow to penetrate the interior spaces, bringing nature and greenery to the office space and creating a pure ambience.

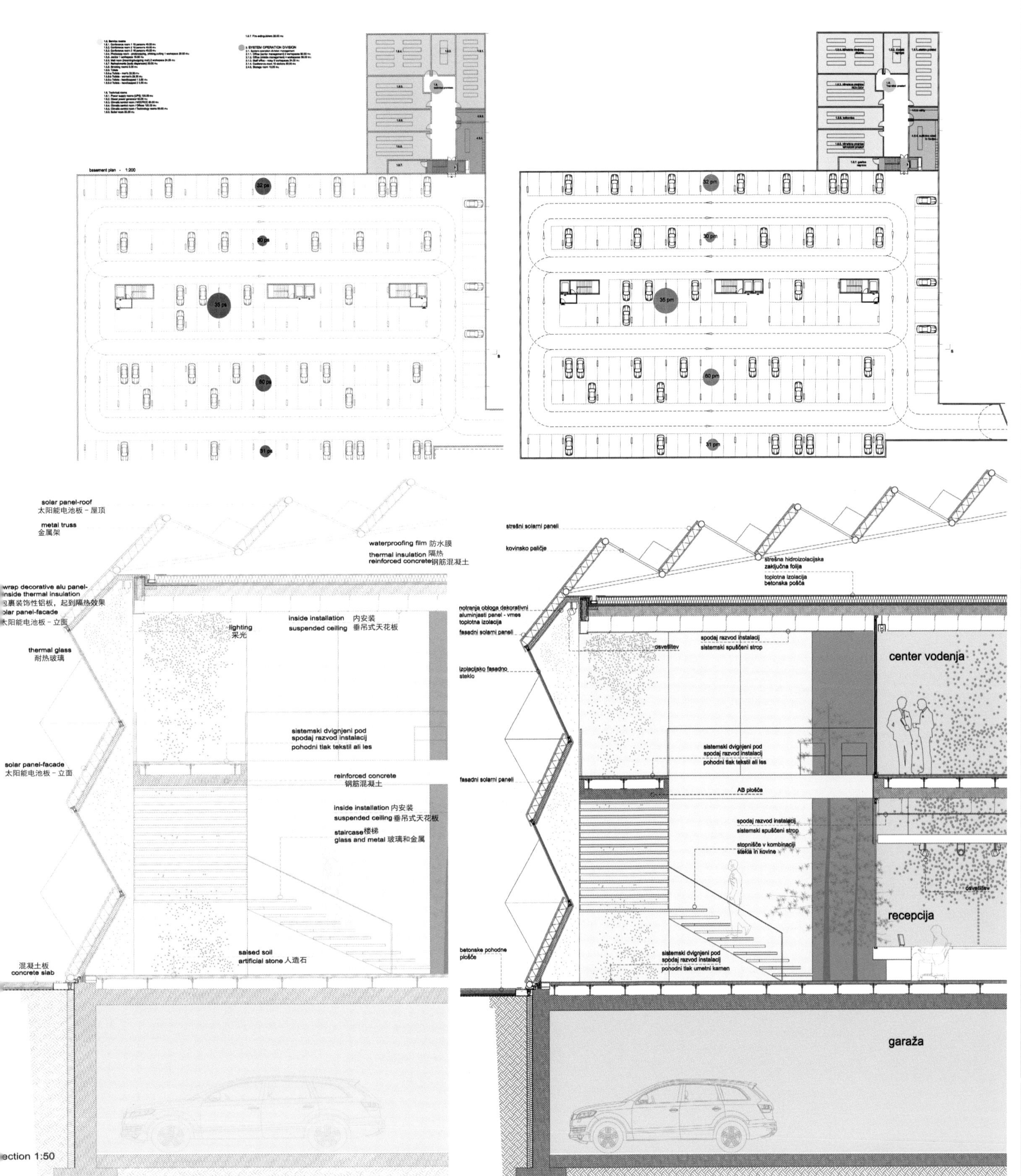

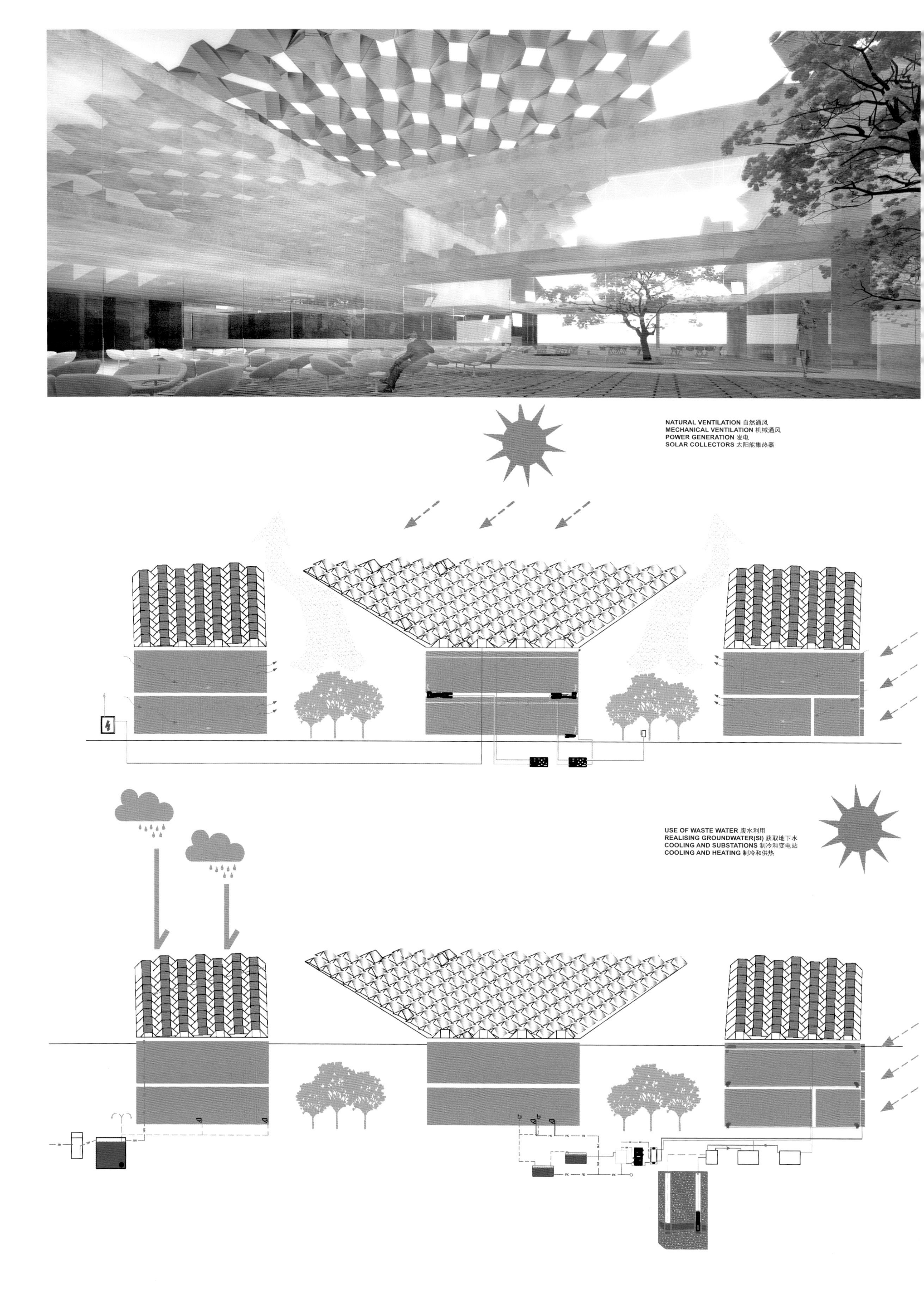
NATURAL VENTILATION 自然通风
MECHANICAL VENTILATION 机械通风
POWER GENERATION 发电
SOLAR COLLECTORS 太阳能集热器
USE OF WASTE WATER 废水利用
REALISING GROUNDWATER(SI) 获取地下水
COOLING AND SUBSTATIONS 制冷和变电站
COOLING AND HEATING 制冷和供热

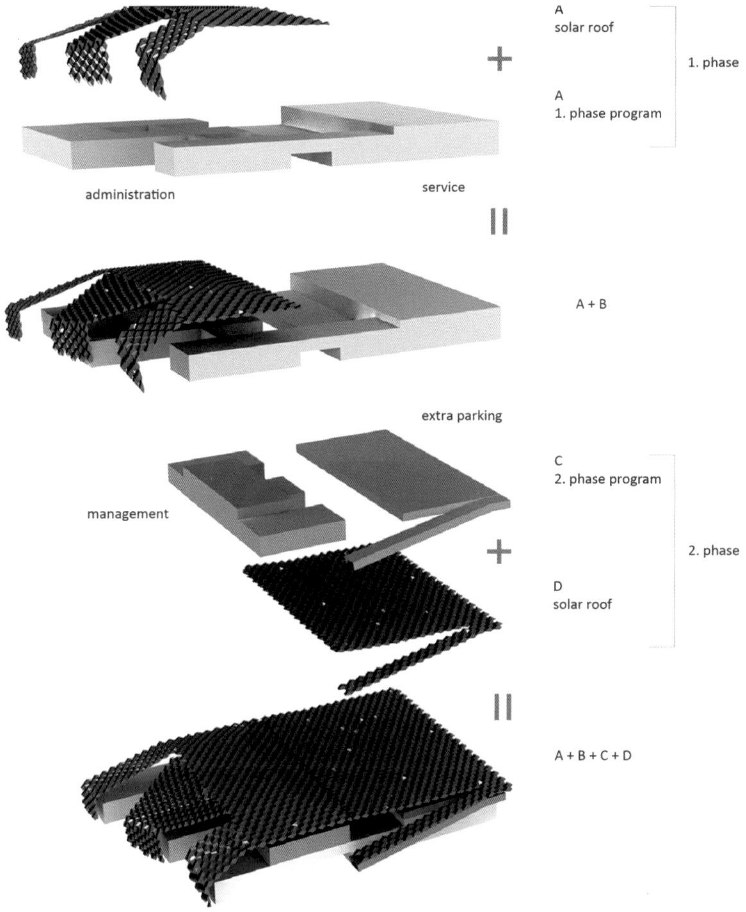

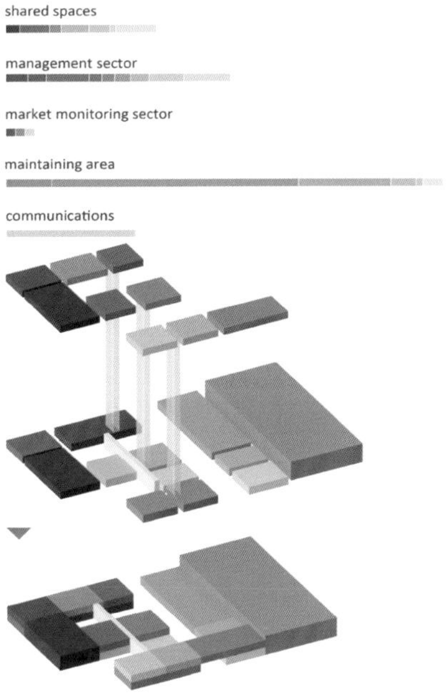

SMT

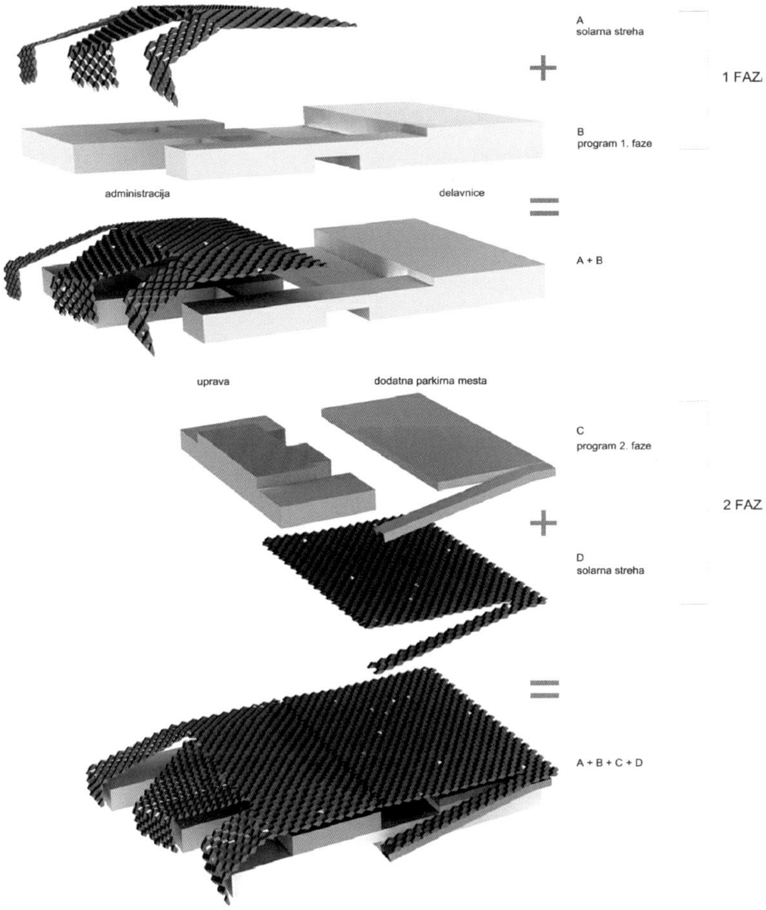

A
solarna streha

B
program 1. faze

1 FAZ.

administracija

delavnice

A + B

uprava

dodatna parkirna mesta

C
program 2. faze

D
solarna streha

2 FAZ.

A + B + C + D

A
solar roof

A
1. phase program

1. phase

administration

service

A + B

extra parking

management

C
2. phase program

D
solar roof

2. phase

A + B + C + D

shared spaces

management sector

market monitoring sector

maintaining area

communications

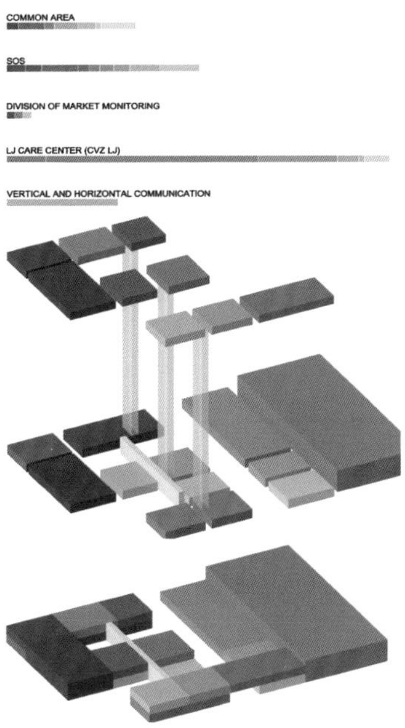

COMMON AREA

SOS

DIVISION OF MARKET MONITORING

LJ CARE CENTER (CVZ LJ)

VERTICAL AND HORIZONTAL COMMUNICATION

Potential: CO2 Neutral Building Operation 潜能：CO2 中性建筑
Energy Balance-Office (6,734m²), Storage and Garage(5,000m²)
能量平衡办公室（6 734m²），储藏室和车库（5 000m²）

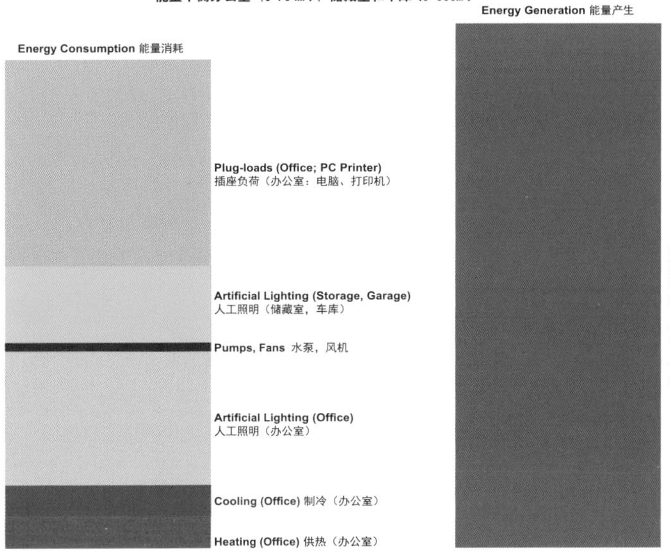

Energy Consumption 能量消耗

Plug-loads (Office; PC Printer)
插座负荷（办公室：电脑、打印机）

Artificial Lighting (Storage, Garage)
人工照明（储藏室，车库）

Pumps, Fans 水泵，风机

Artificial Lighting (Office)
人工照明（办公室）

Cooling (Office) 制冷（办公室）

Heating (Office) 供热（办公室）

Energy Generation 能量产生

Photovoltaic cells 光伏电池：
8300m²

奥地利格拉茨 Crossroad 办公楼

Crossroad Office

设计单位：OFIS 建筑事务所
项目地址：奥地利格拉茨
基地面积：2 060 ㎡
总建筑面积：2 528 ㎡
设计团队：Rok Oman　Spela Videcnik
Janez Martincic　Andrej Gregoric
Jan Celeda　Will Gibson
Carlos Garcia　Zuzana Chupacova
Andrej Kacera　Aquilino Fernandez Lopez
Sara Barriocanal　Radek Toman
Jan Smejkal

Designed by: OFIS Arhitekti
Location: Graz, Austria
Site Area: 2,060 m^2
Gross Floor Area: 2,528 m^2
Design Team: Rok Oman, Spela Videcnik, Janez Martincic, Andrej Gregoric, Jan Celeda, Will Gibson, Carlos Garcia, Zuzana Chupacova, Andrej Kacera, Aquilino Fernandez Lopez, Sara Barriocanal, Radek Toman, Jan Smejkal

CROSSROAD OFFICES Crossroad 办公楼
Navigation: Lechgasse, Graz, Austria 导航：奥利地格拉茨 Lechgasse

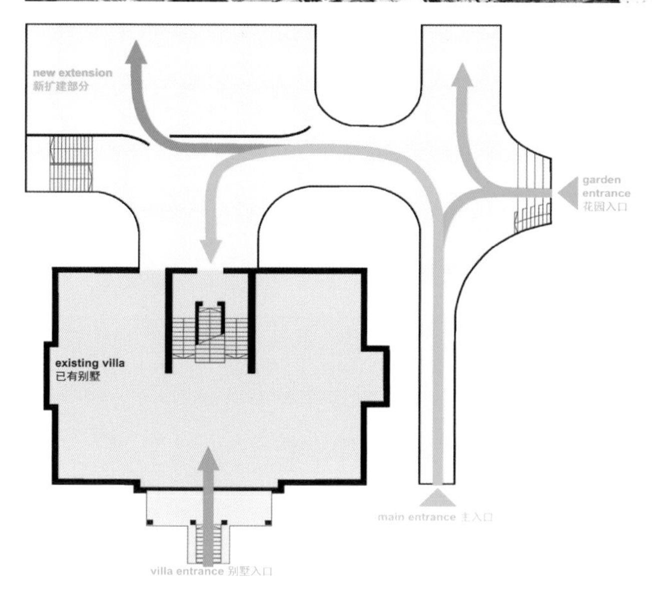

项目概况

方案旨在对原有别墅进行扩展，构建一个可举行讲座和会议等活动的多功能建筑。该方案的重点在于对原有别墅进行整修以及对三个小型会议厅进行扩展。

设计特色

建筑形态源于从场址到主要目的地的逻辑联系。新建的部分靠近南侧的街道和东侧的地下停车场，好像别墅周围的一个十字路口调节器。

新的建筑位于别墅的后面，玻璃中庭下是一个绿色的庭院，这些户外空间构成一个外部花园，延伸到展览空间，同时，它们也是空间与空间之间的隔断与衔接，营造了动感、舒缓的环境。

沿着主街道的缓坡或是车库入口前的楼梯可进入建筑内部。大厅环绕着绿色天井，是一个多功能的空间，可用于接待、就餐、展览等活动。大厅与原有建筑相连，也是通往大型的会议室和一层中央电梯的入口。

会议厅的布局比较灵活，在必要的时候，可以将两个会议厅合并起来。大厅内光照充足，而且可以观赏庭院和公园的景色。

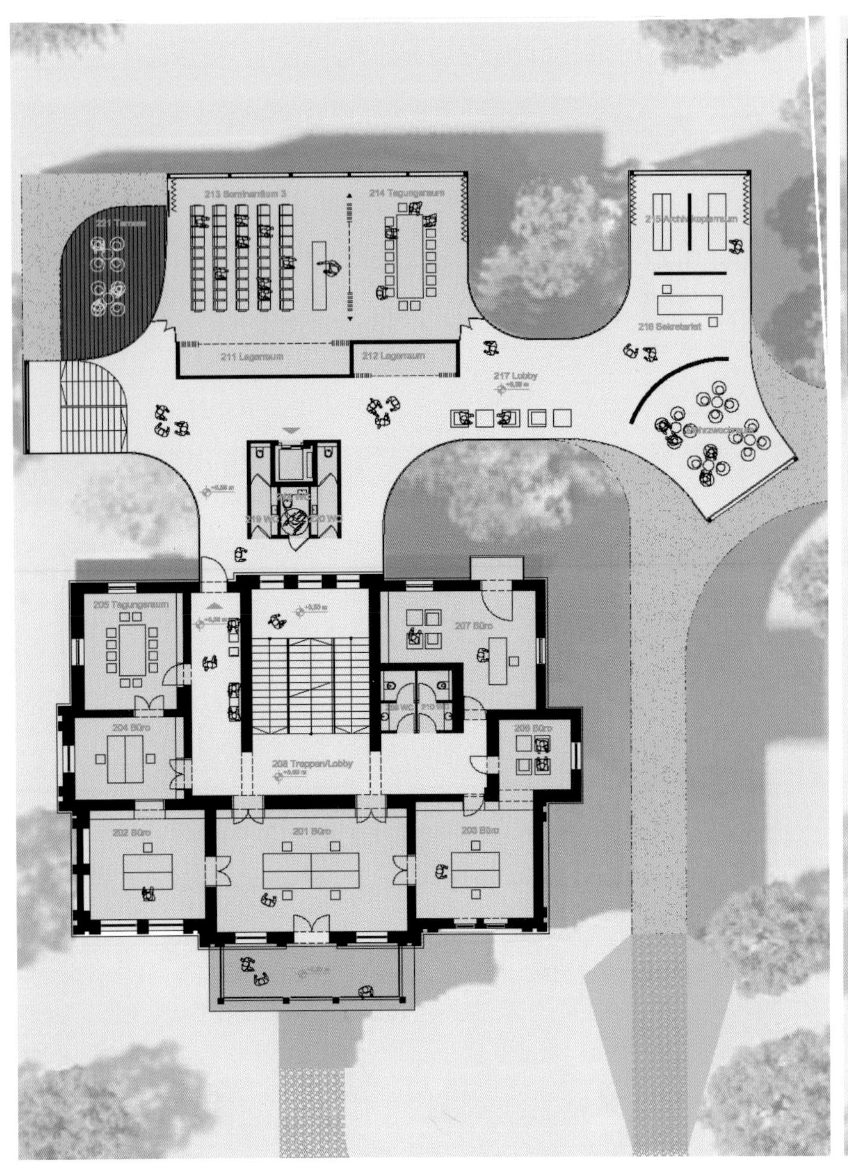

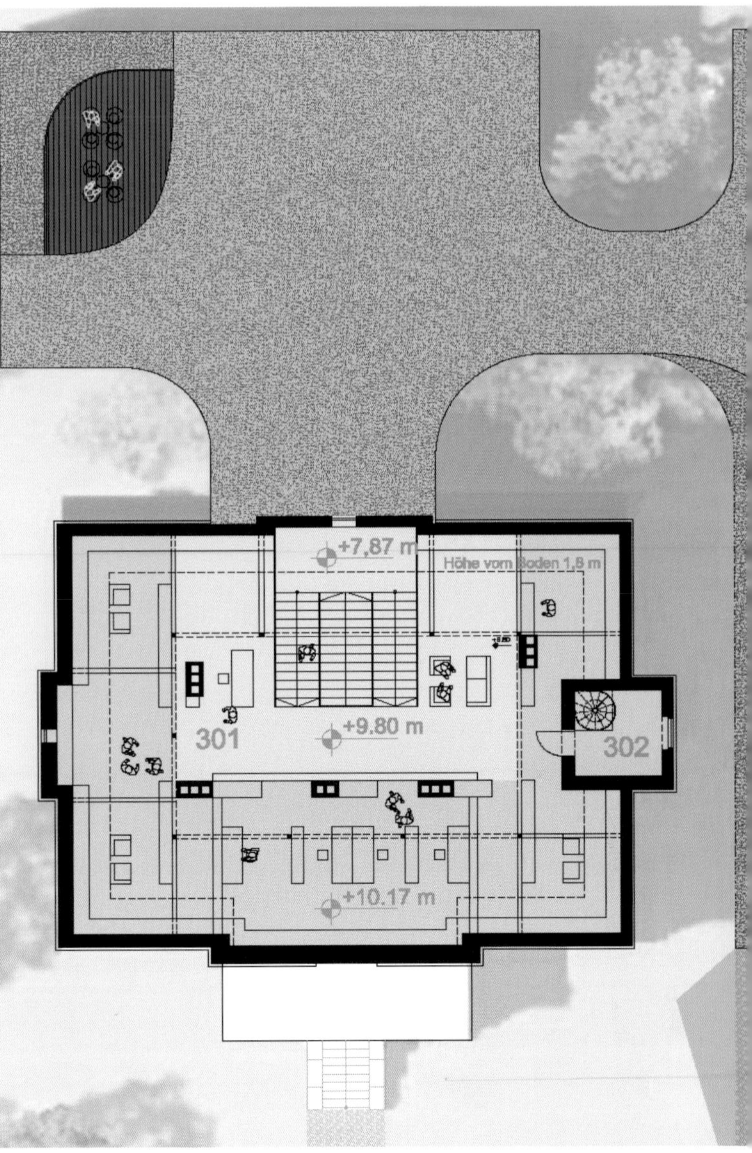

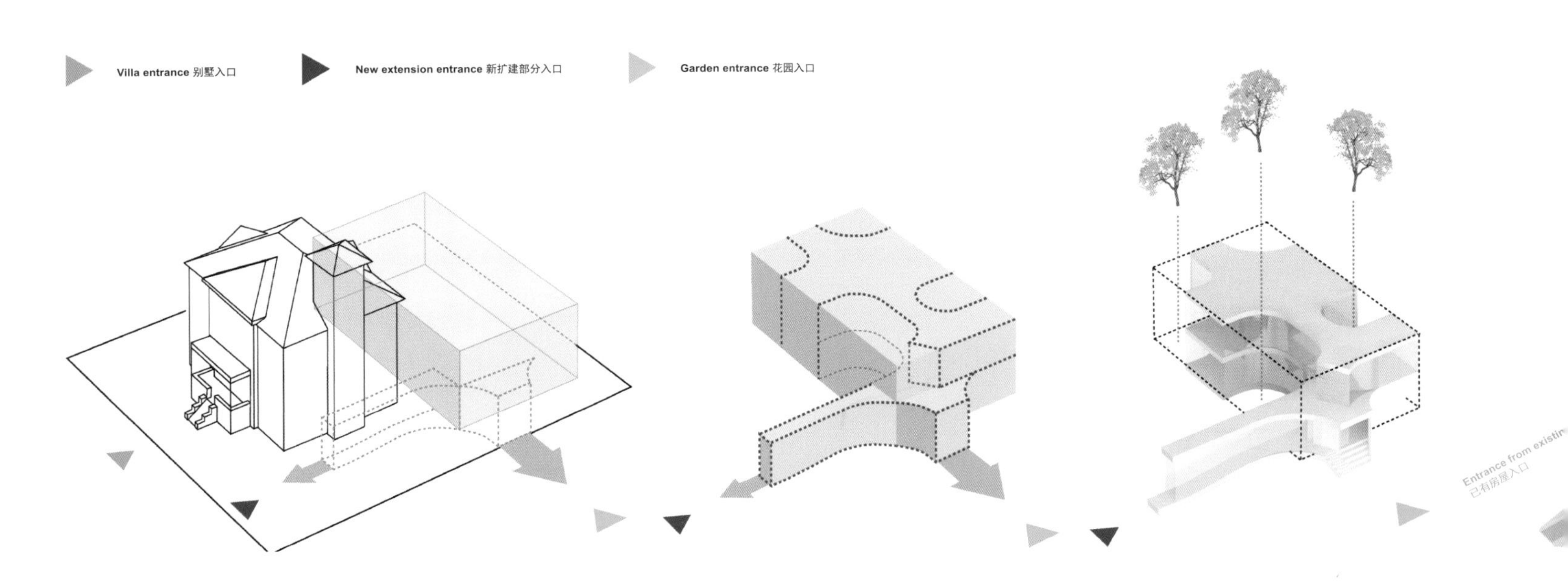

Profile

The concept proposal is an extension to the existing Villa to create a multifunctional building for renting the space for lectures and congress activities. The brief proposed a renovation of the existing Villa and an extension with 3 small congress halls.

Design Feature

The actual form derives from the main logical directions on the site to the main destinations. The new extension acts as a crossroad regulator around the existing Villa and approaches the street on the south side and the underground parking on the east side.

The new volume is positioned behind the existing Villa with cut outs — glass atrium spaces that wrap around a green courtyard. Outside these spaces become an external park that extends into the pavilion. Inside, they form the divisions between the internal spaces and create a dynamic, light and calming atmosphere. One enters the extension through a gentle slope from the main street or through a staircase that fronts the garage entrance. The lobby weaves around the green atria and offers entrances to large conference rooms and a central elevator/staircase core that leads to the first floor. It is also connected with the existing historical Villa. The lobby is a multipurpose space that can work for receptions and other events as a place for catering, exhibition or informal presentations.

Congress halls can be joined if necessary. Both rooms have the possibility of daily light with attractive views to the park and the new green atrium that create calm and intimate atmosphere.

-2.64 m
-2.42 m
003
002
011
010
012
004
001
013
009
-1.90 m
005
006
008
007

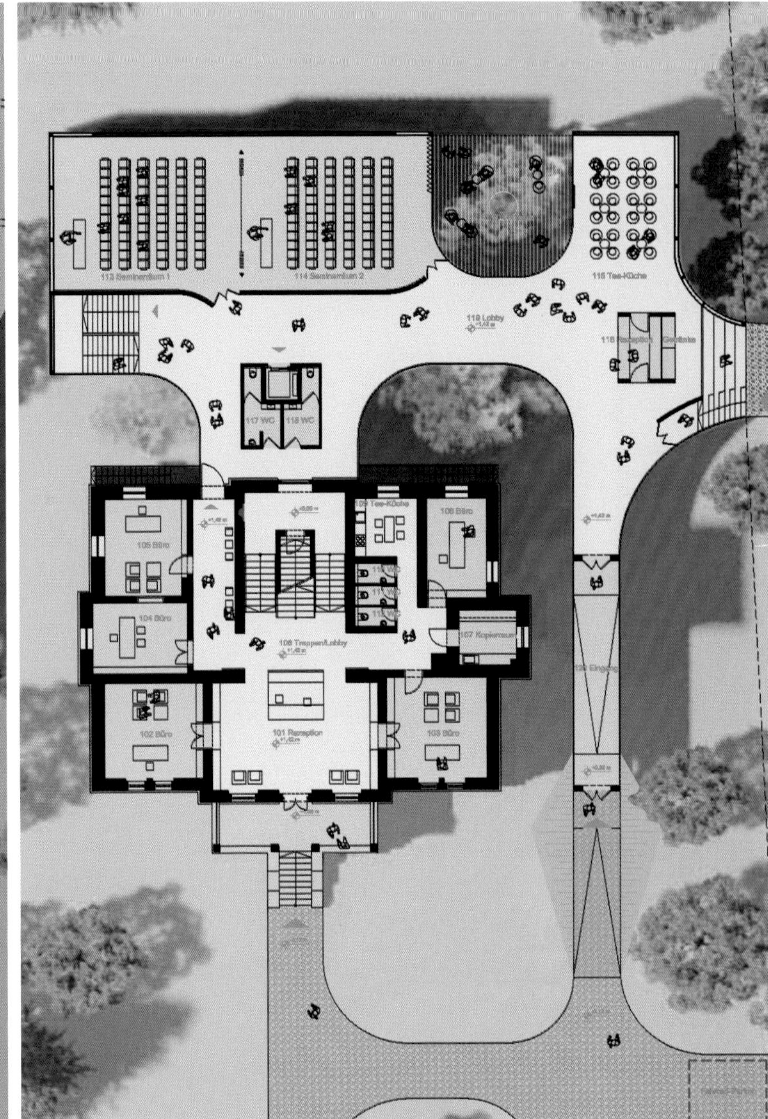

113 Seminarium 1
114 Seminarium 2
115 Tee-Küche
119 Lobby
117 WC
118 WC
105 Büro
106 Büro
104 Büro
108 Treppen/Lobby
107 Kopierraum
102 Büro
101 Rezeption
103 Büro

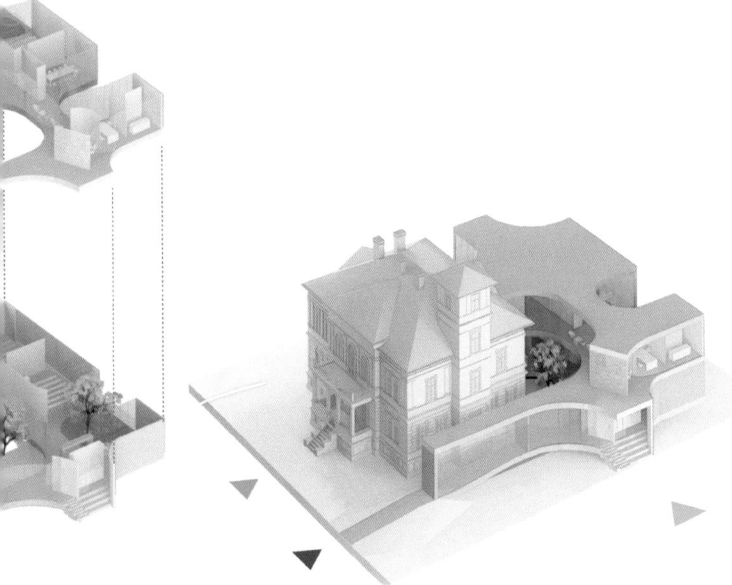

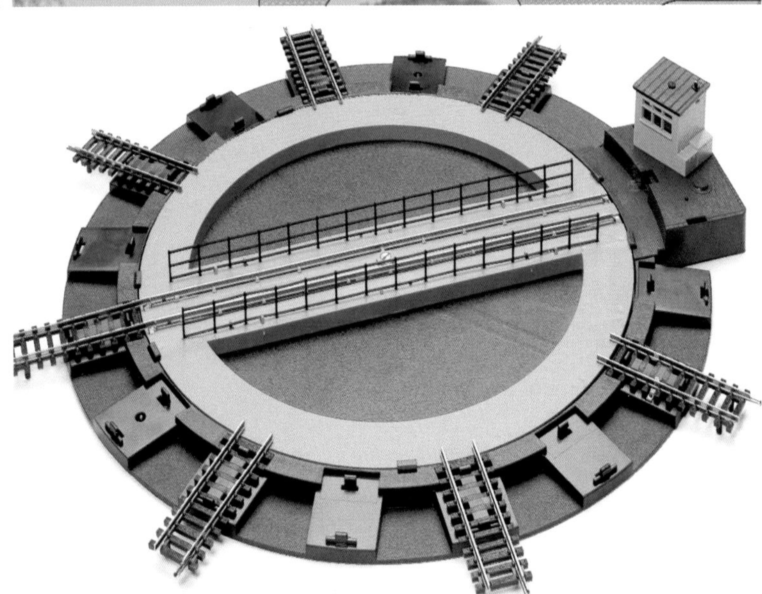

Korridor
Korridor
+5.56 m
+1.42 m
+1.42
-0.13 m
-0.13 m
± 0.00 m
+5.56
+0.62

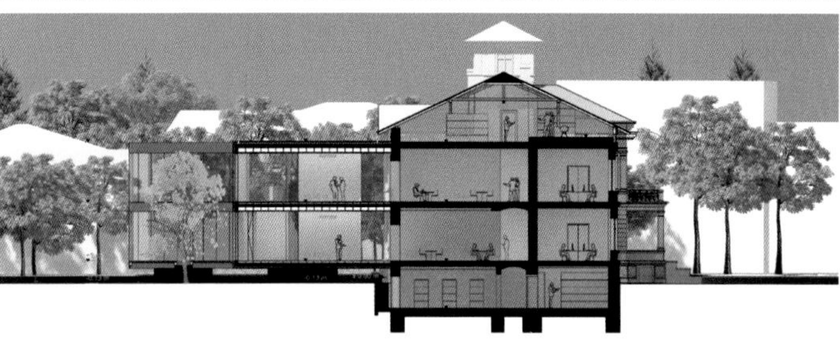

Rezeption
Terrase

丹麦哥本哈根 LM 海港大门

LM Harbor Gateway

设计单位：斯蒂文·霍尔建筑师事务所
开发单位：ATP Ejendomme
哥本哈根港务发展局
项目地址：丹麦哥本哈根
建筑面积：58 018 ㎡
设计团队：Steven Holl　Noah Yaffe
Chris McVoy　Rashid Satti
Justin Allen　Esin Erez
Runar Halldorsson　Fiorenza Matteoni

Designed by: Steven Holl Architects
Client: ATP Ejendomme; CPH City and Port Development
Location: Copenhagen, Denmark
Building Area: 58,018 m^2
Design Team: Steven Holl, Noah Yaffe,
Chris McVoy, Rashid Satti,
Justin Allen, Esin Erez,
Runar Halldorsson, Fiorenza Matteoni

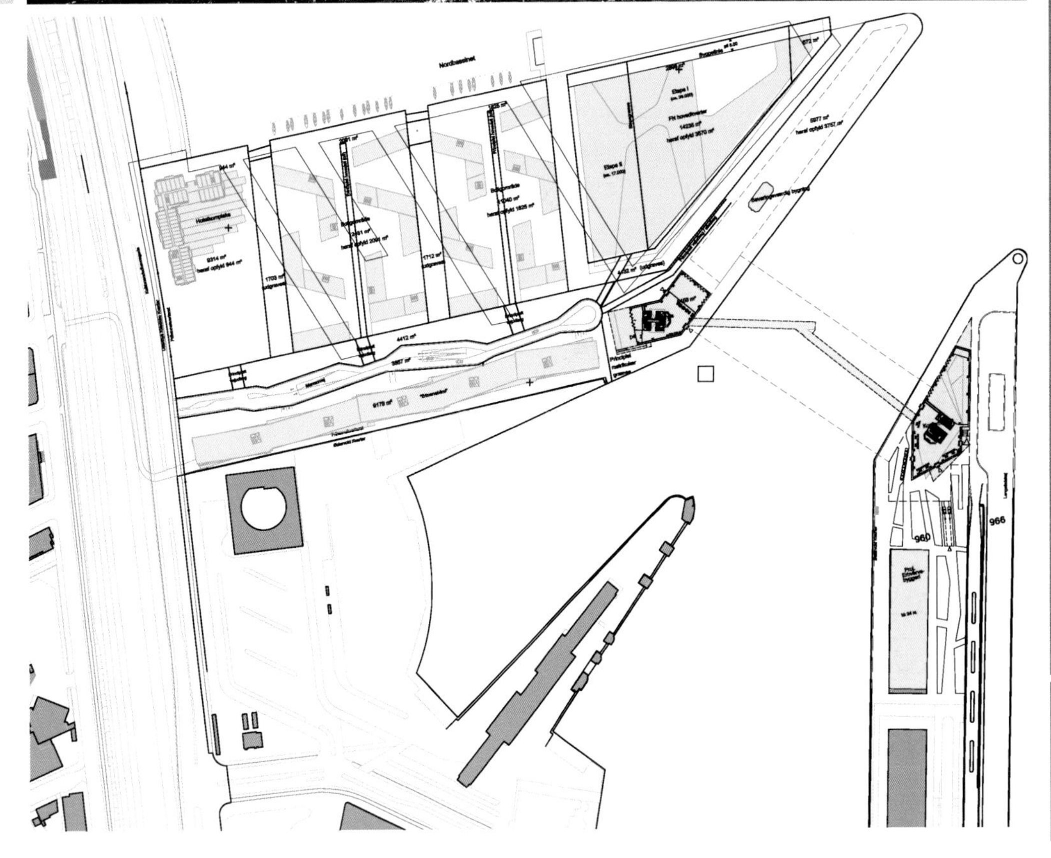

项目概况

这是斯蒂文·霍尔建筑师事务所为由哥本哈根港务发展局与 ATP Ejendomme 共同组织的哥本哈根滨水地区 LM Project 港口设计大赛提出的设计方案。设计旨在在哥本哈根的 Marmormolen 地区建立一座与 Langelinikaj 连接的海港大门，创建一个全新的、连贯的哥本哈根海港城，使之成为最具魅力的建筑景观。

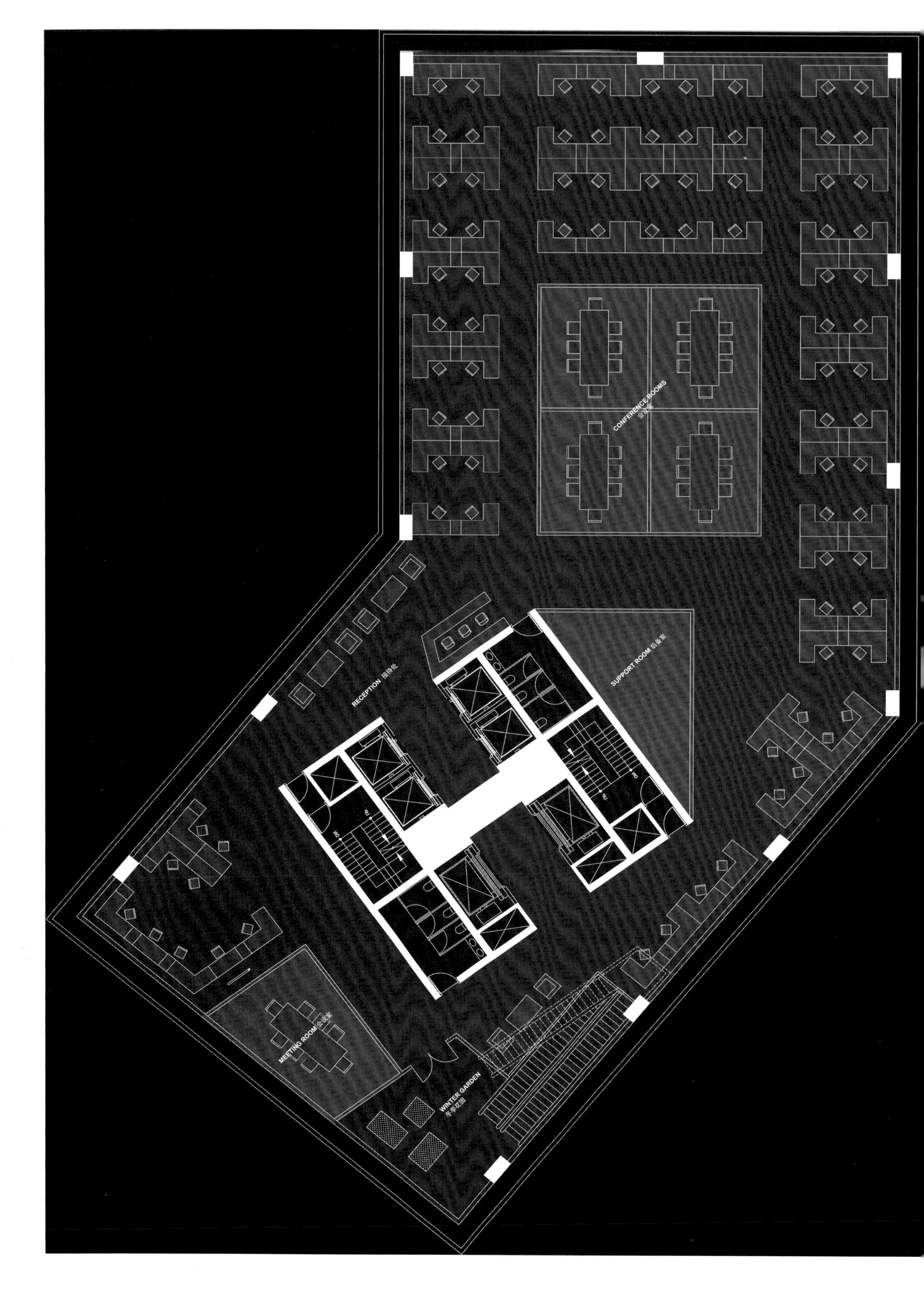
CONFERENCE ROOMS
会议室
RECEPTION 接待处
SUPPORT ROOM 后备室
MEETING ROOM 会议室
WINTER GARDEN
冬季花园

CONFERENCE ROOM 会议室
MEETING ROOM 会议室
RECEPTION 接待处
SUPPORT ROOM 后备室
MEETING ROOM 会议室
CONFERENCE ROOM 会议室
CONFERENCE ROOM 会议室

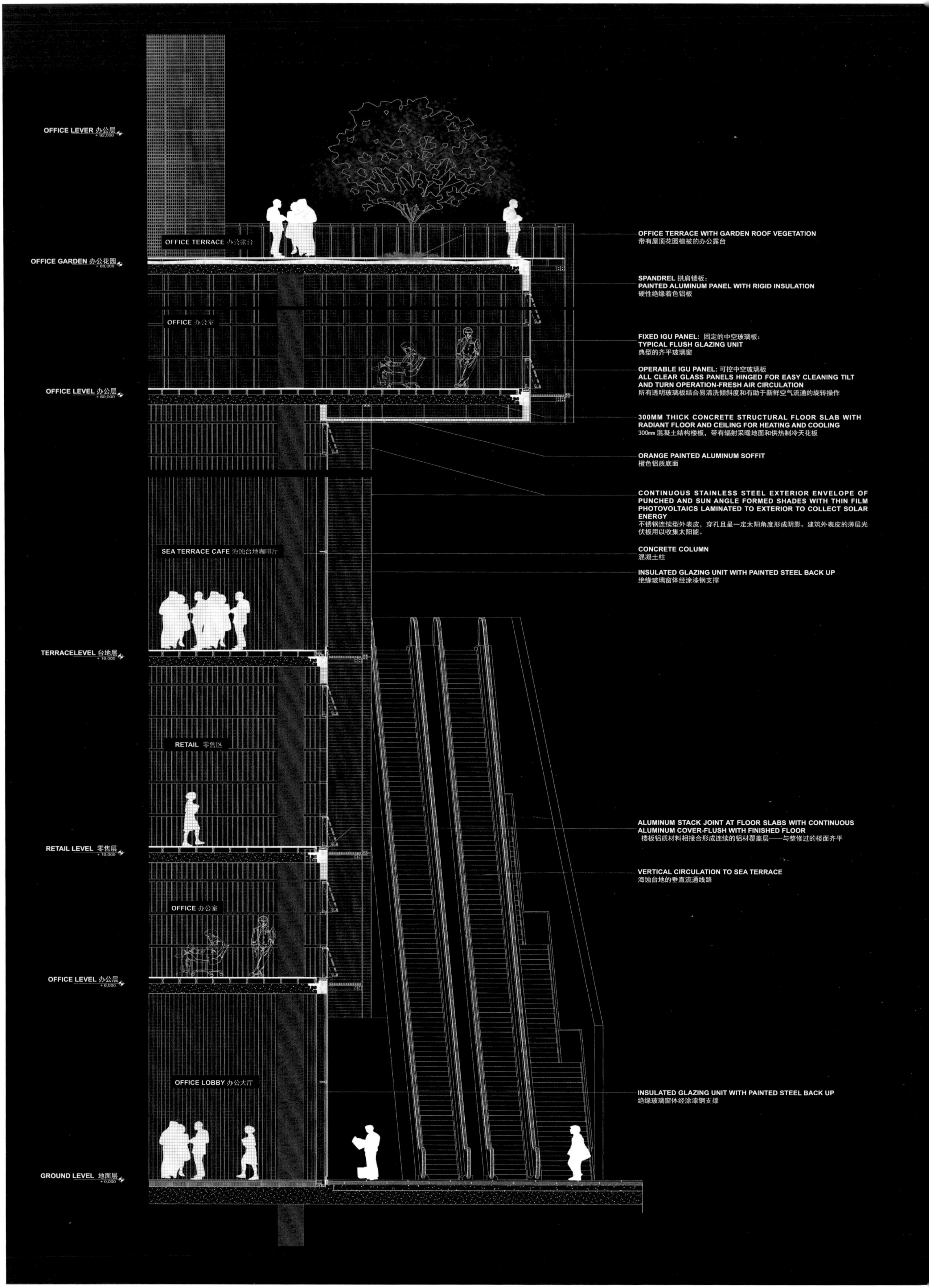

OFFICE LEVER 办公层
+ 92,000

OFFICE TERRACE 办公露台

OFFICE GARDEN 办公花园
+ 88,000

OFFICE 办公室

OFFICE LEVEL 办公层
+ 60,000

SEA TERRACE CAFE 海蚀台地咖啡厅

TERRACELEVEL 台地层
+ 16,000

RETAIL 零售区

RETAIL LEVEL 零售层
+ 10,000

OFFICE 办公室

OFFICE LEVEL 办公层
+ 6,000

OFFICE LOBBY 办公大厅

GROUND LEVEL 地面层
+ 0,000

OFFICE TERRACE WITH GARDEN ROOF VEGETATION
带有屋顶花园植被的办公露台

SPANDREL 拱肩镂板：
PAINTED ALUMINUM PANEL WITH RIGID INSULATION
硬性绝缘着色铝板

FIXED IGU PANEL: 固定的中空玻璃板：
TYPICAL FLUSH GLAZING UNIT
典型的齐平玻璃窗

OPERABLE IGU PANEL: 可控中空玻璃板
ALL CLEAR GLASS PANELS HINGED FOR EASY CLEANING TILT AND TURN OPERATION-FRESH AIR CIRCULATION
所有透明玻璃板结合易清洗倾斜度和有助于新鲜空气流通的旋转操作

300MM THICK CONCRETE STRUCTURAL FLOOR SLAB WITH RADIANT FLOOR AND CEILING FOR HEATING AND COOLING
300mm 混凝土结构楼板，带有辐射采暖地面和供热制冷天花板

ORANGE PAINTED ALUMINUM SOFFIT
橙色铝质底面

CONTINUOUS STAINLESS STEEL EXTERIOR ENVELOPE OF PUNCHED AND SUN ANGLE FORMED SHADES WITH THIN FILM PHOTOVOLTAICS LAMINATED TO EXTERIOR TO COLLECT SOLAR ENERGY
不锈钢连续型外表皮，穿孔且呈一定太阳角度形成阴影。建筑外表皮的薄层光伏板用以收集太阳能。

CONCRETE COLUMN
混凝土柱

INSULATED GLAZING UNIT WITH PAINTED STEEL BACK UP
绝缘玻璃窗体经涂漆钢支撑

ALUMINUM STACK JOINT AT FLOOR SLABS WITH CONTINUOUS ALUMINUM COVER-FLUSH WITH FINISHED FLOOR
楼板铝质材料相接合形成连续的铝材覆盖层——与整修过的楼面齐平

VERTICAL CIRCULATION TO SEA TERRACE
海蚀台地的垂直流通线路

INSULATED GLAZING UNIT WITH PAINTED STEEL BACK UP
绝缘玻璃窗体经涂漆钢支撑

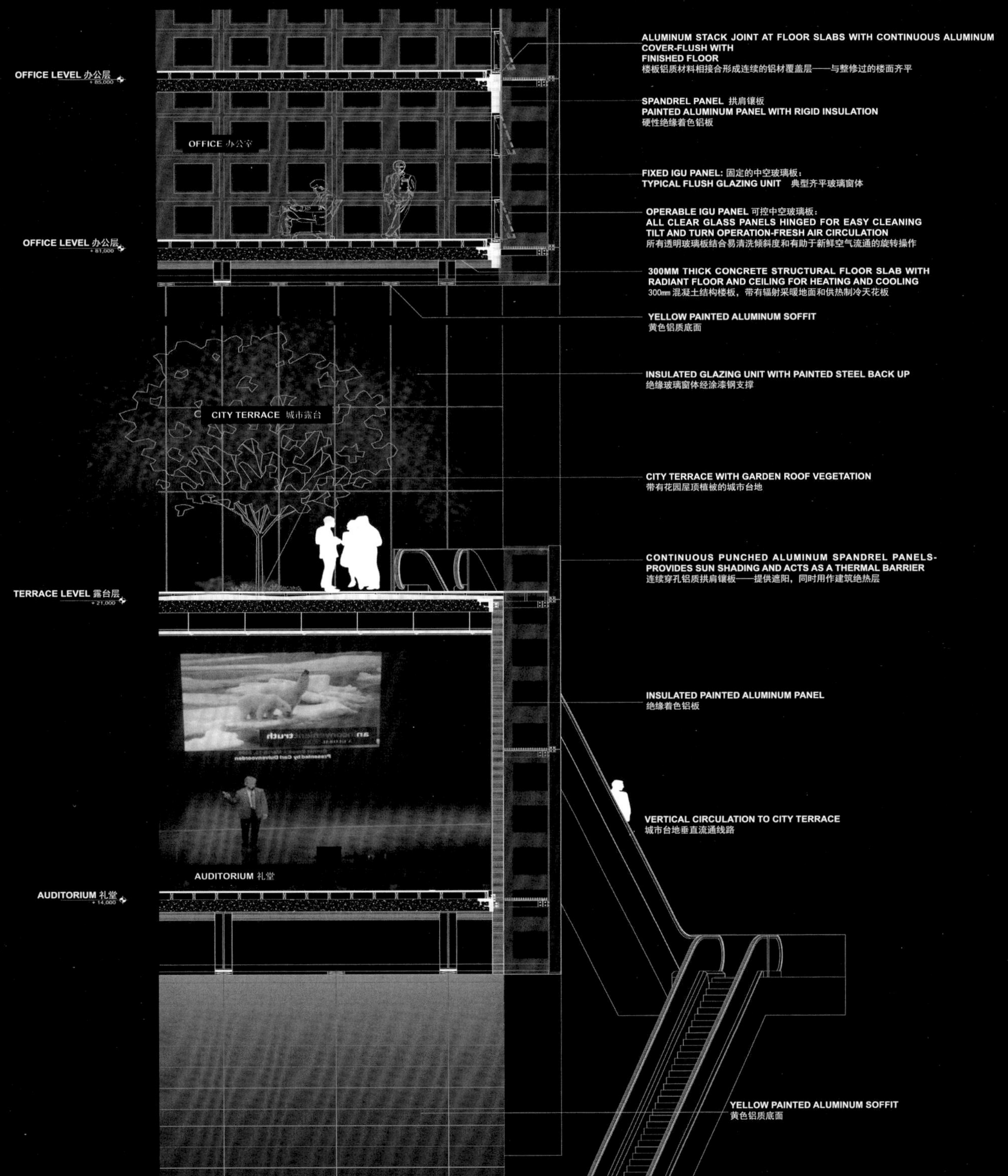

OFFICE LEVEL 办公层 + 85,000

OFFICE 办公室

OFFICE LEVEL 办公层 + 81,000

CITY TERRACE 城市露台

TERRACE LEVEL 露台层 + 21,000

AUDITORIUM 礼堂

AUDITORIUM 礼堂 + 14,000

OFFICE LOBBY 办公大厅

GROUND LEVEL 地面层 + 0,000

ALUMINUM STACK JOINT AT FLOOR SLABS WITH CONTINUOUS ALUMINUM COVER-FLUSH WITH FINISHED FLOOR
楼板铝质材料相接合形成连续的铝材覆盖层——与整修过的楼面齐平

SPANDREL PANEL 拱肩镶板
PAINTED ALUMINUM PANEL WITH RIGID INSULATION
硬性绝缘着色铝板

FIXED IGU PANEL: 固定的中空玻璃板：
TYPICAL FLUSH GLAZING UNIT 典型齐平玻璃窗体

OPERABLE IGU PANEL 可控中空玻璃板：
ALL CLEAR GLASS PANELS HINGED FOR EASY CLEANING
TILT AND TURN OPERATION-FRESH AIR CIRCULATION
所有透明玻璃板结合易清洗倾斜度和有助于新鲜空气流通的旋转操作

300MM THICK CONCRETE STRUCTURAL FLOOR SLAB WITH RADIANT FLOOR AND CEILING FOR HEATING AND COOLING
300mm 混凝土结构楼板，带有辐射采暖地面和供热制冷天花板

YELLOW PAINTED ALUMINUM SOFFIT
黄色铝质底面

INSULATED GLAZING UNIT WITH PAINTED STEEL BACK UP
绝缘玻璃窗体经涂漆钢支撑

CITY TERRACE WITH GARDEN ROOF VEGETATION
带有花园屋顶植被的城市台地

CONTINUOUS PUNCHED ALUMINUM SPANDREL PANELS-PROVIDES SUN SHADING AND ACTS AS A THERMAL BARRIER
连续穿孔铝质拱肩镶板——提供遮阳，同时用作建筑绝热层

INSULATED PAINTED ALUMINUM PANEL
绝缘着色铝板

VERTICAL CIRCULATION TO CITY TERRACE
城市台地垂直流通线路

YELLOW PAINTED ALUMINUM SOFFIT
黄色铝质底面

INSULATED GLAZING UNIT WITH PAINTED STEEL BACK UP
绝缘玻璃窗体经涂漆钢支撑

设计特色

方案包含了两栋各占港口一边的办公大楼以及一条连接两楼的高达 65 米的人行天桥。建筑师在桥两端的摩天大楼上各修建一个缆索桥，由于场地的几何形状，这两座悬索桥相遇时呈一定角度，似乎是在海港上方相握的两只手。这座桥梁给两个塔楼提供了一条连接通道，是一个公众瞭望台。

项目采用了多种先进的可持续方案以突出丹麦是世界领先的新能源利用国之一的地位。两栋大楼采用了高性能的玻璃幕墙，幕墙带有太阳能光伏屏，既能收集太阳能，又能起到遮阳的作用。另一方面，幕墙具备反射光的性能，可为办公空间提供最优的自然采光条件。

楼板中装有辐射供暖系统，天花板中装有辐射冷却系统，这些海水加热和制冷系统与幕墙中的太阳能光伏屏相连，可以为建筑供暖、供热。楼板和天花板上安装的窗口也可为建筑内部提供自然通风。人行天桥的顶部装有风力涡旋机，可将风能转化为电能。

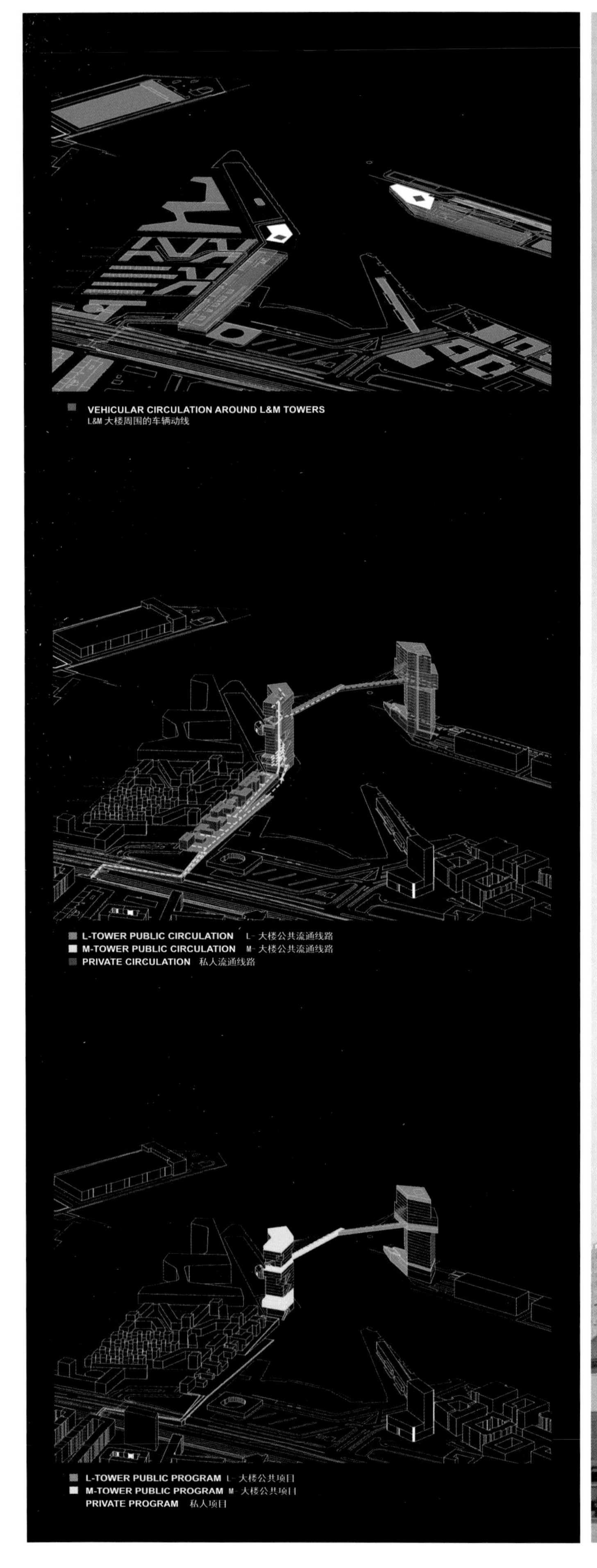

VEHICLE
DROP-OFF
下客区

RETAIL
零售区

OFFICE LOBBY
办公大厅

RETAIL
零售区

BRIDGE BOUNDARY
桥梁边界

BUILDING BOUNDARY
建筑边界

ESCALATOR UP TO
TERRACE LEVEL
通往露台层的扶梯

BRIDGE BOUNDARY
桥梁边界

RETAIL
零售区

BUILDING BOUNDARY
建筑边界

ESCALATOR UP TO
TERRACE LEVEL
通往露台层的扶梯

VEHICLE
DROP-OFF
下客区

GALLERY/RETAIL
走廊 / 零售区

CAFE 咖啡厅
KITCHEN 厨房

LOBBY
大厅

M-TOWER GROUND
LEVEL ENTRY
M- 大楼地面层入口

M-TOWER GROUND
LEVEL ENTRY
M- 大楼地面层入口

L-TOWER GROUND
LEVEL ENTRY
L- 大楼地面层入口

CAFE 咖啡厅

LOBBY
大厅

KITCHEN 厨房

RESTAURANT
餐厅

L-TOWER GROUND
LEVEL ENTRY
L- 大楼地面层入口

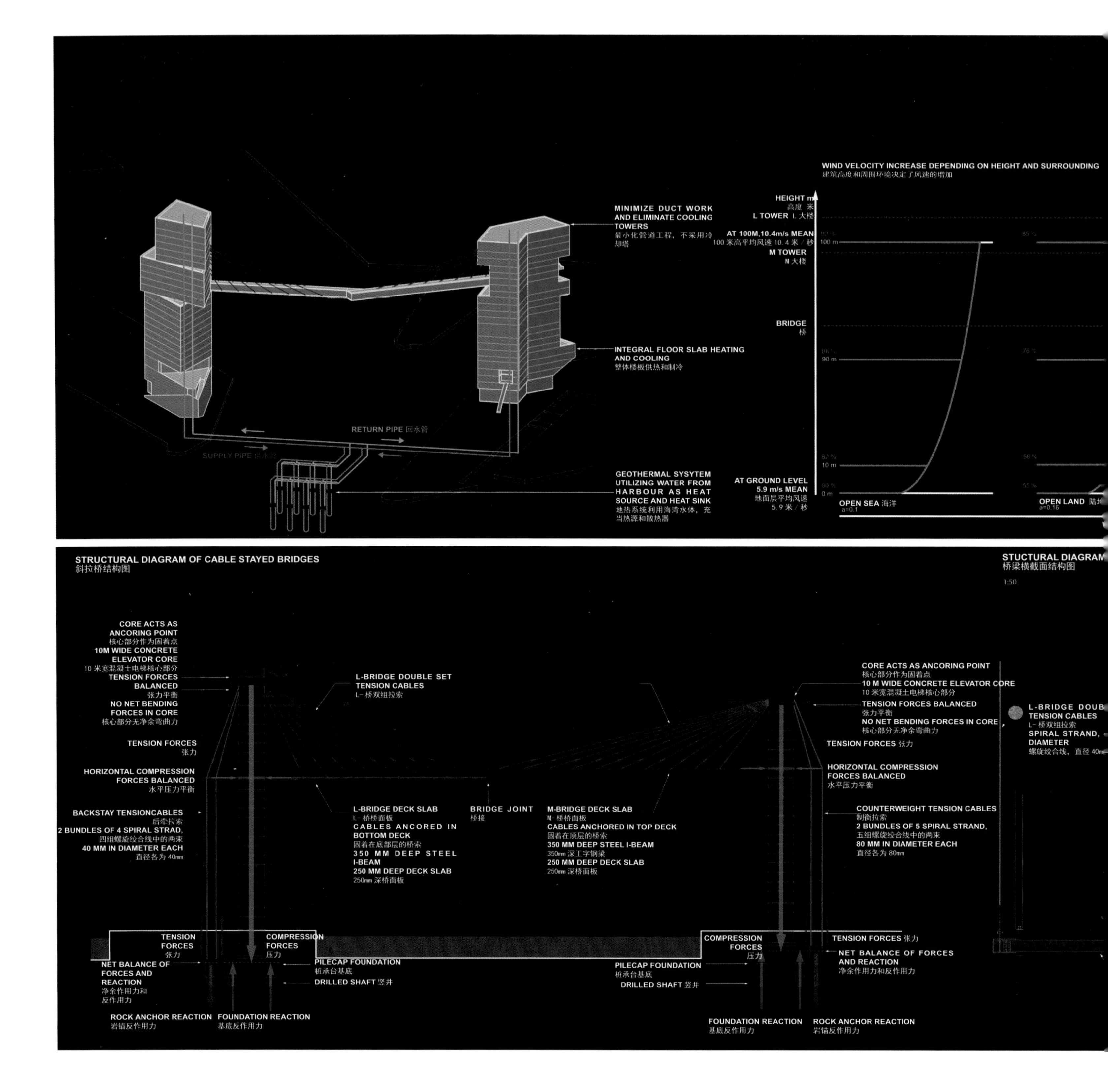

Profile

The design scheme of Steven Holl Architects is proposed for LM Harbor design competition which is organized by ATP Ejendomme and CPH City and Port Development. The project aims to build a harbor gateway in Copenhagen connecting Marmormolen tower and Langeliniekaj tower. It strives to create a brand new while continuous harbor city Copenhagen and make an extremely charming building landscape.

Design Feature

The scheme involves two office towers at both ends of the harbor as well as a 65-meter passageway connecting the two. Each office tower carries its own cable-stay bridge that is a public passageway between the two piers. Due to the site geometry, these bridges

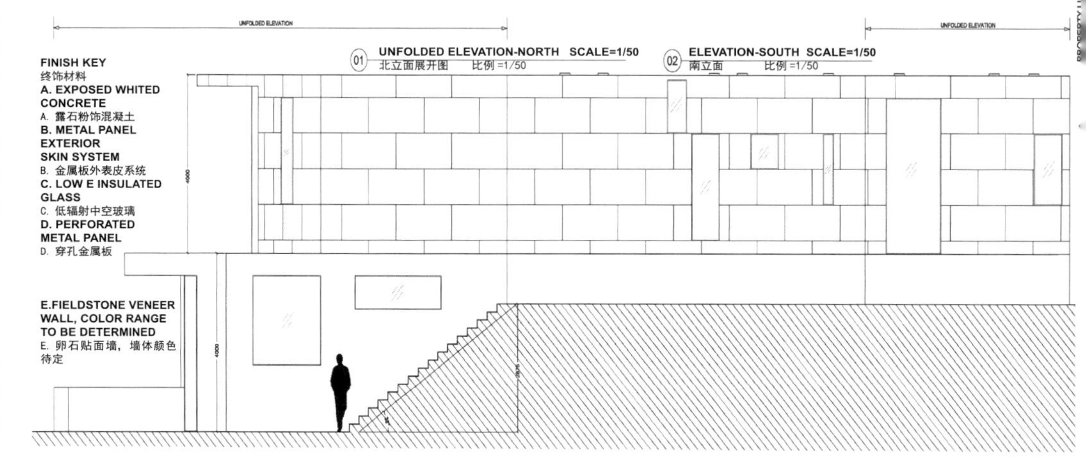

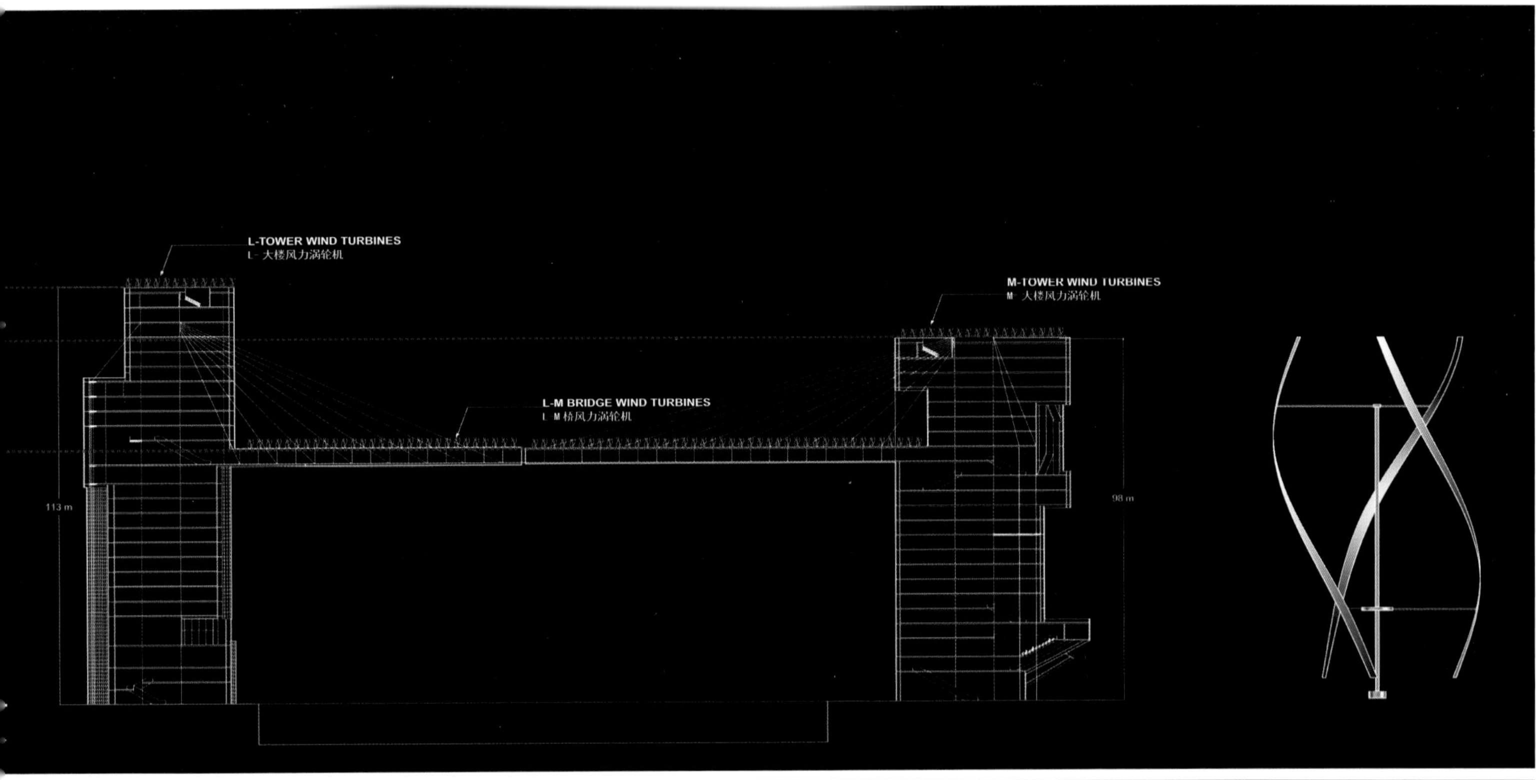
L-TOWER WIND TURBINES
L- 大楼风力涡轮机
M-TOWER WIND TURBINES
M- 大楼风力涡轮机
L-M BRIDGE WIND TURBINES
L-M 桥风力涡轮机
113 m
98 m

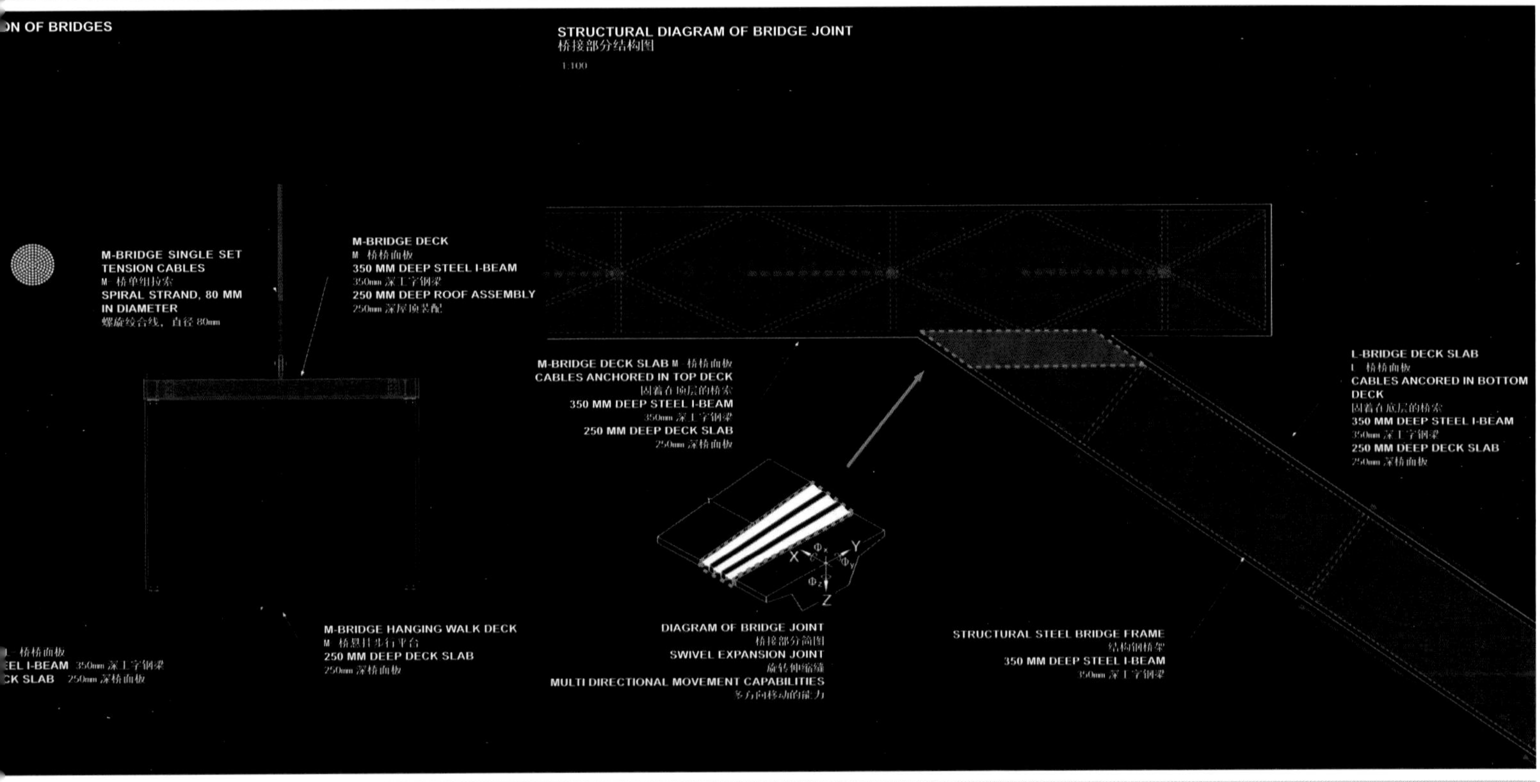
ON OF BRIDGES
STRUCTURAL DIAGRAM OF BRIDGE JOINT
桥接部分结构图
1:100
M-BRIDGE SINGLE SET TENSION CABLES
M- 桥单组拉索
SPIRAL STRAND, 80 MM IN DIAMETER
螺旋绞合线，直径 80mm
M-BRIDGE DECK
M- 桥桥面板
350 MM DEEP STEEL I-BEAM
350mm 深工字钢梁
250 MM DEEP ROOF ASSEMBLY
250mm 深屋顶装配
M-BRIDGE DECK SLAB M- 桥桥面板
CABLES ANCHORED IN TOP DECK
固着在顶层的桥索
350 MM DEEP STEEL I-BEAM
350mm 深工字钢梁
250 MM DEEP DECK SLAB
250mm 深桥面板
L-BRIDGE DECK SLAB
L- 桥桥面板
CABLES ANCORED IN BOTTOM DECK
固着在底层的桥索
350 MM DEEP STEEL I-BEAM
350mm 深工字钢梁
250 MM DEEP DECK SLAB
250mm 深桥面板
X Y Z
桥桥面板
EEL I-BEAM 350mm 深工字钢梁
CK SLAB 250mm 深桥面板
M-BRIDGE HANGING WALK DECK
M- 桥悬挂步行平台
250 MM DEEP DECK SLAB
250mm 深桥面板
DIAGRAM OF BRIDGE JOINT
桥接部分简图
SWIVEL EXPANSION JOINT
旋转伸缩缝
MULTI DIRECTIONAL MOVEMENT CAPABILITIES
多方向移动的能力
STRUCTURAL STEEL BRIDGE FRAME
结构钢桥架
350 MM DEEP STEEL I-BEAM
350mm 深工字钢梁

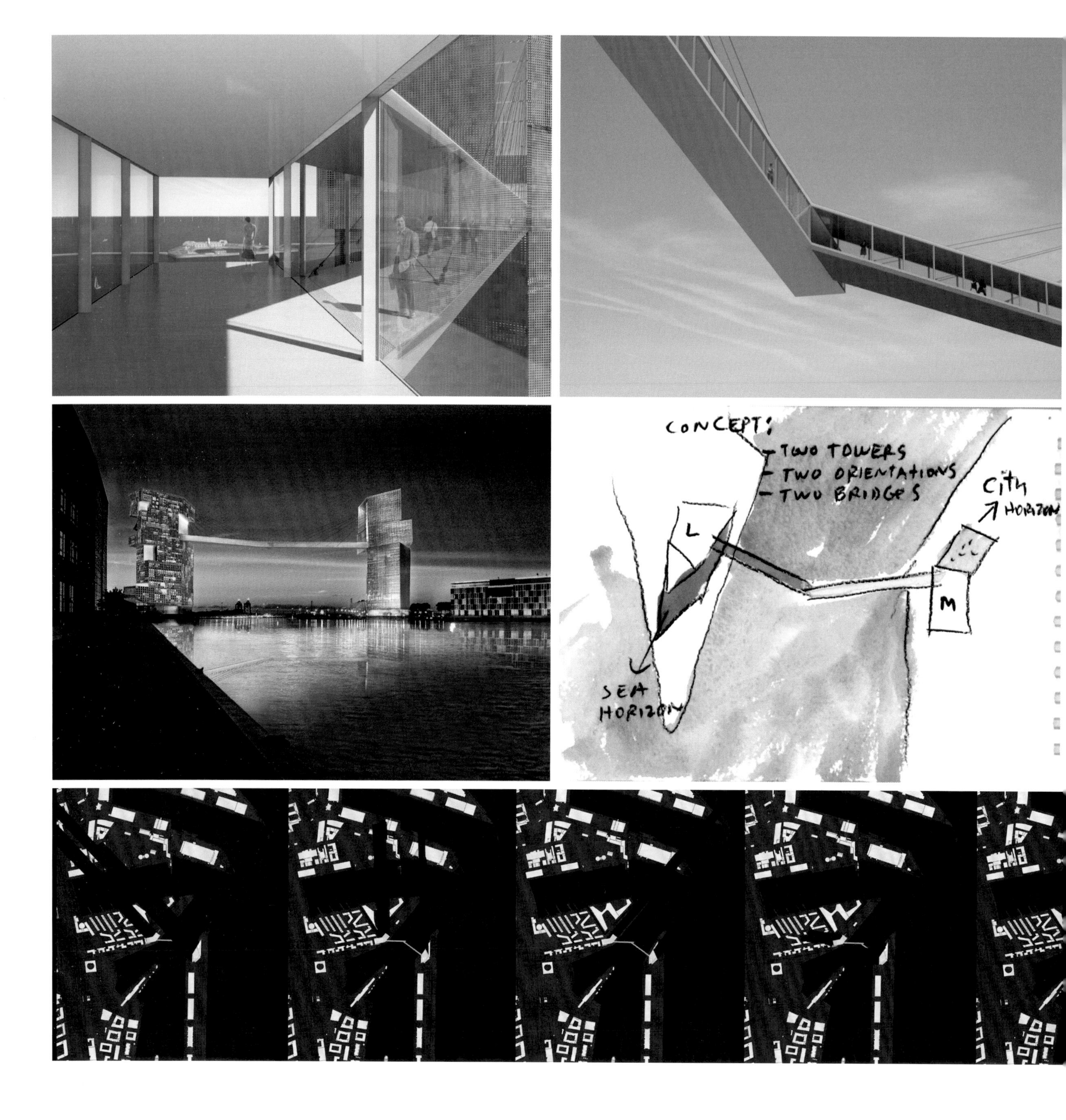

meet at an angle, joining like a handshake over the harbor. The bridge offers a passageway and a public terrace.

The project utilizing a variety of progressive sustainable solutions to ensure this important international landmark is rooted in Denmark's identity as one of the world leaders in alternative energy. Both towers have high performance glass curtain walls with a veil of solar screen made of photovoltaics; collecting the sun's energy while shading. Optimum natural light is provided to all offices due to the reflective light performance of the screens.

There is a seawater heating/cooling system with radiant heating in the floor slabs and radiant cooling in the ceiling. Natural ventilation is provided on every floor with windows opening at the floor level and ceiling level for maximum air circulation. Wind turbines line the top of the pedestrian bridge roof providing all electricity for lighting the public spaces.

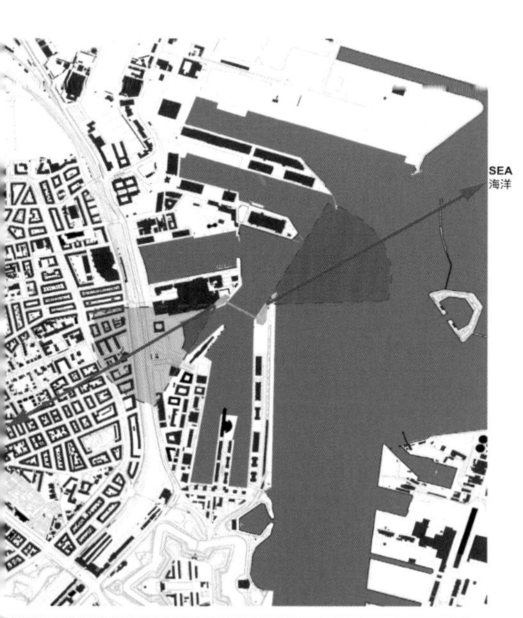

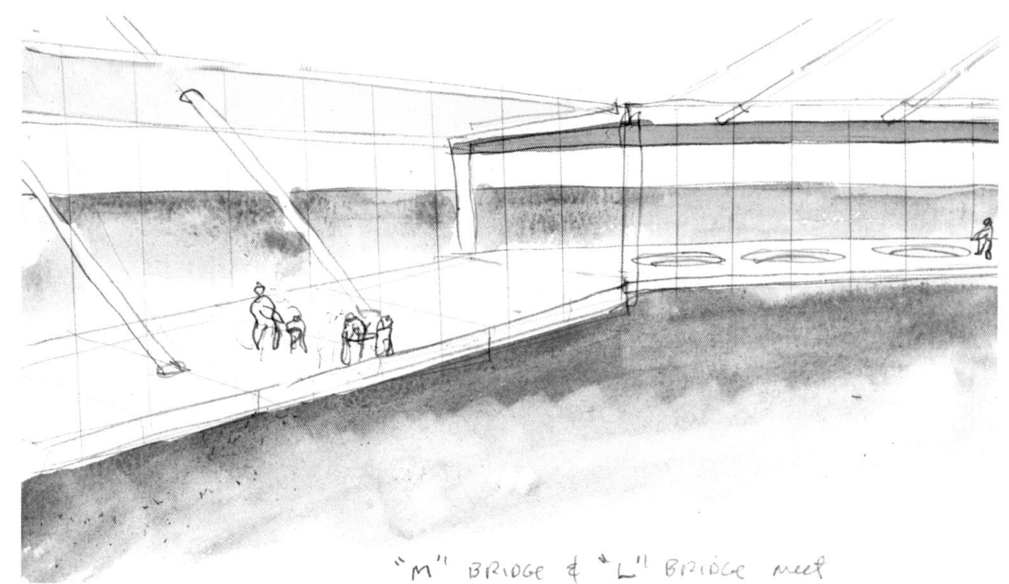

PUBLIC CIRCULATION:

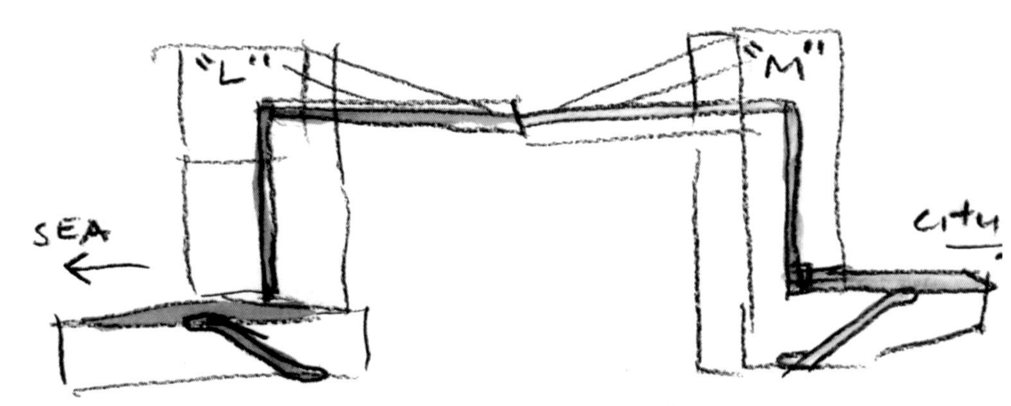

PUBLIC ESCALATORS TO SEA & City TERRACES
LARGE ELEVATORS TO BRIDGE

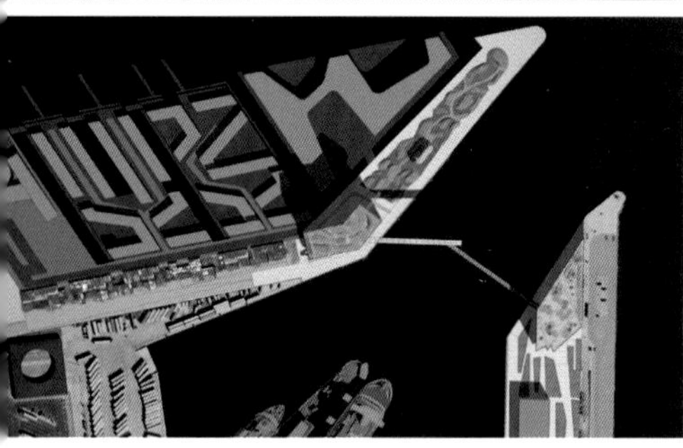

天津 TEDA MSD H2 低碳示范楼

TEDA MSD H2 Low Carbon Building

设计单位：阿特金斯 ATKINS
开发商：天津泰达发展有限公司
项目地址：中国天津市
总建筑面积：21 158 ㎡

Designed by: Atkins
Client: Tianjin TEDA
Location: Tianjin, China
Total Floor Area: 21,158 m^2

五大体系 24项技术

项目概况

项目位于天津泰达经济技术开发区现代服务产业园区内，是一栋兼顾技术研究与科技推广的、具有示范性功能的低碳建筑。

设计特色

设计不仅以理性的方式探索了新材料和新技术的应用问题，而且塑造了现代、高科技的建筑形象，实现了技术与建筑形态的完美结合。同时，项目也可能成为世界上首个同时通过中国三星认证、日本CASBEE认证、英国BREEAM认证、美国LEED认证四项绿色建筑认证的低碳建筑。

建筑及结构
屋顶绿化
幕墙系统
建筑材料
可再生能源利用
太阳能光伏发电
太阳能热水
自然光利用（水平光管）
自然光利用（垂直光管）
地源热泵
电气及照明系统
LED灯源
人工照明与光导照明的结合
楼宇自动控制终端
变频电梯与电梯能源再生
给排水系统
中水回用应用
节水型洁具
空调通风系统
毛细管空调系统
变频技术应用
诱导风机
室内通风温度监测
室内气体浓度监测
地板送风空调系统
冷热源
带热回收的热泵型溶液

•建筑及结构
•可再生能源利用
•电器及照明系统
•给排水系统
•空调通风系统

总节能耗 **30%**

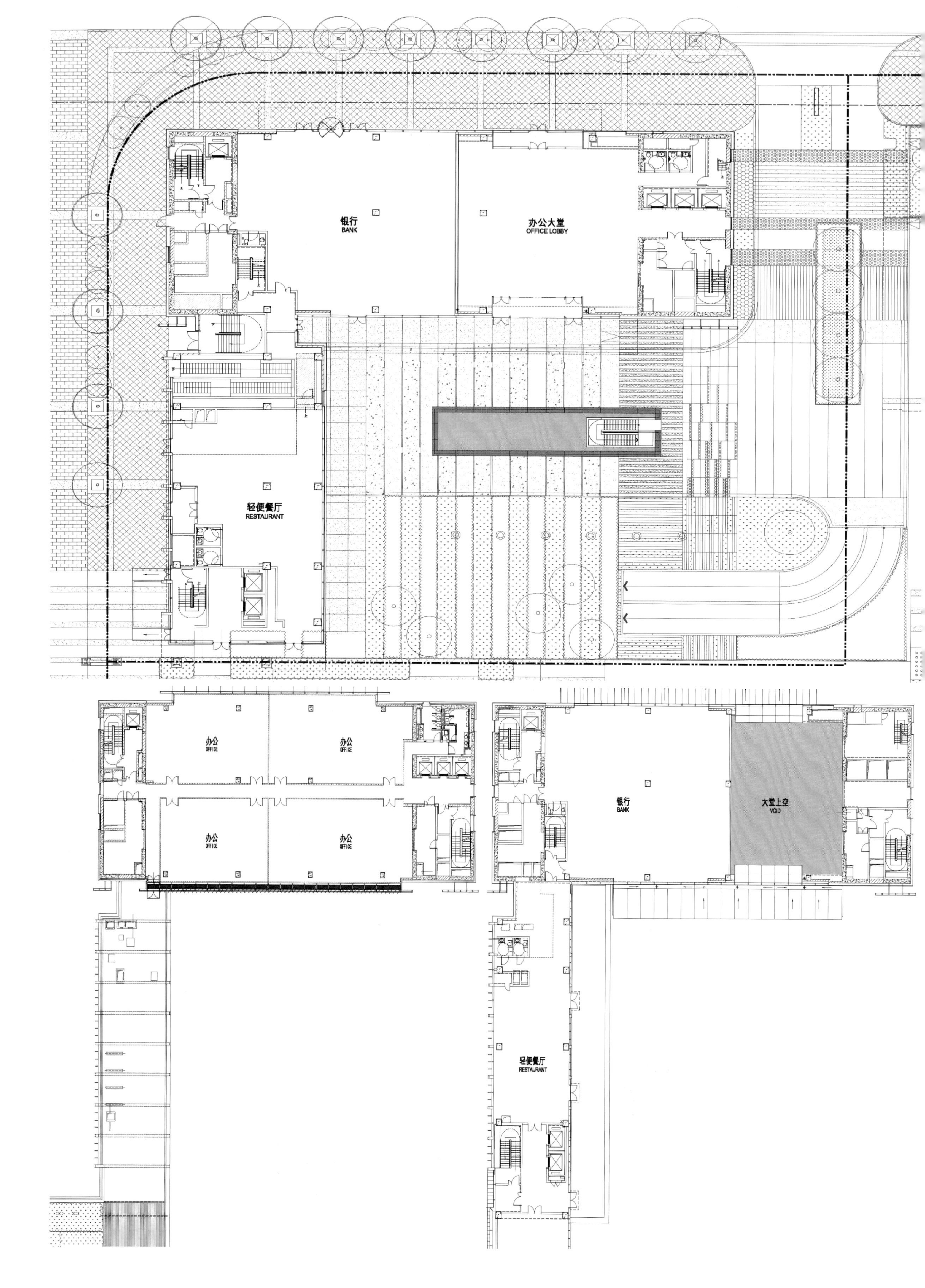

银行
BANK
办公大堂
OFFICE LOBBY
轻便餐厅
RESTAURANT
办公
OFFICE
办公
OFFICE
办公
OFFICE
办公
OFFICE
银行
BANK
大堂上空
VOID
轻便餐厅
RESTAURANT

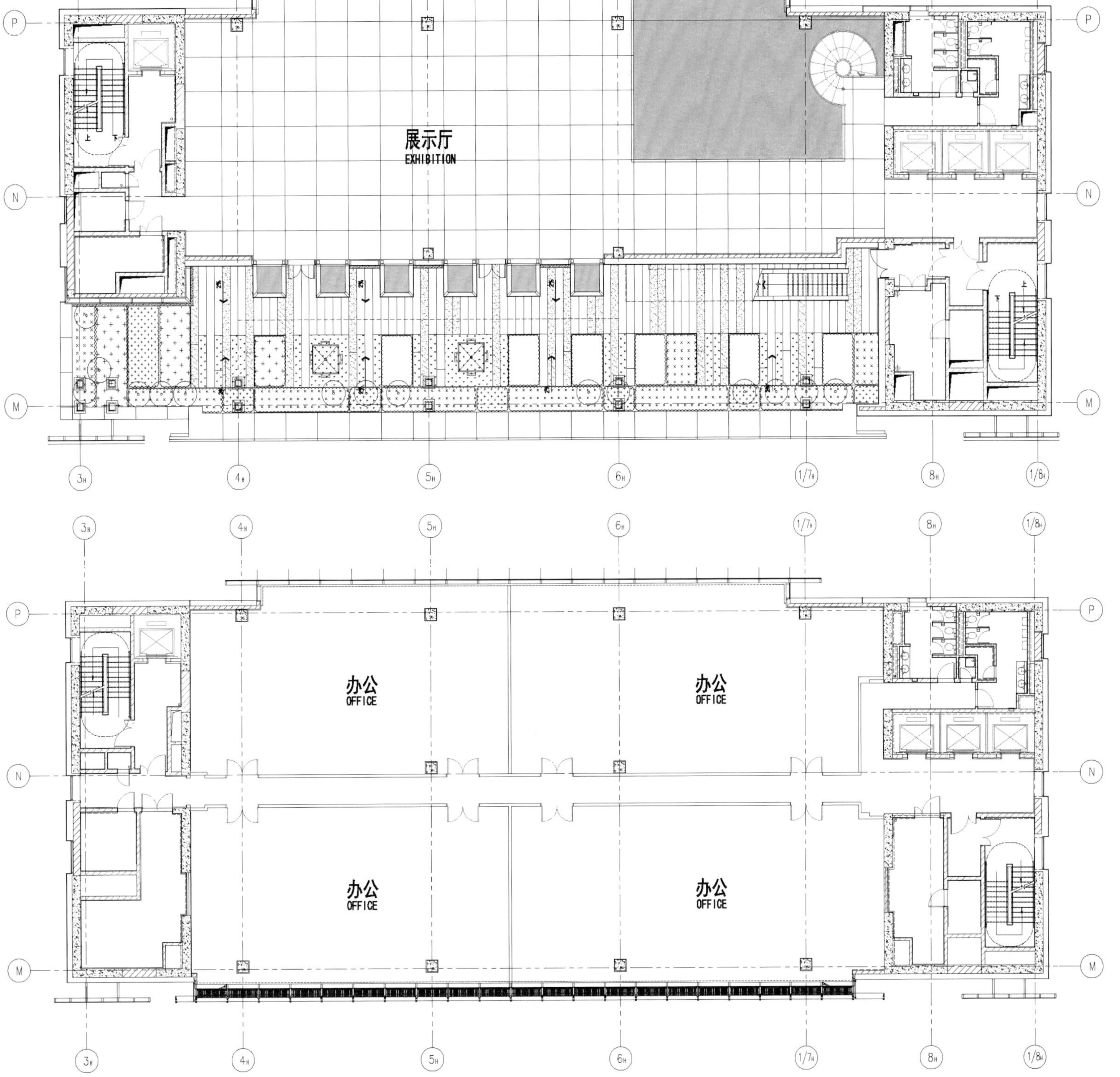

基于该建筑物的复合体量和独特的高性能幕墙体系，设计团队将其理念形象地称为“生态三明治”，即将因功能需要而不同的立面元素从南到北分层设置。南立面最外层为光伏玻璃，独立于建筑主体结构之外，3层以上的南立面设计为双层呼吸式幕墙；北立面为双中空玻璃幕墙，充分考虑了低碳节能的要求。

设计采用了多种绿色技术，包括最基础的雨水中水回收利用技术以及最先进的光伏技术、太阳能利用技术，既低碳节能，又可营造舒适的内部环境。建筑采用的低碳技术主要包括五个体系：建筑及结构、可再生能源利用、电器及照明系统、给排水系统、空调通风系统。

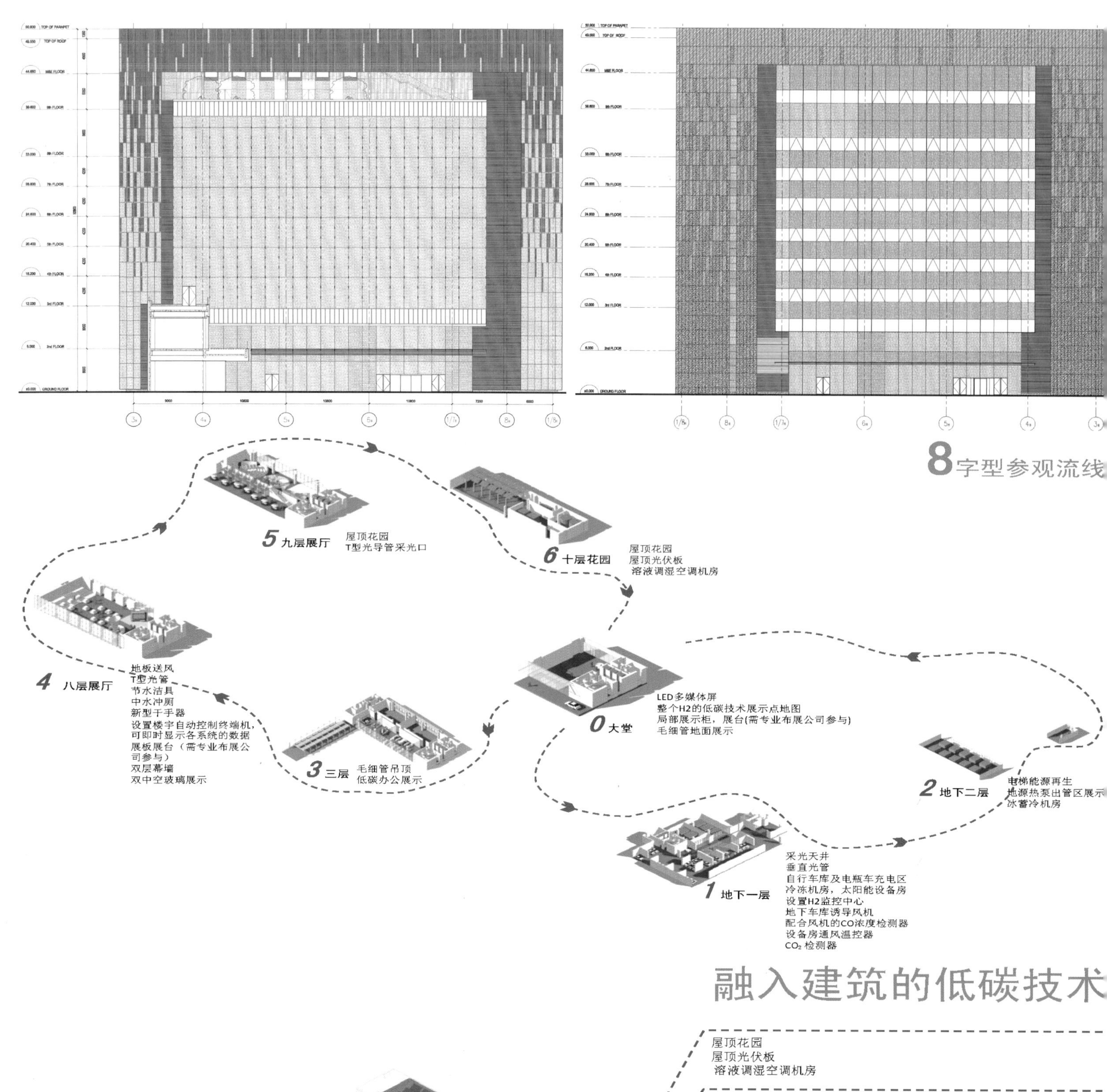
8字型参观流线
5 九层展厅
屋顶花园
T型光导管采光口
6 十层花园
屋顶花园
屋顶光伏板
溶液调湿空调机房
4 八层展厅
地板送风
T型光管
节水洁具
中水冲厕
新型干手器
设置楼宇自动控制终端机，可即时显示各系统的数据
展板展台（需专业布展公司参与）
双层幕墙
双中空玻璃展示
3 三层
毛细管吊顶
低碳办公展示
0 大堂
LED多媒体屏
整个H2的低碳技术展示点地图
局部展示柜，展台(需专业布展公司参与)
毛细管地面展示
2 地下二层
电梯能源再生
地源热泵出管区展示
冰蓄冷机房
1 地下一层
采光天井
垂直光管
自行车库及电瓶车充电区
冷冻机房，太阳能设备房
设置H2监控中心
地下车库诱导风机
配合风机的CO浓度检测器
设备房通风温控器
CO2 检测器

融入建筑的低碳技术

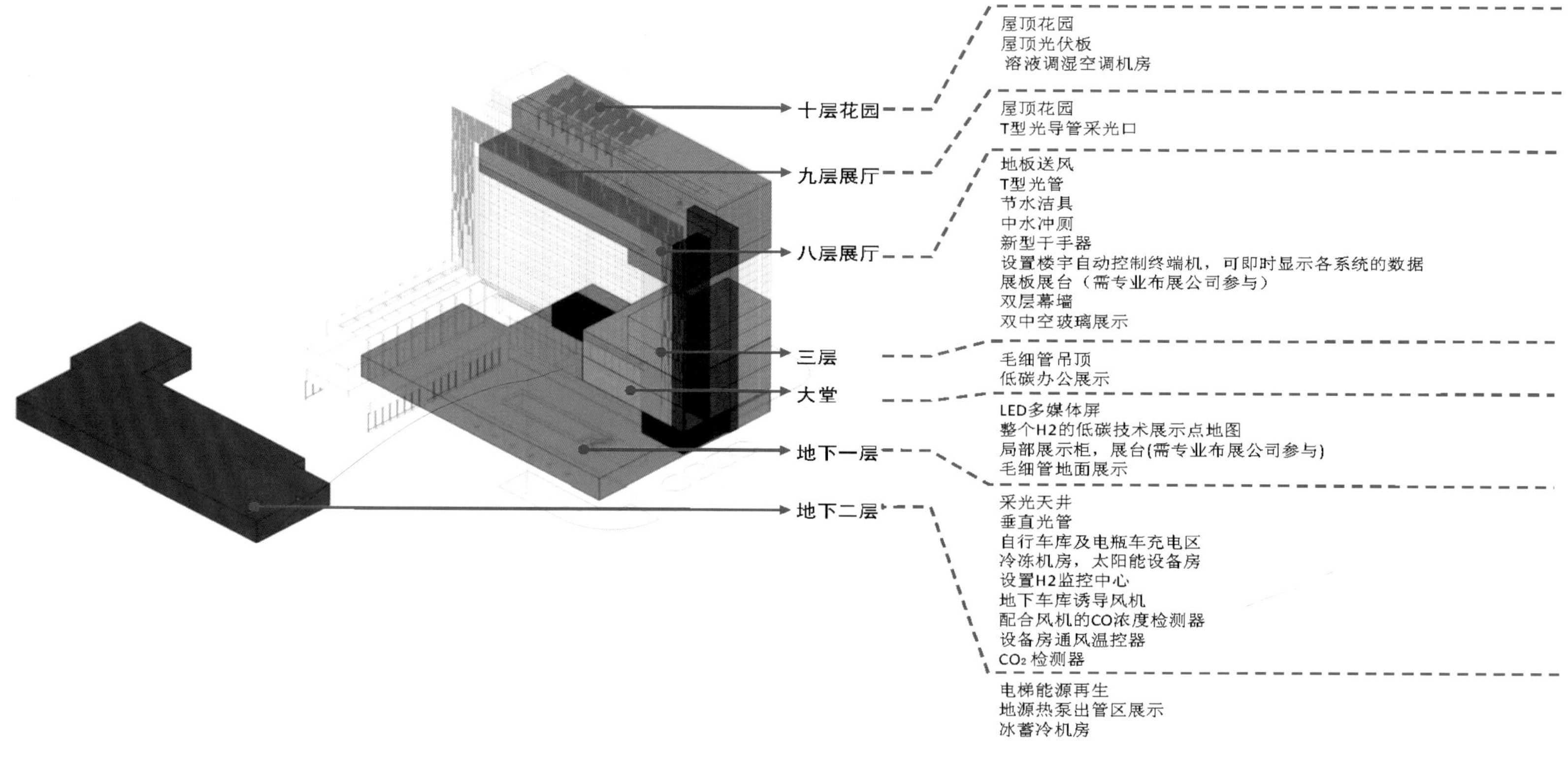
十层花园
屋顶花园
屋顶光伏板
溶液调湿空调机房
九层展厅
屋顶花园
T型光导管采光口
八层展厅
地板送风
T型光管
节水洁具
中水冲厕
新型干手器
设置楼宇自动控制终端机，可即时显示各系统的数据
展板展台（需专业布展公司参与）
双层幕墙
双中空玻璃展示
三层
毛细管吊顶
低碳办公展示
大堂
LED多媒体屏
整个H2的低碳技术展示点地图
局部展示柜，展台(需专业布展公司参与)
毛细管地面展示
地下一层
采光天井
垂直光管
自行车库及电瓶车充电区
冷冻机房，太阳能设备房
设置H2监控中心
地下车库诱导风机
配合风机的CO浓度检测器
设备房通风温控器
CO2 检测器
地下二层
电梯能源再生
地源热泵出管区展示
冰蓄冷机房

Profile

Located at Tianjin TEDA Modern Service District, H2 low carbon building was conceived as a demonstration project and research platform for green building technologies.

Design Feature

As a technological showcase, the project not only explores the feasibility of new building materials and systems, it is an attempt to integrate low carbon design with a smart, attractive architectural expression. H2 low carbon building is poised to be the first low carbon project in the world to be accredited for four green building standards: China 3 Stars, CASBEE, BREEAM and LEED.

Due to its layered, slim rectilinear appearance and performance-inspired facades, the project has been nicknamed "Ecological Sandwich" by the design team. An array of photovoltaic cells is positioned in front of double-skinned façade on south elevation, while the north façade utilizes triple-glazing curtain wall panels. Service cores are located at the east and west ends of the building.

Incorporated green building initiatives include simple technique such as rainwater harvesting to relatively complex photovoltaic system for solar energy. The aim is to create a energy saving, comfortable indoor building environment through 5 aspects of architectural and engineering designs — building structure & architectural envelope, renewable energy, electrical & lighting design, plumbing & sanitary engineering, HVAC engineering.

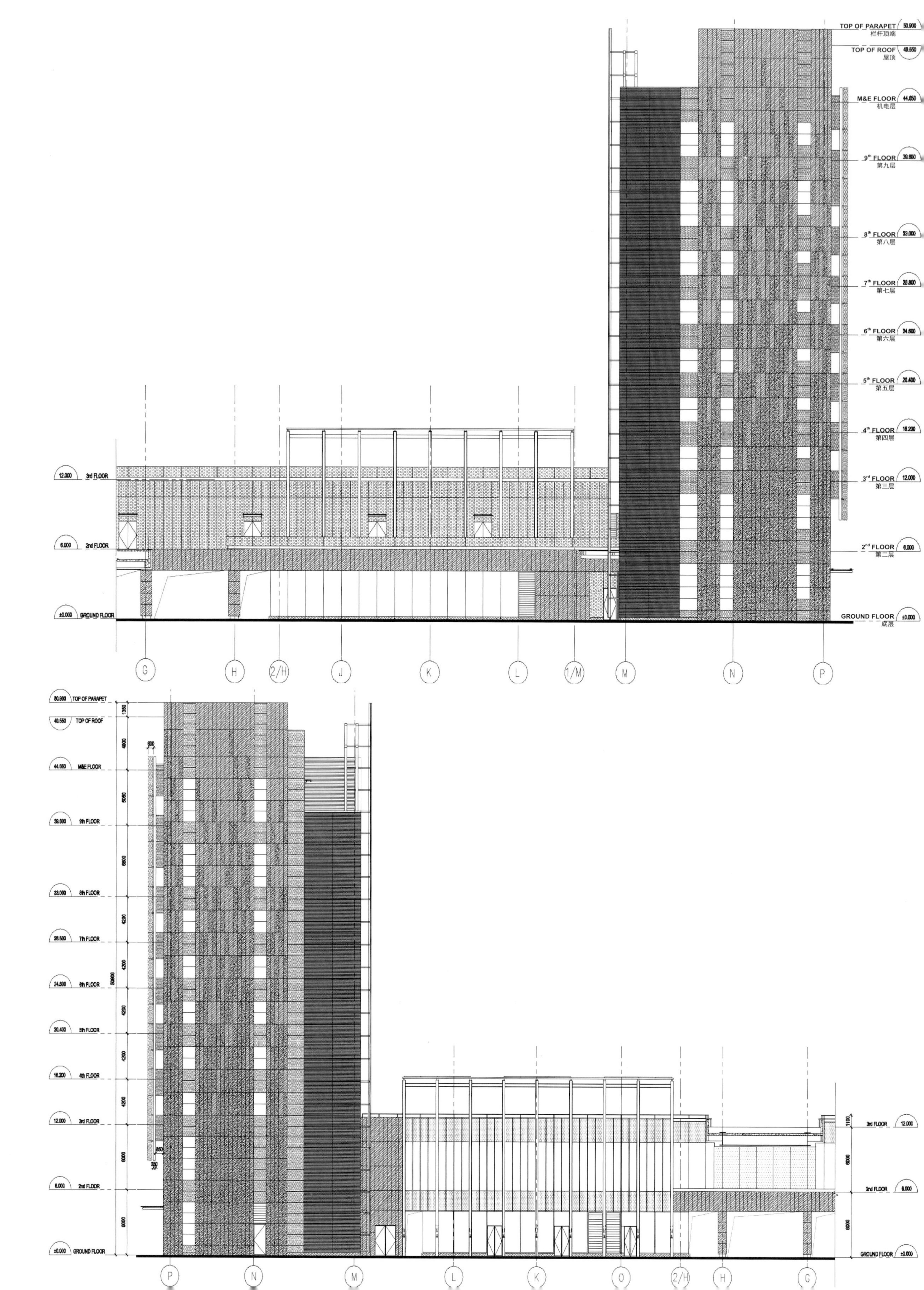

TOP OF PARAPET 栏杆顶端 50.900
TOP OF ROOF 屋顶 49.550
M&E FLOOR 机电层 44.650
9th FLOOR 第九层 39.600
8th FLOOR 第八层 33.000
7th FLOOR 第七层 28.800
6th FLOOR 第六层 24.600
5th FLOOR 第五层 20.400
4th FLOOR 第四层 16.200
3rd FLOOR 第三层 12.000
2nd FLOOR 第二层 6.000
GROUND FLOOR 底层 ±0.000
G
H
2/H
J
K
L
1/M
M
N
P

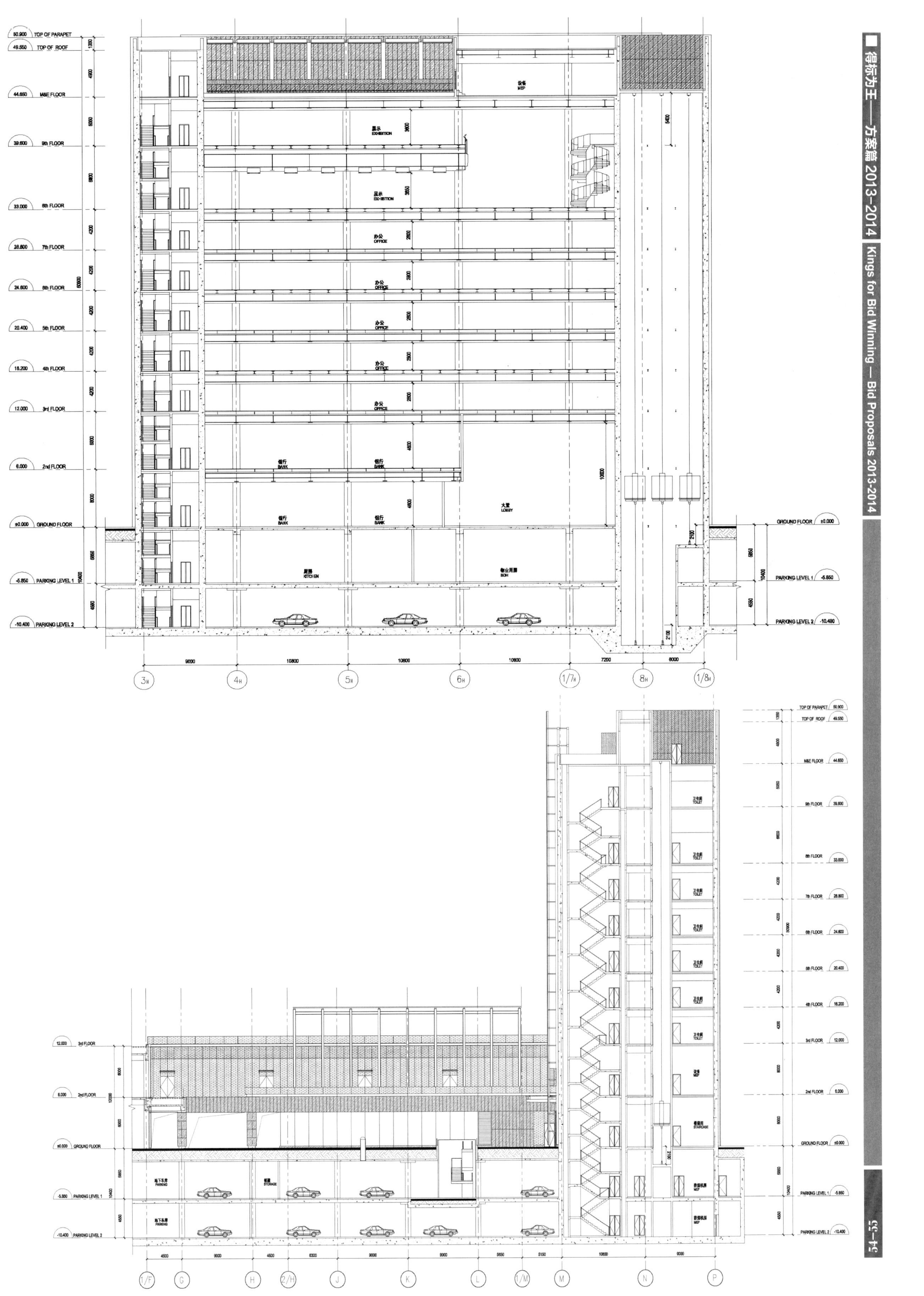
50.900 TOP OF PARAPET
49.550 TOP OF ROOF
44.650 M&E FLOOR
39.600 9th FLOOR
33.000 8th FLOOR
28.800 7th FLOOR
24.600 6th FLOOR
20.400 5th FLOOR
16.200 4th FLOOR
12.000 3rd FLOOR
6.000 2nd FLOOR
±0.000 GROUND FLOOR
-5.850 PARKING LEVEL 1
-10.400 PARKING LEVEL 2
设备 MEP
展示 EXHIBITION
办公 OFFICE
银行 BANK
大堂 LOBBY
厨房 KITCHEN
物业用房 BOH
卫生间 TOILET
楼梯间 STAIRCASE
地下车库 PARKING
储藏 STORAGE
9000
10800
7200
6000
3H
4H
5H
6H
1/7H
8H
1/8H
1/F
G
H
2/H
J
K
L
1/M
M
N
P

医疗建筑 Medical Building

挪威贝尔根 Haraldsplass 医院扩展

Haraldsplass Hospital Extension

设计单位：C.F.Møller Architects
项目地址：挪威贝尔根
项目面积：10 000 ㎡

Designed by: C.F. Møller
Location: Bergen, Norway
Area: 10,000 m²

项目概况

这是 C.F.Møller Architects 为 Haraldsplass 医院设计的新住院大楼的获奖方案。设计移除了传统的医院走廊，用开放的公共空间和高效的后勤设施来替代，设计了一个全新类别的医院项目。

建筑设计

Haraldsplass 医院建于 1986 年，配备有 184 个床位。新的建筑占地 10 000 平方米，将在建筑的三层空间内新增 108 个床位，同时还将建造一个可容纳 400 个车位的地下停车场。

这栋新的竹园大楼没有长长的走廊，与传统的医院建筑形成鲜明的对比。病房围绕着两个大型的有顶的天井分布，为设置两个不同类型的公共空间预留了空间，其中一个是配备了接待区、咖啡吧、商店和休息区的公共空间；另外一个则是只为患者及其访客服务的较为私密的公共区。

天井可以将日光渗透到建筑内部，建筑内部茂盛的植被，如种植在水池中的竹子、草坪、花卉以及藤蔓类植物，确保了良好的室内环境。这栋新的建筑沿着 Møllendalselven 河分布，形成了一个呈一定角度的立面，这使所有的患者都可以享有朝向山谷和城市的视野。

项目因其生态友好性而凸显出来，因为相对总建筑面积来说，建筑外立面的尺度很小。同时，通过改善建筑的通风设备和减少已有建筑的热量损耗，这栋新的住院大楼可以达到“被动屋”的标准。

HAUKELAND SKOLE
ELVEPARK
TRENINGS-
LØYPEN
BORETTSLAGS-
HAGEN
SØSTRENES
BORETTSLAG
SANSEHAGEN
VARELEVERING/
RENOVASJON
TAKHAGEN
AKUTT-
MOTTAK
AKUTT-
MOTTAKET
SMU
FORPLASS
NY GANGBRO
ELVE-
PLASSEN
ADKOMST-
TORGET
RESEPSJONS-
TORGET
MELLOM-
ROMMET
EKSIST.
SYKEHUS
STRIPE
HAGEN
KAFÉ-
HAGEN
MEDITASJONS-
HAGEN
ELVEPARK
NY GANGBRO
KANTINE
EKSISTERENDE
ALLÉHAGE
RUINHAGEN
SJELESORG-
SENTERET
OG SØSTER
HJEMMET
KANTINE-
HAGEN
P-PLASS
UTSIKTEN
MOHNS HOVEDGÅRD
SYKEPLEIER-
HØYSKOLE
REGINES
MINNE
EKSISTERENDE
FJELLHAGE
EKSISTERENDE
SKOLEHAGE
ULRIKSDAL
HELSEPARK
eks gangbro
Asfalt gangarealer
Sykehusbygg HDS
Plen
Hekk
Mur
Taktil ledelinje
Asfalt kjøreareal
Plassdekke
Uberørt
Eks. trær
Trapp/amfi
Nye koter
Grusdekke
Grønt tak sengebygg
Busker/plantefelt
Nye trær
Benk
Utebord og stoler
N
0
50
100
150
Horisontal skala
1:500

Luftrom
Stillerom
Arbeidsrom
Arbeidsstasjon
Kontor
WC Pers.
Skyllerom
Forrom 6,1 m²
Bad 5,6 m²
Isolat 16,5 m²
Enerom 16,5 m²
10 m
UTSNITT SENGEPOST_1: 50 (1: 100 i A3)
DETALJE SENGEPOST_1: 50 (1: 100 i A3)
PLAN U1_1:200 (1: 400 i A3)
PLAN 1_1:200 (1: 400 i A3)
PLAN 1
PLAN 3

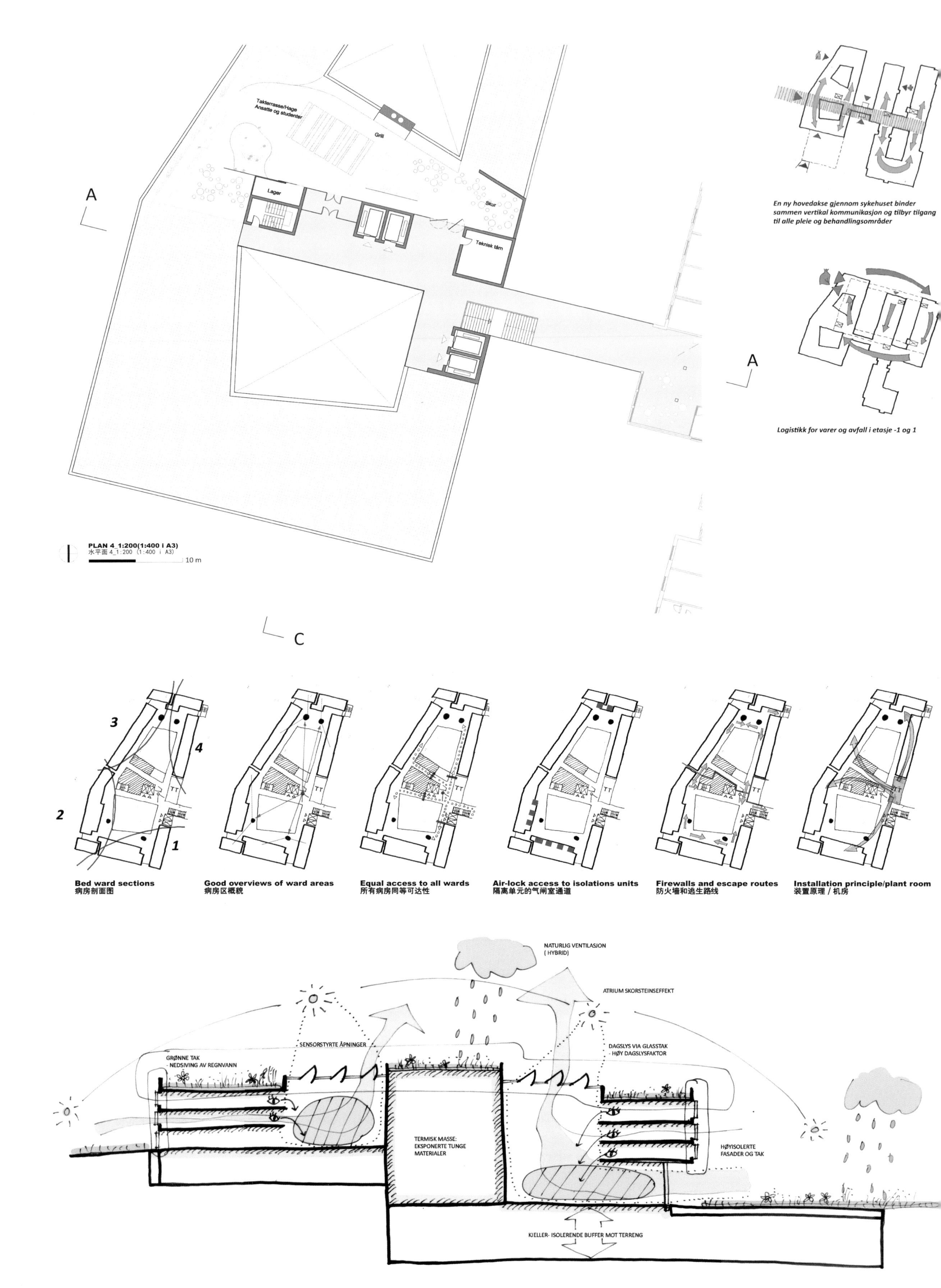
Takterrasse/Hage
Ansatte og studenter
Grill
Lager
Skur
Teknisk tårn
A
A
C
PLAN 4_1:200(1:400 i A3)
水平面 4_1:200（1:400 i A3）
10 m
En ny hovedakse gjennom sykehuset binder sammen vertikal kommunikasjon og tilbyr tilgang til alle pleie og behandlingsområder
Logistikk for varer og avfall i etasje -1 og 1
1
2
3
4
Bed ward sections
病房剖面图
Good overviews of ward areas
病房区概貌
Equal access to all wards
所有病房同等可达性
Air-lock access to isolations units
隔离单元的气闸室通道
Firewalls and escape routes
防火墙和逃生路线
Installation principle/plant room
装置原理 / 机房
NATURLIG VENTILASJON
(HYBRID)
ATRIUM SKORSTEINSEFFEKT
SENSORSTYRTE ÅPNINGER
GRØNNE TAK
- NEDSIVING AV REGNVANN
DAGSLYS VIA GLASSTAK
- HØY DAGSLYSFAKTOR
TERMISK MASSE:
EKSPONERTE TUNGE
MATERIALER
HØYISOLERTE
FASADER OG TAK
KJELLER- ISOLERENDE BUFFER MOT TERRENG

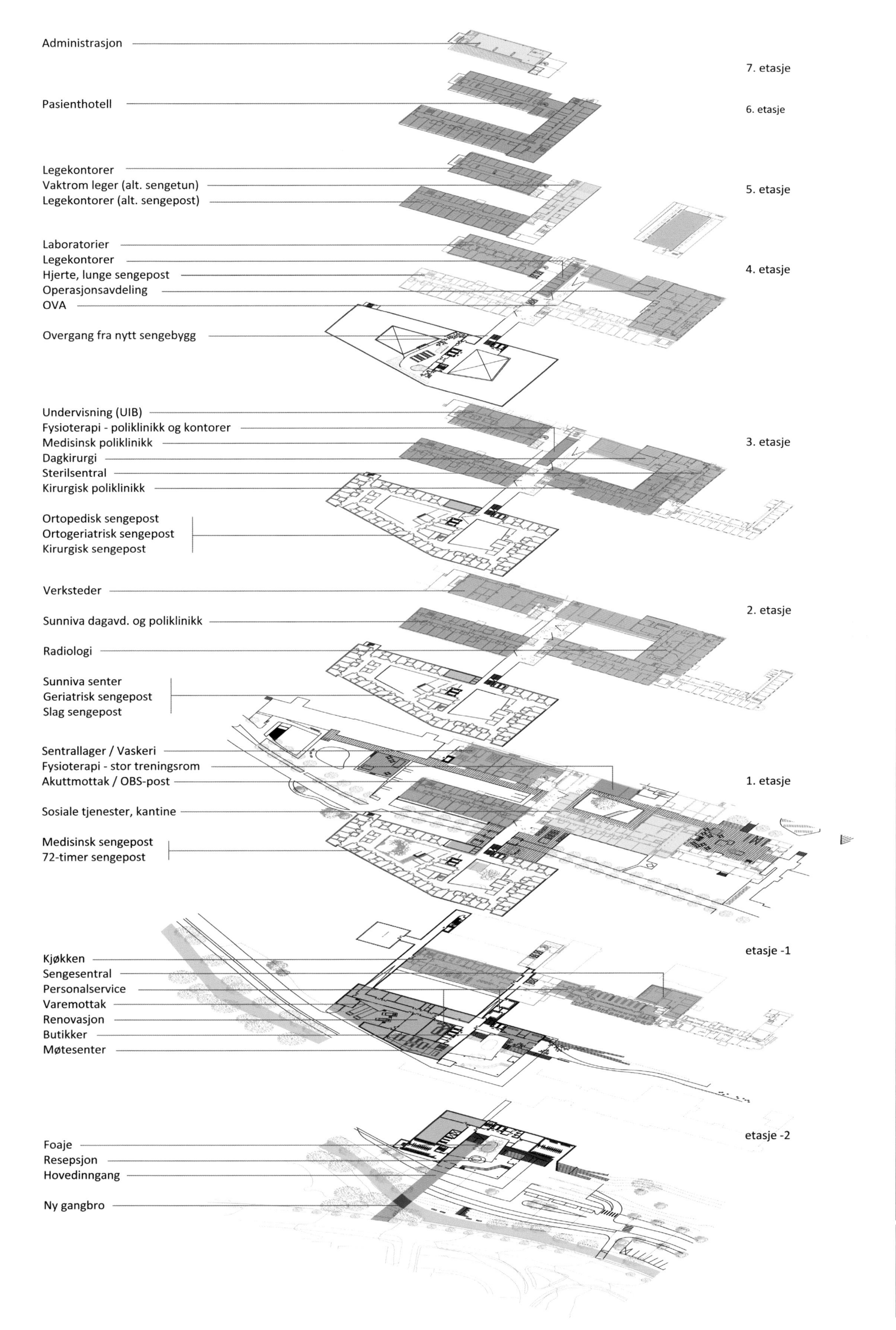
Administrasjon
7. etasje
Pasienthotell
6. etasje
Legekontorer
Vaktrom leger (alt. sengetun)
Legekontorer (alt. sengepost)
5. etasje
Laboratorier
Legekontorer
Hjerte, lunge sengepost
Operasjonsavdeling
OVA
4. etasje
Overgang fra nytt sengebygg
Undervisning (UIB)
Fysioterapi - poliklinikk og kontorer
Medisinsk poliklinikk
Dagkirurgi
Sterilsentral
Kirurgisk poliklinikk
3. etasje
Ortopedisk sengepost
Ortogeriatrisk sengepost
Kirurgisk sengepost
Verksteder
Sunniva dagavd. og poliklinikk
2. etasje
Radiologi
Sunniva senter
Geriatrisk sengepost
Slag sengepost
Sentrallager / Vaskeri
Fysioterapi - stor treningsrom
Akuttmottak / OBS-post
1. etasje
Sosiale tjenester, kantine
Medisinsk sengepost
72-timer sengepost
Kjøkken
etasje -1
Sengesentral
Personalservice
Varemottak
Renovasjon
Butikker
Møtesenter
Foaje
etasje -2
Resepsjon
Hovedinngang
Ny gangbro

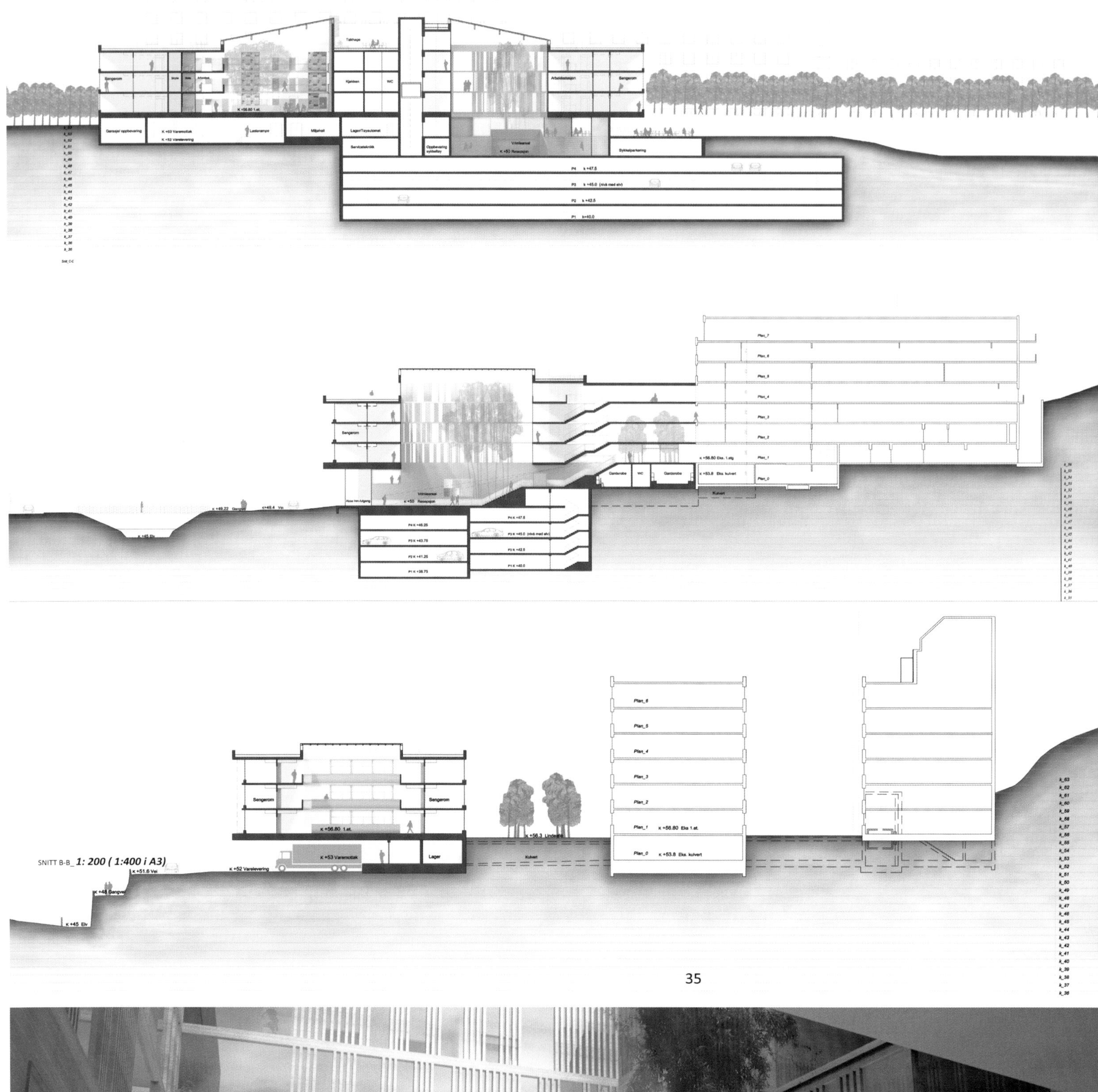
Sengerom
Sengerom
κ +56.80 1.et.
κ +53 Varemottak
Lager
SNITT B-B_ 1: 200 (1:400 i A3)
κ +51.6 Vei
κ +52 Varelevering
κ +48 Gangvei
κ +45 Elv
Kulvert
Plan_6
Plan_5
Plan_4
Plan_3
Plan_2
Plan_1 κ +56.80 Eks 1.et.
Plan_0 κ +53.8 Eks. kulvert

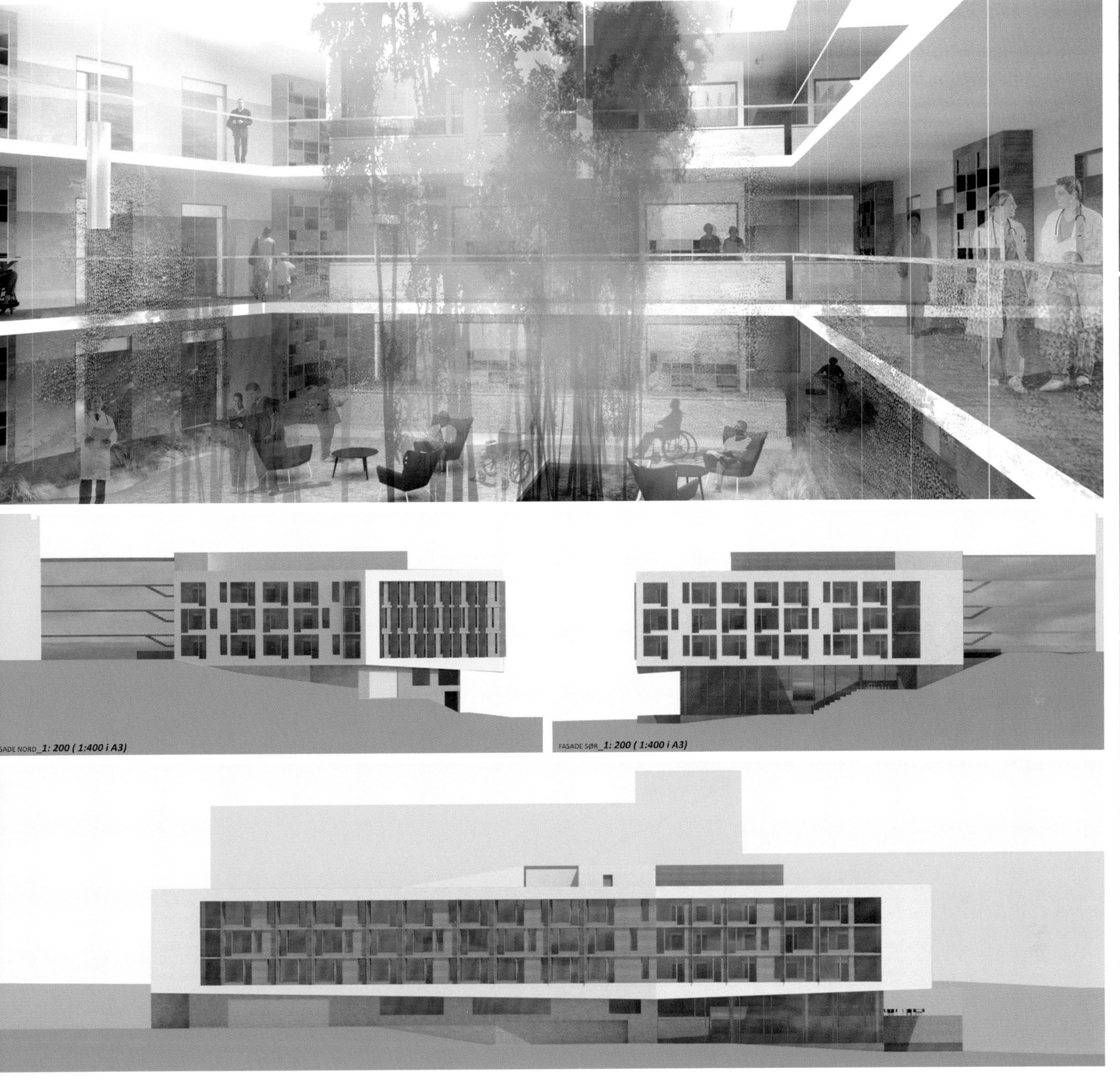

Profile

This is the winning proposal by C. F. Møller Architects for a new ward building of Haraldsplass Hospital. Replacing the traditional hospital corridors with open common areas and efficient logistics, the project has designed a whole new kind of hospital.

Architectural Design

Haraldsplass Hospital was built in 1986 and has approximately 184 beds. The new building will cover 10,000 square meters and give the hospital a further 108 beds on three storeys. There will also be new underground parking facilities for approximately 400 cars.

In stark contrast to traditional hospital buildings, there are no long corridors. The wards are located around two large covered atria, which provide the setting for two different kinds of common areas: a public arrivals area with a reception, café, shop and seating area, and a more private space for patients and their guests only.

The atria draw daylight into the building, where lush vegetation with bamboo plants in water pools and a bed of grass, flowers and creeping plants help to ensure a good indoor climate. All patients will have access to views of the valley and the city, as the new building follows the course of the Møllendalselven River, with an angled facade.

The project has also been highlighted as being very eco-friendly, because the facade size is small relative to the gross area. By taking new approaches to ventilation and reusing waste heat from the existing hospital, the new ward can achieve "passive house" standard.

北京新安贞医院

Beijing New Anzhen Hospital

设计单位：澳大利亚 SDG 设计集团
开发商：北京安贞医院
北京宏福集团
内蒙古广厦集团
项目地址：中国北京市
总用地面积：60 000 m²
总建筑面积：90 000 m²
建筑密度：15%
绿化率：60%
容积率：1.5
设计团队：聂建鑫　王新强　张春红　胡乃昆
叶伟聪　卢　松　翟　莹　钱　捷

Designed by: Shine Design Group Pty. Ltd.
Developer: Beijing Anzhen Hospital; Beijing Hongfu Group;
Inner Mongolia Guangsha Group
Location: Beijing, China
Site Area: 60,000 m²
Floor Area: 90,000 m²
Building Density: 15%
Greening Ratio: 60%
Plot Ratio: 1.5
Design Team: Nie Jianxin, Wang Xinqiang,Zhang Chunhong,
Hu Naikun, Ye Weicong, Lu Song, Di Ying, Qian Jie

项目概况

北京新安贞医院位于北京中轴线北侧的昌平区温都水城大社区内，交通便利，社区功能配套完善。项目定位为面向中高端患者的，集医、教、研、培为一体，以心血管为主的大专科小综合三甲综合医院。

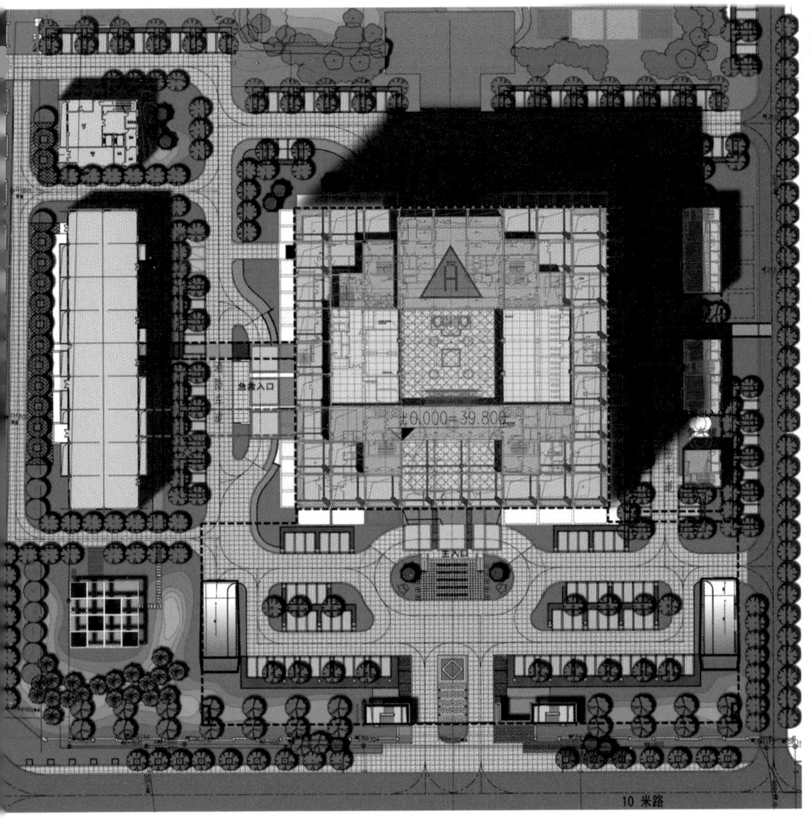

总平面图

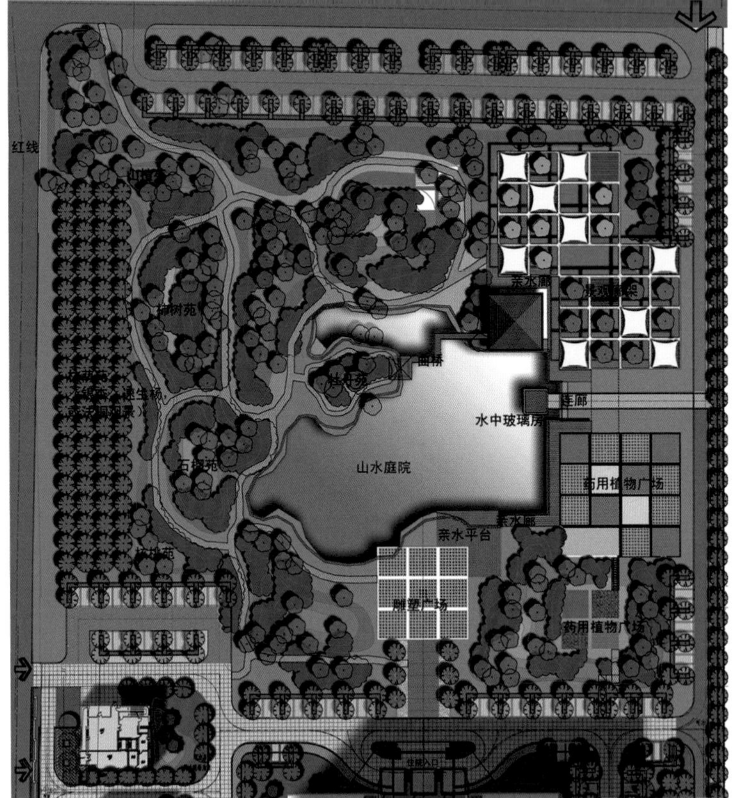

北区平面图

中药房 10
中药房 9
8
NO. 8
7
6
服务台

建筑设计

为了节约土地，为未来发展预留空间，设计采用了与常规医院水平三段式布局（门诊、医技、病房）截然不同的竖向处理方式，设计了一个长、宽、高各72米的“建筑魔方”立方体。北面的大片空地则规划成一个有湖、有山、有丛林的户外大公园。

在一个建筑立方体内涵盖复杂的医院功能区，如何处理好医患分流、洁污分流、各部室既相对独立又相互衔接等问题是此项目的重点。设计师采用了建筑核心概念，在立方体内设计了上下两个共享空间：下层共享空间（1-5层）分布着门诊；上层共享空间（9-17层）布置为病房；中间被6、7、8层（分别是ICU、手术层、设备层）隔开，只留一个8米见方的光井将上下空间联系起来。

这样的空间布局使下层共享空间形成内环以病患为主、外环以医护人员为主的交通流线，使病人走动距离最少；上层共享空间则是外环以病患为主、内环以医护人员为主，使医护巡行距离最短，效率最高。

白天最繁忙的门诊位于下层共享空间，自然光线十分有限，设计通过引进太阳能追踪器和太阳折射板，成功解决了下层空间的自然光利用问题。

收费取药
超市
1F
综合服务台
咨询处 General Counter

Profile

Beijing New Anzhen Hospital is located in Wendu Water Town community of Changping District, the north side of the central axis of Beijing City. It has convenient transportation and complete functional facilities. The project is supposed to be a comprehensive Top Three Hospital that gears to middle and critical patients, integrates medical treatment, education, research and training, and specializes in cardiovascular disease.

Architectural Design

To save land and reserve space for future development, the design has taken a distinctive vertical layout different from the ternary form (outpatient service, medical technologies and wards) of common hospitals. It has erected a "Cube" building with length, width and height of 72 meters. A large open space in the north is planned as a large outdoor park embracing lake, mountain and jungles.

The building cube contains complex functional areas. Issues of this project, such as separation of doctors and patients, separation of the clean and the polluted, and mutually independent and interrelated connection of offices, are of great importance. Designers have adopted the concept of building core. There are two shared space in the cube: lower shared space (1-5 storey) is for outpatient service; upper shared space (9-17 storey) is for wards; they are separated by the 6, 7, 8 storey (floors for ICU, surgery and equipment), only reserving a 8-meter square light well to connect the lower and upper spaces.

The inner ring of lower shared space puts patients in the center while the external ring is dominated by medical workers. Such spatial layout minimizes walking distance for patients. The external ring of upper shared space puts patients in the center while the inner ring is dominated by medical workers. It ensures a minimum ward round distance for medical workers and the highest efficiency.

The busiest outpatient department is located in the lower shared space. Since the natural lighting is limited, sun seeker and solar deflector are utilized, which has managed to solve the problem of natural lighting in the lower space.

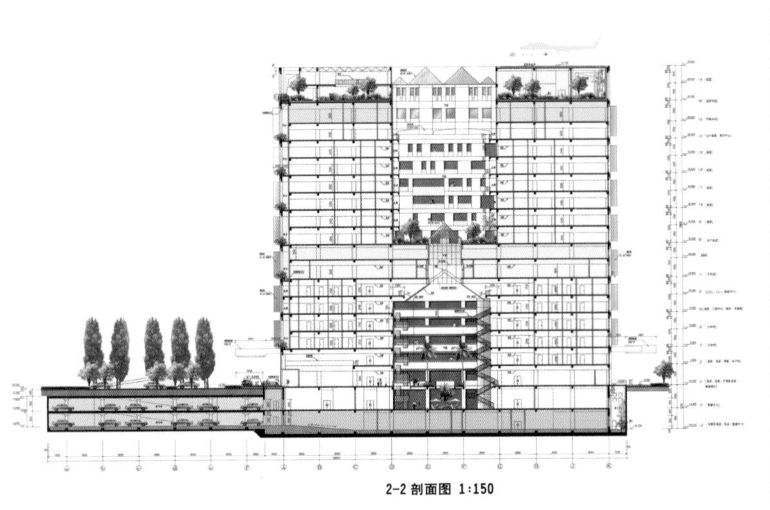

2-2 剖面图 1:150

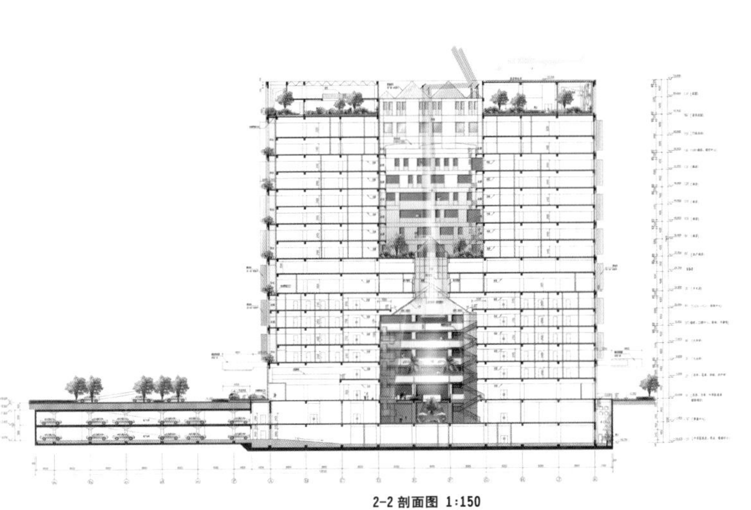

2-2 剖面图 1:150

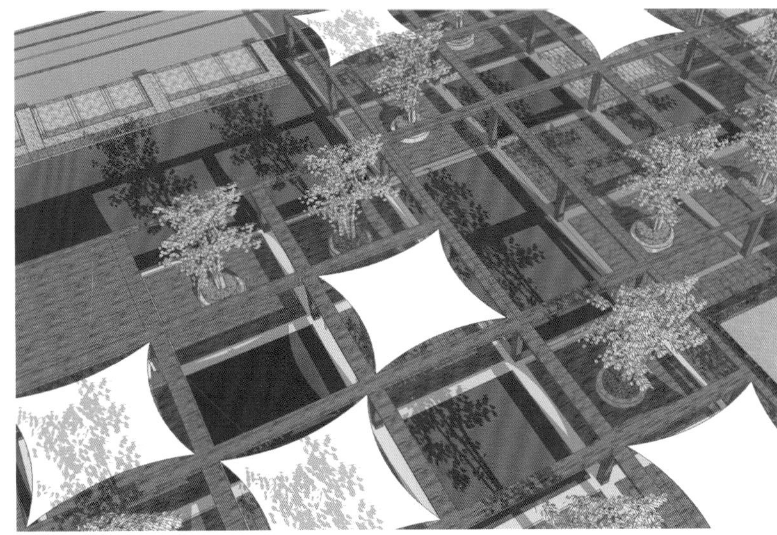

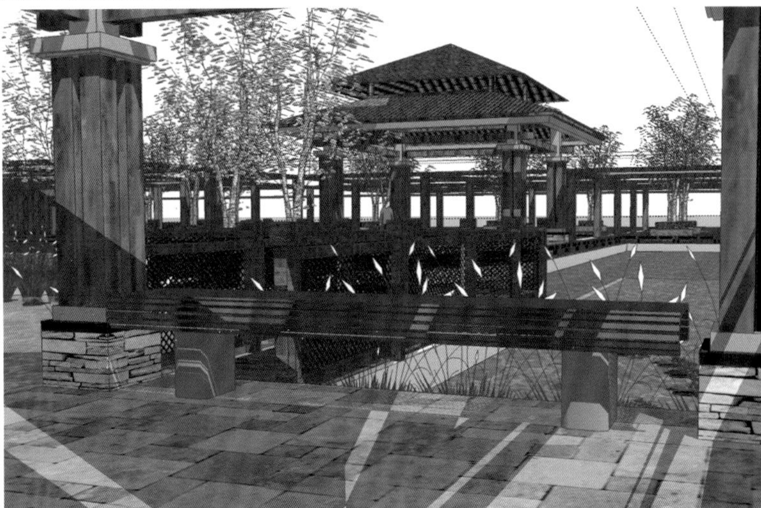

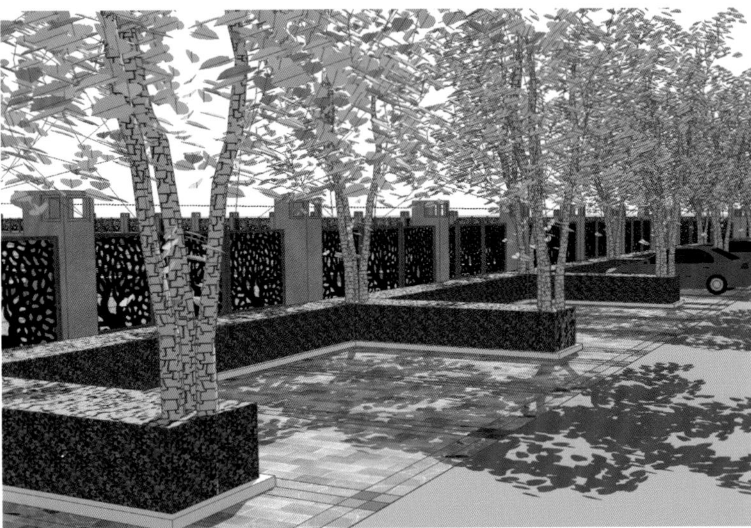

伊朗拉什特 Pars 综合医院

Pars General Hospital, Rasht

设计单位：New Wave Architecture
项目地址：伊朗拉什特
项目面积：23 000 m²
绿化率：70%

Designed by: New Wave Architecture
Location: Rasht, Iran
Area: 23,000 m²
Greening Ratio: 70%

项目概况

项目位于拉什特市 Goslar 地区一个 17 400 平方米的地块上，这一场址为构建一个综合医院提供了良好环境。Pars 医院规划面积达 23 000 平方米，配备了 120 个达到国际最新医疗保健标准的病房床位，可为整个区域服务。

建筑设计

项目旨在为医院建筑提供一种新的模式，同时改变伊朗医院建筑的常规形态。设计不仅回应了物理规划的需求以及客户的要求，也设想了一个统一的建筑形式，有效地推动了这个区域的发展。

设计将几个不同高度的体量安置在混凝土结构的基座上，突出了这个综合医疗机构的特征。中庭将建筑整合起来，也可将自然光线渗透到不同的功能区，而门廊则使室内空间与周围的环境形成了一种微妙的、相互作用的视觉联系。

建筑沿着东西方向伸展，顺应了当地的气候条件。白色的石灰华天然洞石和玻璃是建筑的主要材料，这两种材料与外墙上的木质结构一起，成为 Guilan 地区建筑的象征。各部门的内表面将由抗菌效果均匀的楼板覆盖层包裹，这是为医疗机构特别设计的。

Profile

A 17,400 square meters site in Goslar district of Rasht provides an appropriate context for designing and constructing a General Hospital of Rasht. Pars hospital is a 23,000 square meters design, occupied by 120 beds in line with latest international health care standards.

Architectural Design

The project aims to show a new mode of hospital building while at the same time break the conventional mode of the hospital buildings in Iran. The project is trying to not only present responds to physical planning requirement and client's demands, but also approach to a unified form and leaves effective influence on this region of the city.

Composing several blocks in three different heights built on the basis of concrete structure features a comprehensive medical infrastructure, moreover, being integrated via a transitive atrium that pervades and provides natural light accessibility in different departments, while porches create a subtle and interactive visual communication between inside and surrounding environment.

The building is extended along east to west to be adopted with appropriate climatic direction in the context. White travertine stone and glass as the main material are applied in combination of wooden texture panels on the exterior walls as a sort of emblematic of vernacular architecture in Guilan. Interior surfaces of departments will be covered by anti-bacterial homogeneous floor-covering, exclusively designed for health care facilities.

INTER DEPARTMENTAL RELATIONSHIPS 跨部门联系

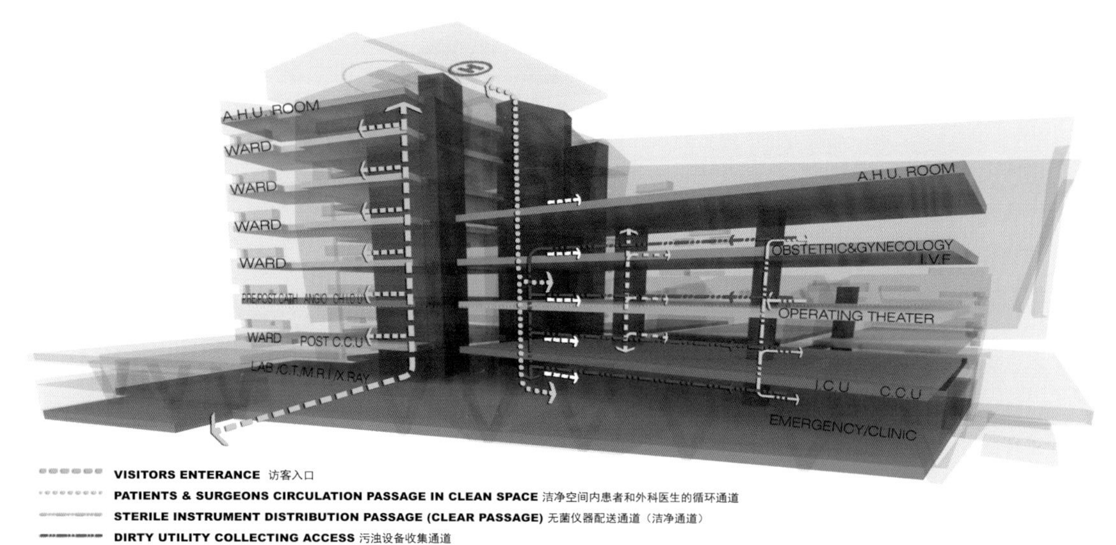

VISITORS ENTERANCE 访客入口
PATIENTS & SURGEONS CIRCULATION PASSAGE IN CLEAN SPACE 洁净空间内患者和外科医生的循环通道
STERILE INSTRUMENT DISTRIBUTION PASSAGE (CLEAR PASSAGE) 无菌仪器配送通道（洁净通道）
DIRTY UTILITY COLLECTING ACCESS 污浊设备收集通道
STAFF ACCESSIBILITY 员工可达性
TRANSFER ACCESS FROM HELI-PAD TO ACCIDENT & EMERGENCY DEPARTMENT & OPERATING THEATER 从直升机场通往事故＆急诊室＆手术室转移通道

A.H.U ROOM
OBSTETRIC&GYNECOLOGY I.V.F
C.S.R
OPERATING THEATER
I.C.U
C.C.U
EMERGENCY/CLINIC

STERILE INSTRUMENT DISTRIBUTION MAIN PASSAGE
无菌仪器主要配送通道

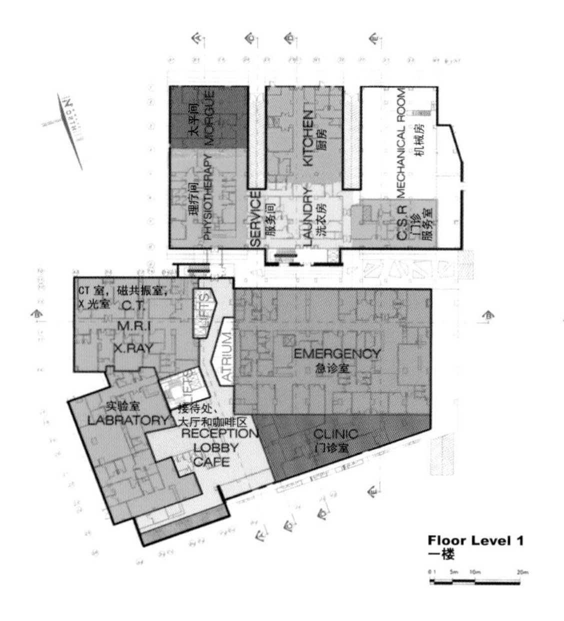

Floor Level 1
一楼

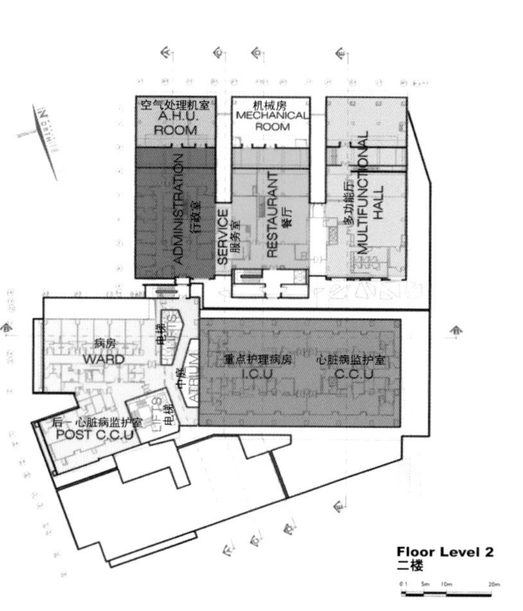

Floor Level 2
二楼

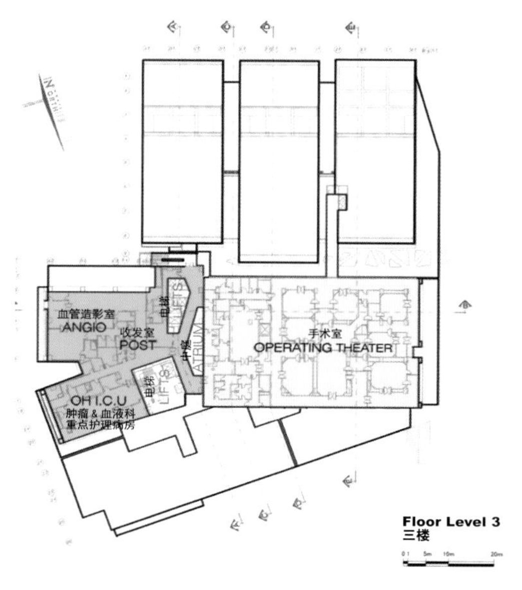

Floor Level 3
三楼

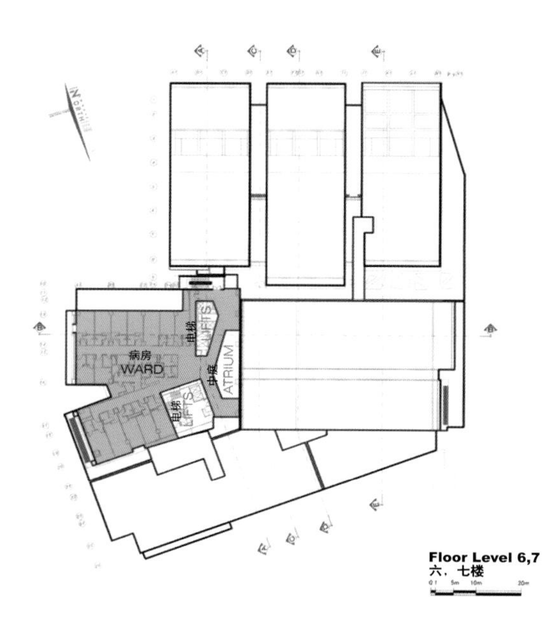

Floor Level 6,7
六、七楼

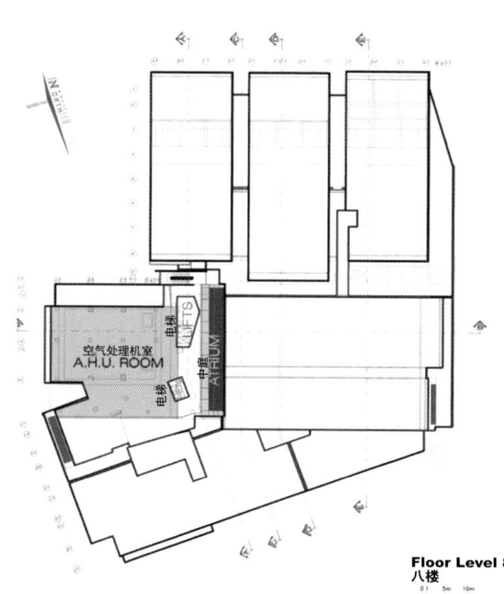

Floor Level 8
八楼

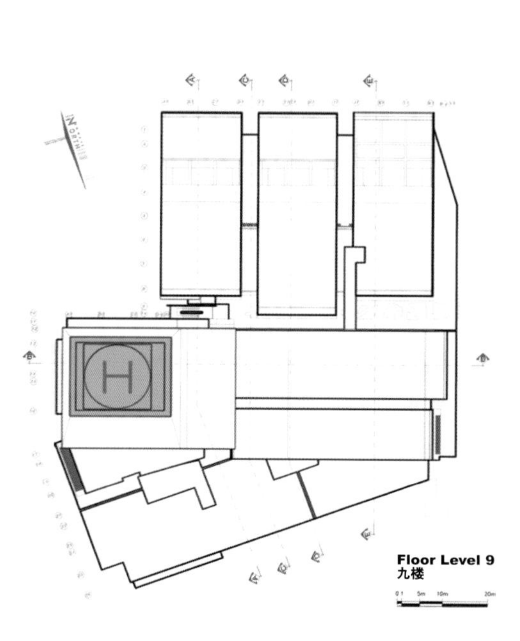

Floor Level 9
九楼

INTER-DEPARTMENTAL RELATIONSHIPS 跨部门联系

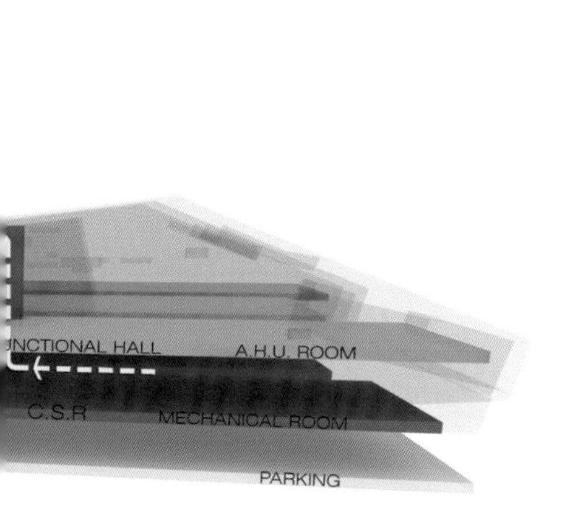

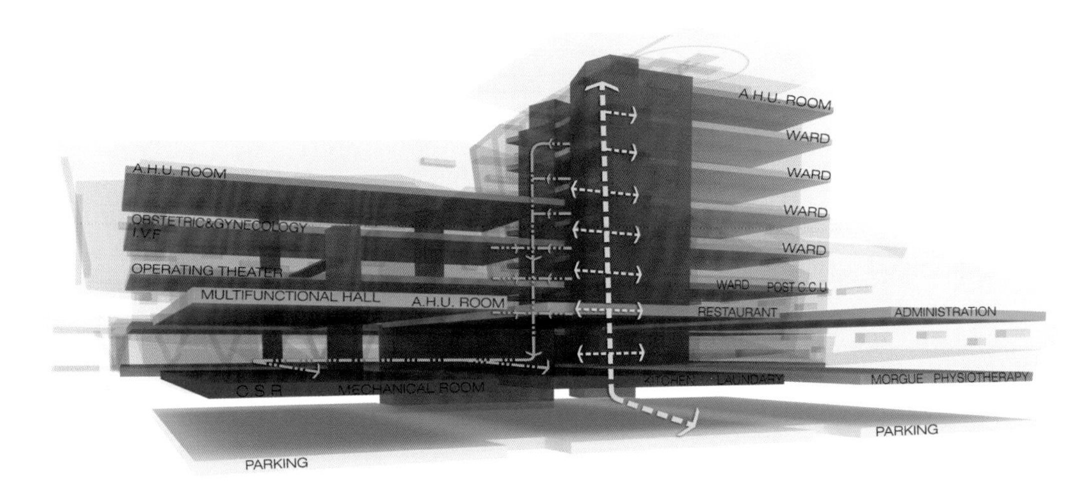

STAFF ACCESSIBILITY 员工可达性

DIRTY UTILITY COLLECTING ACCESS 污浊设备收集通道

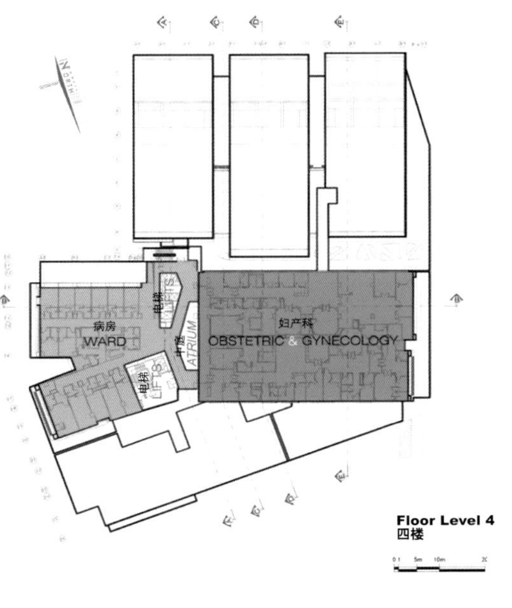

Floor Level 4
四楼

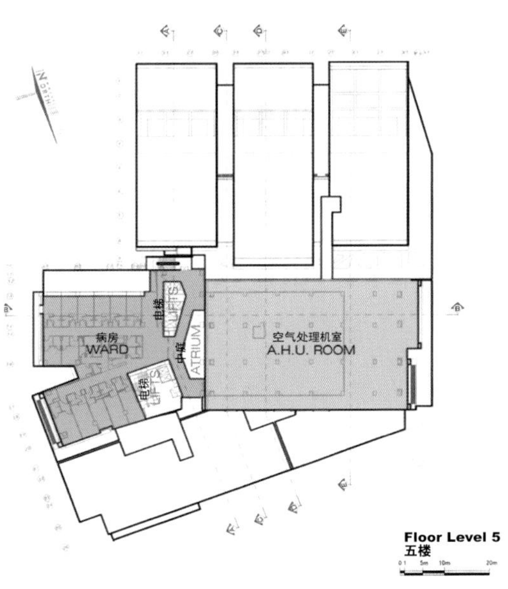

Floor Level 5
五楼

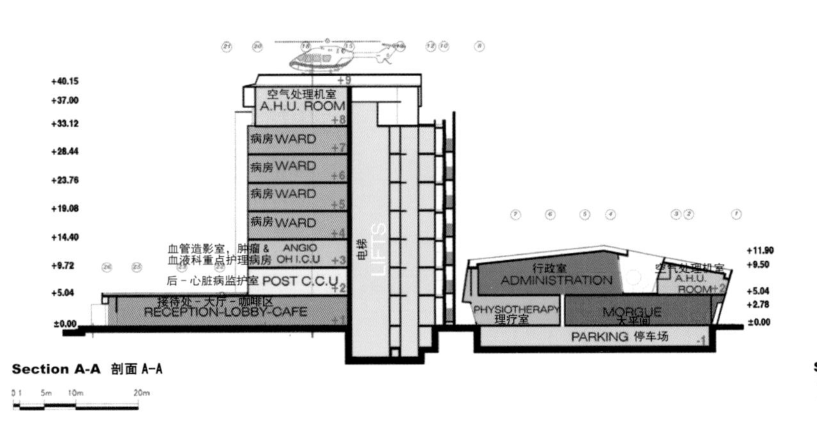

Section A-A 剖面 A-A

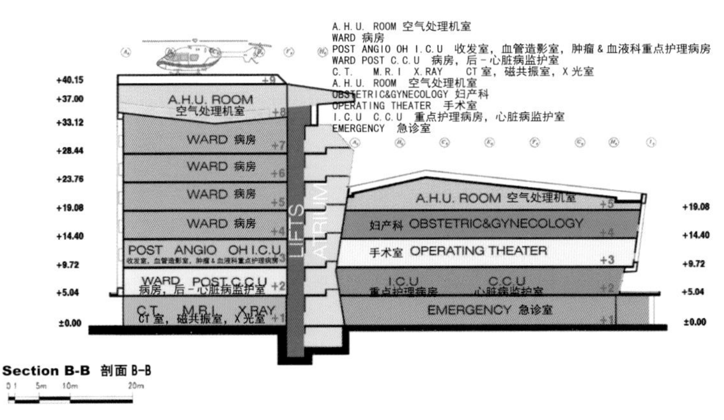

Section B-B 剖面 B-B

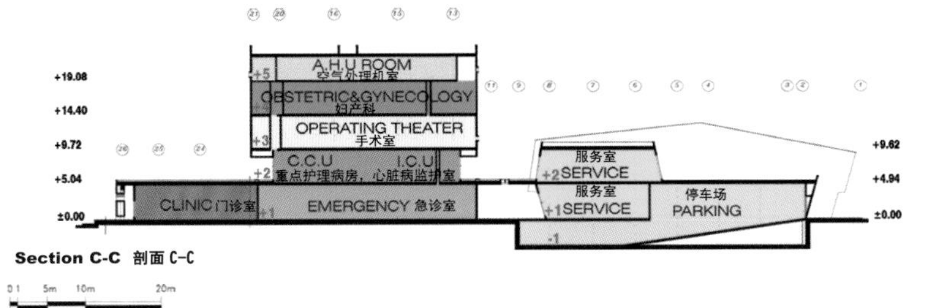

Section C-C 剖面 C-C

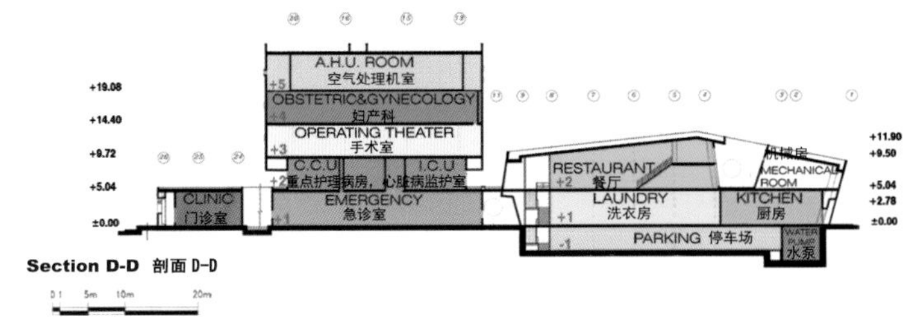

Section D-D 剖面 D-D

冰岛雷克雅未克 Landspitali 大学医院

Landspitali University Hospital of Iceland

设计单位： C.F.Møller Architects

开发单位： 冰岛卫生和社会保障部

项目地址： 冰岛雷克雅未克

项目面积： 161 500 ㎡

Designed by: *C.F. Møller*

Client: *Ministry of Health and Social Security, Iceland*

Location: *Reykjavik, Iceland*

Area: *161,500 m²*

项目概况

项目是一个占地 161 500 平方米的综合设施，其中 130 000 平方米的空间将用作纯粹的医院功能区。Landspitali 大学医院将成为一个理想的国立医院，设计将把医院设计和针对性治疗的现代化需求与简约却清晰的建筑设计联系起来。

设计理念

设计的主要理念在于构建一个与常规医疗建筑截然不同的建筑。设计师为这个国立大学医院高品质的设计提供了清晰的视野，这一视野将在住院大楼得到加强。

设计特色

设计旨在为冰岛的人们提供安全保障和社会福利。除了传统的检查和治疗区，这个项目的设计将设置可供患者、家属以及来访者聚集的节点，这些空间将设置在住院大楼周围、主要的大厅内、主入口处以及门诊部和主楼的一层大厅。

大部分病房的患者都可以欣赏周边的自然景观。设计旨在构建一栋医疗建筑，这一建筑将有效规划和集中治疗的需求与为人们提供一个与自然融合的娱乐休闲场所的愿景结合起来。

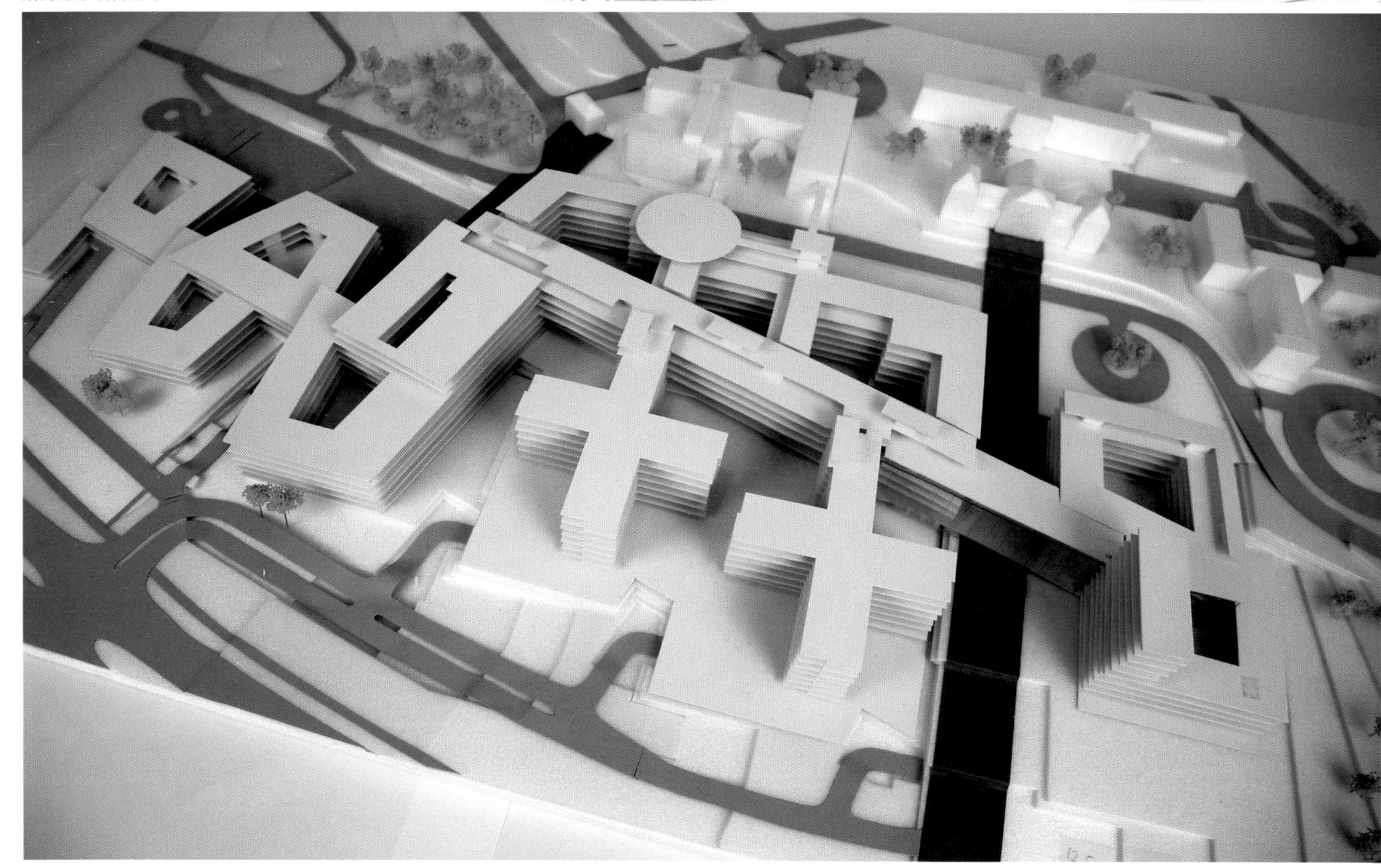

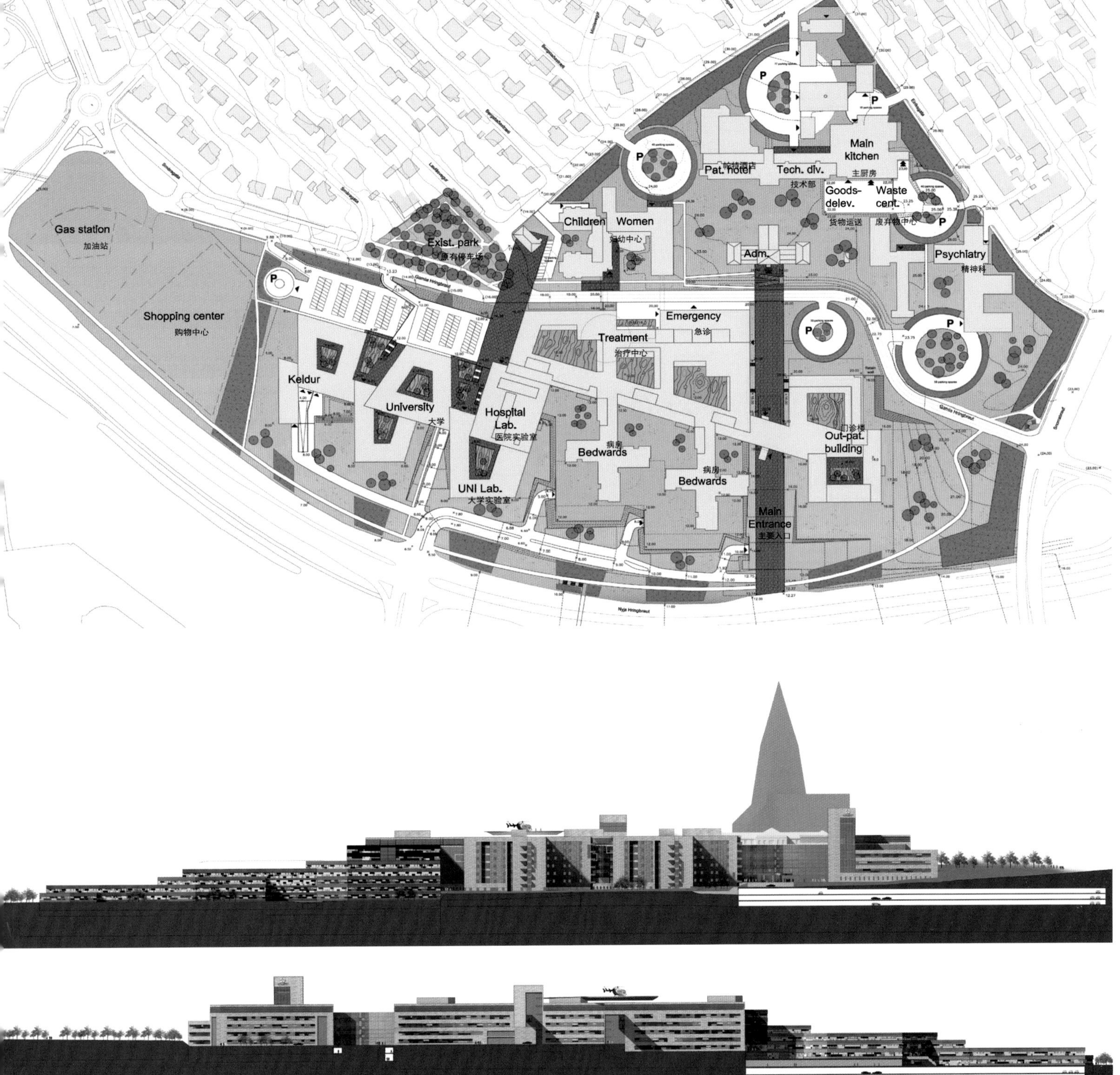
Gas station
加油站
Shopping center
购物中心
Exist. park
原有停车场
Keldur
University
大学
Hospital
Lab.
医院实验室
UNI Lab.
大学实验室
Children
Women
妇幼中心
Treatment
治疗中心
Emergency
急诊
病房
Bedwards
病房
Bedwards
Pat. hotel
Tech. div.
技术部
Main
kitchen
主厨房
Goods-
delev.
货物运送
Waste
cent.
废弃物中心
Adm.
Psychiatry
精神科
门诊楼
Out-pat.
building
Main
Entrance
主要入口
P

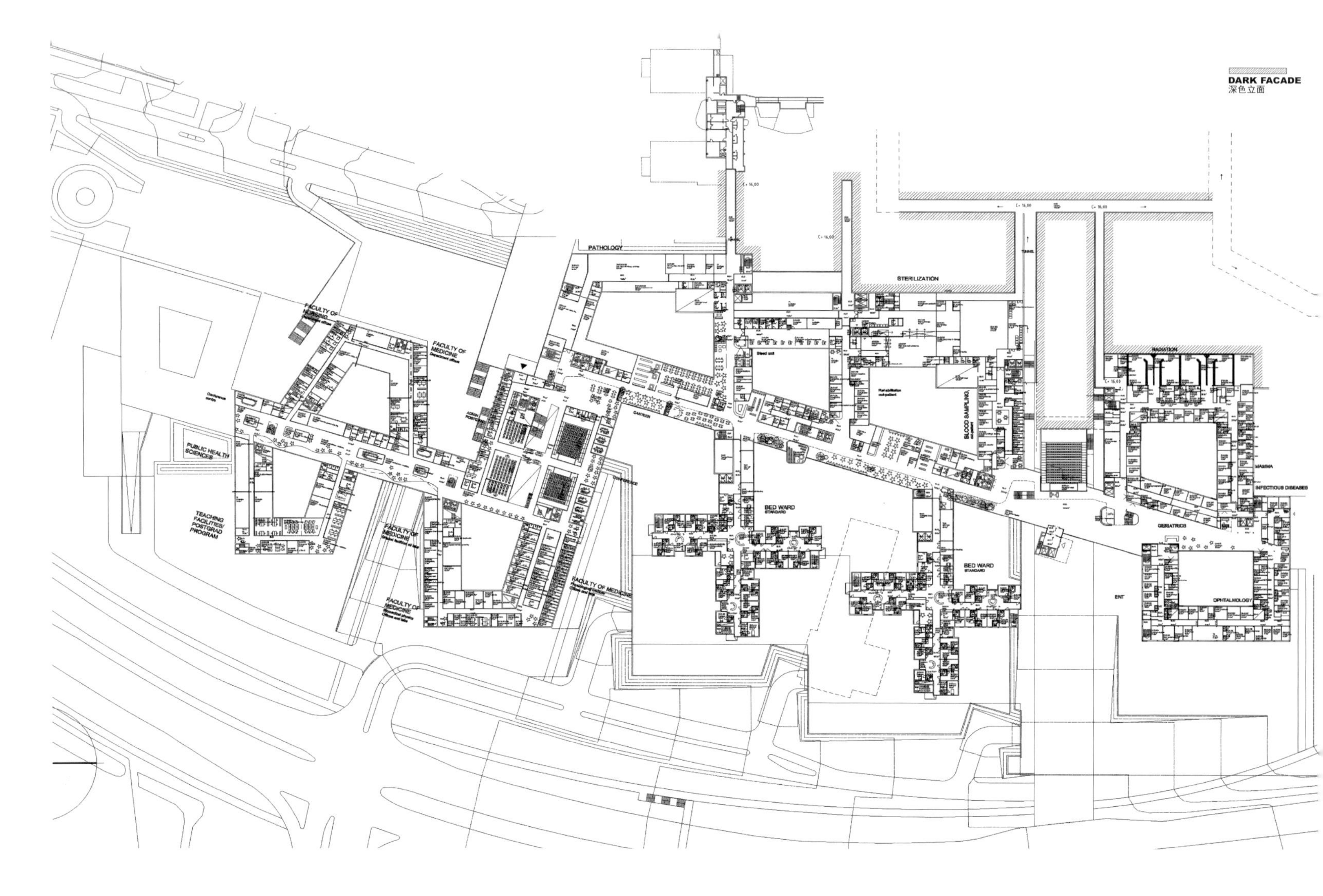

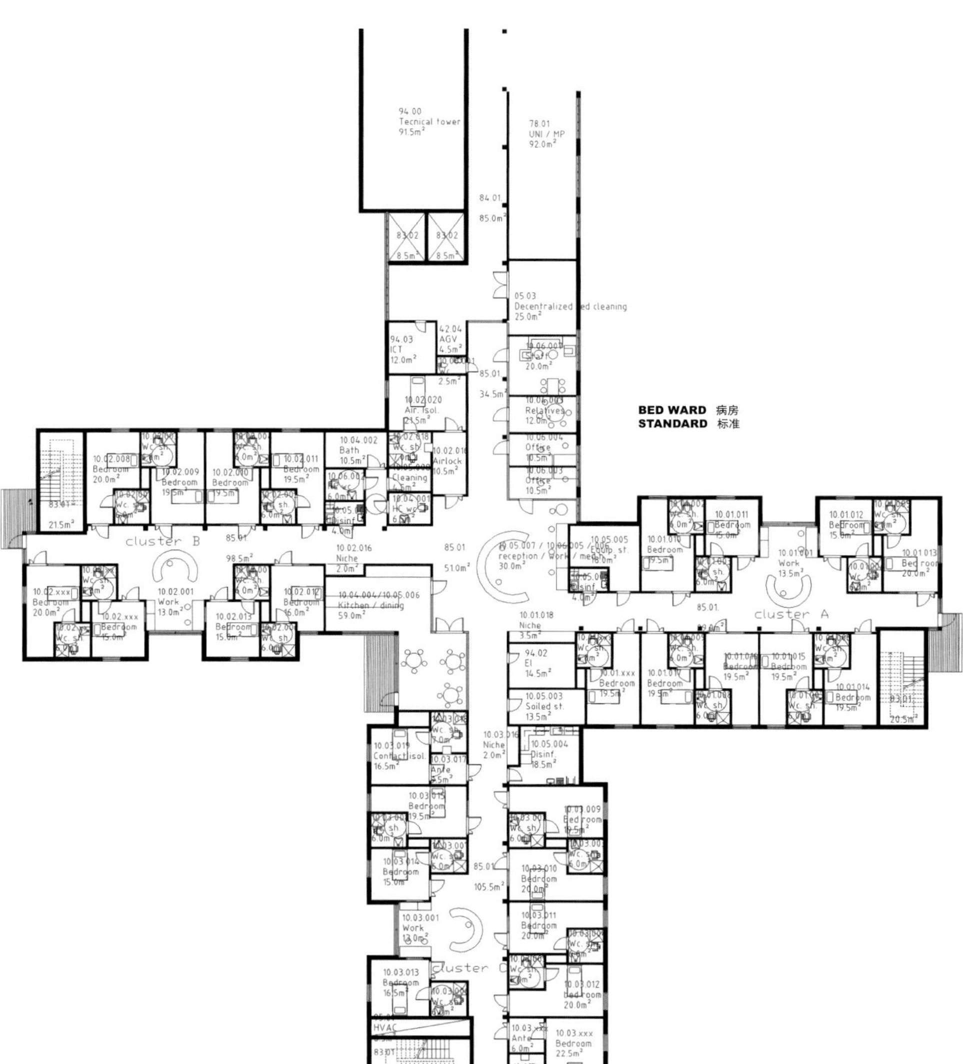

Profile

The project is for a complex covering 161,500 square meters, of which 130,000 square meters will be devoted to pure hospital functions. The University Hospital will be a visionary national hospital in which modern demands towards hospital planning and targeted treatment will be combined with simple and clear architecture.

Design Concept

The main idea for the NUH project has been to form a hospital that has no resemblance to common institutional buildings. It has been a clear vision to provide a design quality to the NUH which is focusing on the patient.

Design Feature

The intention of NUH is to supply safety and wellbeing for the public of Iceland. In addition to the traditional areas for examination and treatment, NUH is designed with locations where patients, families and visitors can get together. These areas are placed near the bed wards as well as in the main lobby area, near the main entrance and in the lobby on the first floor within the outpatient building and the main building.

Inpatients have full view to the surrounding nature from most beds. The project aims to create a university hospital where demands for efficient planning and focused treatment go along with a strong wish to provide human with rest and recreation space in connection with the surrounding environment.

DARK FACADE
深色立面

FACULTY OF NURSING
CLINICAL LABS
BED WARD STANDARD
RHEUMATOLOGY
PAIN
GYN
UROLOGY
FACULTY OF MEDICINE

OP I 手术室一
手术室一 + 手术室二
OP I + OP II
手术室二
OP II
OP I 手术室一
PRE/POST I
前 / 后 1
PRE/POST II
前 / 后 2
手术室二
OP II
干预
INTERVENTION
手术室二
OP II
日间手术
DAY SURGERY
临床检验室
CLINICAL LABS.
临床检验室
CLINICAL LABS.
临床检验室
CLINICAL LABS.
病房
BED WARD
STANDARD
病房
BED WARD
STANDARD
P

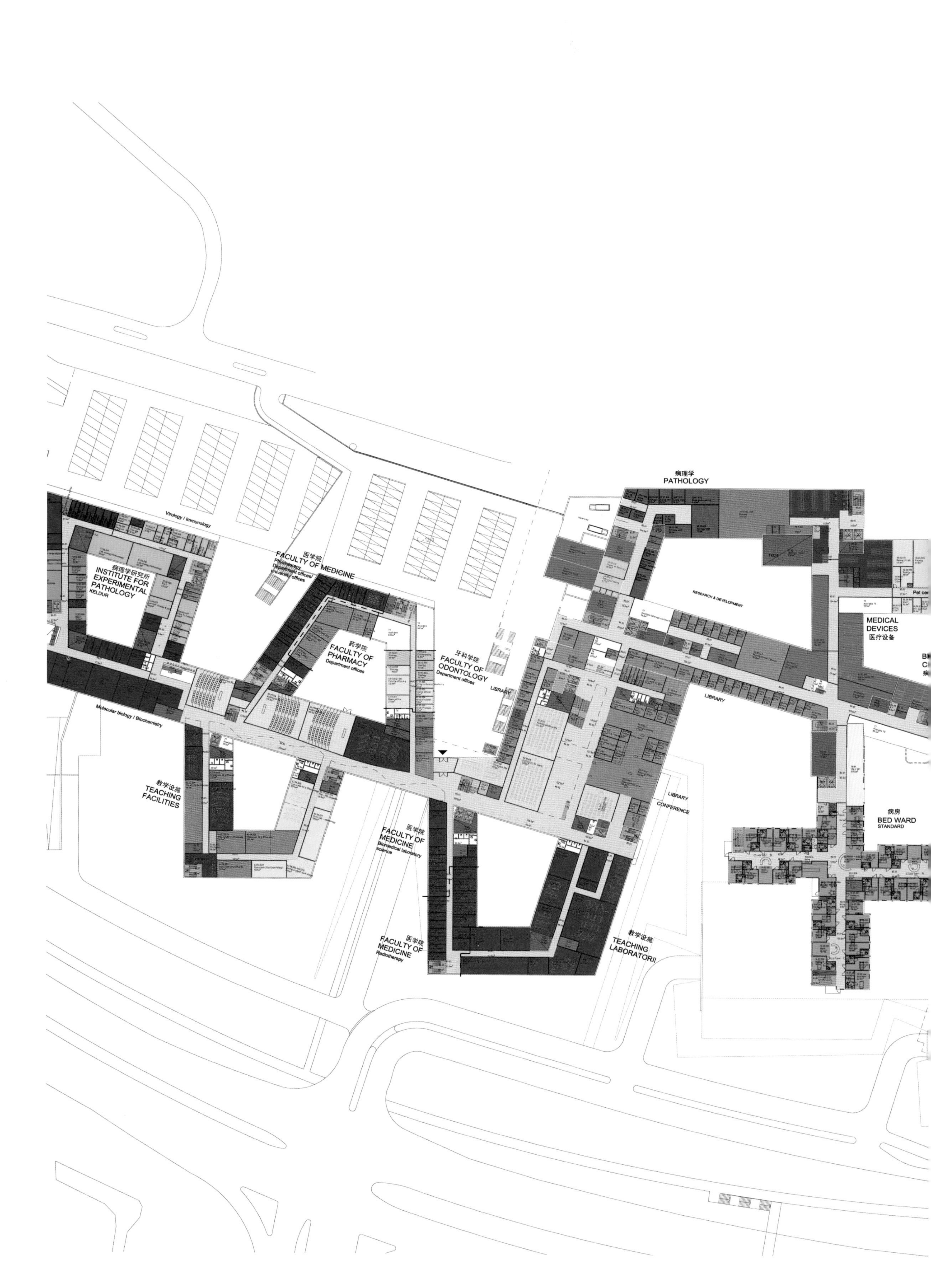

病理学
PATHOLOGY
Virology / Immunology
医学院
FACULTY OF MEDICINE
病理学研究所
INSTITUTE FOR EXPERIMENTAL PATHOLOGY
KELDUR
药学院
FACULTY OF PHARMACY
Department offices
牙科学院
FACULTY OF ODONTOLOGY
Department offices
LIBRARY
RESEARCH & DEVELOPMENT
MEDICAL DEVICES
医疗设备
Molecular biology / Biochemistry
教学设施
TEACHING FACILITIES
医学院
FACULTY OF MEDICINE
Biomedical laboratory science
医学院
FACULTY OF MEDICINE
Radiotherapy
教学设施
TEACHING LABORATORII
LIBRARY
CONFERENCE
病房
BED WARD
STANDARD

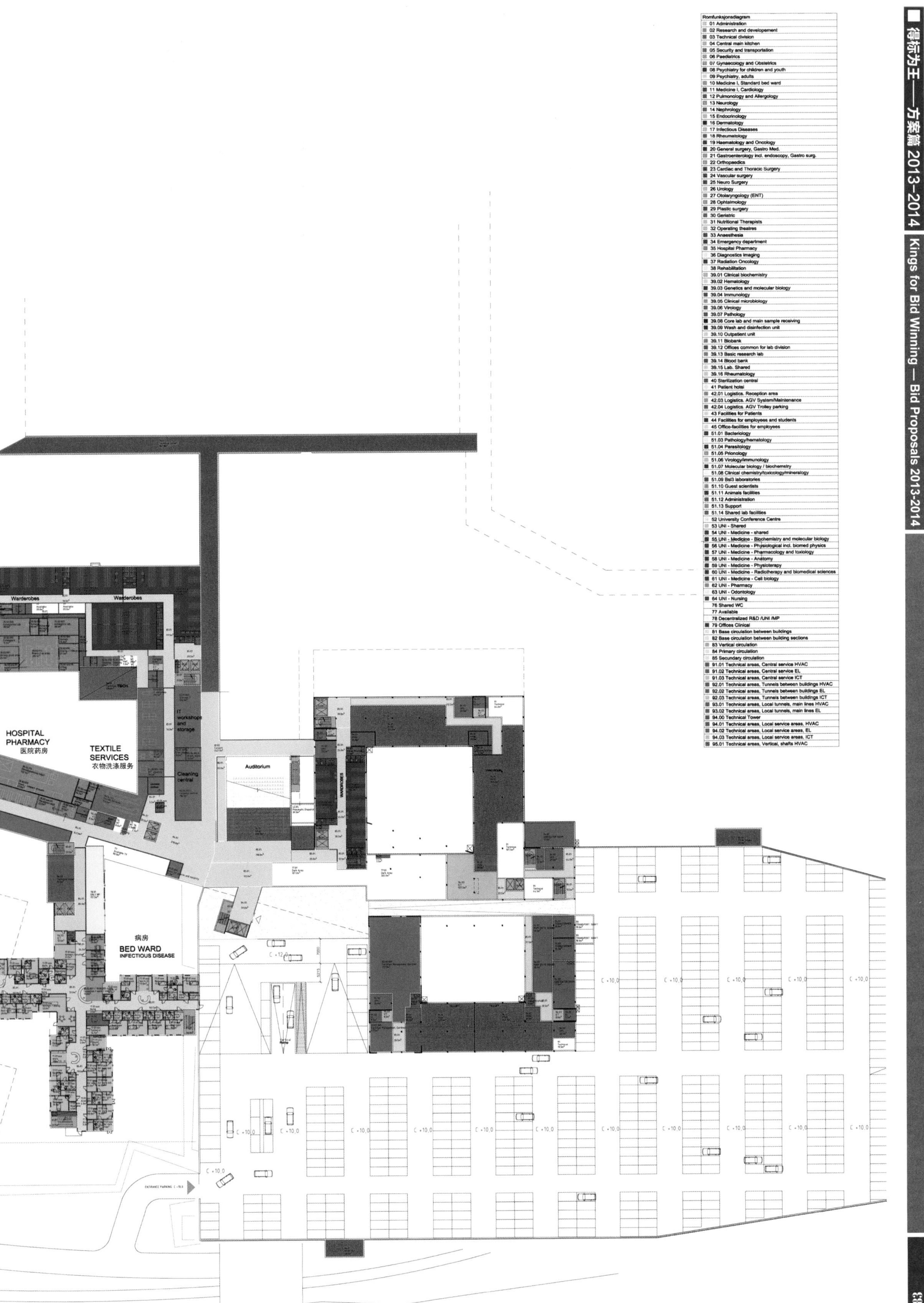
Romfunksjonsdiagram
01 Administration
02 Research and developement
03 Technical division
04 Central main kitchen
05 Security and transportation
06 Paediatrics
07 Gynaecology and Obstetrics
08 Psychiatry for children and youth
09 Psychiatry, adults
10 Medicine I, Standard bed ward
11 Medicine I, Cardiology
12 Pulmonology and Allergology
13 Neurology
14 Nephrology
15 Endocrinology
16 Dermatology
17 Infectious Diseases
18 Rheumatology
19 Haematology and Oncology
20 General surgery, Gastro Med.
21 Gastroenterology incl. endoscopy, Gastro surg.
22 Orthopaedics
23 Cardiac and Thoracic Surgery
24 Vascular surgery
25 Neuro Surgery
26 Urology
27 Otolaryngology (ENT)
28 Ophtalmology
29 Plastic surgery
30 Geriatric
31 Nutritional Therapists
32 Operating theatres
33 Anaesthesia
34 Emergency department
35 Hospital Pharmacy
36 Diagnostics imaging
37 Radiation Oncology
38 Rehabilitation
39.01 Clinical biochemistry
39.02 Hematology
39.03 Genetics and molecular biology
39.04 Immunology
39.05 Clinical microbiology
39.06 Virology
39.07 Pathology
39.08 Core lab and main sample receiving
39.09 Wash and disinfection unit
39.10 Outpatient unit
39.11 Biobank
39.12 Offices common for lab division
39.13 Basic research lab
39.14 Blood bank
39.15 Lab. Shared
39.16 Rheumatology
40 Sterilization central
41 Patient hotel
42.01 Logistics. Reception area
42.03 Logistics. AGV System/Maintenance
42.04 Logistics. AGV Trolley parking
43 Facilities for Patients
44 Facilities for employees and students
45 Office-facilities for employees
51.01 Bacteriology
51.03 Pathology/hematology
51.04 Parasitology
51.05 Prionology
51.06 Virology/immunology
51.07 Molecular biology / biochemstry
51.08 Clinical chemistry/toxicology/mineralogy
51.09 Bsl3 laboratories
51.10 Guest scientists
51.11 Animals facilities
51.12 Administration
51.13 Support
51.14 Shared lab facilities
52 University Conference Centre
53 UNI - Shared
54 UNI - Medicine - shared
55 UNI - Medicine - Biochemistry and molecular biology
56 UNI - Medicine - Physiological incl. biomed physics
57 UNI - Medicine - Pharmacology and toxiology
58 UNI - Medicine - Anatomy
59 UNI - Medicine - Physioterapy
60 UNI - Medicine - Radiotherapy and biomedical sciences
61 UNI - Medicine - Cell biology
62 UNI - Pharmacy
63 UNI - Odontology
64 UNI - Nursing
76 Shared WC
77 Available
78 Decentralized R&D /UNI /MP
79 Offices Clinical
81 Base circulation between buildings
82 Base circulation between building sections
83 Vertical circulation
84 Primary circulation
85 Secundary circulation
91.01 Technical areas, Central service HVAC
91.02 Technical areas, Central service EL
91.03 Technical areas, Central service ICT
92.01 Technical areas, Tunnels between buildings HVAC
92.02 Technical areas, Tunnels between buildings EL
92.03 Technical areas, Tunnels between buildings ICT
93.01 Technical areas, Local tunnels, main lines HVAC
93.02 Technical areas, Local tunnels, main lines EL
94.00 Technical Tower
94.01 Technical areas, Local service areas, HVAC
94.02 Technical areas, Local service areas, EL
94.03 Technical areas, Local service areas, ICT
95.01 Technical areas, Vertical, shafts HVAC
Wardrobes
Wardrobes
HOSPITAL PHARMACY
医院药房
TEXTILE SERVICES
衣物洗涤服务
IT workshops and storage
Cleaning central
Auditorium
WARDROBES
病房
BED WARD
INFECTIOUS DISEASE
ENTRANCE PARKING

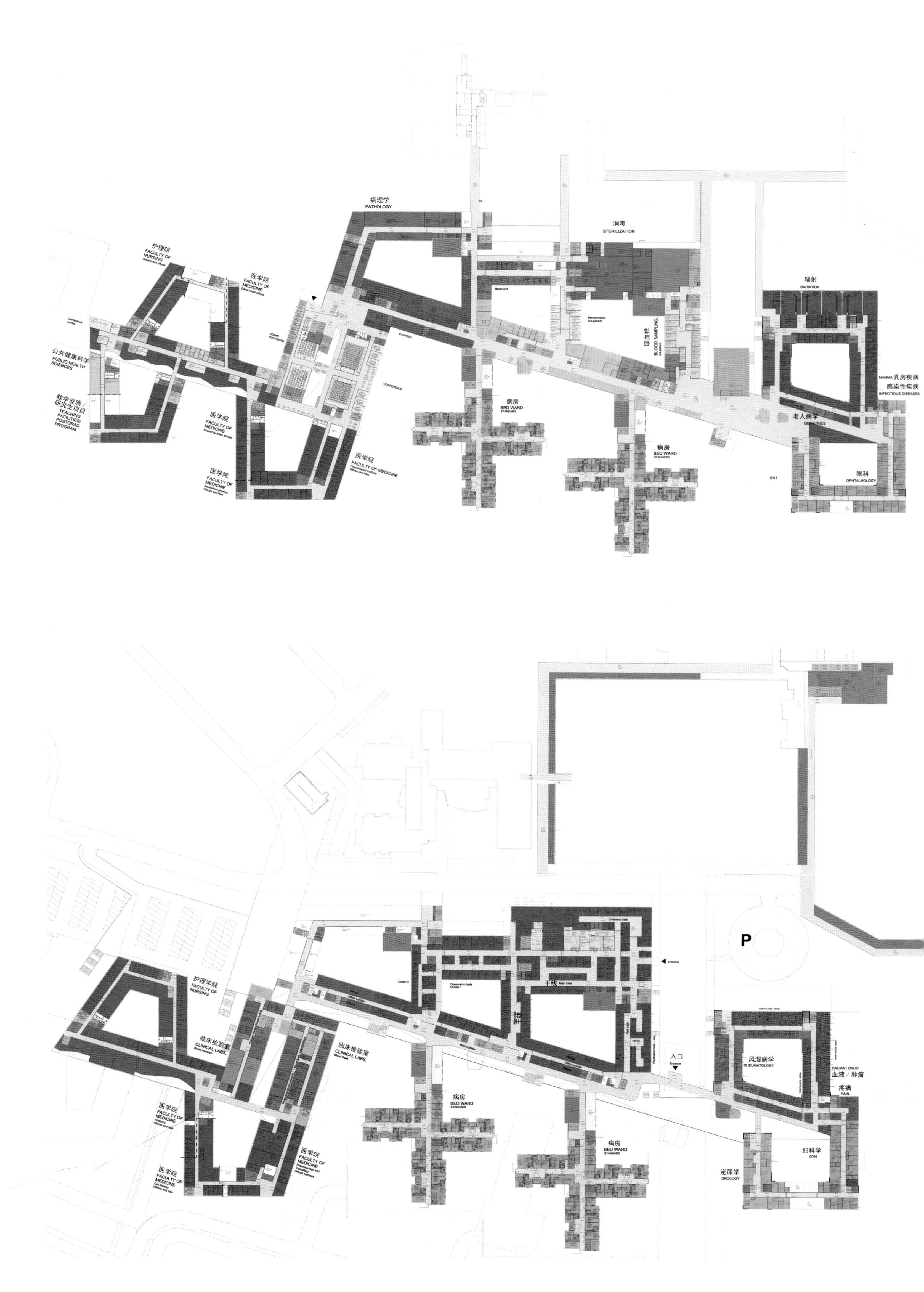

病理学
PATHOLOGY
消毒
STERILIZATION
护理院
FACULTY OF NURSING
医学院
FACULTY OF MEDICINE
辐射
RADIATION
公共健康科学
PUBLIC HEALTH SCIENCES
取血样
BLOOD SAMPLING
教学设施/研究生项目
TEACHING FACILITIES/ POSTGRAD PROGRAM
MAMMA 乳房疾病
感染性疾病
INFECTIOUS DISEASES
病房
BED WARD
医学院
FACULTY OF MEDICINE
老人病学
GERIATRICS
医学院
FACULTY OF MEDICINE
医学院
FACULTY OF MEDICINE
眼科
OPHTALMOLOGY
P
护理学院
FACULTY OF NURSING
干线
临床检验室
CLINICAL LABS.
临床检验室
CLINICAL LABS.
入口
风湿病学
RHEUMATOLOGY
HAEMA / ONCO
血液/肿瘤
疼痛
PAIN
病房
BED WARD
医学院
FACULTY OF MEDICINE
医学院
FACULTY OF MEDICINE
病房
BED WARD
妇科学
GYN
泌尿学
UROLOGY
医学院
FACULTY OF MEDICINE

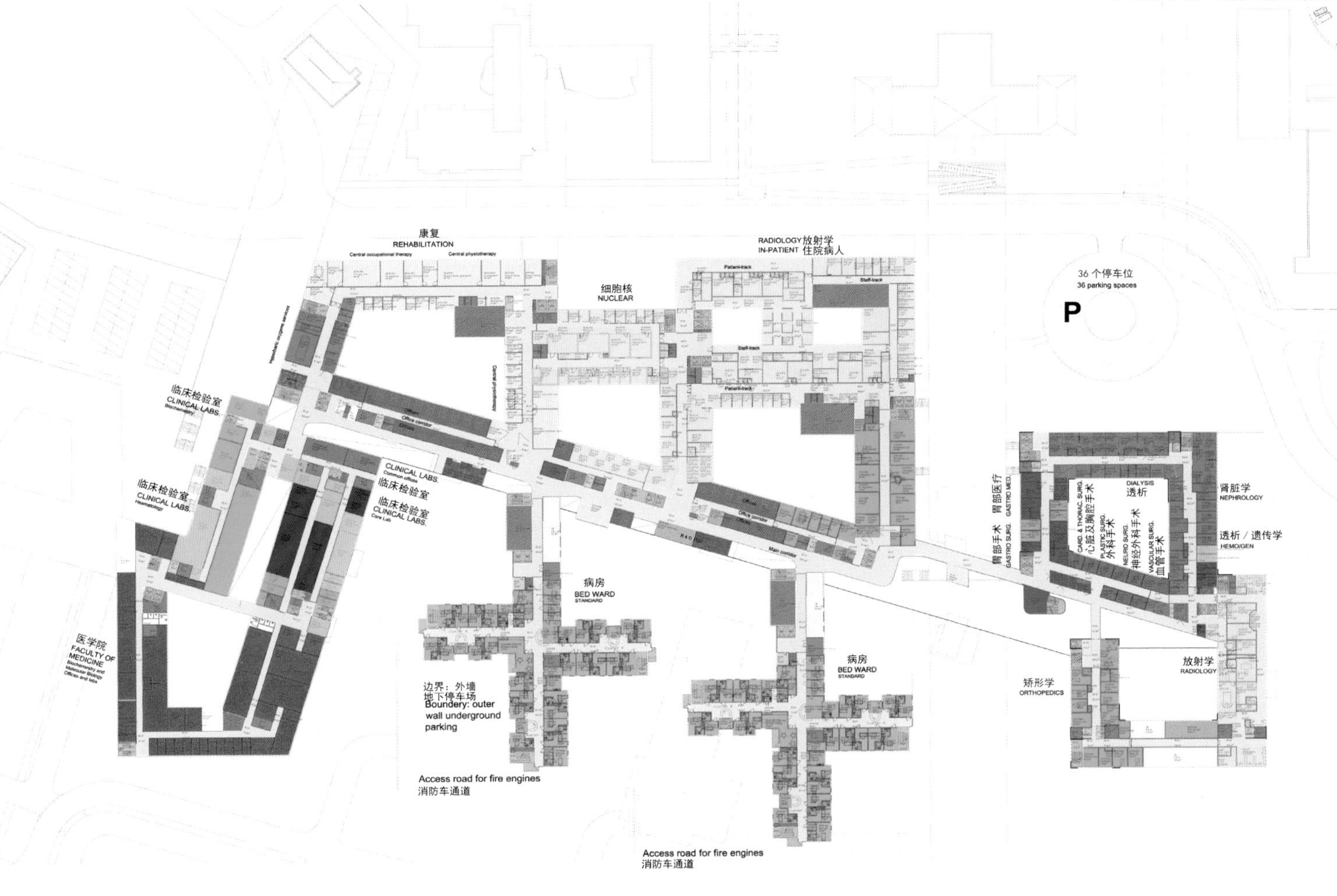
康复
REHABILITATION
RADIOLOGY 放射学
IN-PATIENT 住院病人
细胞核
NUCLEAR
36 个停车位
36 parking spaces
P
临床检验室
CLINICAL LABS.
CLINICAL LABS.
临床检验室
临床检验室
CLINICAL LABS.
临床检验室
CLINICAL LABS.
肾脏学
NEPHROLOGY
透析
DIALYSIS
透析／遗传学
HEMO/GEN
胃部手术
GASTRO SURG.
胃部医疗
GASTRO MED.
心脏及胸腔手术
CARD. & THORAC. SURG.
外科手术
PLASTIC SURG.
神经外科手术
NEURO SURG.
血管手术
VASCULAR SURG.
病房
BED WARD
病房
BED WARD
医学院
FACULTY OF MEDICINE
边界：外墙
地下停车场
Boundery: outer wall underground parking
矫形学
ORTHOPEDICS
放射学
RADIOLOGY
Access road for fire engines
消防车通道
Access road for fire engines
消防车通道

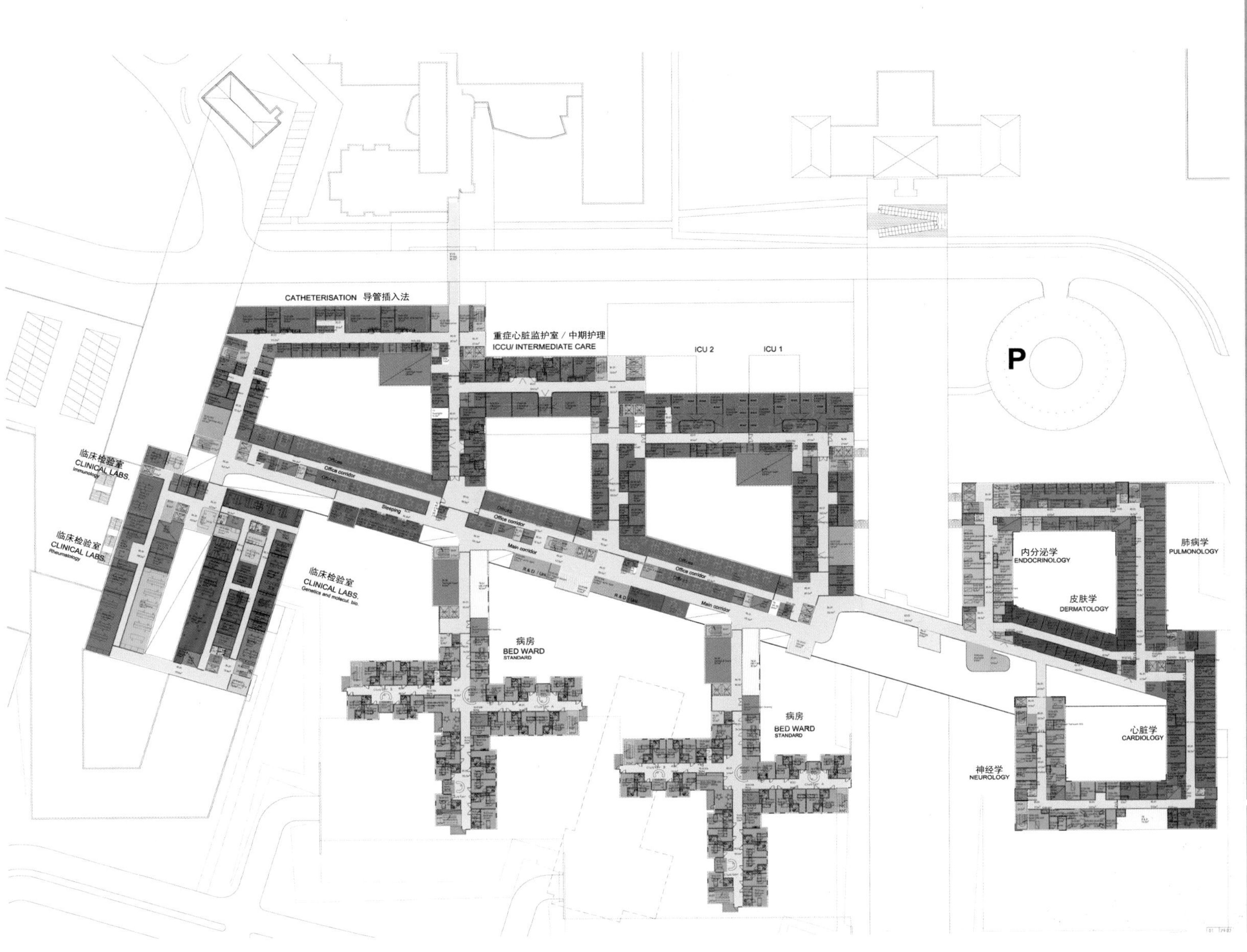
CATHETERISATION 导管插入法
重症心脏监护室／中期护理
ICCU/ INTERMEDIATE CARE
ICU 2
ICU 1
P
临床检验室
CLINICAL LABS.
临床检验室
CLINICAL LABS.
临床检验室
CLINICAL LABS.
内分泌学
ENDOCRINOLOGY
肺病学
PULMONOLOGY
皮肤学
DERMATOLOGY
病房
BED WARD
病房
BED WARD
心脏学
CARDIOLOGY
神经学
NEUROLOGY

丹麦新奥尔胡斯大学医院

New University Hospital in Aarhus

设计单位：C.F.Møller Architects
合作单位：Cubo Arkitekter A/S
开发单位：Region Mid-Jutland
项目地址：丹麦奥尔胡斯
总用地面积：970 000 ㎡
总建筑面积：400 000 ㎡

Designed by: C.F. Møller
Collaboration: Cubo Arkitekter A/S
Client: Region Mid-Jutland
Location: Aarhus, Denmark
Site Area: 970,000 m^2
Gross Floor Area: 400,000 m^2

项目概况

这将是丹麦历史上规模最大的医院建设项目，建成后的大学医院相当于丹麦的一个省级城镇。项目将建设在现有的奥尔胡斯大学医院 Skejby 之上，形成一个独立的医院综合体。

设计理念

新奥尔胡斯大学医院将展现未来医院的物理模型：一个以患者为中心，以技术创新、治疗环境和工作实践为导向的医院，它将为未来医院的发展指明方向。

建筑设计

项目包括一个两层的治疗中心、门诊部、一般的临床科室以及作为分散的单元的病房楼。这个大型的医院综合体好似一个城镇，附近社区、街道、广场分明，为构建一个动态的多样化绿色城区提供了基础。项目在整体布局上更趋向于分散式，而不是趋向于紧凑，这在一定程度上加强了已有建筑的人性化特征。

The Forum，是医院的中央大道，具有最外向的功能。这个富有特色的、宽敞的公共区域占据了医院综合体的核心位置，从这里，可通过开放式的拱廊通往专业社区。东北部的拱廊可通往各个已有的主入口，这些主入口很自然地并入综合体新的交通系统中；南部则是一个通往画廊的过渡区。

SKEJ BAR
Arhus Universitets Hospital

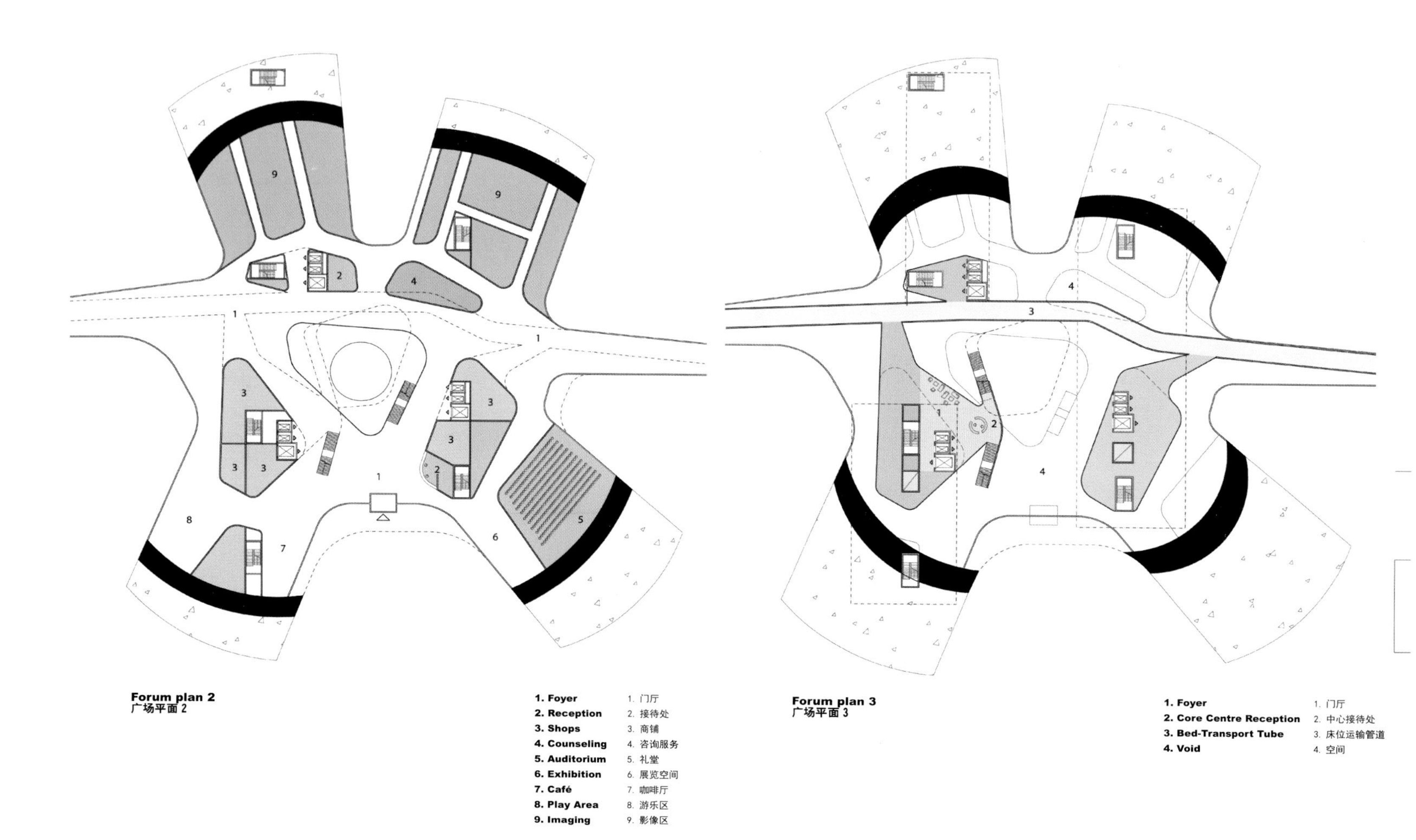

Forum plan 2
广场平面 2

1. Foyer	1. 门厅
2. Reception	2. 接待处
3. Shops	3. 商铺
4. Counseling	4. 咨询服务
5. Auditorium	5. 礼堂
6. Exhibition	6. 展览空间
7. Café	7. 咖啡厅
8. Play Area	8. 游乐区
9. Imaging	9. 影像区

Forum plan 3
广场平面 3

1. Foyer	1. 门厅
2. Core Centre Reception	2. 中心接待处
3. Bed-Transport Tube	3. 床位运输管道
4. Void	4. 空间

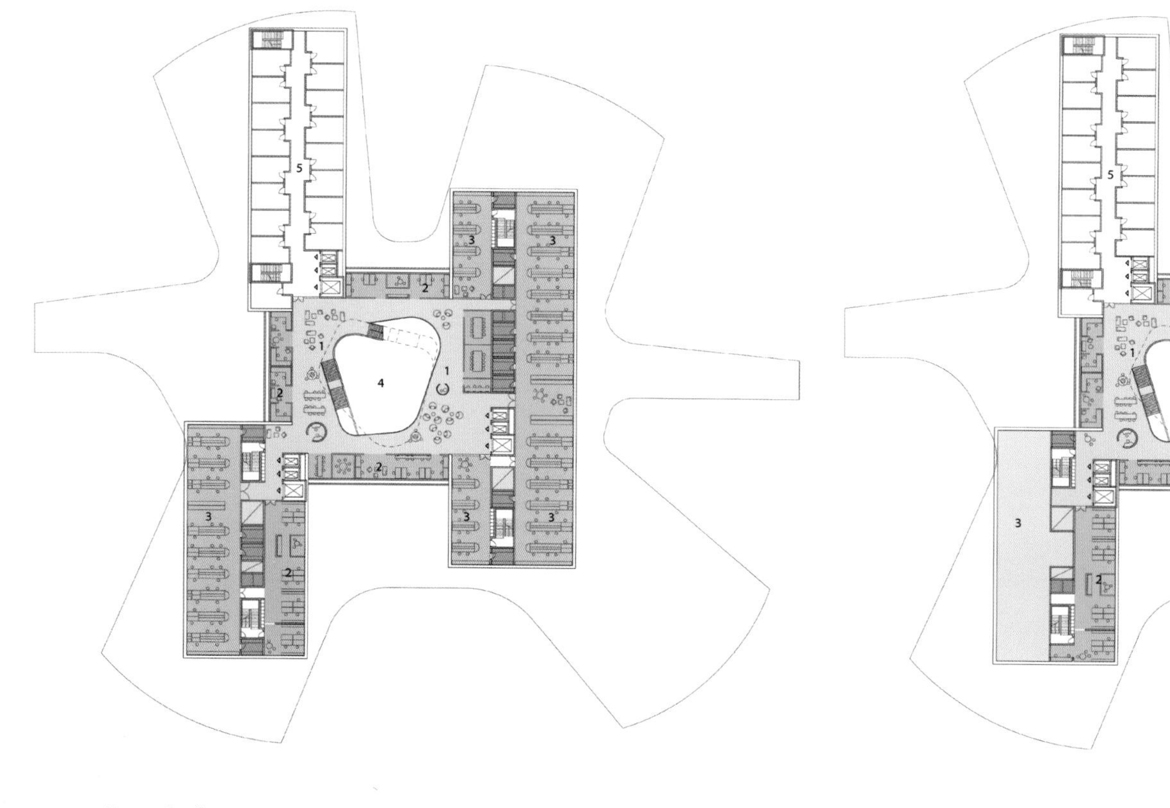

Forum plan 4
广场平面 4

1. Foyer	1. 门厅
2. Offices	2. 办公室
3. Laboratories	3. 实验室
4. Void	4. 空间
5. Patient Hotel	5. 患者酒店

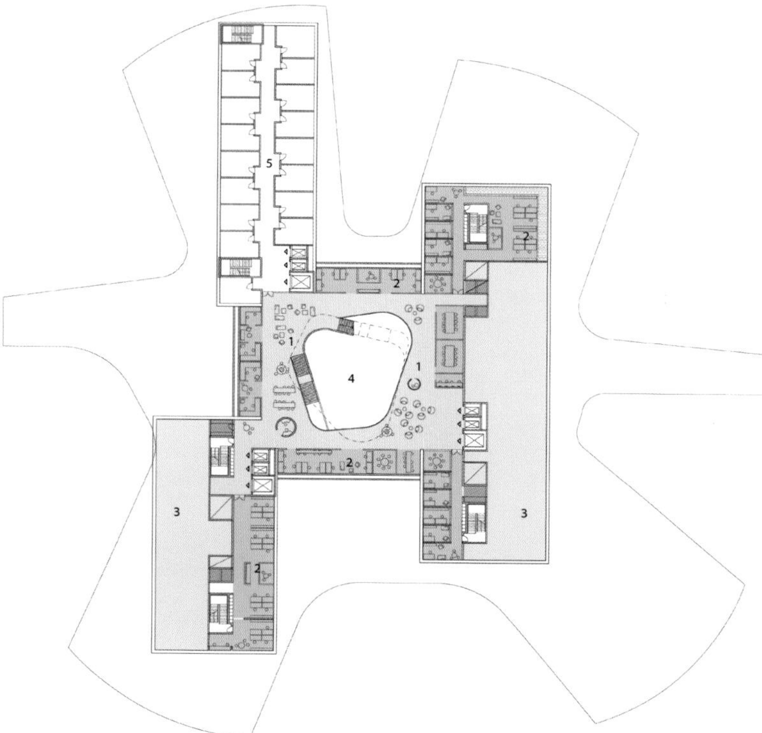

Forum plan 5
广场平面 5

1. Foyer	1. 门厅
2. Offices	2. 办公室
3. Plant (technical floor)	3. 机房（技术层）
4. Void	4. 空间
5. Patient Hotel	5. 患者酒店

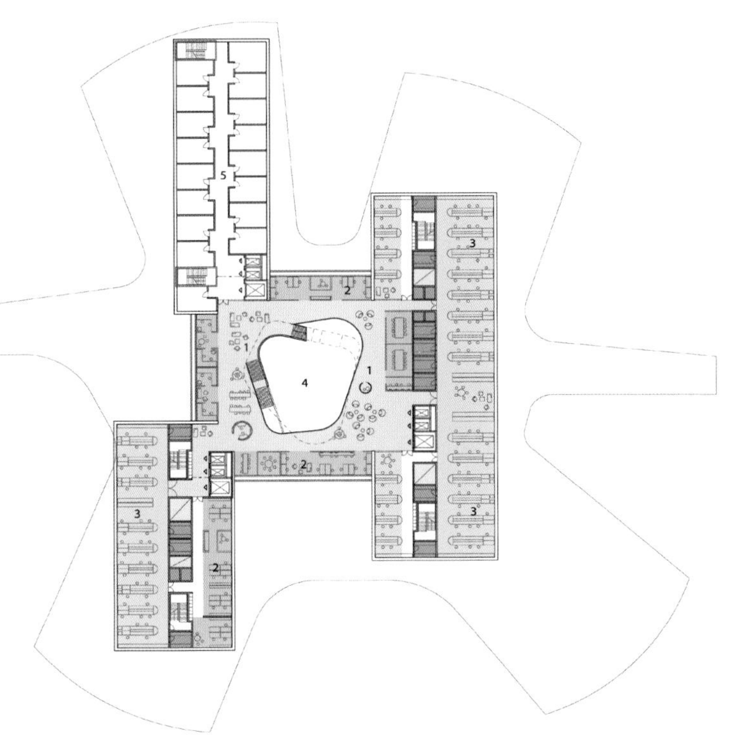

Forum plan 6
广场平面 6

1. Foyer	1. 门厅
2. Offices	2. 办公室
3. Laboratories	3. 实验室
4. Void	4. 空间
5. Patient Hotel	5. 患者酒店

Forum plan 7
广场平面 7

1. Foyer	1. 门厅
2. Offices	2. 办公室
3. Laboratories	3. 实验室
4. Void	4. 空间
5. Plant (technical floor)	5. 机房（技术层）
6. Café	6. 咖啡厅
7. Terrace	7. 露台

1. CENTRAL RECEPTION/LOUNGE
1. 中心接待处 / 休息室

2. COMMON 'SQUARE'
2. 公共"广场"

3. UNIVERSITY-OFFICES
3. 大学 - 办公室

4. VOID/VIEW TO ARCADE
4. 空间 / 视线朝向拱廊方向

5. LOUNGE
5. 休息室

6. CONNECTING WALKWAY
6. 连接通道

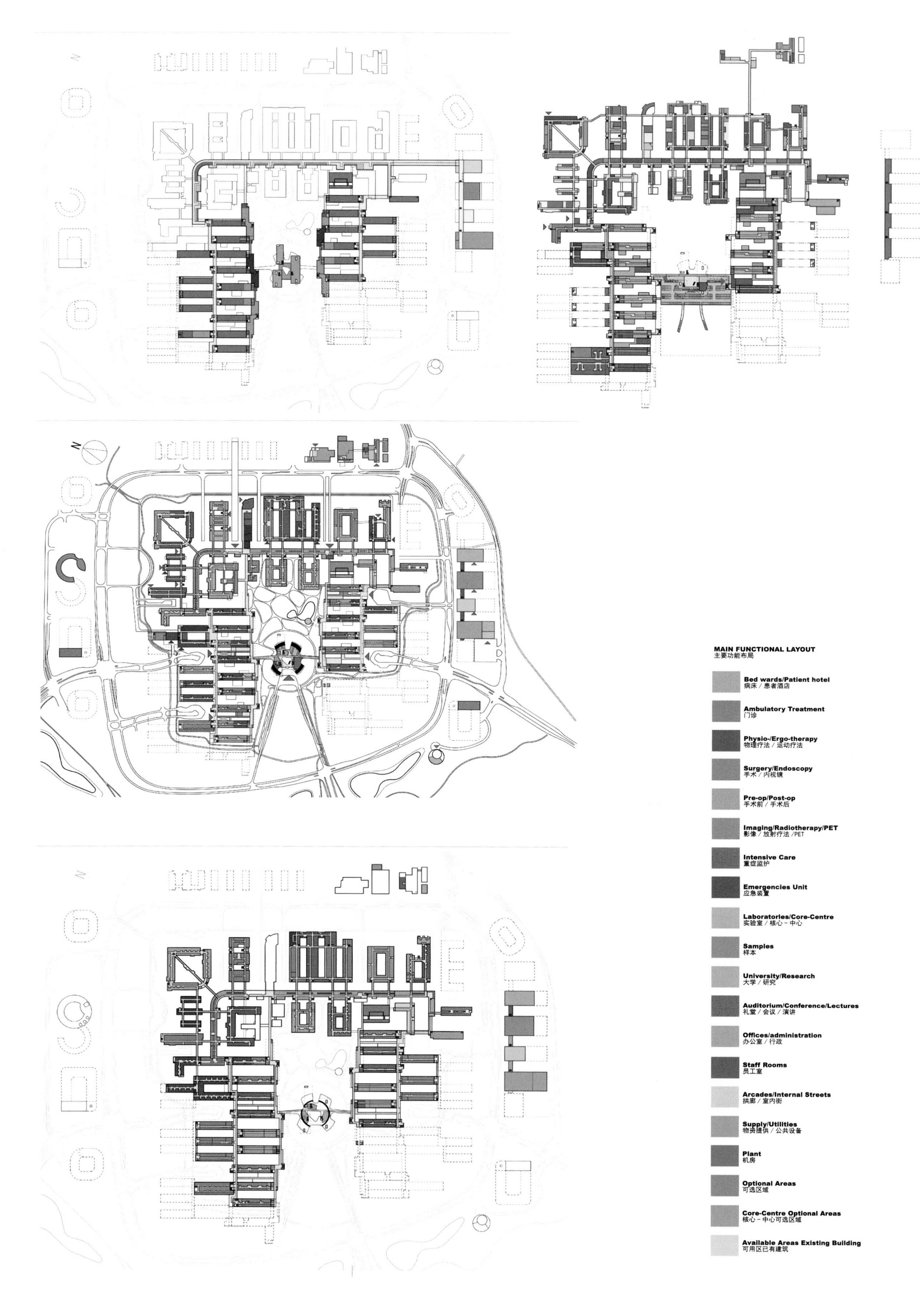
N
N
N
MAIN FUNCTIONAL LAYOUT
主要功能布局
Bed wards/Patient hotel
病床 / 患者酒店
Ambulatory Treatment
门诊
Physio-/Ergo-therapy
物理疗法 / 运动疗法
Surgery/Endoscopy
手术 / 内视镜
Pre-op/Post-op
手术前 / 手术后
Imaging/Radiotherapy/PET
影像 / 放射疗法 /PET
Intensive Care
重症监护
Emergencies Unit
应急装置
Laboratories/Core-Centre
实验室 / 核心 – 中心
Samples
样本
University/Research
大学 / 研究
Auditorium/Conference/Lectures
礼堂 / 会议 / 演讲
Offices/administration
办公室 / 行政
Staff Rooms
员工室
Arcades/Internal Streets
拱廊 / 室内街
Supply/Utilities
物资提供 / 公共设备
Plant
机房
Optional Areas
可选区域
Core-Centre Optional Areas
核心 – 中心可选区域
Available Areas Existing Building
可用区已有建筑

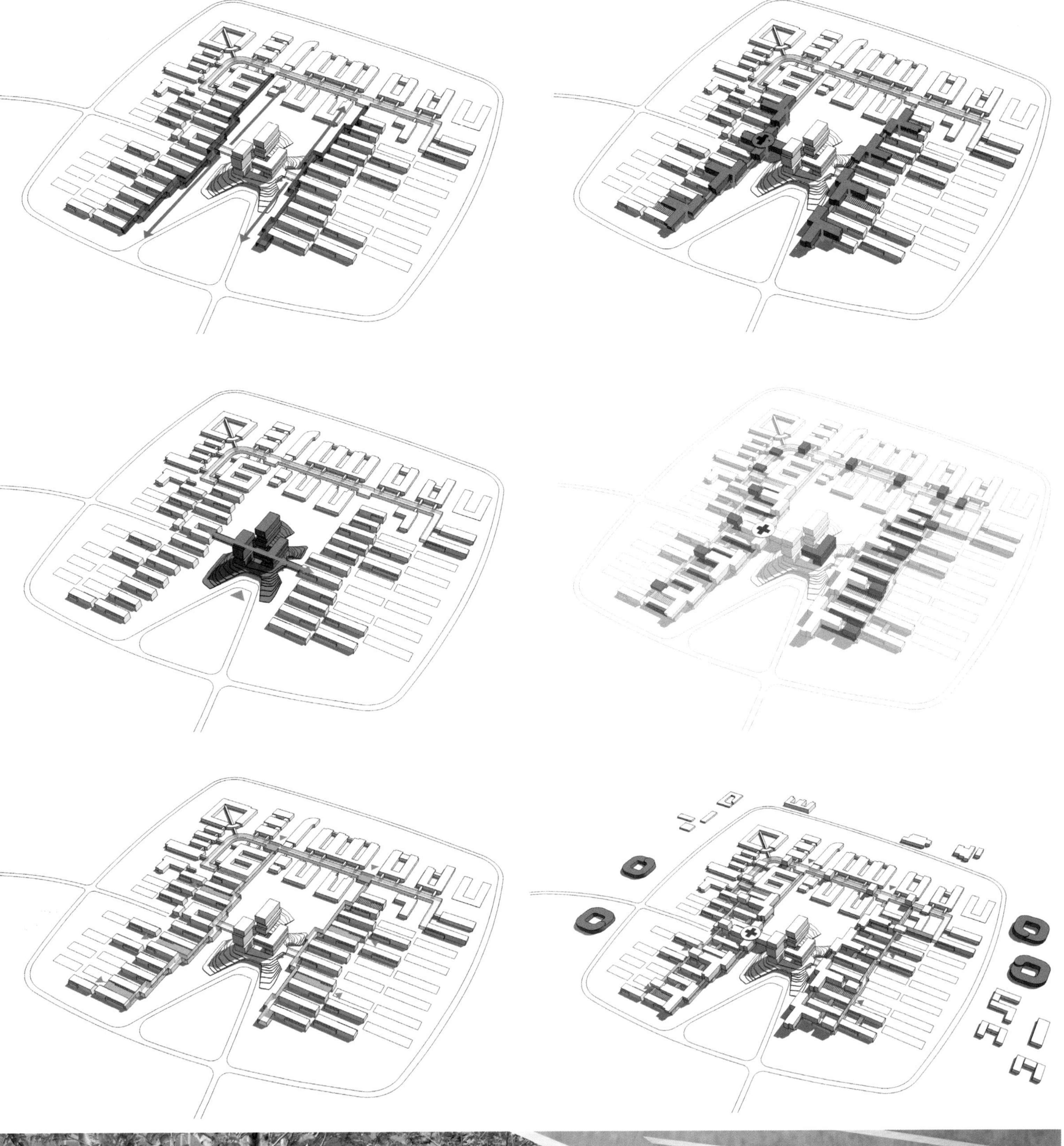

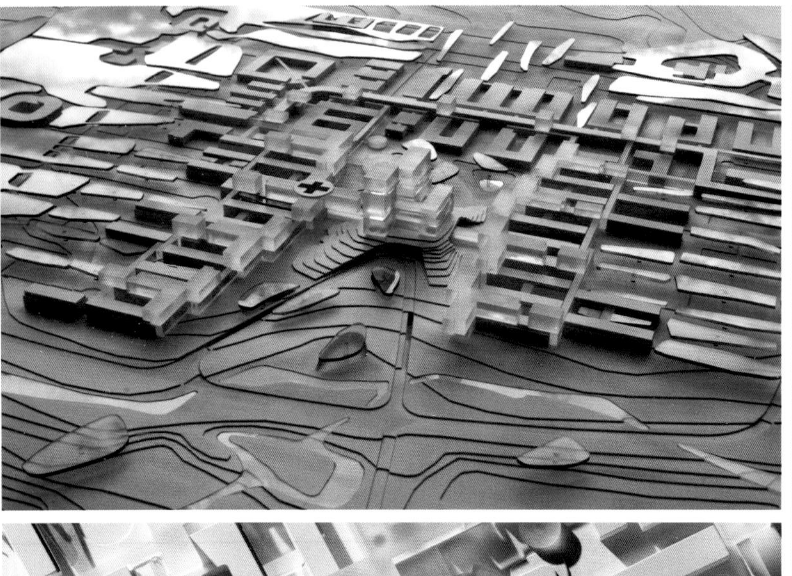

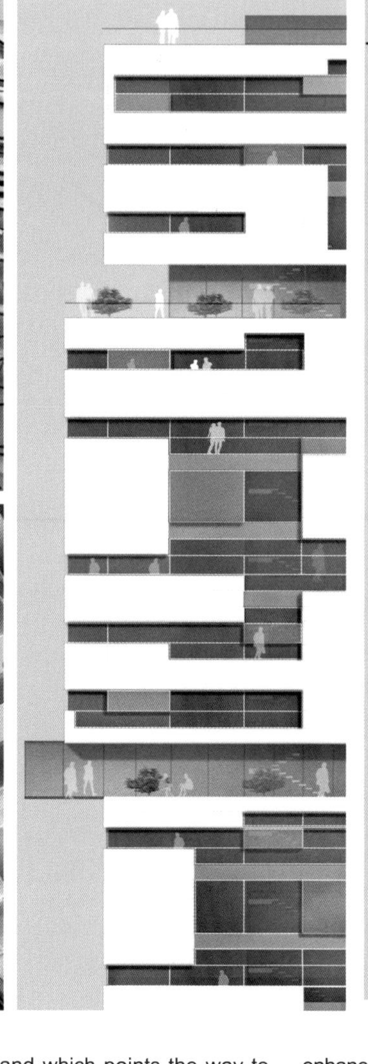

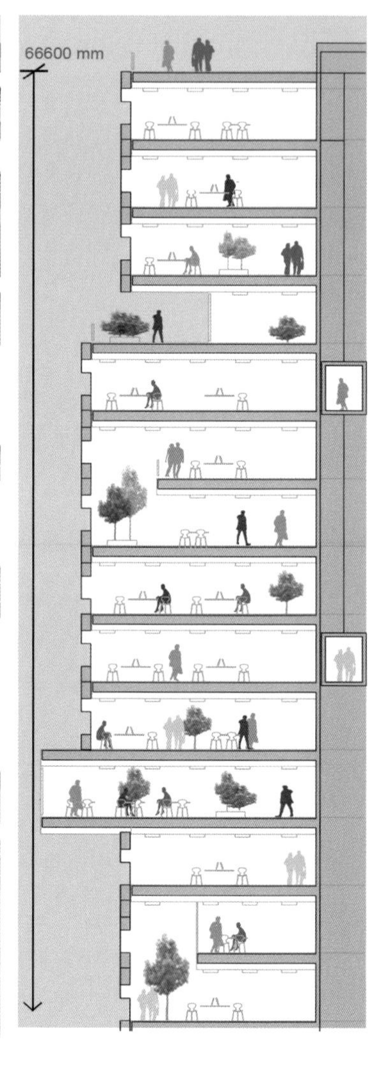

Profile

The largest hospital construction project in the history of Denmark, the New University Hospital in Aarhus, will be built onto the existing Aarhus University Hospital, Skejby, to form a single hospital complex. The resulting New University Hospital will be the size of a Danish provincial town.

Design Concept

The New University Hospital in Aarhus will represent a physical model of the hospital of the future: a patient-centred hospital founded on concepts of the healing environment, technological innovation and health-promoting surroundings, and which points the way to future hospital architecture.

Architectural Design

The project includes a two-storey treatment centre, with out-patient departments, the common clinical department and ward buildings as decentralised units. The giant hospital complex is also organised like a town, with a hierarchy of neighbourhoods, streets and squares which provide a foundation for a diverse and dynamic "green" urban area. The overall style of the New University Hospital in Aarhus thus favours the decentralised over a more compact building character, in order to retain and enhance the humane character and identity of the existing building.

The Forum, with the most outwardly-directed functions is the most important thoroughfare: a distinctive, capacious public space which occupies a central position as the complex's natural heart. From here, circulation routes will branch out towards the professional communities via open arcades. The arcades to the north-east lead all the way to the existing main entrances, which have naturally been incorporated into the complex's new traffic system, while to the south they provide a transition zone to galleries.

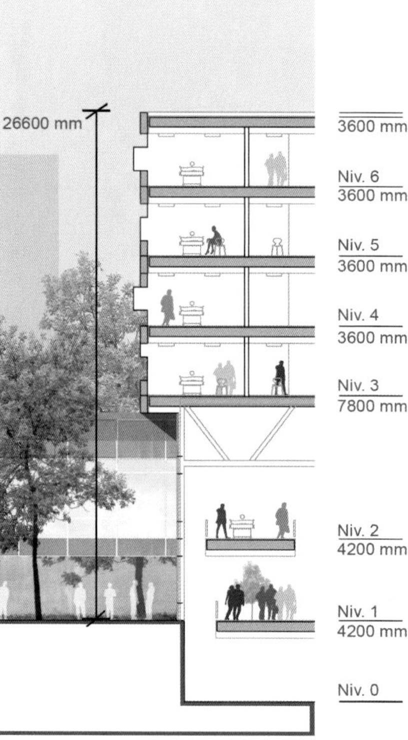

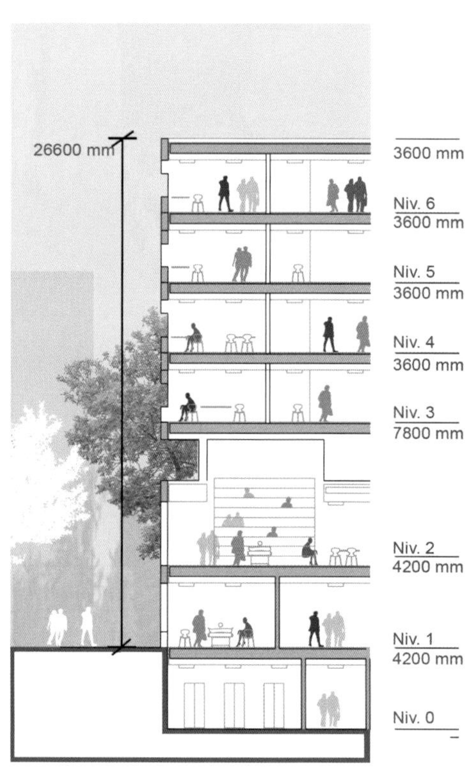

Plan 4
Teknik / Logistik
Plan 3
Behandling
Plan 2
Ambulatorium
Plan 1
Universitet / Teknik
GÅRDHAVE
Plan 8
Senge
Plan 7
Senge
Plan 6
Senge
Plan 5
Senge / Behandling
Plan 8
Kontor
Plan 7
Kontor
Plan 6
Kontor
Plan 5
Kontor

丹麦 Viborg 地区医院新急性治疗中心

New Acute Treatment Center of Viborg Regional Hospital

设计单位：C.F.Møller Architects
开发商：Region Midtjylland
项目地址：丹麦 Viborg
项目面积：22 000 ㎡

Designed by: C.F. Møller
Client: Region Midtjylland
Location: Viborg, Denmark
Area: 22,000 m²

项目概况

Viborg 地区医院的扩建大楼新增了一个急救中心和一个通向这栋综合设施的高度可见的主入口。这个项目还构建了一个更具吸引力的外观，并建立了医院与城市以及 Søndersø 湖自然保护区的联系。

建筑设计

设计旨在为患者和家属营造一个人性化的医疗环境，在这里，建筑各部分的方位和导向系统清晰可辨。开放通透的接待区既确定了项目的场所感，同时也提供了观赏室外景观的视野。各主要功能区之间的路径很短，既有利于患者的健康，同时也为员工提供了一个高效的工作环境。

项目方案依据康复建筑标准设计，实现了自然光照的优化利用。建筑材料和建筑外表皮保证了良好的室内氛围、室内气候和声学效果。设计采用的可持续性策略能够减少能源消耗和资源浪费，并将达到丹麦 2015 年的节能标准。

单独的病房单元采用了标准化设计，这便于结构的重组和功能的转换。同样的，立面的设计采用了可替换性的体量和通透的模块，这使得立面的建造不会干扰到医院的运作。这些措施使病房具备了极好的灵活性，并符合了建筑的扩展策略，使这个综合建筑在未来能够满足新的要求。

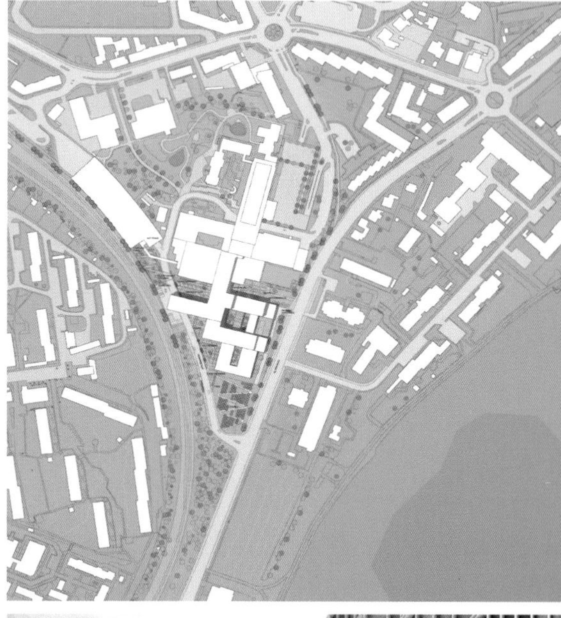

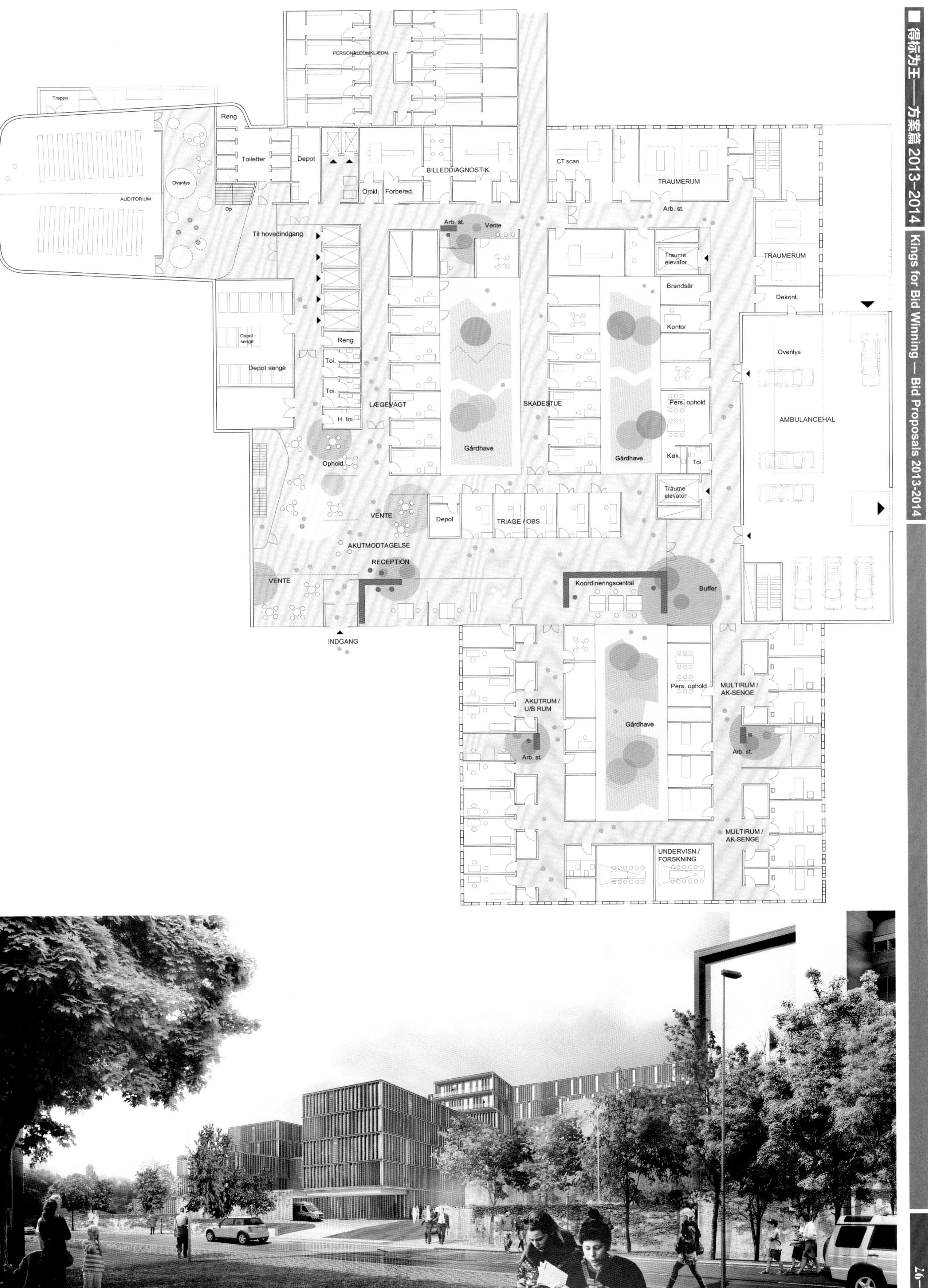
Trappe
Reng.
Toiletter
Depot
BILLEDDIAGNOSTIK
Omkl.
Forbered.
CT scan.
TRAUMERUM
Arb. st.
Ovenlys
AUDITORIUM
Op
Til hovedindgang
Vente
Traume elevator
TRAUMERUM
Brandsår
Dekont.
Kontor
Depot - senge
Depot senge
Reng.
Toi.
Toi.
H. toi
LÆGEVAGT
SKADESTUE
Pers. ophold
Ovenlys
AMBULANCEHAL
Gårdhave
Gårdhave
Køk.
Toi
Ophold
Traume elevator
VENTE
Depot
TRIAGE / OBS.
AKUTMODTAGELSE
RECEPTION
VENTE
Koordineringscentral
Buffer
INDGANG
AKUTRUM / U/B RUM
Pers. ophold
MULTIRUM / AK-SENGE
Gårdhave
Arb. st.
Arb. st.
MULTIRUM / AK-SENGE
UNDERVISN./ FORSKNING

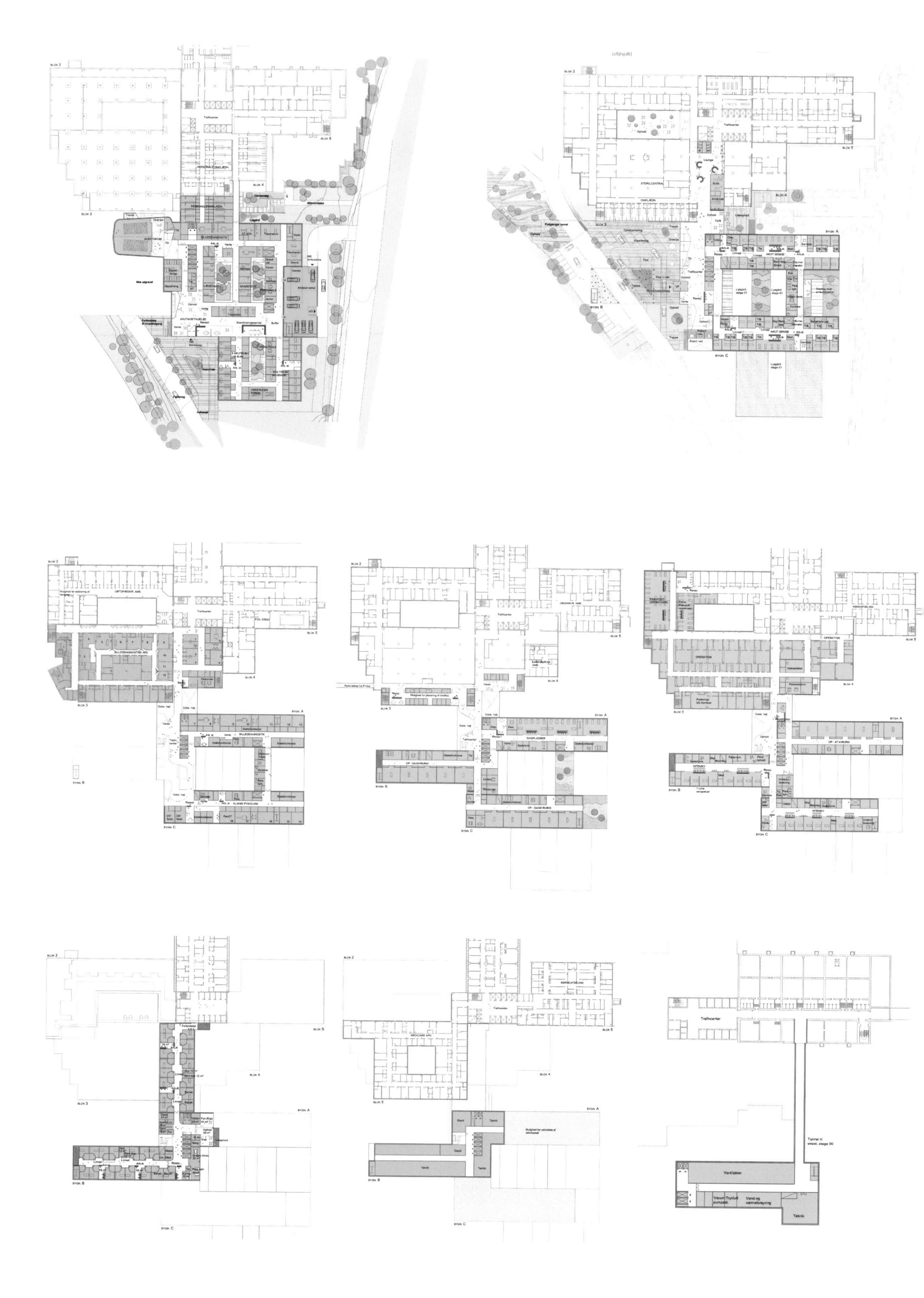

ccf@sj.dk
Trafikcenter
Ventilation
Teknik
Tunnel til eksist. etage 00
Vand og varmeforsyning

LÆGEVAGTEN

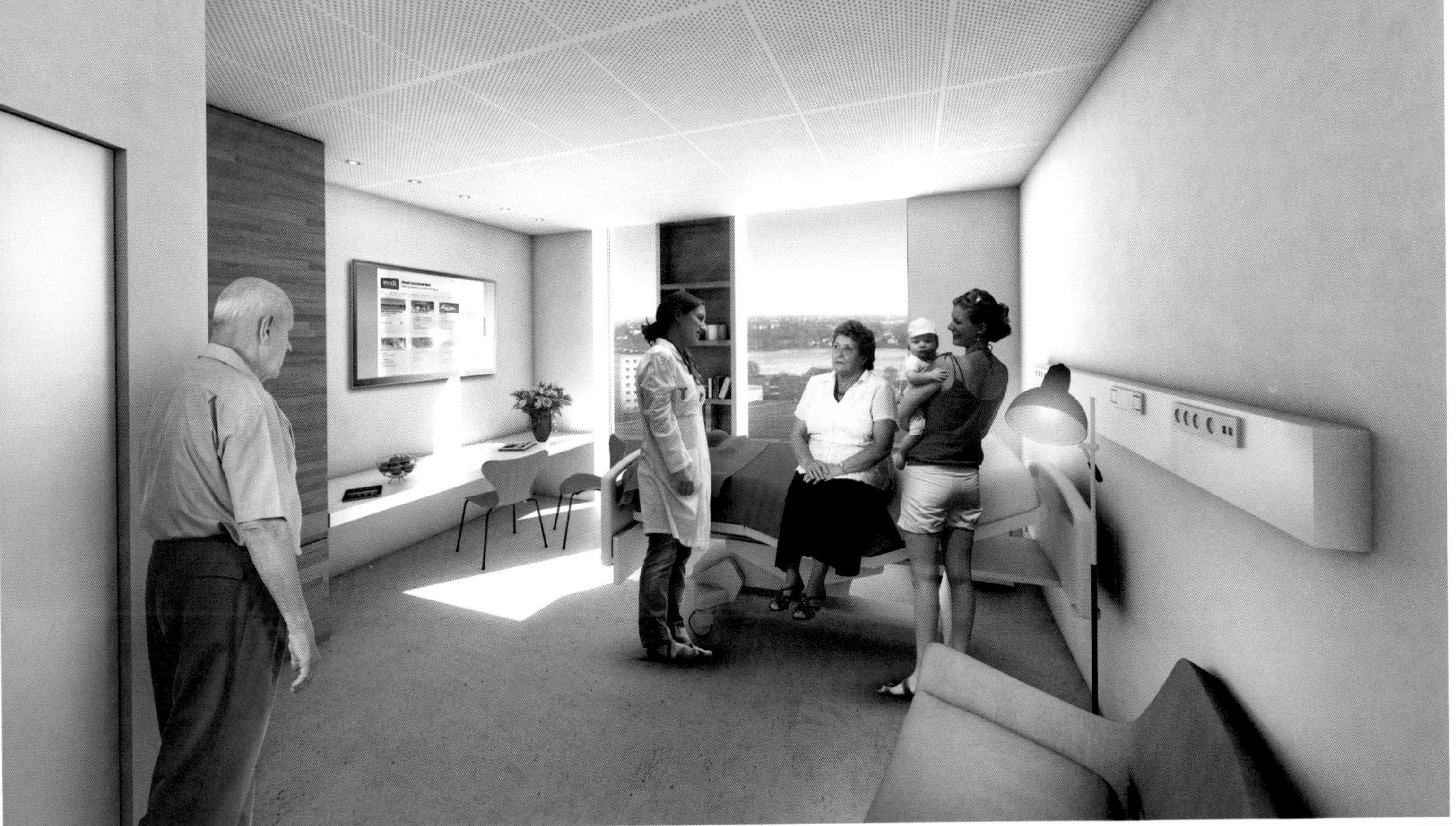

AKUTSENGE

INFORMATION

Arb. st.
Standard U/B rum
AMBULATORIE
Special U/B rum
Special U/B rum
Depot
Toi.
Møde
Venteomr. børn
Multirum
Vente
H. toi.
Reng.
Udeophold
Skylle-rum
Rent depot
Standard U/B rum
Bifunk.
Special U/B rum
Medicin
Depot
Undervisn/forsk. 32 m²
Linned
Arb. st.
AMBULATORIE
Reception
Pers. ophold 40 m²
Toi.
Bifunk.
Special U/B rum
Standard U/B rum
Special U/B rum
Special U/B rum
Kontor 12 m²
Køkken
Arb. st.
Linned
Linned
Linned
Skylle-rum
Rent depot
Arb. st.
Bariatrisk sengestue
SENGEAFSNIT
Bariatrisk sengestue
Depot
Udstyr 21 m²
Fysio- og ergo-terapi 44 m²
Kontor 16 m²
Linned
Samtale / møde 16 m²
Toi.
Ophold 59 m²
H. toi.
Køkken
Reng.
Udeophold
Skylle-rum 12 m²
Rent depot 12 m²
Depot 13 m²
Linned
Medicin 10 m²
Undervisn/forsk. 32 m²
Foldevæg
Linned
Linned
Linned
Arb. st.
Arb. st.
SENGEAFSNIT
Reception
Pers. ophold 40 m²
Toi.
18 m²
18 m²
Bariatrisk sengestue
Bariatrisk sengestue
Kontor 12 m²
Køkken

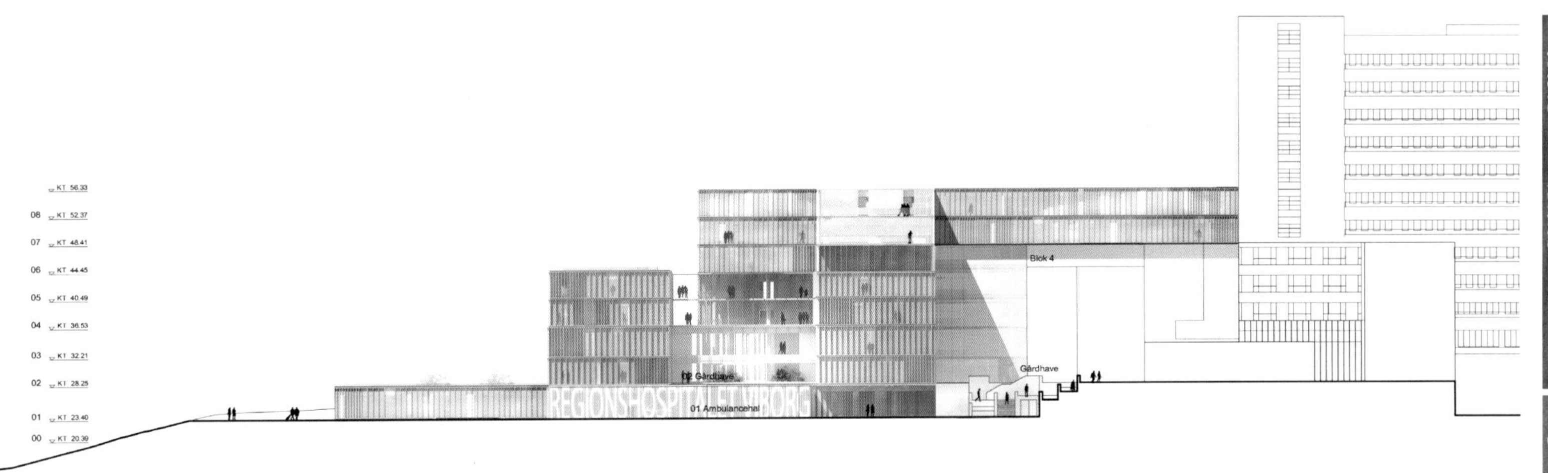

FACADEOPSTALT MOD ØST

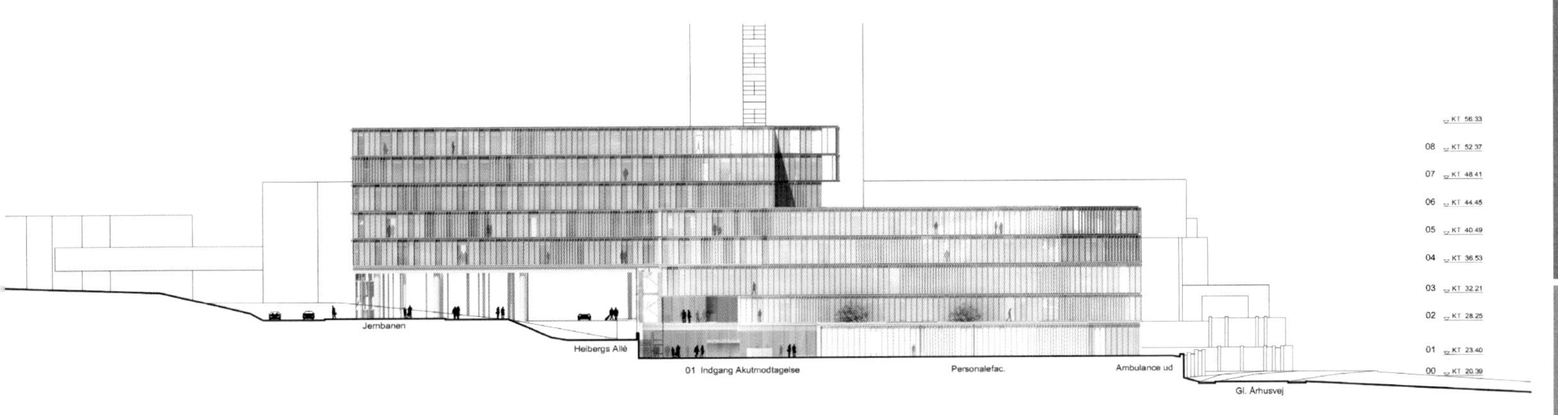

FACADEOPSTALT MOD SYD

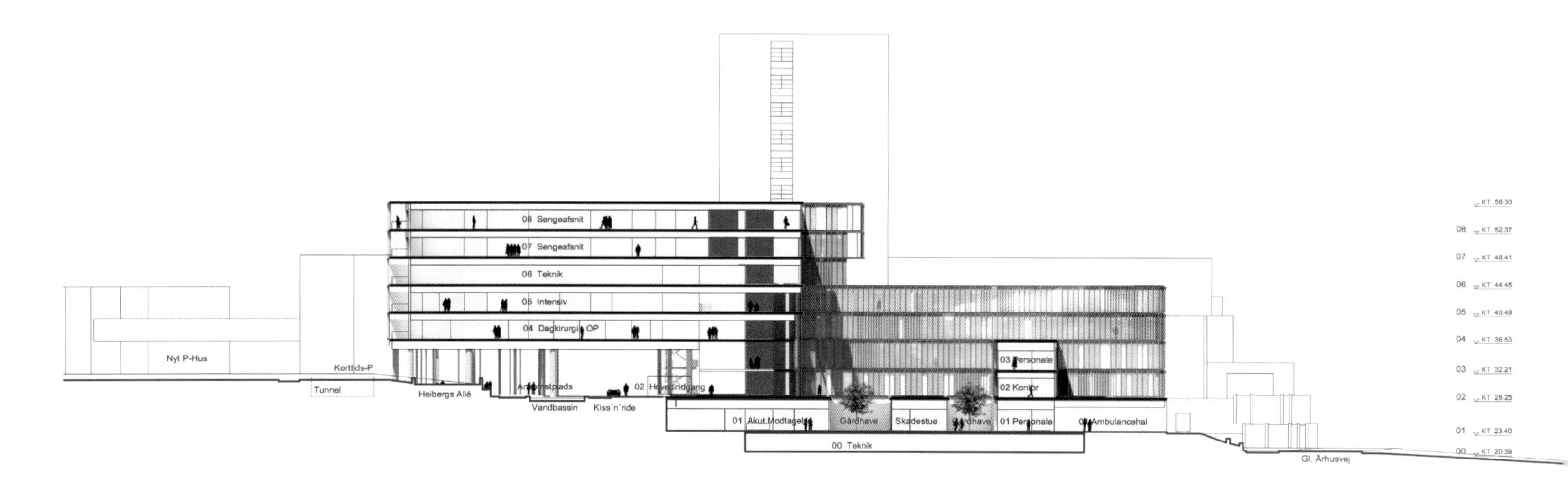

08 Sengeafsnit
07 Sengeafsnit
06 Teknik
05 Intensiv
04 Dagkirurgi OP
03 Personale
02 Kontor
02 Hovedindgang
01 Akut Modtagelse
Gårdhave
Skadestue
01 Personale
Ambulancehal
00 Teknik
Nyt P-Hus
Korttids-P
Tunnel
Helbergs Allé
Vandbassin
Kiss´n´ride
Gl. Århusvej

SNIT **AA**

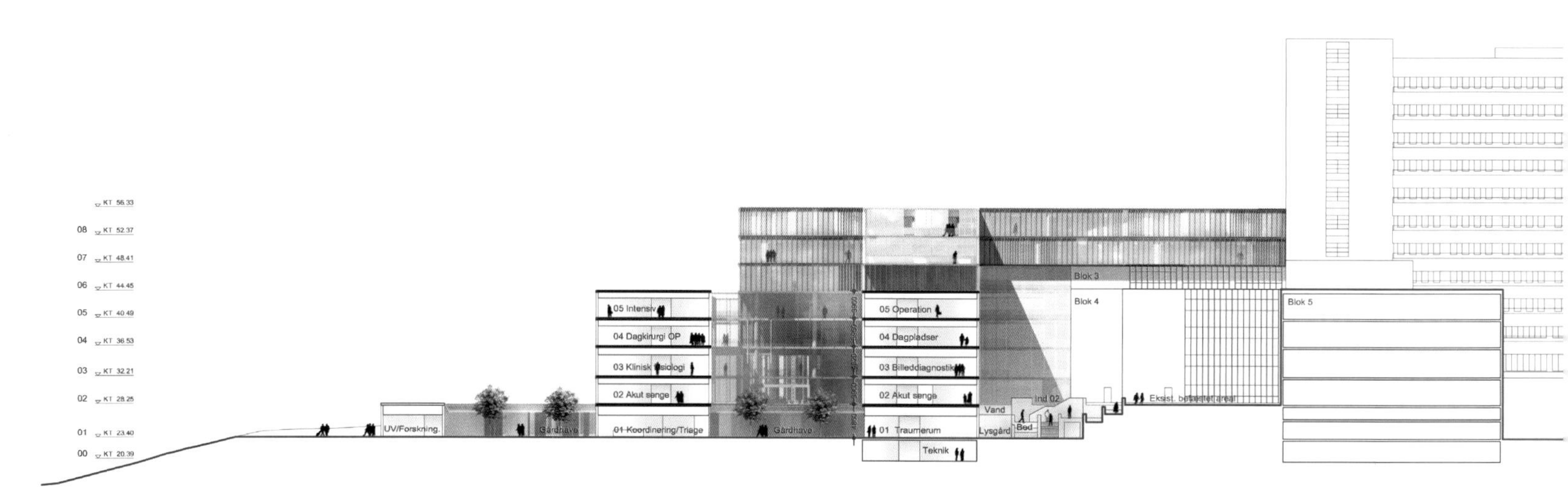

05 Intensiv
04 Dagkirurgi OP
03 Klinisk fysiologi
02 Akut senge
01 Koordinering/Triage
05 Operation
04 Dagpladser
03 Billeddiagnostik
02 Akut senge
01 Traumerum
Teknik
UV/Forskning
Gårdhave
Blok 3
Blok 4
Blok 5
Vand
Lysgård
Bed

SNIT **CC**

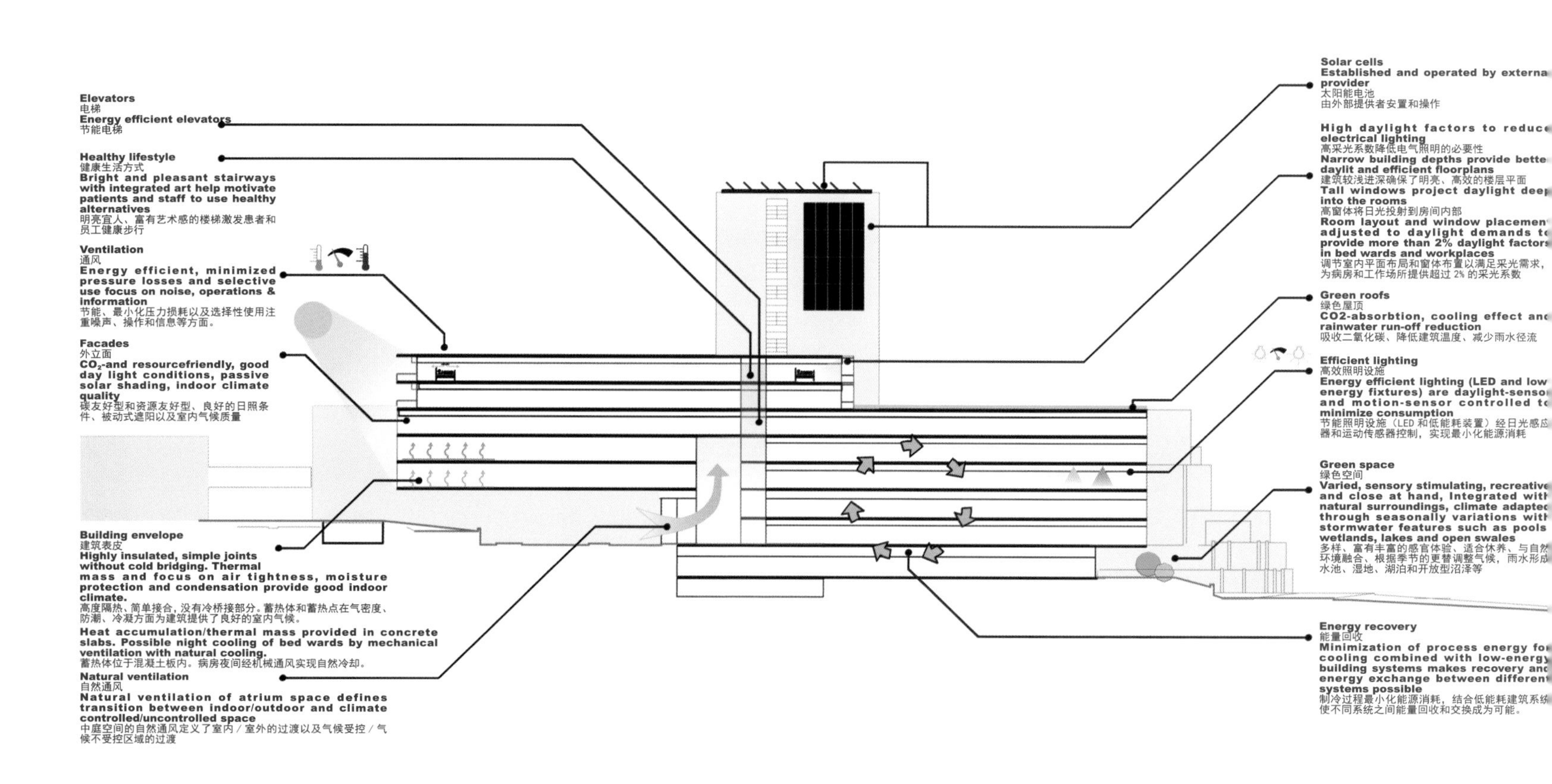

Elevators
电梯
Energy efficient elevators
节能电梯
Healthy lifestyle
健康生活方式
Bright and pleasant stairways with integrated art help motivate patients and staff to use healthy alternatives
明亮宜人、富有艺术感的楼梯激发患者和员工健康步行
Ventilation
通风
Energy efficient, minimized pressure losses and selective use focus on noise, operations & information
节能、最小化压力损耗以及选择性使用注重噪声、操作和信息等方面。
Facades
外立面
CO_2- and resourcefriendly, good day light conditions, passive solar shading, indoor climate quality
碳友好型和资源友好型、良好的日照条件、被动式遮阳以及室内气候质量
Building envelope
建筑表皮
Highly insulated, simple joints without cold bridging. Thermal mass and focus on air tightness, moisture protection and condensation provide good indoor climate.
高度隔热、简单接合，没有冷桥接部分。蓄热体和蓄热点在气密度、防潮、冷凝方面为建筑提供了良好的室内气候。
Heat accumulation/thermal mass provided in concrete slabs. Possible night cooling of bed wards by mechanical ventilation with natural cooling.
蓄热体位于混凝土板内。病房夜间经机械通风实现自然冷却。
Natural ventilation
自然通风
Natural ventilation of atrium space defines transition between indoor/outdoor and climate controlled/uncontrolled space
中庭空间的自然通风定义了室内 / 室外的过渡以及气候受控 / 气候不受控区域的过渡
Solar cells
Established and operated by externa
provider
太阳能电池
由外部提供者安置和操作
High daylight factors to reduce electrical lighting
高采光系数降低电气照明的必要性
Narrow building depths provide bette daylit and efficient floorplans
建筑较浅进深确保了明亮、高效的楼层平面
Tall windows project daylight dee into the rooms
高窗体将日光投射到房间内部
Room layout and window placemen adjusted to daylight demands t provide more than 2% daylight facto in bed wards and workplaces
调节室内平面布局和窗体布置以满足采光需求，为病房和工作场所提供超过 2% 的采光系数
Green roofs
绿色屋顶
CO2-absorbtion, cooling effect an rainwater run-off reduction
吸收二氧化碳、降低建筑温度、减少雨水径流
Efficient lighting
高效照明设施
Energy efficient lighting (LED and low energy fixtures) are daylight-senso and motion-sensor controlled t minimize consumption
节能照明设施（LED 和低能耗装置）经日光感应器和运动传感器控制，实现最小化能源消耗
Green space
绿色空间
Varied, sensory stimulating, recreative and close at hand, Integrated wit natural surroundings, climate adapte through seasonally variations wit stormwater features such as pools wetlands, lakes and open swales
多样、富有丰富的感官体验、适合休养、与自然环境融合、根据季节的更替调整气候，雨水形成水池、湿地、湖泊和开放型沼泽等
Energy recovery
能量回收
Minimization of process energy fo cooling combined with low-energ building systems makes recovery an energy exchange between differen systems possible
制冷过程最小化能源消耗，结合低能耗建筑系统使不同系统之间能量回收和交换成为可能。

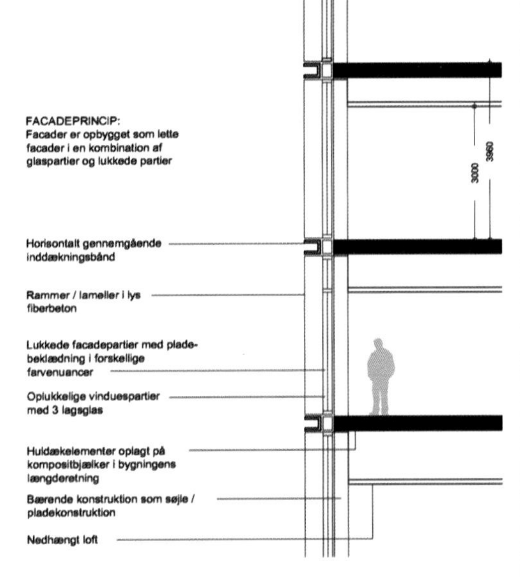

Profile

The Viborg Regional Hospital's new extension building adds both a new acute treatment centre and a highly visible new main entrance to the complex. It also creates a more inviting exterior, and connects the hospital to the city and to the Søndersø Lake nature reserve.

Architectural Design

The aim is to create a protective atmosphere for patients and relatives, where orientation and wayfinding is easy. Open and transparent reception areas provide a sense of place as well as views to the exterior landscape. Short distances between the main functions are a benefit to the patient's wellbeing, and also secure an efficient working environment for the staff.

The designs focusing on healing architecture include optimal use of natural daylight, and use of materials and surfaces which provide fine atmosphere, good indoor climate and pleasant acoustics in the spaces. The sustainability strategy reduces waste and energy consumption, and targets the ambitious future Danish 2015 energy standards.

Individual rooms and units are planned to standardized dimensions, to allow for easy reconfigurations and changes in use. Similarly, the facades are planned to allow the interchangeability of solid and transparent modules without interrupting the hospitals operations. These measures ensure full flexibility for the wards, and are combined with a future extension strategy to allow the entire complex to be adapted to new demands.

丹麦哥本哈根 VIlla Vita 癌症中心

Villa Vita Cancer Center, Copenhagen, Denmark

设计单位： C.F.Møller Architects
开发单位： 哥本哈根市和丹麦癌症协会
项目地址： 丹麦哥本哈根
项目面积： 1 800 ㎡

Designed by: *C.F. Møller*
Client: *Municipality of Copenhagen and the Danish Cancer Society*
Location: *Copenhagen, Denmark*
Area: *1,800 m^2*

项目概况

Villa Vita 癌症中心是为癌症患者提供的一个咨询服务中心，在这里，癌症患者可以得到专业的治疗或帮助，也可以加入团体或参加课程培训。

建筑设计

该中心旨在呈现一个独特的连贯性建筑，建筑良好的连贯性使之在任意一个角度都可以被轻易识别。为了消除人们对这一机构的负面情绪，为病人和员工提供安全感，设计将建筑的内部空间分割成多个小单元，这些小单元由中央公共空间相连。

门厅和公共空间占据了建筑的中心位置。建筑内包含了公共空间和社交空间，既为患者和员工提供了充足的空间进行社交聚会，同时也使之成为了一个充满生机和活力的空间。这一空间涵盖了医院的基本功能，整个区域由半透明的墙壁照亮，并向内部景观庭院开放。

中心区域在功能上进行了清晰明确的划分，使各专业职能得到集中。小型的咨询室和大型的活动空间分布较为自由，在整个空间内形成了开放、流动的序列。

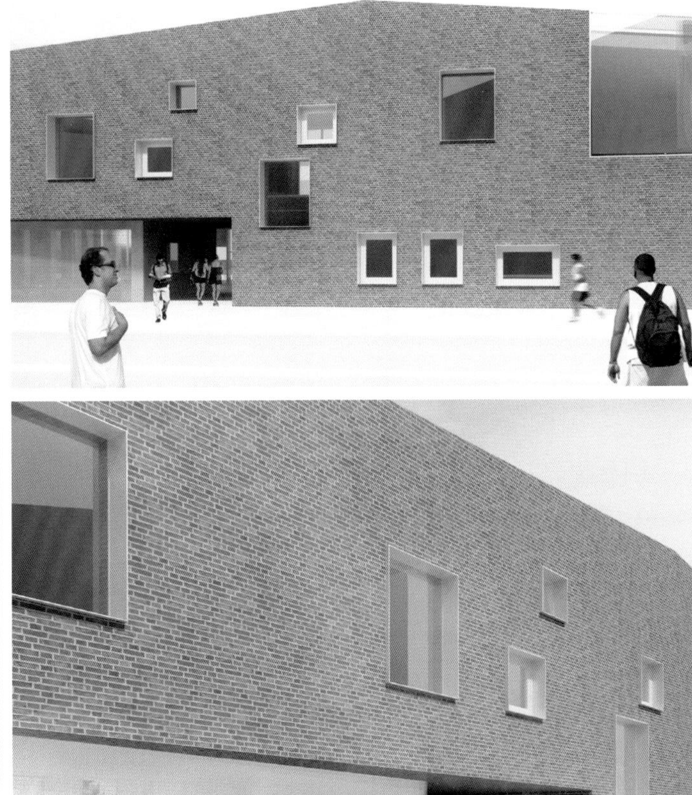

BASEMENT 地下室
gorss area 480m² 总面积 480m²
GROUND FLOOR 底层
gorss area 990m² 总面积 990m²
1st.FLOOR 一层
gorss area 960m² 总面积 960m²
COMMON AREAS 公共区
OFFICES/STAFF 办公室 / 员工
COUNCELLING & TREATMENT 咨询 / 治疗
EXERCISE ROOMS 健身房
EDUCATION ROOMS 教育室
KNOWLEDGE CENTRE 知识中心
STAFF ROOMS 教研室
KÆLDER 1:200 Brutto
480 m2
FØRSTE SAL 1:200 Brutto
880 m2

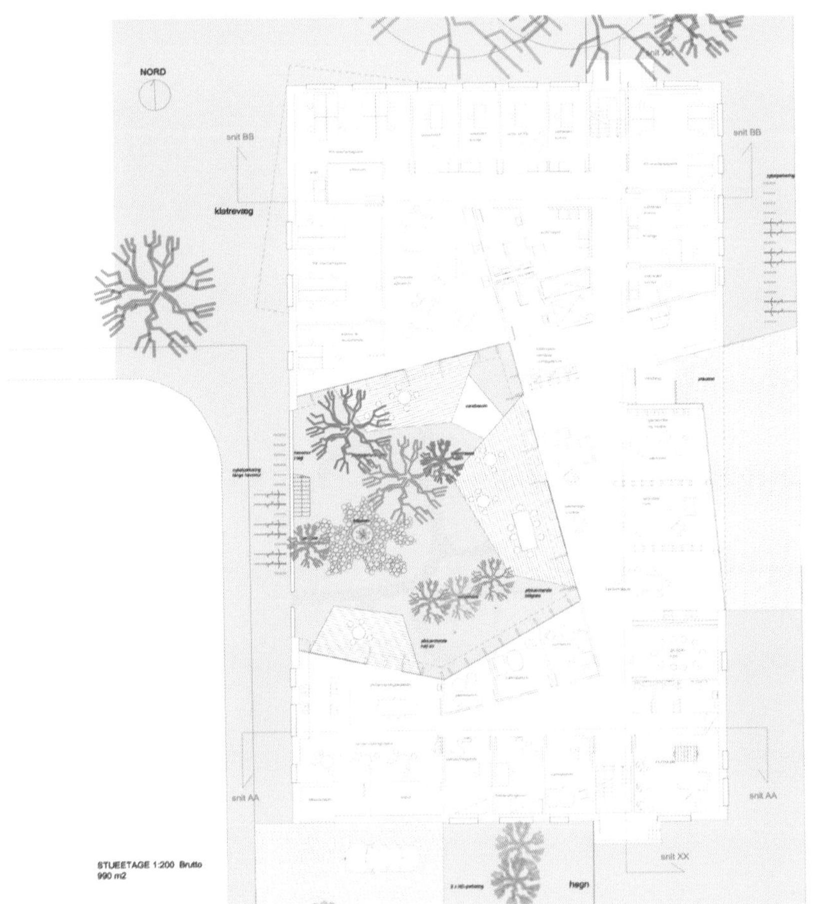

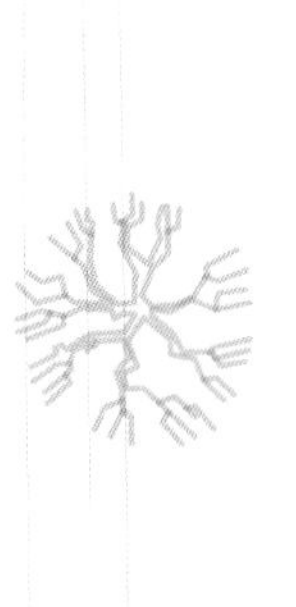

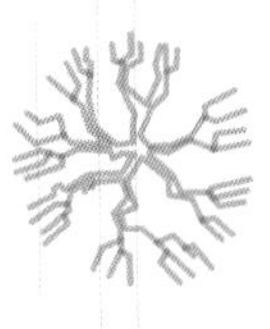

Profile

"Villa Vita" is a counselling facility for cancer patients, where they can receive advice and support, therapy and exercise, and participate in groups and courses.

Architectural Design

The centre is designed to appear as a distinct and coherent building, recognizable from all sides of the site. To avoid an institutional feel, and ensure the necessary sense of secureness for patients and staff, the interior layout is broken down into smaller units linked by central social spaces.

A foyer and common area forms the centrepiece, designed with niches and social areas, to make it a welcoming and lively space with enough room for social gatherings. All the primary functions can be accessed from this space, which is lit by a semi-transparent wall and opens up to an interior landscaped courtyard.

The various departments of the centre are distinctly demarcated, enabling professional concentration in each one, whereas small counselling rooms and larger activity spaces are freely distributed within the centre to create an open and fluid sequence of spaces through the entire building.

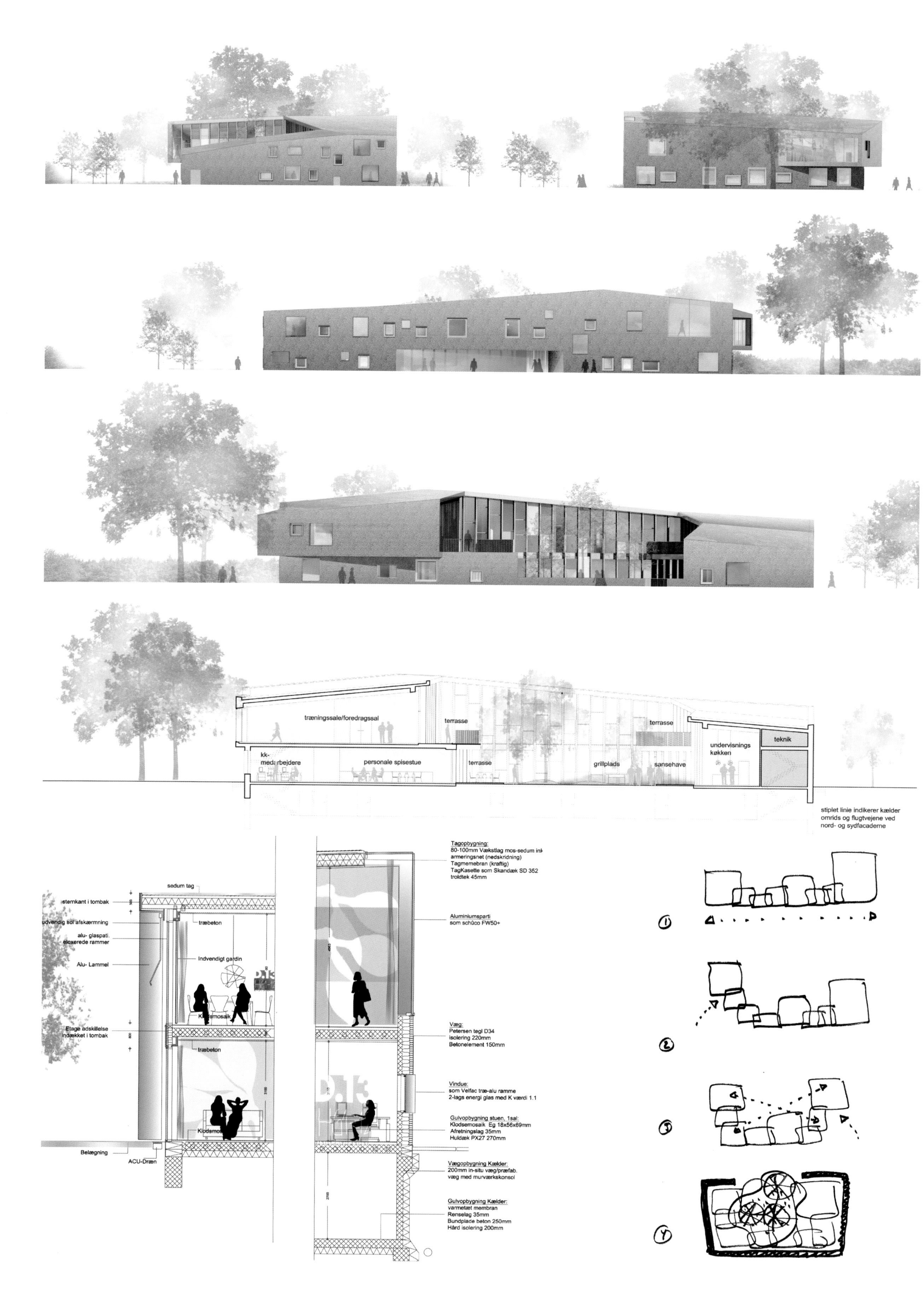

træningssale/foredragssal
terrasse
terrasse
kk-
medarbejdere
personale spisestue
terrasse
grillplads
sansehave
undervisnings
køkken
teknik
stiplet linie indikerer kælder
omrids og flugtvejene ved
nord- og sydfacaderne
sedum tag
sternkant i tombak
udvendig sol afskærmning
alu- glaspati.
eloxerede rammer
Alu- Lammel
træbeton
Indvendigt gardin
Etage adskillelse
indækket i tombak
træbeton
Belægning
ACU-Dræn
Tagopbygning:
80-100mm Vækstlag mos-sedum ink
armeringsnet (nedskridning)
Tagmemebran (kraftig)
TagKasette som Skandæk SD 352
troldtek 45mm
Aluminiumsparti
som schûco FW50+
Væg:
Petersen tegl D34
isolering 220mm
Betonelement 150mm
Vindue:
som Velfac træ-alu ramme
2-lags energi glas med K værdi 1.1
Gulvopbygning stuen, 1sal:
Klodsemosaik Eg 18x56x69mm
Afretningslag 35mm
Huldæk PX27 270mm
Vægopbygning Kælder:
200mm in-situ væg/præfab.
væg med murværkskonsol
Gulvopbygning Kælder:
varmetæt membran
Renselag 35mm
Bundplade beton 250mm
Hård isolering 200mm

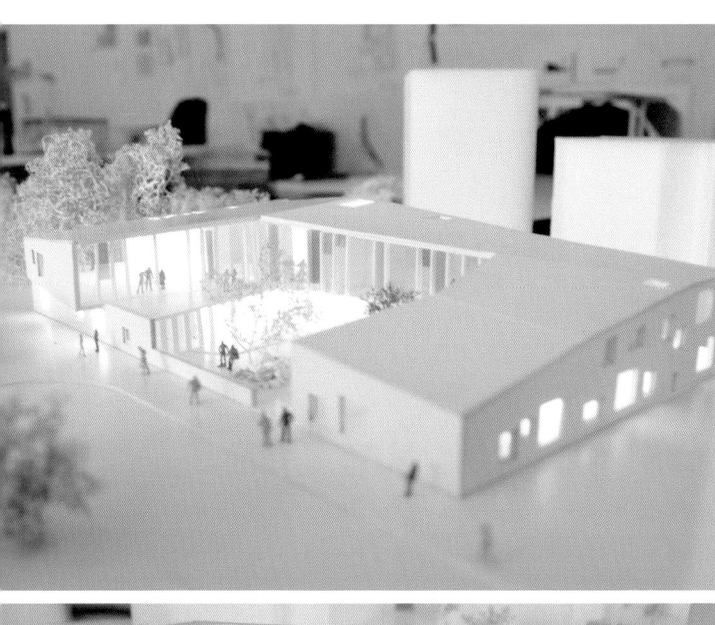

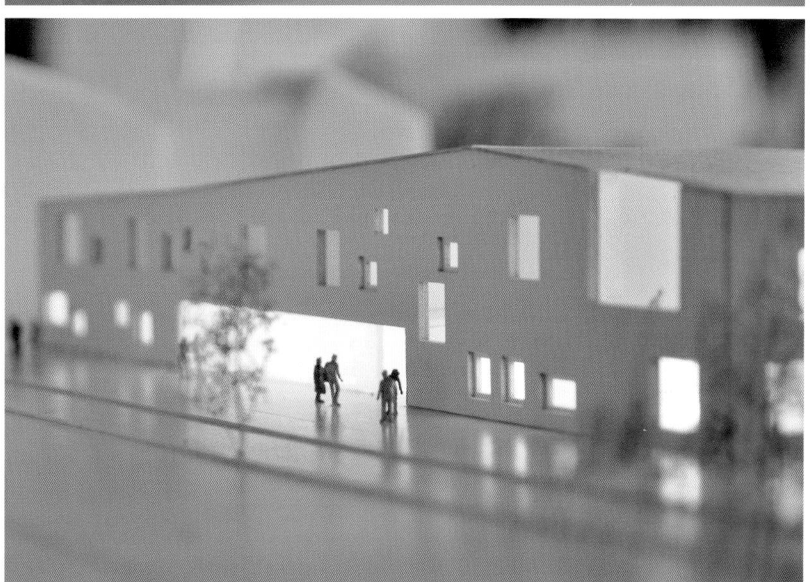

克罗地亚斯普利特私人医疗中心

Polyclinic St

设计单位：3LHD
开发商：斯普利特市 Krupa d.o.o.
项目地址：克罗地亚斯普利特
用地面积：2 086 ㎡
设计团队：Sasa Begovic　Marko Dabrovic
Tatjana Grozdanic Begovic　Silvije Novak
Koraljka Brebric Kleoncic　Sanja Jasika
Ines Vlahovic　Josko Kotula
Dragana Simic

Designed by: 3LHD
Client: City of Split, Krupa d.o.o.
Location: Split, Croatia
Site Area: 2,086 m²
Project Team: Sasa Begovic, Marko Dabrovic, Tatjana Grozdanic Begovic, Silvije Novak, Koraljka Brebric Kleoncic, Sanja Jasika, Ines Vlahovic, Josko Kotula, Dragana Simic

1.Garage entrance 1. 车库入口
2.Entrance 2. 入口

Ground floor plan
底层平面图

1.Garage entrance 1. 车库入口
2.Entrance 2. 入口
3.Pharmacy 3. 药房
4.Reception 4. 接待处
5.Caffe bar 5. 咖啡酒吧
6.Restaurant 6. 餐厅
7.Shop 7. 商店

Ground floor plan
底层平面图

项目概况

这个项目是 3LHD 在克罗地亚斯普利特某比赛中赢得的私人医疗中心附属设施建设项目。这座医疗中心围绕在 Firule 地区的一个历史遗迹周边，靠近当地现有的一家综合医院，靠海的自然环境以及新鲜的空气都为这个项目的建设提供了良好的条件。

建筑设计

依照参赛规则，设计方案旨在最大化实现技术和设备的实用性。设计师将服务、门诊、诊所和实验室等功能区间进行垂直规划，所有的公共场所都设在地下室、地面层和二楼，门诊和行政区在以上层面，并规划有地下车库。

建筑的外观和入口进行了横向装饰，覆盖在建筑表面的“绷带型”条形构件是一种先进的光线调节器，在建筑外表上形成一层保护膜，可以有效地遮挡阳光。这一设计灵感来自于斯普利特现代建筑风格——伊沃萨纳拉迪奇的“Briss Soleil”。

所有楼层的外墙和半公共场所都有庭院或是长廊相连，这些庭院或长廊将诊所隔断或连接。面向医院的内走廊种有绿植，为这个靠海的空间提供了一个舒适的地中海式的环境。

Profile

3LHD got the first prize in competing for subsidiary facilities construction of this private medical center in Split, Croatia. Close surrounding and historical site of Firule area are one of the most enjoyable living, working and recreation environments of the city of Split. Extraordinary location for the polyclinic is one of its greatest advantages. Placement near the existing hospital complex on Firule close to the sea and fresh air gives it even more importance and value.

8.Polyclinc-practice 8. 综合医院 - 实践
9.Office 9. 办公室
10.Drugstore 10. 药房
11.Hair salon 11. 发廊
12.Beauty salon 12. 美容院

First floor plan
第一层平面图

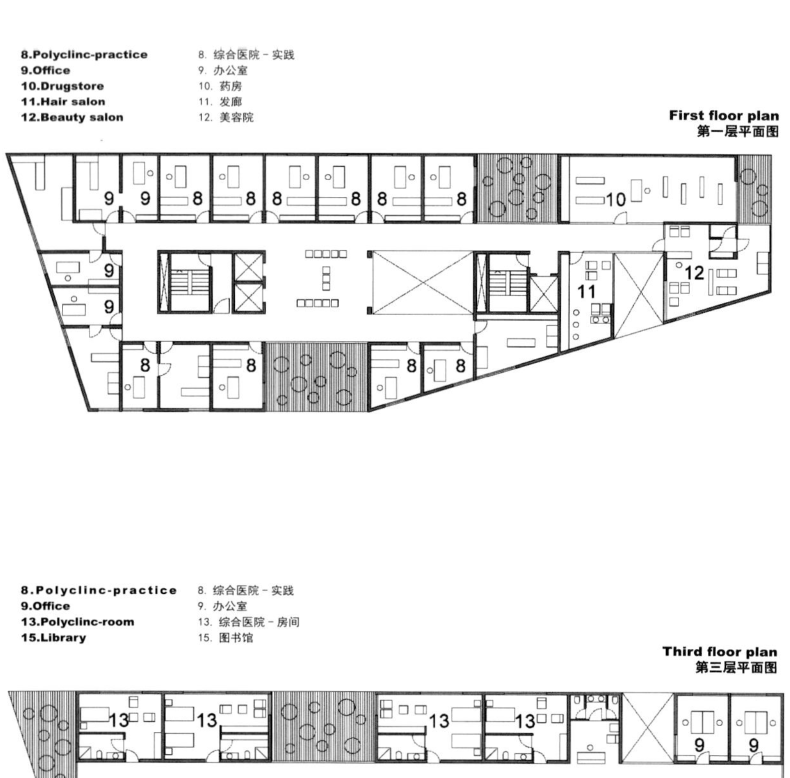

8.Polyclinc-practice 8. 综合医院 - 实践
9.Office 9. 办公室
13.Polyclinc-room 13. 综合医院 - 房间
14.Stomatology 14. 口腔科

Second floor plan
第二层平面图

8.Polyclinc-practice 8. 综合医院 - 实践
9.Office 9. 办公室
13.Polyclinc-room 13. 综合医院 - 房间
15.Library 15. 图书馆

Third floor plan
第三层平面图

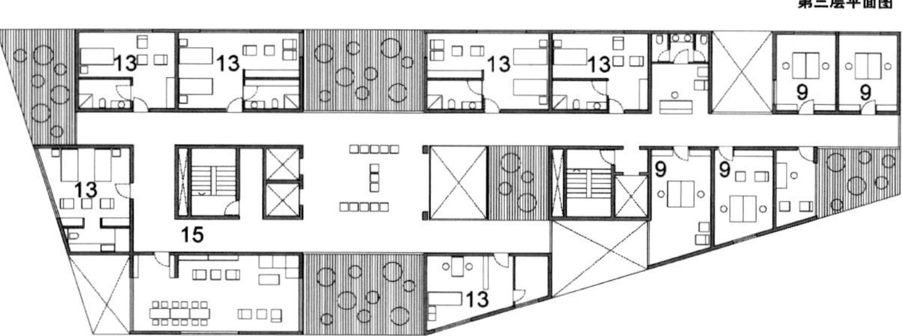

9.Polyclinc-room 9. 综合医院 - 房间
13.Office 13. 办公室

Fourth floor plan
第四层平面图

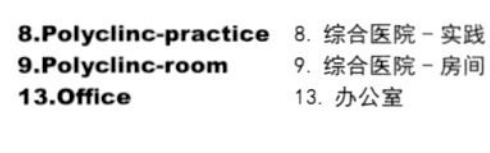
8.Polyclinc-practice 8. 综合医院 - 实践
9.Polyclinc-room 9. 综合医院 - 房间
13.Office 13. 办公室

Fifth floor plan
第五层平面图

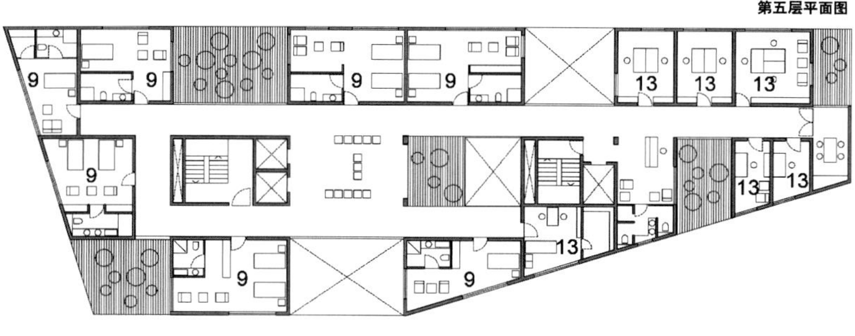

27.Garage 27. 车库
28.Utility room 28. 杂物间

Basement plan
地下室平面图

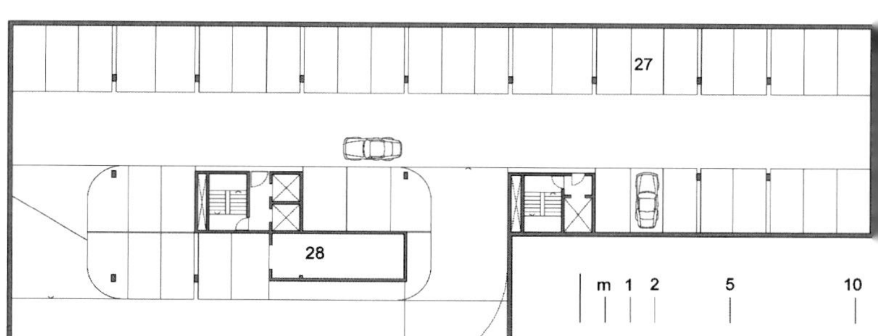

16.MR 16. 机械房
17.CT 17. CT 室
18.X ray 18. X- 光室
19.US 1 19. US 1
20.US 2 20. US 2
21.US 3 21. US 3
22.Densitometry 22. 密度计量学
23.Mammogram 23. 乳房 X 线照片
24.Farmacy store 24. 药店
25.Diagnostics 25. 诊断学
26.Polyclinic laboratory 26. 综合医院实验室

Basement-planv
地下室平面图

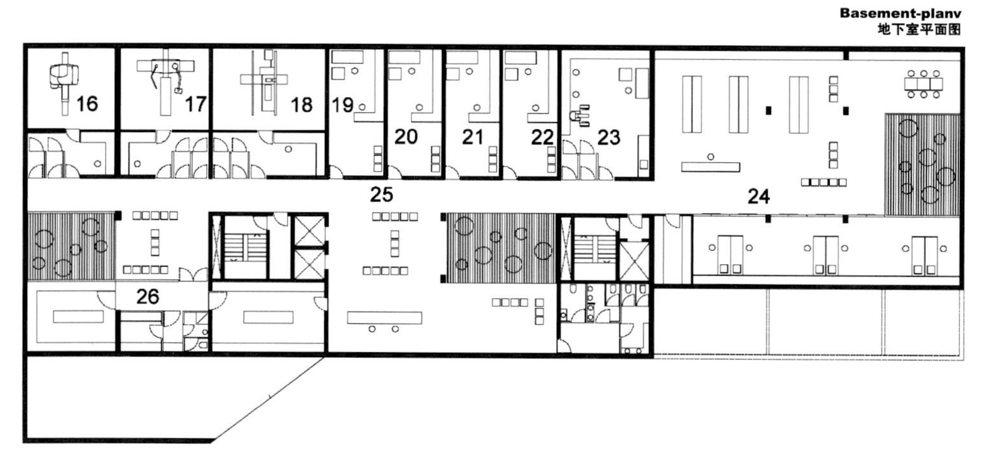

Architectural Design

Project of such determinate program and function requires maximal pragmatism and respect to the competition program, as well as maximum of technical and technological equipment and functionality. It was necessary to develop acquired functions vertically in order to place all the services, diagnostics, practices and laboratories as requested. All public spaces are located in the basement, ground and first floors, patient clinic and administration on the upper floors and garage in the underground.

The main visual element of the building and accent of the whole house is the facade envelope made of horizontal bands. It covers the house like bandages protecting a patient. Like a membrane which protects the building from the sun, it is a sophisticated light regulator that conceals as much as reveals. The design is inspired by one of the most recognizable elements of modern Split's architecture — Ivo Radić's "briss soleil".

They divide and connect spaces of the polyclinic, and instead of creating an introvert hospital corridor, they provide a relaxing zone filled with Mediterranean greenery, wooden terraces with a view to the sea and the surrounding islands.

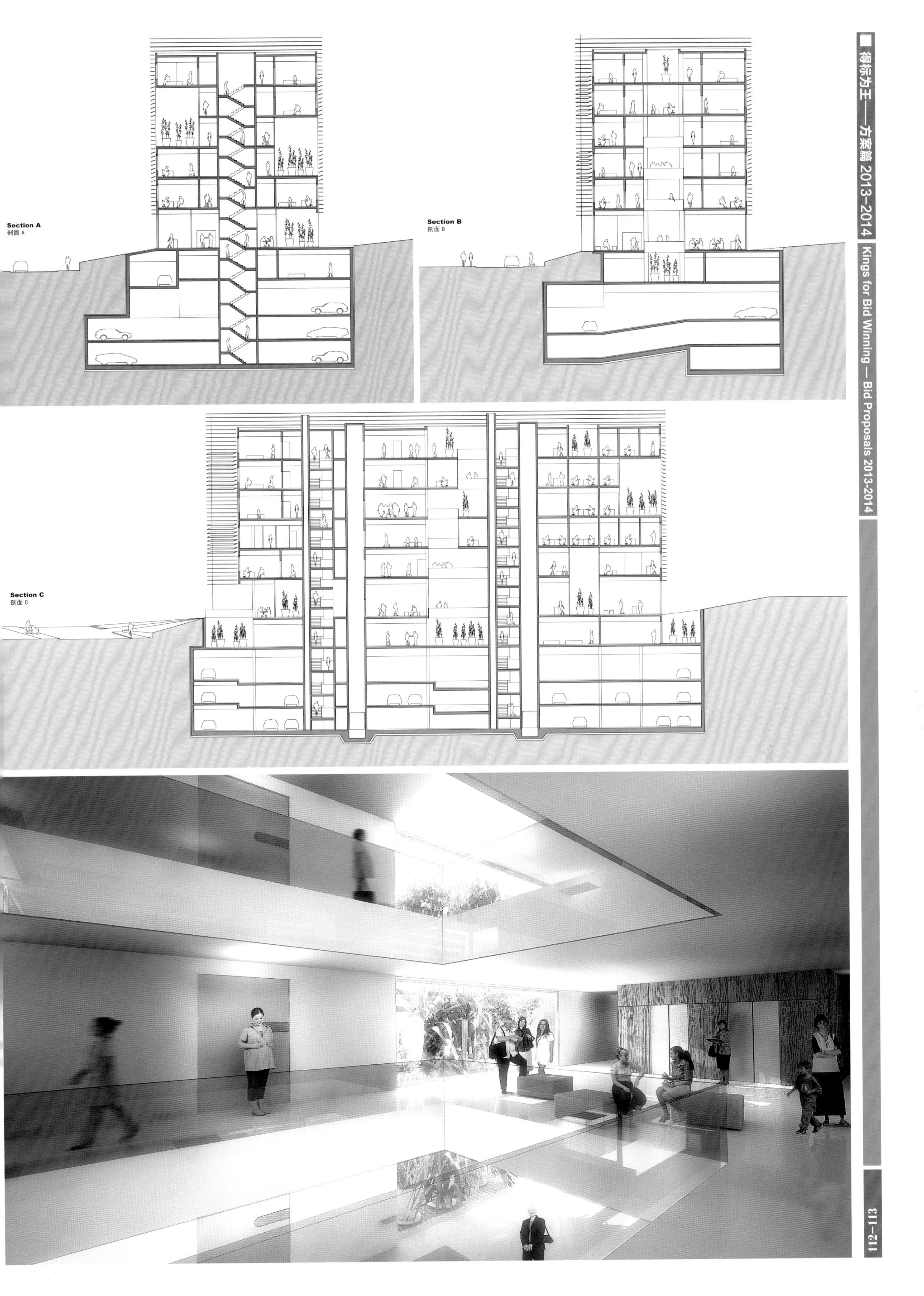
Section A
剖面 A
Section B
剖面 B
Section C
剖面 C

学校建筑 School Building

奥地利维也纳新应用艺术大学

The New University of Applied Arts

设计单位：蓝天组
开发商：BIG Bundesimmobiliengesellschaft Vienna
项目地址：奥地利维也纳
场地面积：7 060 ㎡
设计团队：Wolf D. Prix　Karolin Schmidbaur
Helmut Holleis　Friedrich Hähle
Jan Brosch　Laura Ghitta
Tom Hindelang　Daniela Kröhnert
Heimo Math　Daniel Moral
Martin Mostböck　Andrea Müllner
Arndt Prager　Anja Sorger
Paul Hoszowski　Sebastian Buchta
Steven Ma　Cynthia Sanchez Morales
摄影：Markus Pillhofer

Designed by: COOP HIMMELB(L)AU
Client: BIG Bundesimmobiliengesellschaft Vienna
Location: Vienna, Austria
Site Area: 7,060 m^2
Project Team: Wolf D. Prix, Karolin Schmidbaur, Helmut Holleis, Friedrich Hähle, Jan Brosch, Laura Ghitta, Tom Hindelang, Daniela Kröhnert, Heimo Math, Daniel Moral, Martin Mostböck, Andrea Müllner, Arndt Prager, Anja Sorger, Paul Hoszowski, Sebastian Buchta, Steven Ma, Cynthia Sanchez Morales
Photography: Markus Pillhofer

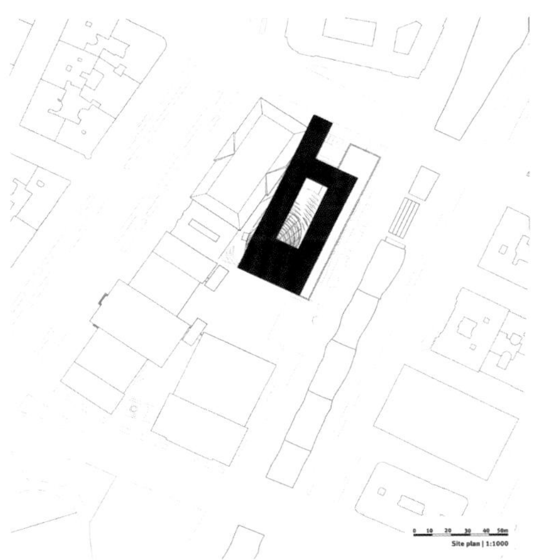

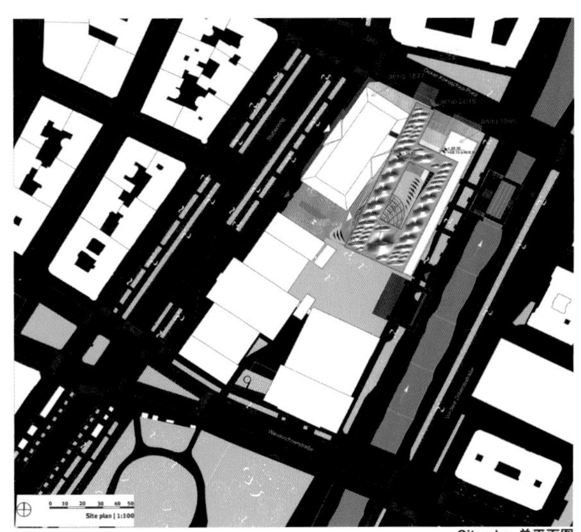

Site plan 总平面图

项目概况

应用艺术大学与应用艺术博物馆、国家歌剧院等建筑一样，都体现了 19 世纪维也纳文化和教育建筑所具有的精神与内涵。这栋新应用艺术大学建筑的设计，综合考虑了建筑的功能性、经济性以及一些特定情况，结合当地已有的两栋（Ferstel 翼楼和 Schwanzer 翼楼），将之建造成为这个城市的第三大建筑。

建筑设计

建筑有着明确的功能分区，行政区位于 Ferstel 翼楼，教育区设置在锥形建筑体中，工作室则分布在 Schwanzer 翼楼。位于锥形建筑第一层的入口大厅与 Schwanzer 翼楼的展示空间和活动空间以及位于地下室的演讲厅和研讨室直接相连，人们可由入口大厅的楼梯直接通往这些空间。Schwanzer 翼楼底层设有餐厅，服务于建筑的其他区间，餐厅设有可直达庭院的独立入口，可作为独立的餐饮企业运作。

设计移除了 Schwanzer 翼楼废弃的核心通道和卫生设施，以营造新的阁楼空间。这一空间可以灵活地分为多个部分，Schwanzer 翼楼的可用楼面面积也得以扩展。Schwanzer 翼楼的庭院大多得以保留，以最大限度维持建筑原有的历史面貌。Ferstel 翼楼的顶部插入了一个棱镜结构，这个扭曲的光棱镜可为翼楼内的休息室提供自然光。

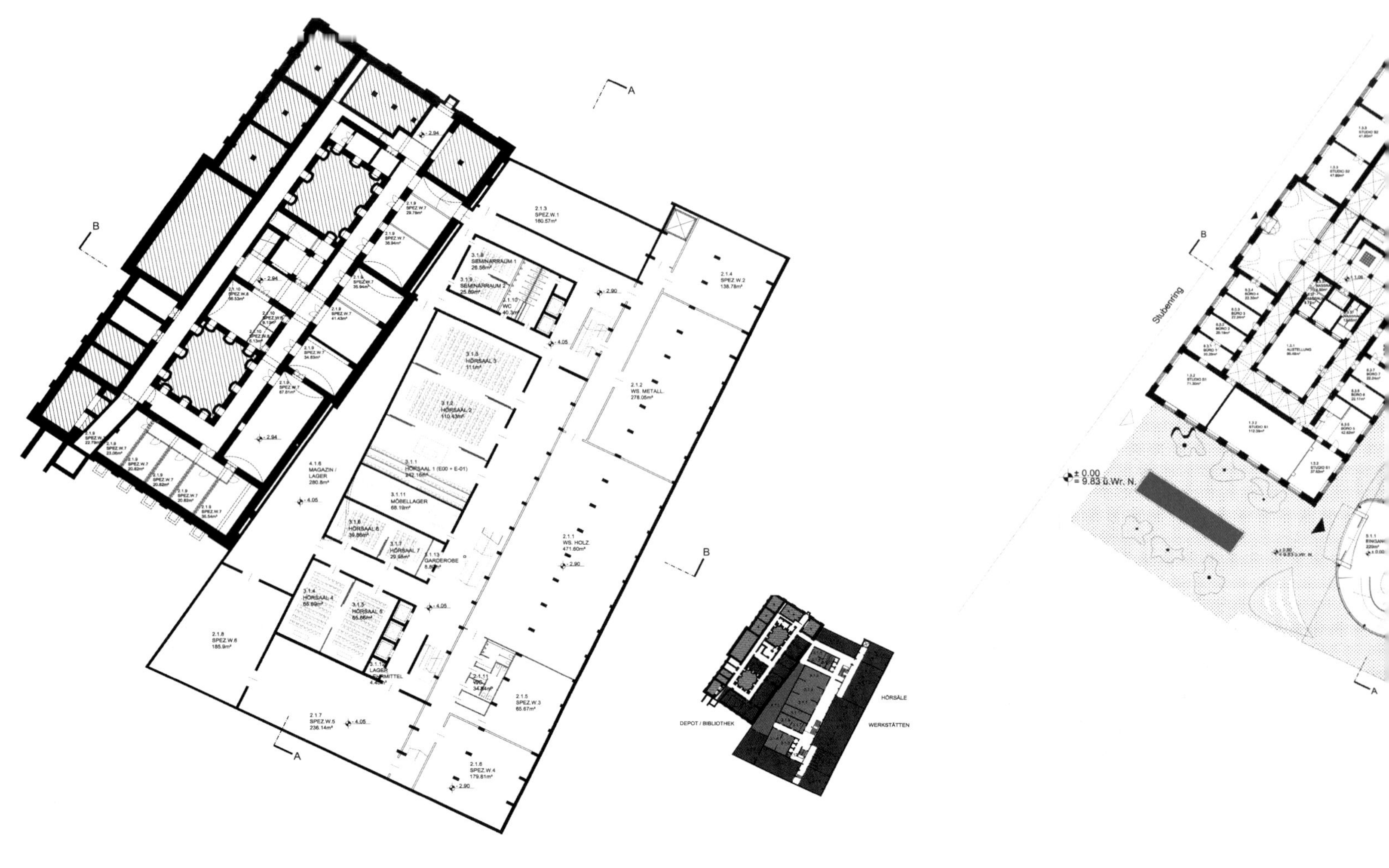

FLOORPLAN -01 负一层平面图

Profile

The New University of Applied Arts, as the Museum for Applied Arts and the State Opera, exemplifies the state of mind reflected by the 19th century cultural and educational buildings in Vienna. The design of the New University of Applied Arts takes functional and economic parameters and the specific situation into consideration. In conjunction with the two existing buildings (Ferstel Wing, Schwanzer Wing), it is rather a third building, which forms a single urban space requiring restructuring.

Architectural Design

The individual functions are clearly assigned to certain spaces: administration (in the Ferstel Wing), education (in the cone) and the studios (in the Schwanzer Wing). The entrance lobby on the first floor of the cone is directly linked to the exhibition and event space on the first floor of the Schwanzer Wing and the lecture halls and seminar rooms in the basement. These spaces are accessible from the entrance hall by staircases. The event space is serviced by the restaurant located on the ground floor of the Schwanzer Wing. The restaurant, which also has a separate entrance to the courtyard, is operated as an independent catering business.

Removal of obsolete access cores and sanitary facilities from the Schwanzer Wing creates new loft spaces that can be freely divided into flexible segments across the entire floor. As a result, the usable floor area in the Schwanzer Wing is expanded. There will be no large-scale construction works on the courtyard side of Schwanzer Wing, leaving intact the protected historical appearance and the courtyard façade. Natural lighting, for the recreational rooms in the Ferstel Wing is provided by the twisted light prism, which is inserted through a corresponding incision in the top floor.

0 5 10 15 20 25m

GROUND FLOOR 00 | 1:500

GROUND FLOOR 底层

FLOORPLAN +01 一层平面图

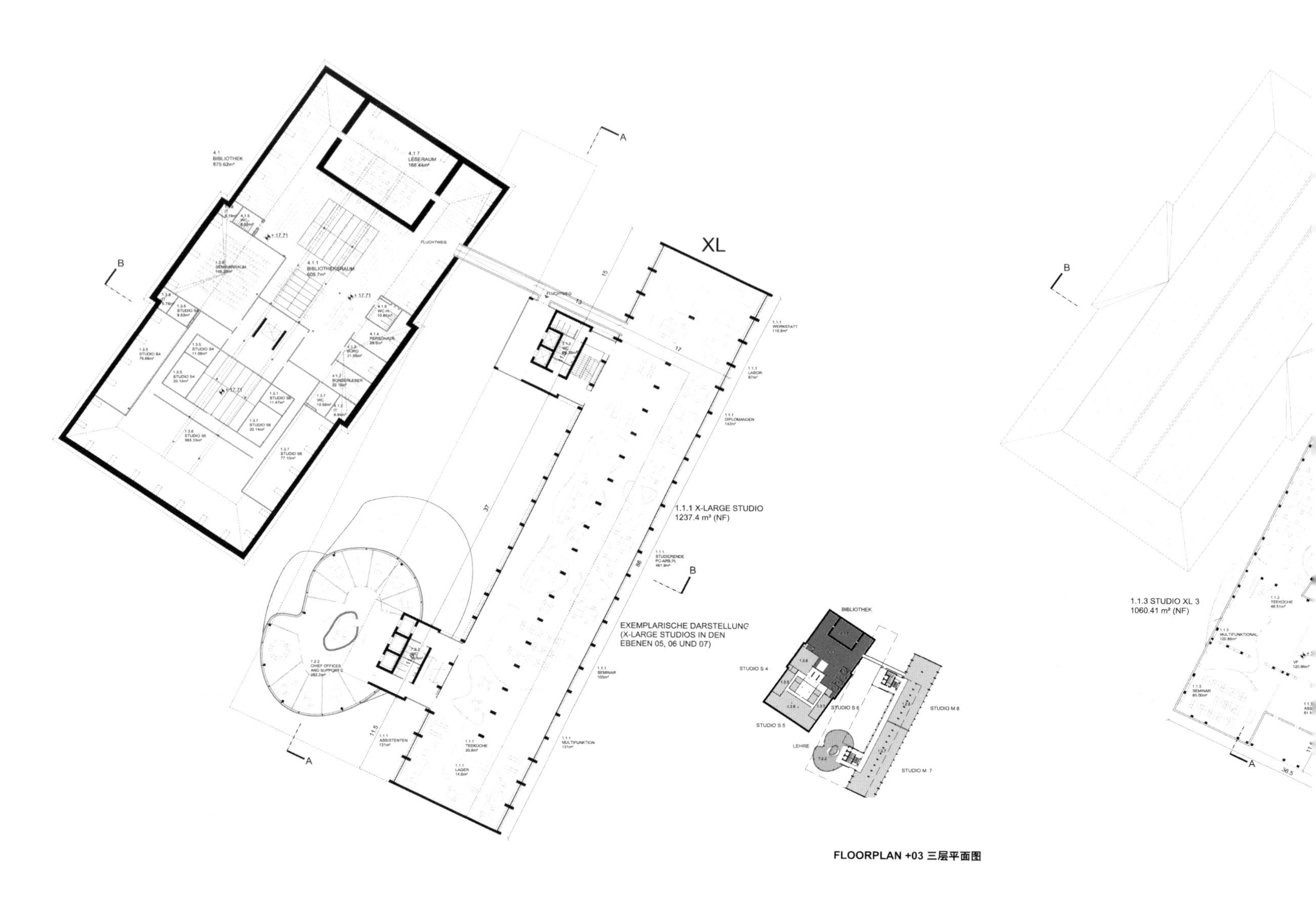

FLOORPLAN +03 三层平面图

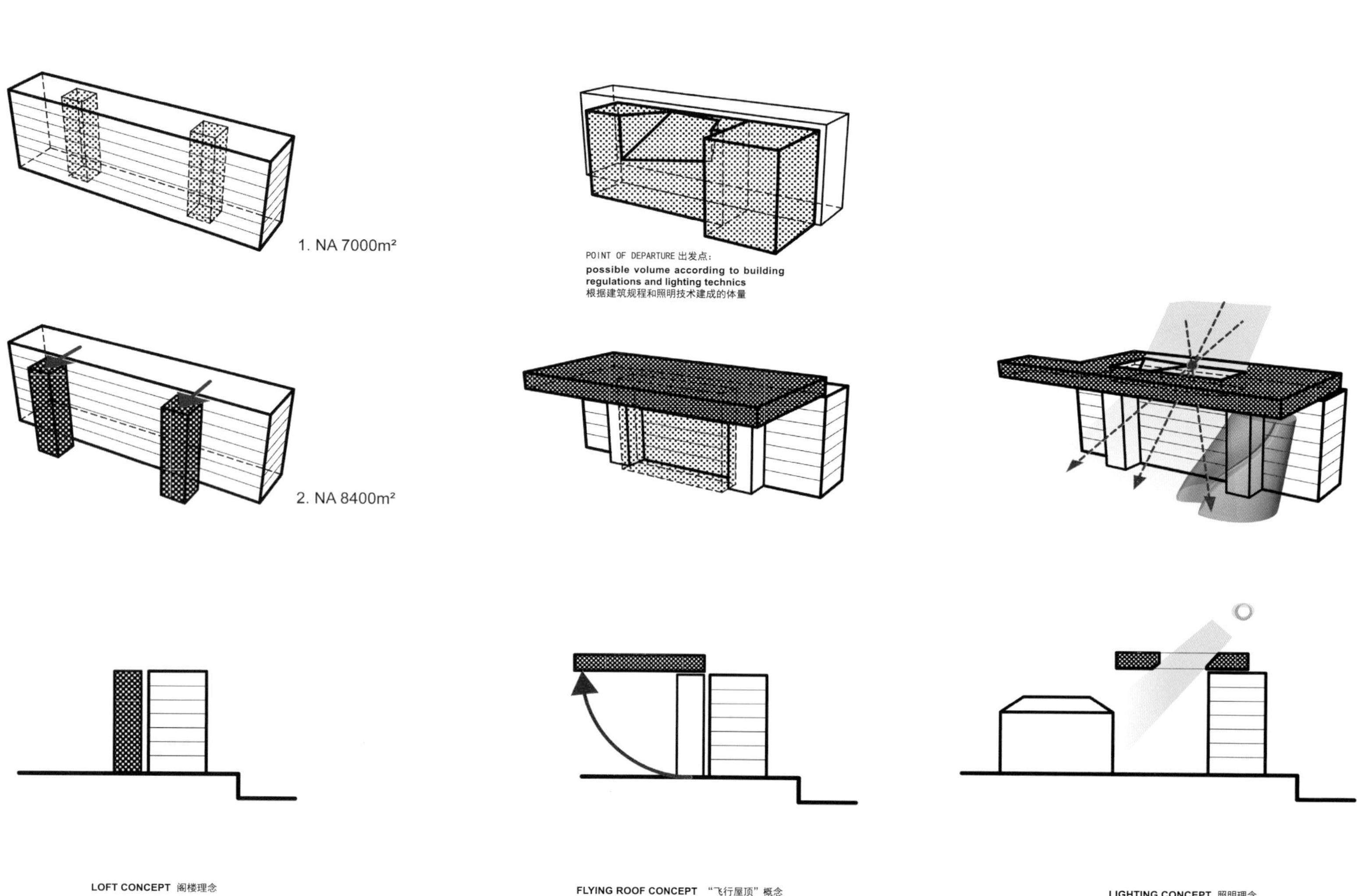

LOFT CONCEPT 阁楼理念

FLYING ROOF CONCEPT “飞行屋顶”概念

LIGHTING CONCEPT 照明理念

CONCEPT 理念

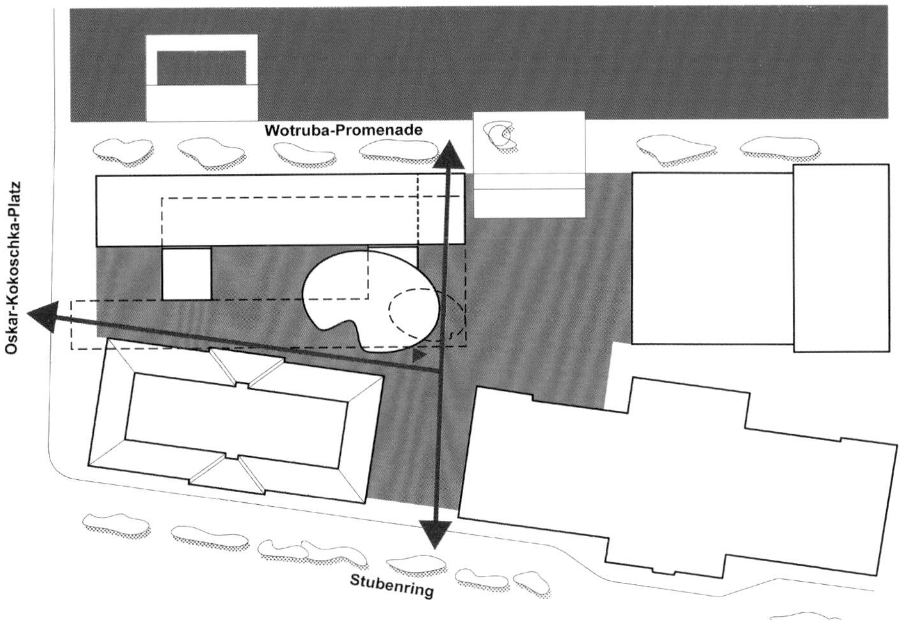

Wotruba-Promenade
Oskar-Kokoschka-Platz
Stubenring

CONCEPT CAMPUS 概念校园

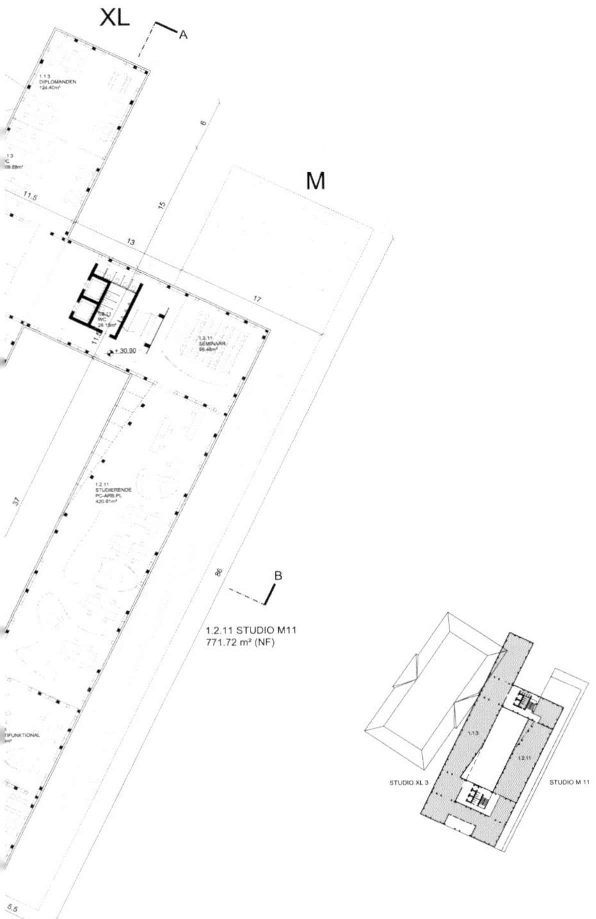

XL
M
1.2.11 STUDIO M11
771.72 m² (NF)
STUDIO XL 3
STUDIO M 11

FLOORPLAN +07 七层平面图

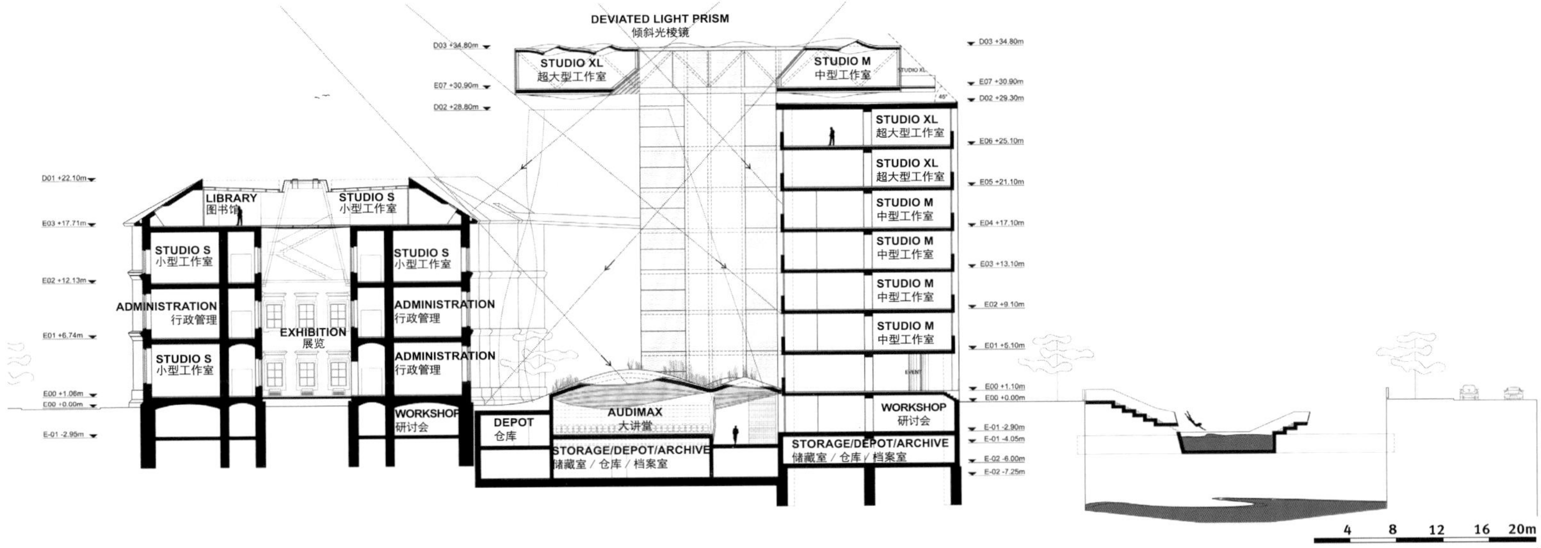

DEVIATED LIGHT PRISM
倾斜光棱镜
STUDIO XL
超大型工作室
STUDIO M
中型工作室
LIBRARY
图书馆
STUDIO S
小型工作室
ADMINISTRATION
行政管理
EXHIBITION
展览
WORKSHOP
研讨会
DEPOT
仓库
AUDIMAX
大讲堂
STORAGE/DEPOT/ARCHIVE
储藏室／仓库／档案室

D03 +34.80m
E07 +30.90m
D02 +29.30m
E06 +25.10m
E05 +21.10m
E04 +17.10m
E03 +13.10m
E02 +9.10m
E01 +5.10m
E00 +1.10m
E00 +0.00m
E-01 -2.90m
E-01 -4.05m
E-02 -6.00m
E-02 -7.25m

4 8 12 16 20m

SECTION A-A 剖面 A-A 1:400

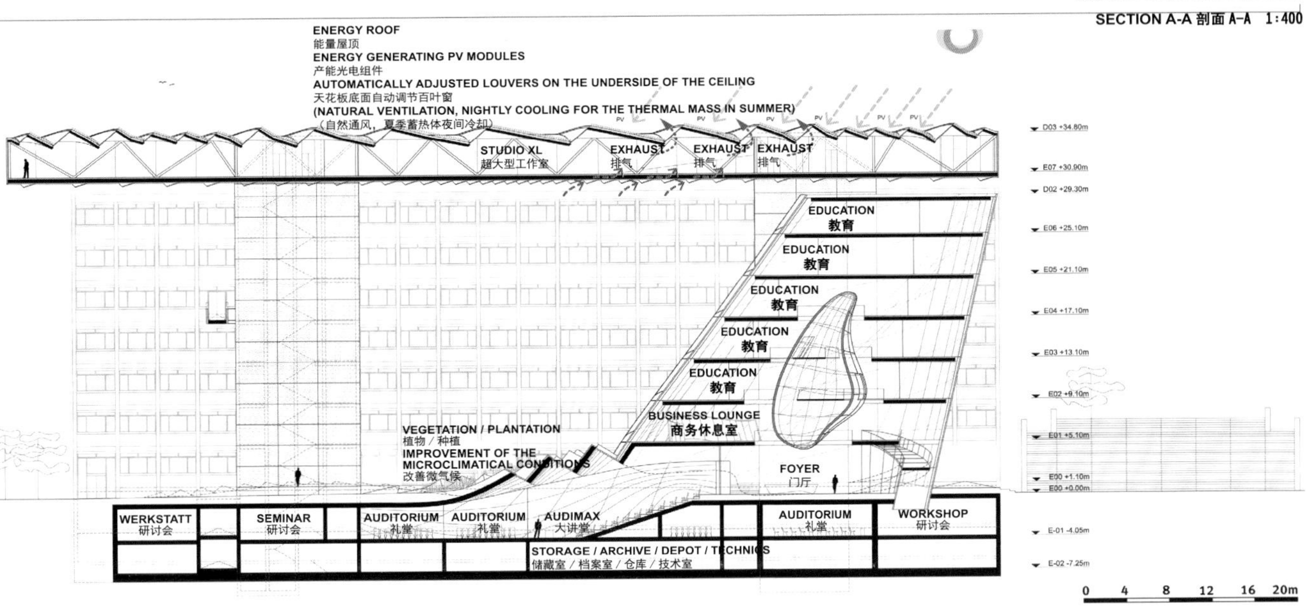

ENERGY ROOF
能量屋顶
ENERGY GENERATING PV MODULES
产能光电组件
AUTOMATICALLY ADJUSTED LOUVERS ON THE UNDERSIDE OF THE CEILING
天花板底面自动调节百叶窗
(NATURAL VENTILATION, NIGHTLY COOLING FOR THE THERMAL MASS IN SUMMER)
（自然通风，夏季蓄热体夜间冷却）

STUDIO XL
超大型工作室
EXHAUST
排气
EDUCATION
教育
BUSINESS LOUNGE
商务休息室
VEGETATION / PLANTATION
植物／种植
IMPROVEMENT OF THE MICROCLIMATICAL CONDITIONS
改善微气候
FOYER
门厅
WERKSTATT
研讨会
SEMINAR
研讨会
AUDITORIUM
礼堂
AUDIMAX
大讲堂
WORKSHOP
研讨会
STORAGE / ARCHIVE / DEPOT / TECHNICS
储藏室／档案室／仓库／技术室

0 4 8 12 16 20m

SECTION B-B 剖面 B-B 1:400

澳大利亚未来教室

Classroom of the Future

设计单位：LAVA
开发单位：Future Proofing School

Designed by: LAVA
Client: Future Proofing School

项目概况

这是LAVA建筑事务所设想的一个未来学习空间，这一空间融合了景观元素，与校园环境建立了直接的联系，适合大规模地定制和推广。

设计理念

一般来说，可移动空间是为解决人口迁移、躲避自然灾害等问题而建立的，这一认知使大多数人认为可移动空间是廉价且粗糙的。LAVA设想的这个未来教室试图改变人们的这一观念，重新定义可持续的、实用的可移动空间，为学生提供生动有趣的教学空间。

建筑设计

这种预制构件环保教室有夹心板材制成的地板和天花板，所有组成部分均由螺丝连接。搭建好后的建筑包括一个屋顶花园，由可以收集能量的太阳能光电板、雨水收集装置等环保系统组成。屋顶花园可以为建筑降温，当电力负荷达到最高值时，流动的水流也可以蒸发降温。此外，教室还配备了先进的通讯系统。

Profile

LAVA's relocatable school is a learning space for the future. This space integrates with the landscape, connects with the school environment, and is suitable for prefabrication and mass customization.

Design Concept

Generally, mobile spaces are the decades old solution to changing demographics, remote community needs, and natural disasters. Unsightly, they are perceived as cheap and unpleasant spaces. This idea is upturned with spaces that are sustainable, practical, cost-effective whilst making learning fun and exciting.

Architectural Design

The prefabricated environment friendly classroom has floor and ceilings made of sandwich panels. All components are connected with simple connections, e.g screws. The building completed shall have a roof garden which contains solar photovoltaic panels and rainwater collection devices. The roof garden cools down the building. When electric load reaches its climax, evaporation of running water also helps to lower temperature. The classrooms are equipped with advance communications systems.

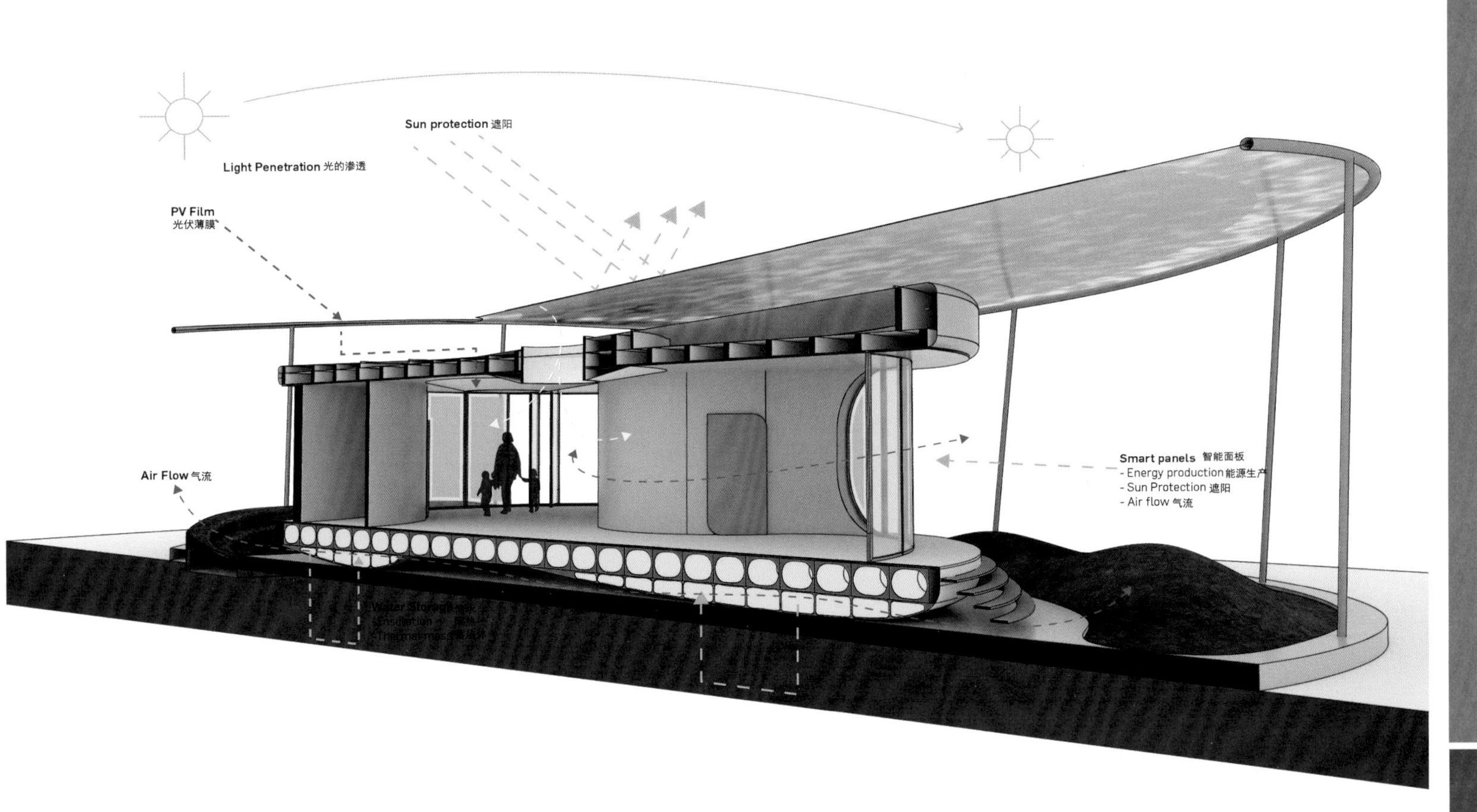
Sun protection 遮阳
Light Penetration 光的渗透
PV Film
光伏薄膜
Air Flow 气流
Smart panels 智能面板
- Energy production 能源生产
- Sun Protection 遮阳
- Air flow 气流

Arid Climate 干燥气候

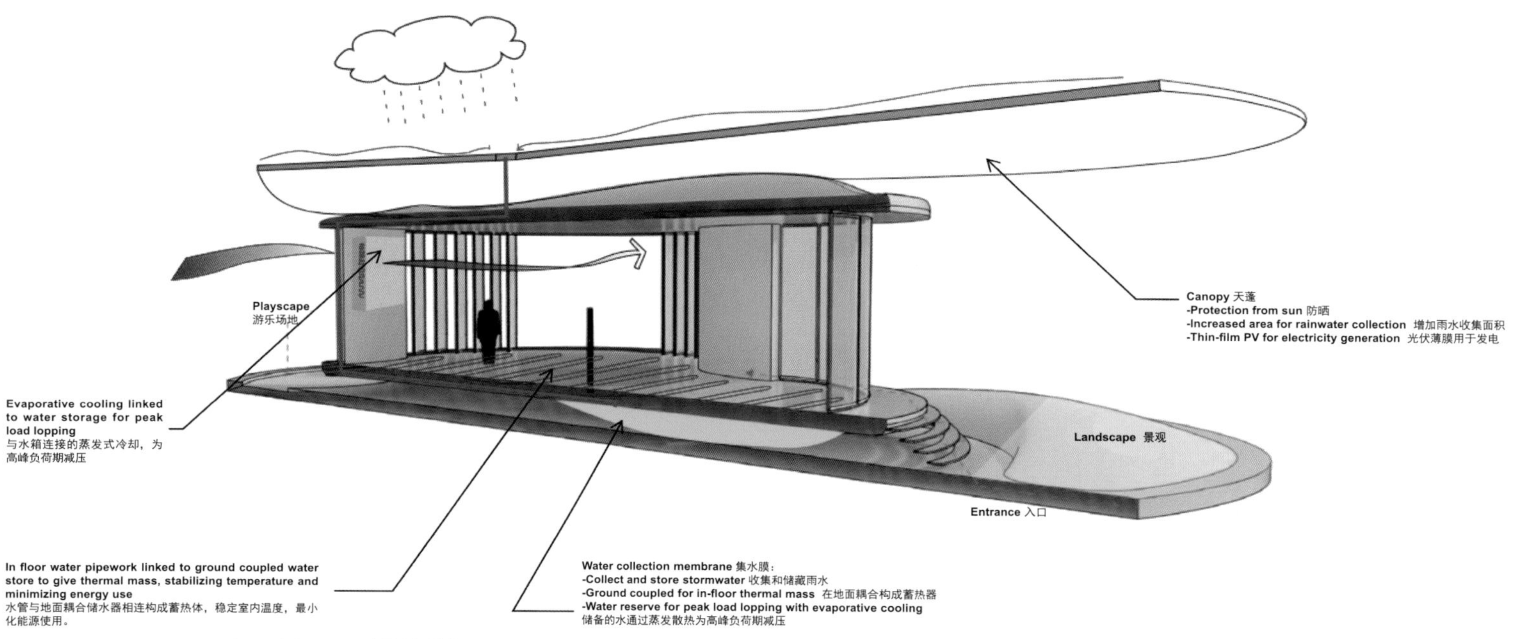

Canopy 天蓬
-Protection from sun 防晒
-Increased area for rainwater collection 增加雨水收集面积
-Thin-film PV for electricity generation 光伏薄膜用于发电
Playscape
游乐场地
Evaporative cooling linked to water storage for peak load lopping
与水箱连接的蒸发式冷却，为高峰负荷期减压
Landscape 景观
Entrance 入口
In floor water pipework linked to ground coupled water store to give thermal mass, stabilizing temperature and minimizing energy use
水管与地面耦合储水器相连构成蓄热体，稳定室内温度，最小化能源使用。
Water collection membrane 集水膜：
-Collect and store stormwater 收集和储藏雨水
-Ground coupled for in-floor thermal mass 在地面耦合构成蓄热器
-Water reserve for peak load lopping with evaporative cooling
储备的水通过蒸发散热为高峰负荷期减压

Facade Solar Access 立面太阳照射入口：

Arid Climate 干燥气候

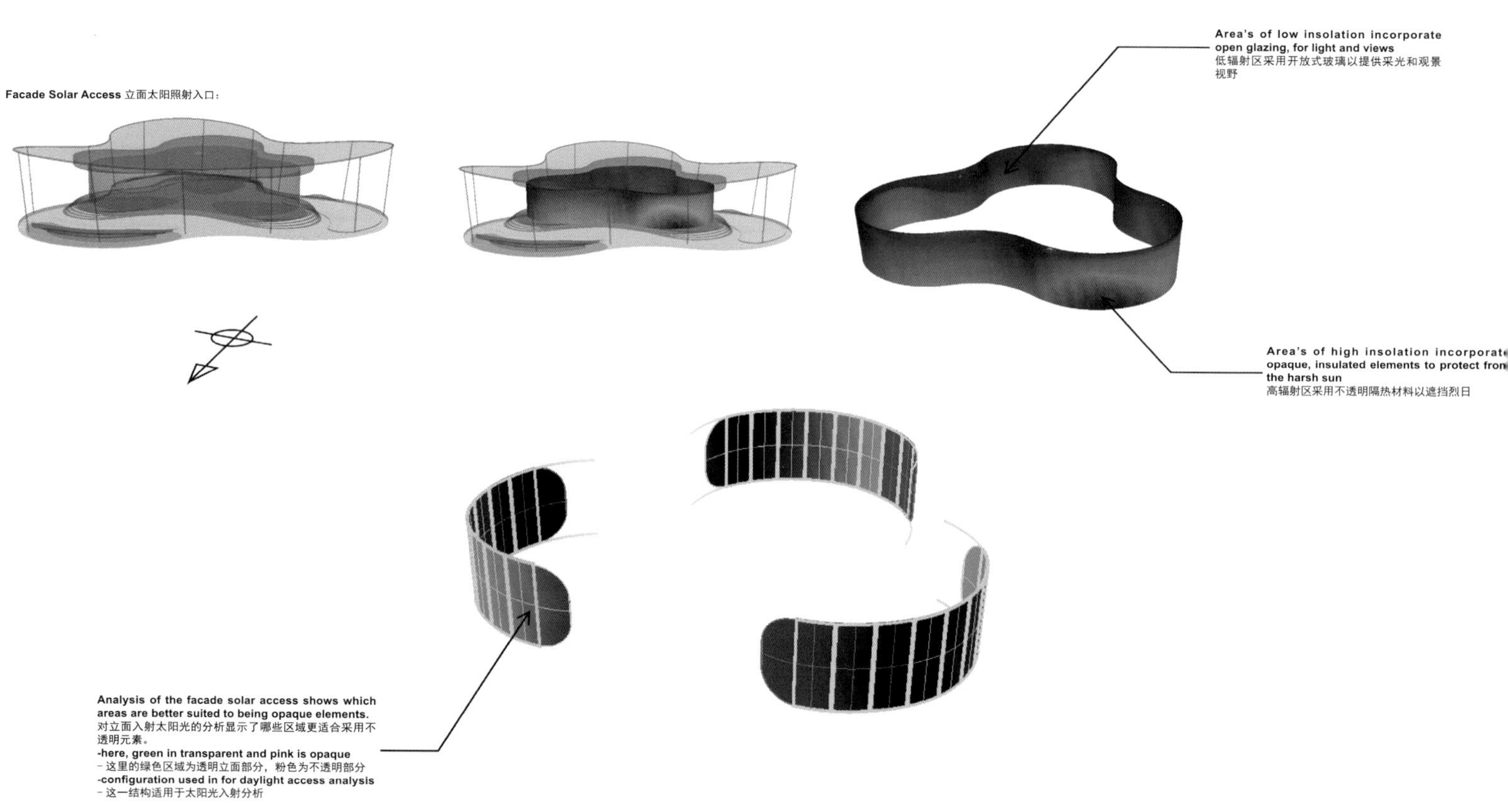

Facade Solar Access 立面太阳照射入口：
Area's of low insolation incorporate open glazing, for light and views
低辐射区采用开放式玻璃以提供采光和观景视野
Area's of high insolation incorporate opaque, insulated elements to protect from the harsh sun
高辐射区采用不透明隔热材料以遮挡烈日
Analysis of the facade solar access shows which areas are better suited to being opaque elements.
对立面入射太阳光的分析显示了哪些区域更适合采用不透明元素。
-here, green in transparent and pink is opaque
- 这里的绿色区域为透明立面部分，粉色为不透明部分
-configuration used in for daylight access analysis
- 这一结构适用于太阳光入射分析

ROOF 屋顶

-Structure: Prefabricated modular timber elements
结构：预制模块化木材元素
-Insulation: Depending on the climate at the specific location, insulation will be provided between timber joists
隔热：根据特定位置的气候，在木搁栅之间进行隔热处理

THREE SERVICE MODULES 三个服务模块

-Walls: Timber structure with lining on either side
墙壁：木材结构，两侧都有内衬

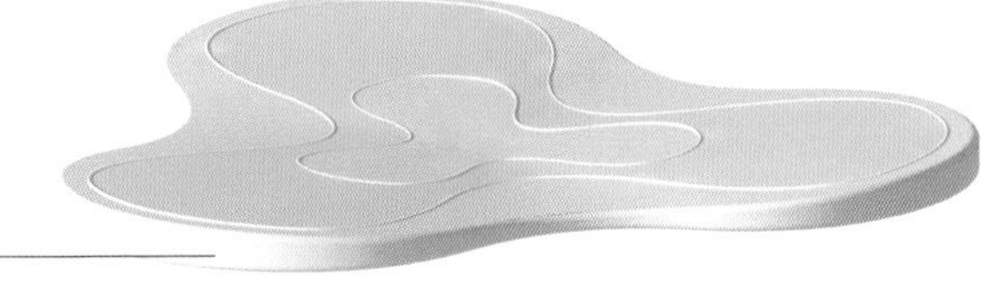

SMART PANELS 智能面板

-Operable windows 可控窗体
-Framework provided for various "smart" infill panels
为各色"智能"嵌板提供的框架

PLINTH 底座

WATER TANKS 水箱
-Rain water will be collected and used for irrigation and various other gray water applications
收集雨水用作灌溉以及其他用途

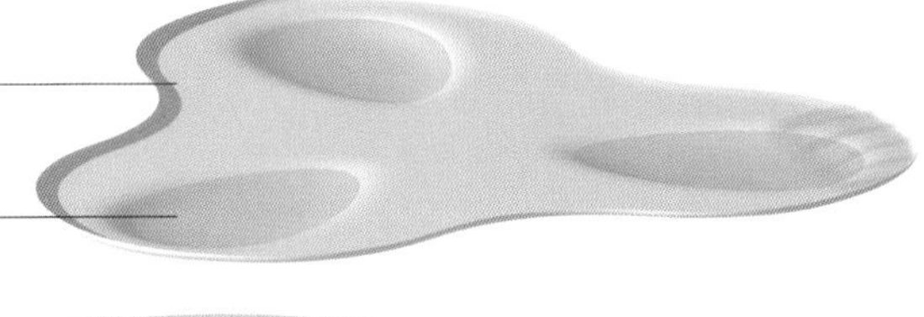

MODULATED LANDSCAPE 经调节的景观

-Outdoor classrooms visually connect to the interior
室外教室在视觉上与室内空间相连
-The design of external areas could be part of individual school projects
外部区域的设计可以是学校项目的一部分

Temperate Climate 温带气候

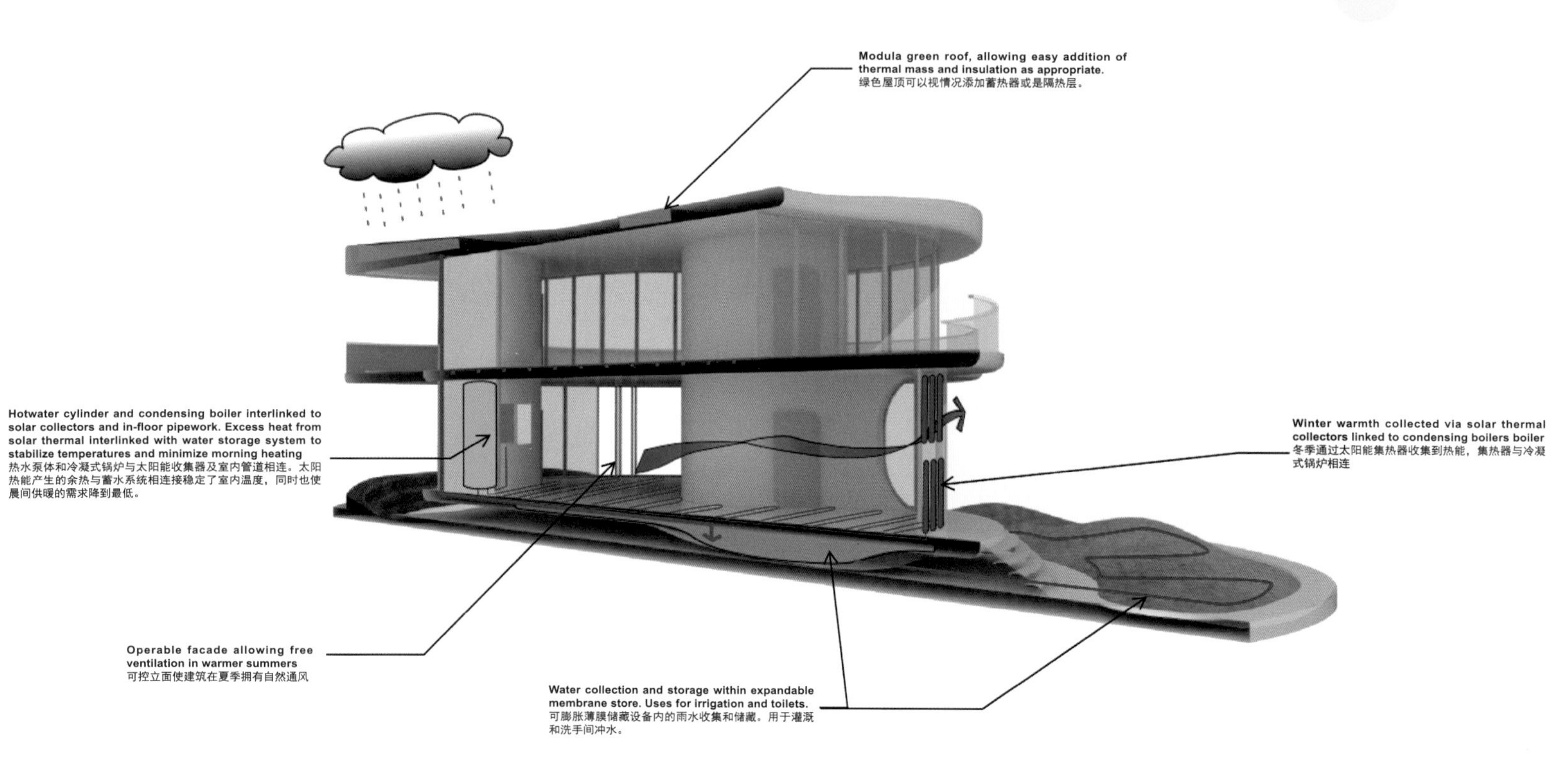
Modula green roof, allowing easy addition of thermal mass and insulation as appropriate.
绿色屋顶可以视情况添加蓄热器或是隔热层。
Hotwater cylinder and condensing boiler interlinked to solar collectors and in-floor pipework. Excess heat from solar thermal interlinked with water storage system to stabilize temperatures and minimize morning heating
热水泵体和冷凝式锅炉与太阳能收集器及室内管道相连。太阳热能产生的余热与蓄水系统相连接稳定了室内温度，同时也使晨间供暖的需求降到最低。
Winter warmth collected via solar thermal collectors linked to condensing boilers boiler
冬季通过太阳能集热器收集到热能，集热器与冷凝式锅炉相连
Operable facade allowing free ventilation in warmer summers
可控立面使建筑在夏季拥有自然通风
Water collection and storage within expandable membrane store. Uses for irrigation and toilets.
可膨胀薄膜储藏设备内的雨水收集和储藏。用于灌溉和洗手间冲水。

Temperate Climate 温带气候

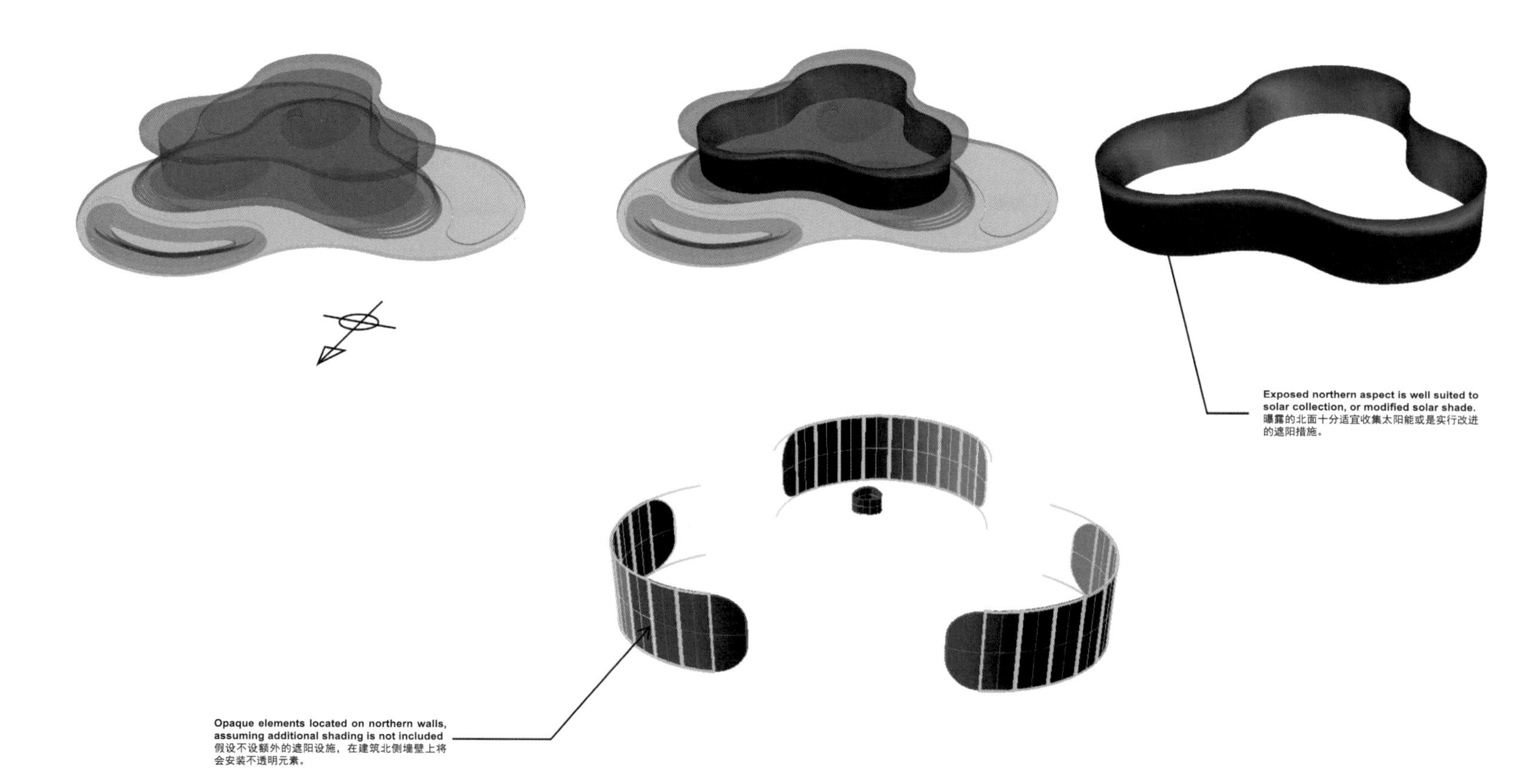
Exposed northern aspect is well suited to solar collection, or modified solar shade.
曝露的北面十分适宜收集太阳能或是实行改进的遮阳措施。
Opaque elements located on northern walls, assuming additional shading is not included
假设不设额外的遮阳设施，在建筑北侧墙壁上将会安装不透明元素。

Temperate Climate 温带气候

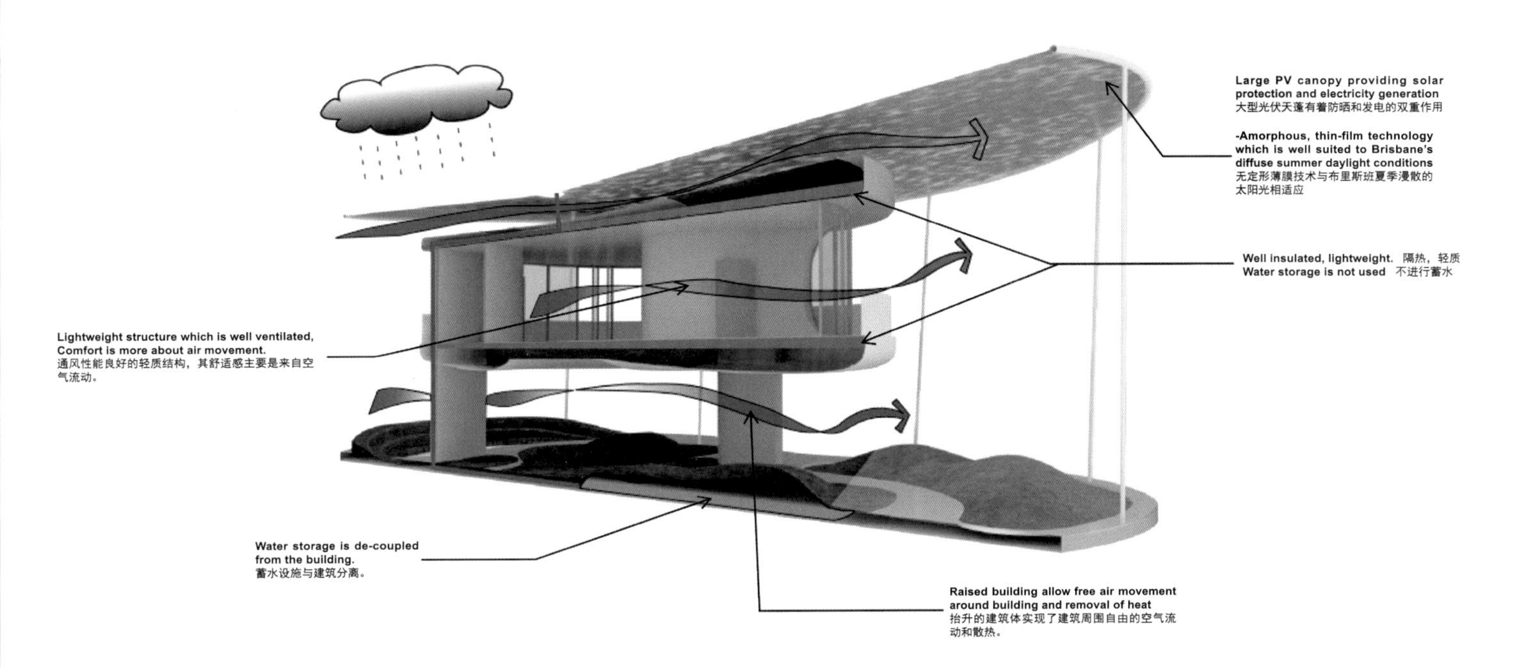
Large PV canopy providing solar protection and electricity generation
大型光伏天蓬有着防晒和发电的双重作用
-Amorphous, thin-film technology which is well suited to Brisbane's diffuse summer daylight conditions
无定形薄膜技术与布里斯班夏季漫散的太阳光相适应
Well insulated, lightweight. 隔热，轻质
Water storage is not used 不进行蓄水
Lightweight structure which is well ventilated, Comfort is more about air movement.
通风性能良好的轻质结构，其舒适感主要是来自空气流动。
Water storage is de-coupled from the building.
蓄水设施与建筑分离。
Raised building allow free air movement around building and removal of heat
抬升的建筑体实现了建筑周围自由的空气流动和散热。

Temperate Climate 温带气候

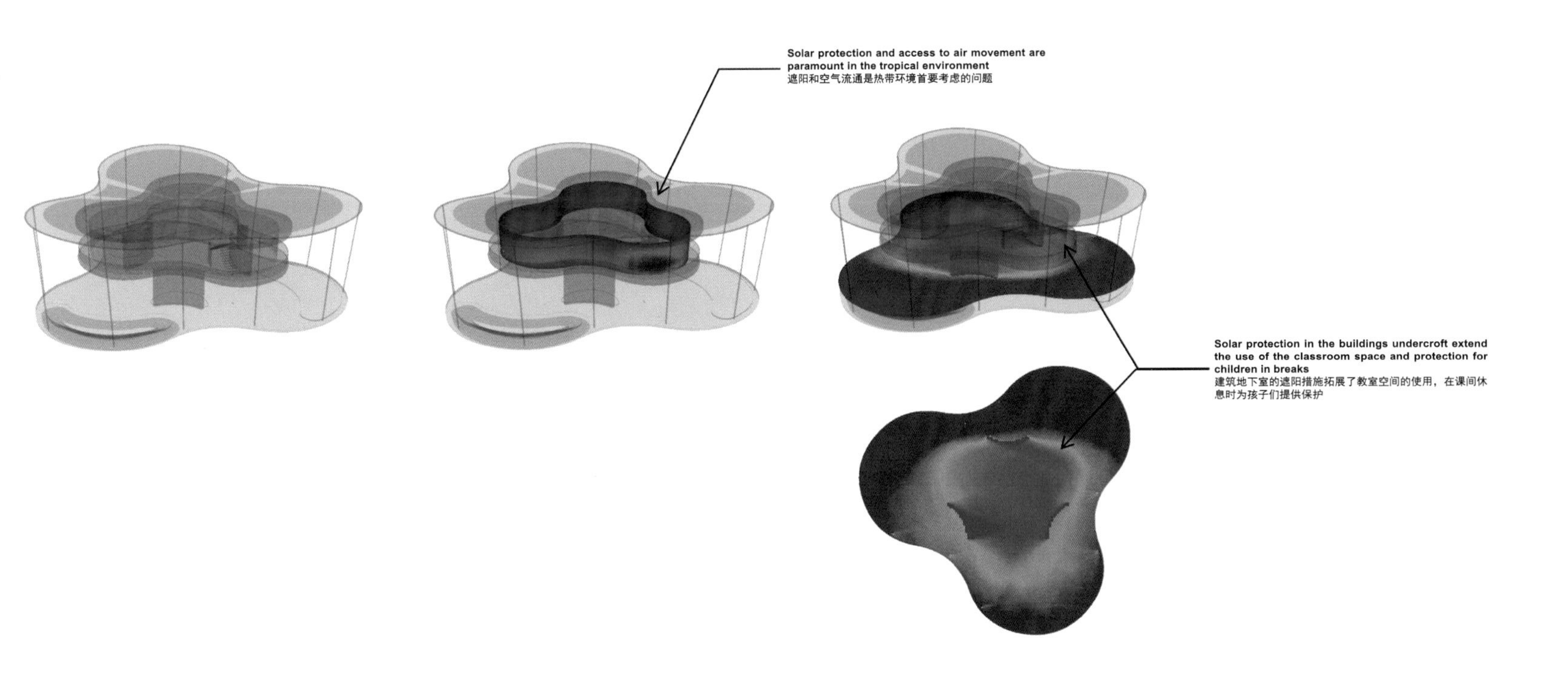
Solar protection and access to air movement are paramount in the tropical environment
遮阳和空气流通是热带环境首要考虑的问题
Solar protection in the buildings undercroft extend the use of the classroom space and protection for children in breaks
建筑地下室的遮阳措施拓展了教室空间的使用，在课间休息时为孩子们提供保护

奥地利阿姆施泰滕校园

Amstetten School Campus

设计单位：Atelier Thomas Pucher

开发单位：Amstettner Schulinfrastrukturentwicklungs GmbH & CoKG

项目地址：奥地利阿姆施泰滕

建筑面积：11 150 ㎡

设计团队：Thomas Pucher　Georg Auinger
Florian Fanta　Klaus Hohsner
Christine Pucher　Robert Lamprecht
Erich Ranegger　Peter Rous
Jan Schrader　Hannes Stöffler

Designed by: Atelier Thomas Pucher

Client: Amstettner Schulinfrastrukturentwicklungs GmbH & CoKG

Location: Amstetten, Austria

Area: 11,150 m^2

Design Team: Thomas Pucher, Georg Auinger, Florian Fanta, Klaus Hohsner, Christine Pucher, Robert Lamprecht, Erich Ranegger, Peter Rous, Jan Schrader, Hannes Stöffler

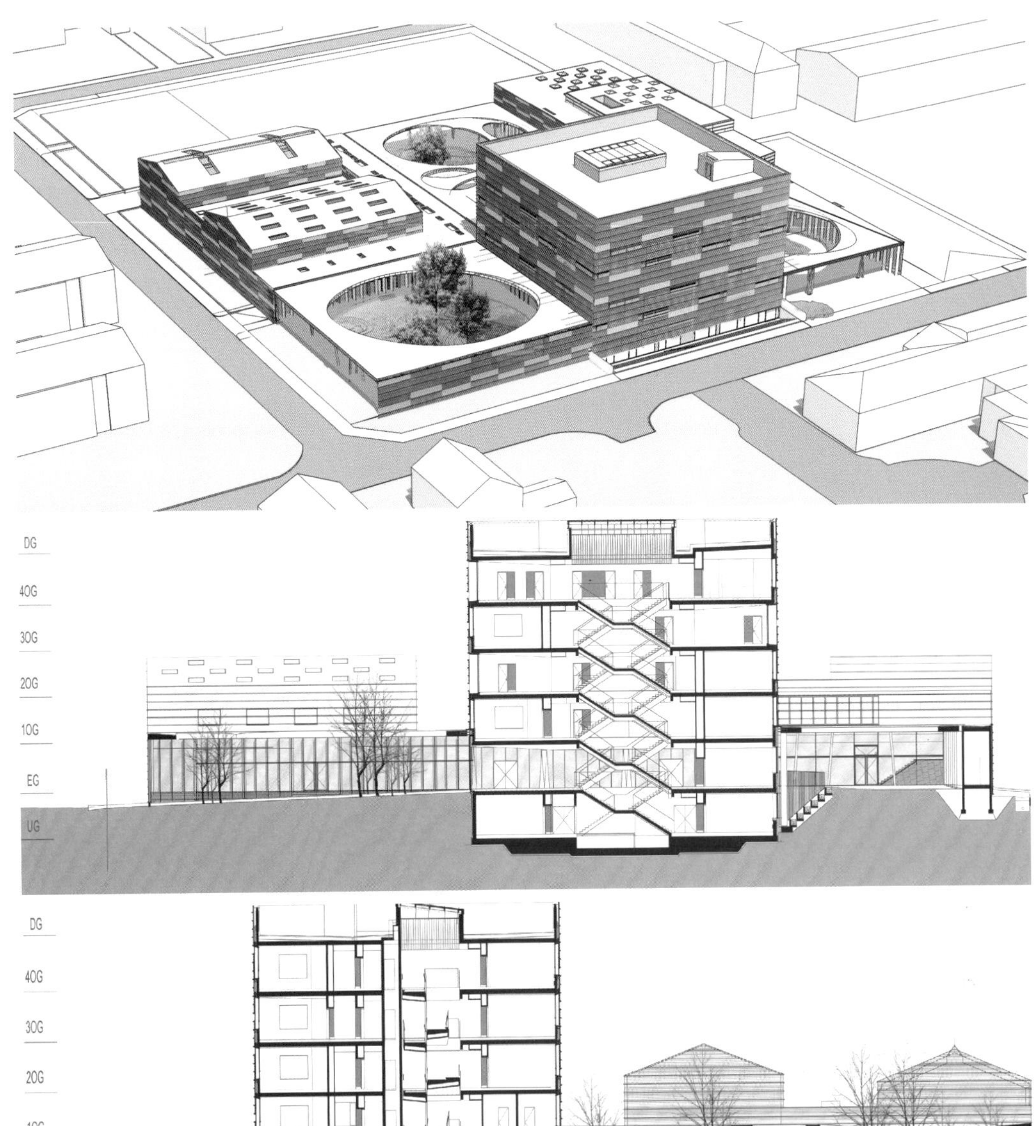

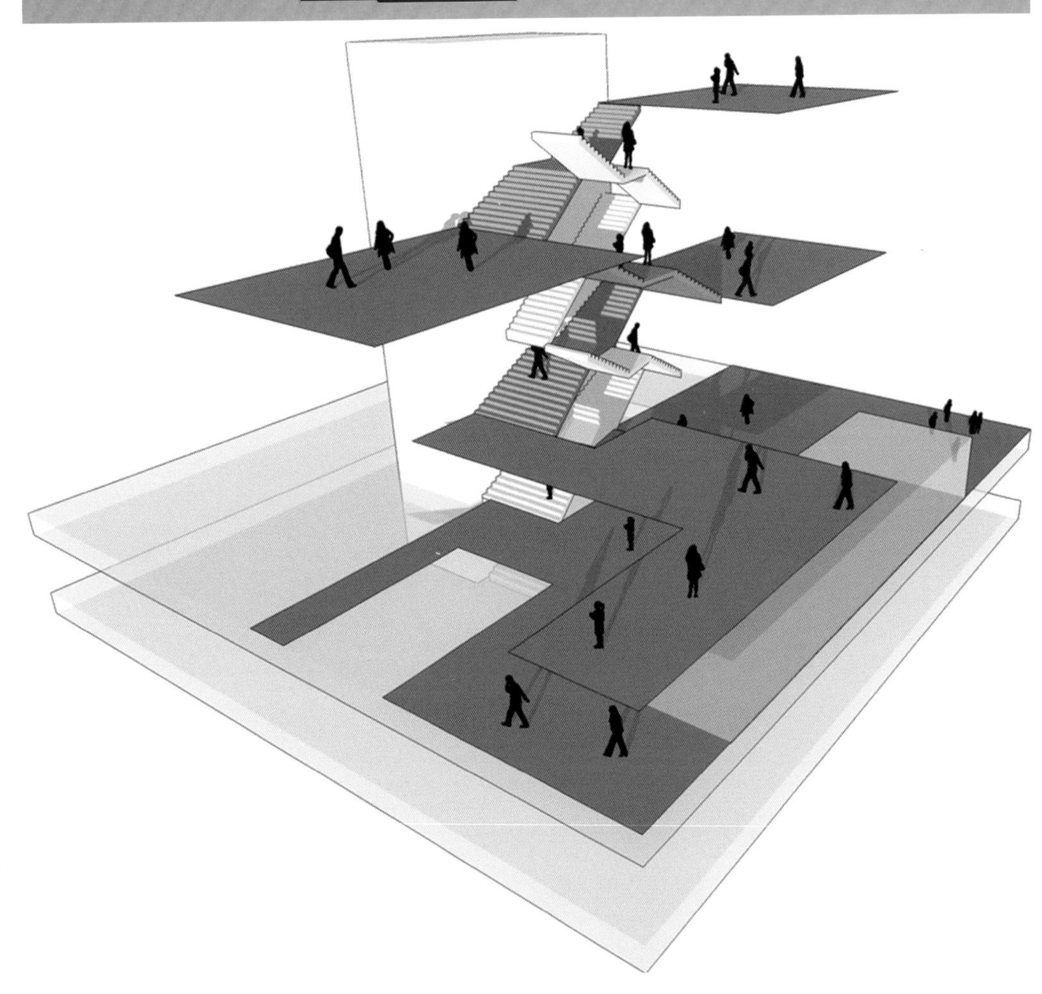

项目概况

这是在 2009 年举办的国际竞赛中获得一等奖的作品，该项目将成为首座获得奥地利“Climate：Active”建筑评级的学校，相当于美国 LEED 认证。项目所在区域存在着不同功能、风格和特色的建筑，且已有两所中学，设计旨在为该区域新建一所音乐学校和一个体育馆。

空间构成

设计的主要概念在于构建各类空间，明确、清晰地组织建筑功能。设计方案包括一个结构紧凑的 5 层立方体建筑，这一建筑在周围环境中拔地而起，成为一座独特的地标式建筑，旁边还有两座较矮的建筑和 3 个庭院，构成了棋盘状的布局。这就如同在城市扩张过程中筑造的一顶王冠，成为周边地区的主要特色。

建筑设计

外围的立面用穿孔金属板包面，既可作为遮阳板，又维持了外墙的连续性，同时也给居于室内空间的人一种微妙的亲切感。庭院从视觉和实际上都连接着各个建筑，成为公众在休息时间聚会的场所，每一座庭院根据朝向和景观设计的差异形成了自己的特征。主要的立方体建筑有一个中庭，该中庭成为可通向建筑各个部分的中轴区域。

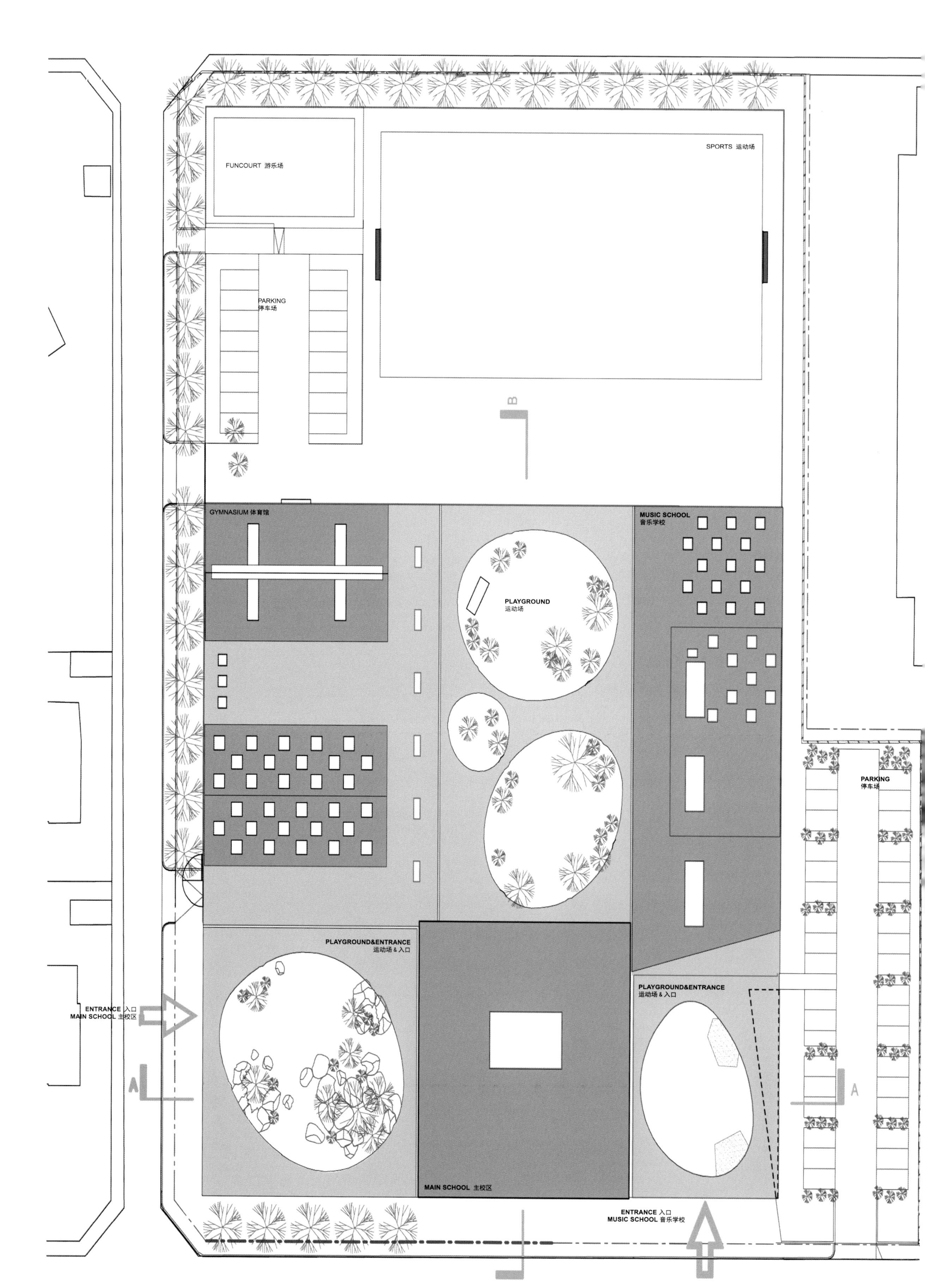
FUNCOURT 游乐场
SPORTS 运动场
PARKING
停车场
B
GYMNASIUM 体育馆
MUSIC SCHOOL
音乐学校
PLAYGROUND
运动场
PARKING
停车场
PLAYGROUND&ENTRANCE
运动场 & 入口
ENTRANCE 入口
MAIN SCHOOL 主校区
A
PLAYGROUND&ENTRANCE
运动场&入口
A
MAIN SCHOOL 主校区
ENTRANCE 入口
MUSIC SCHOOL 音乐学校

Profile

The project which got the first prize in an international open competition in 2009 is one of the first schools to be registered as a green Climate: Active building in Austria, which is the equivalent of the LEED qualification. The surrounding area of the school is characterized by an extreme heterogeneous development, with a wide range of buildings with various functions and styles. The main requirement of the competition was to combine two existing secondary schools in one location and complementing the program with a new music school and an additional gym.

Space Composition

The main concept behind the formal design is the creation of a variety of spaces that can clearly organize the functions of the building. The design comprises a compact five-storey cube which rises in the surroundings as a distinctive landmark, combined with two lower volumes and three courtyards, within a chessboard formation. This strong formal design acts as a stable crown in the urban sprawl, the main characteristic of the surroundings.

Architectural Design

The external facades are clad in perforated metal panels that act as sunscreen and provide both an image of continuity to the outside viewer, and a subtle sense of intimacy in the interior spaces. The courtyards are both physical and visual connections between the different volumes and perfect meeting places for the pupils during the recreational breaks. Each courtyard has a distinct identity, regarding its solar orientation and landscape design. As for the main cube – "the cube of education" – the program is distributed on the five floors around a central atrium. Acting as the main spine, scissor stairs grant direct access to all parts of the buildings.

澳大利亚昆士兰大学高级工程楼

Advanced Engineering Building of the University of Queensland

设计单位：Richard Kirk Architect

Hassell

项目地址：澳大利亚昆士兰州

项目面积：20 000 ㎡

Designed by: Richard Kirk Architect; Hassell

Location: Queensland, Australia

Area: 20,000 m^2

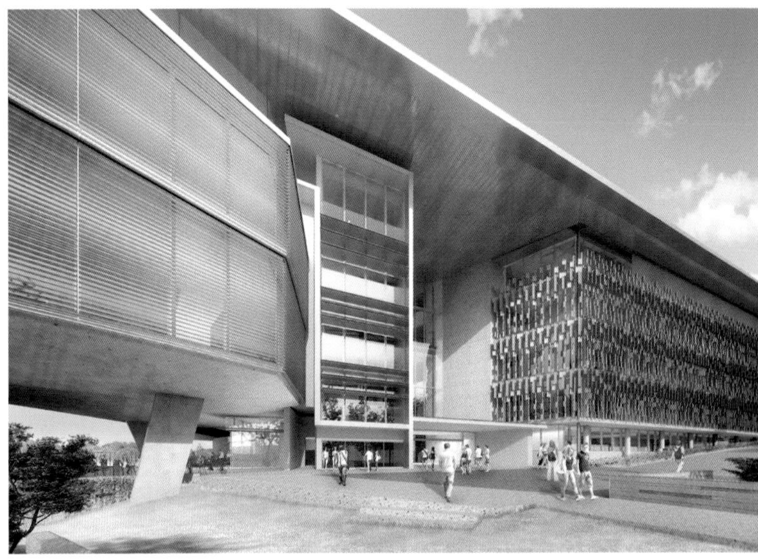

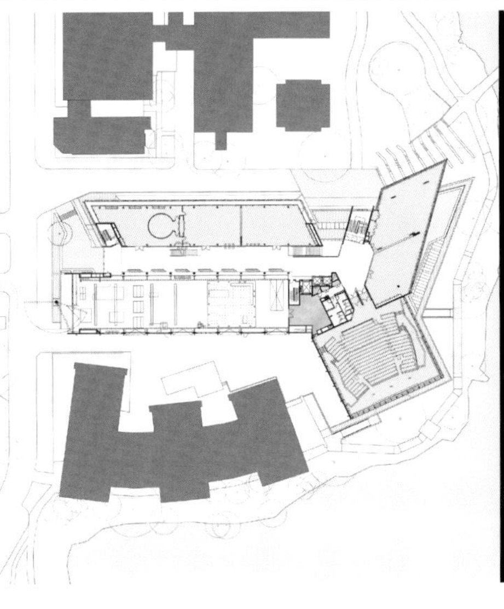

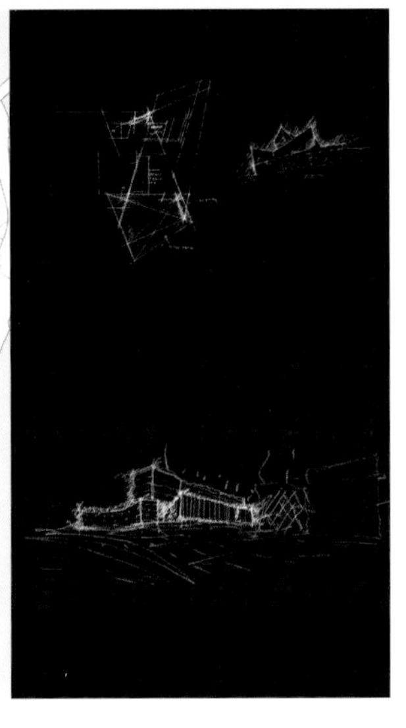

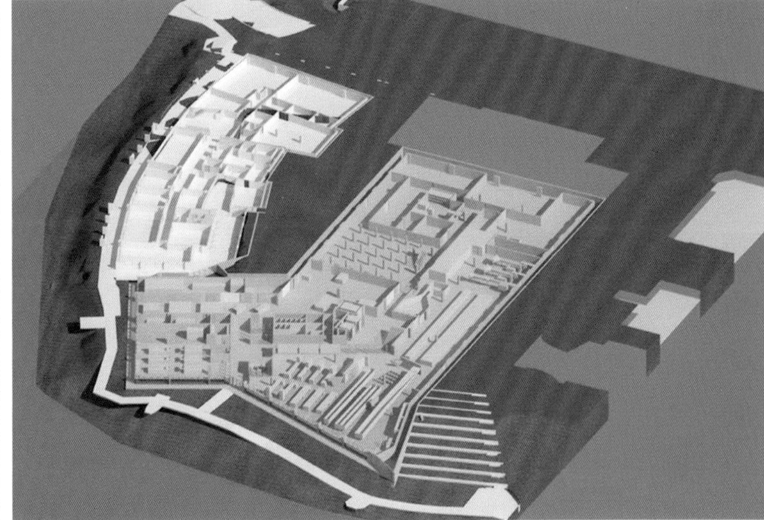

项目概况

昆士兰大学高级工程楼是 Richard Kirk Architect 和 Hassell 联合投标并胜出的项目。这座大楼的建筑将是探索新型教学空间和节能环保教学楼道路上的一座里程碑。

建筑设计

工程楼的设计旨在促进对先进材料和制造工艺的了解，加强学校的创新教育意识。开放式的建筑结构增强了楼内活动的可视性，学生、教工和参观者都有机会看到科研过程和成果。

这座工程楼将会是一个五星级绿色项目，其环保系统包括了错综复杂的空气调节系统、可加强自然通风和扩大自然光源的天井、可调控的智能外墙面、混合模式通风和夜间净化等。这套可持续发展的环保设计方案预期可降低 40% 的能源消耗，将成为昆士兰大学内最具节能环保效应的教学楼。

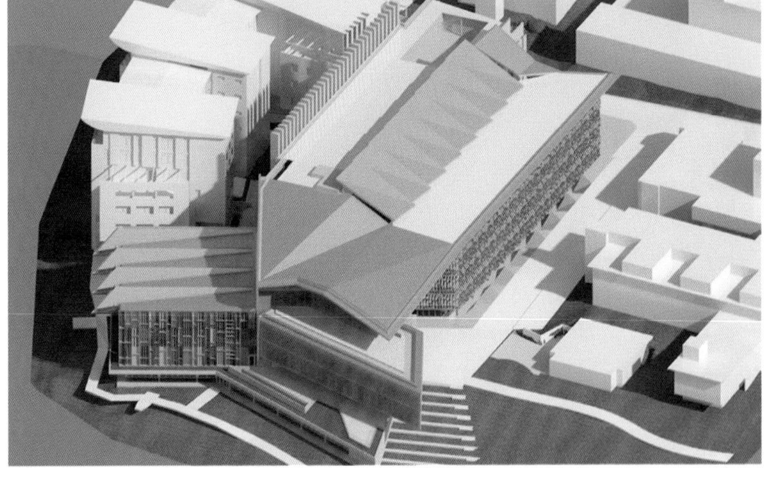

Profile

Advanced Engineering Building (AEB) was awarded to Richard Kirk Architect and Hassell joint venture through a design competition. The AEB will establish a new benchmark for sustainability and explore new possibilities for teaching and learning spaces.

Architectural Design

AEB is designed to stimulate the recognition of advanced materials and manufacturing processes, and enhance the awareness of creative education. The open structure of the building allows a high level of visibility of processes and equipment to visitors and staff alike.

AEB will be a five star green project. Its environmental protection system includes complicated air conditioning system, the atrium to introduce tempered air and light into the centre of the building, a highly efficient façade, mixed mode ventilation, night purging and etc. This sustainable and environment friendly design scheme is expected to achieve a 40% reduction in energy consumption. The AEB will become the most energy-saving and environment friendly teaching building of the University of Queensland.

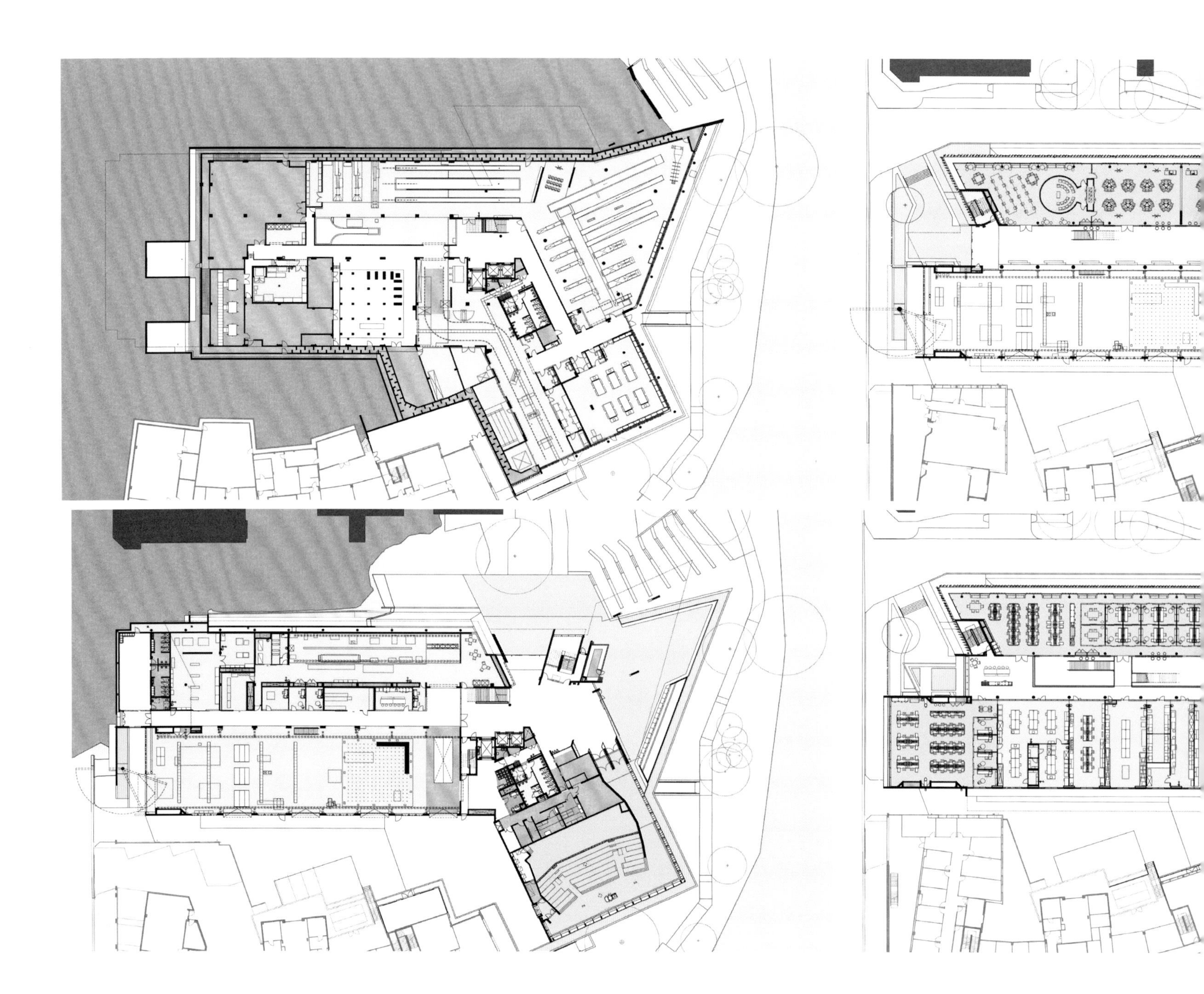

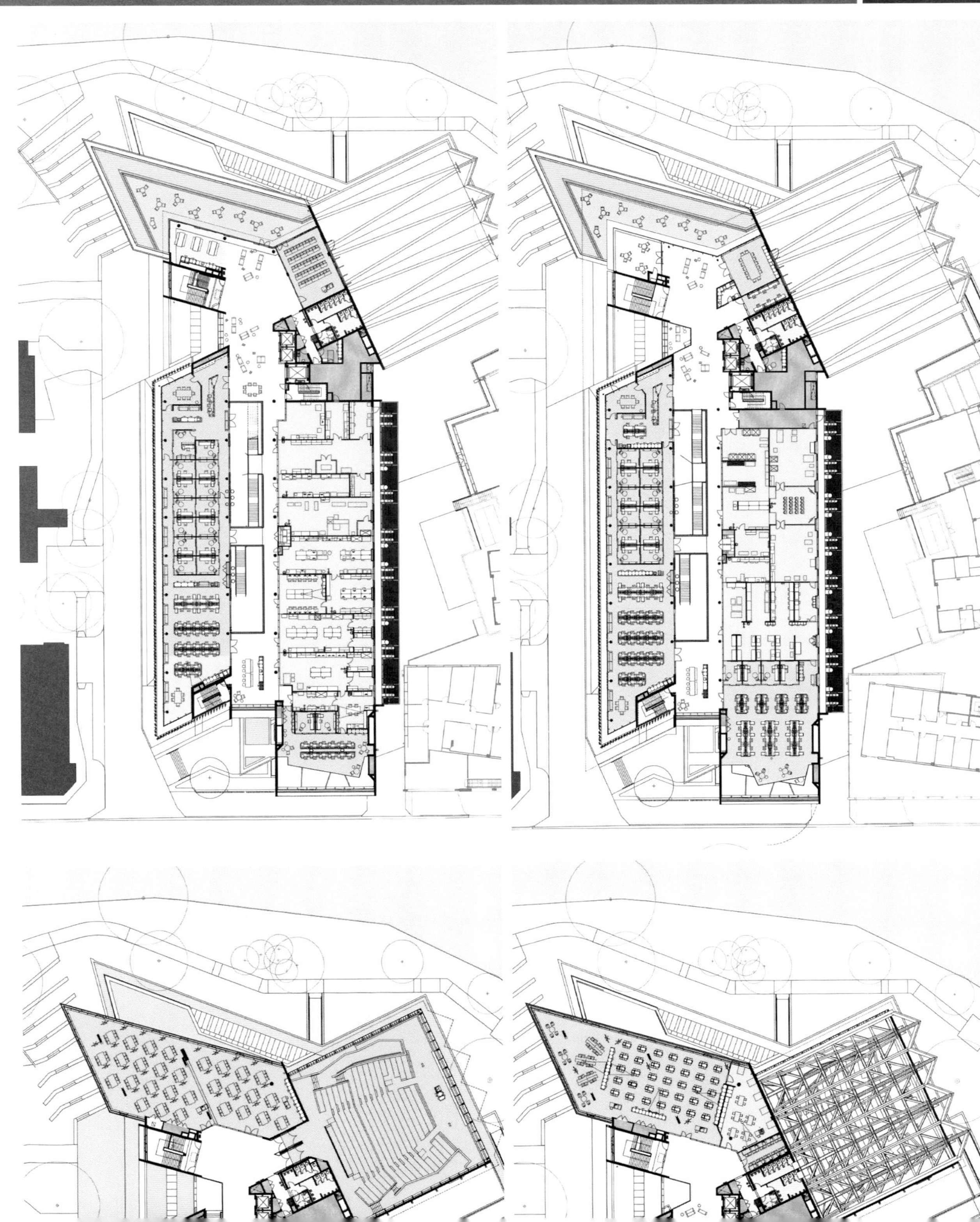

澳大利亚悉尼科技大学主楼

Tower Skin

设计单位：LAVA
项目地址：澳大利亚悉尼

Designed by: LAVA
Location: Sydney, Australia

项目概况

这是对悉尼科技大学的主楼进行外观改造的一个项目，LAVA 提出了一种简单、低成本、高效益的建筑表皮改造方案，即在保持现有结构的前提下，改变建筑的外观，提高建筑的可持续性和内部的舒适度。

设计理念

设计延续了 LAVA 一贯的原则——将最新的数字化制造技术与轻巧的现代建材结合起来，构建可持续发展的公共建筑。方案中，LAVA 提出了“再生皮肤”的概念，这一设想通过对当地环境和原建筑的最小化处理，可高效地提升建筑的外观效果和性能，并简单地应用到其他建筑上。

设计特色

悉尼科技大学主楼室的“再生皮肤”呈半透明的蚕茧状，可以调节建筑内部的小气候。这一表皮可以利用光伏电池产生能量、收集雨水、提高自然光线在建筑物表面的分布效果，并可通过对空气对流的影响，提高建筑物内部自由通风的效果。

现有的建筑表皮上包覆着一层三维、轻量化的高性能复合网，其表面的张力可使材料在轻钢结构中自由伸展。“再生皮肤”中布满了嵌入式的 LED，可发布时装、传媒、通信等多方面的信息，使整个大楼的外表面成为一个智能化的媒体墙和一个可供交流的信息平台。

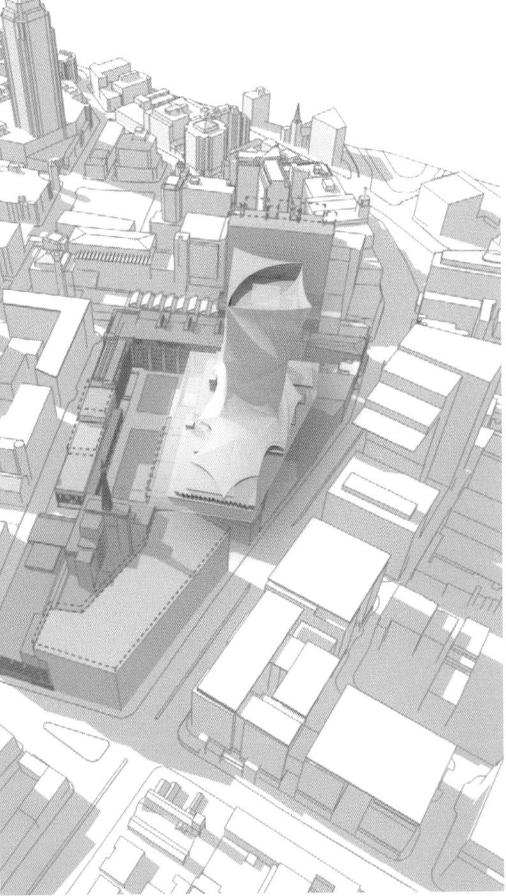

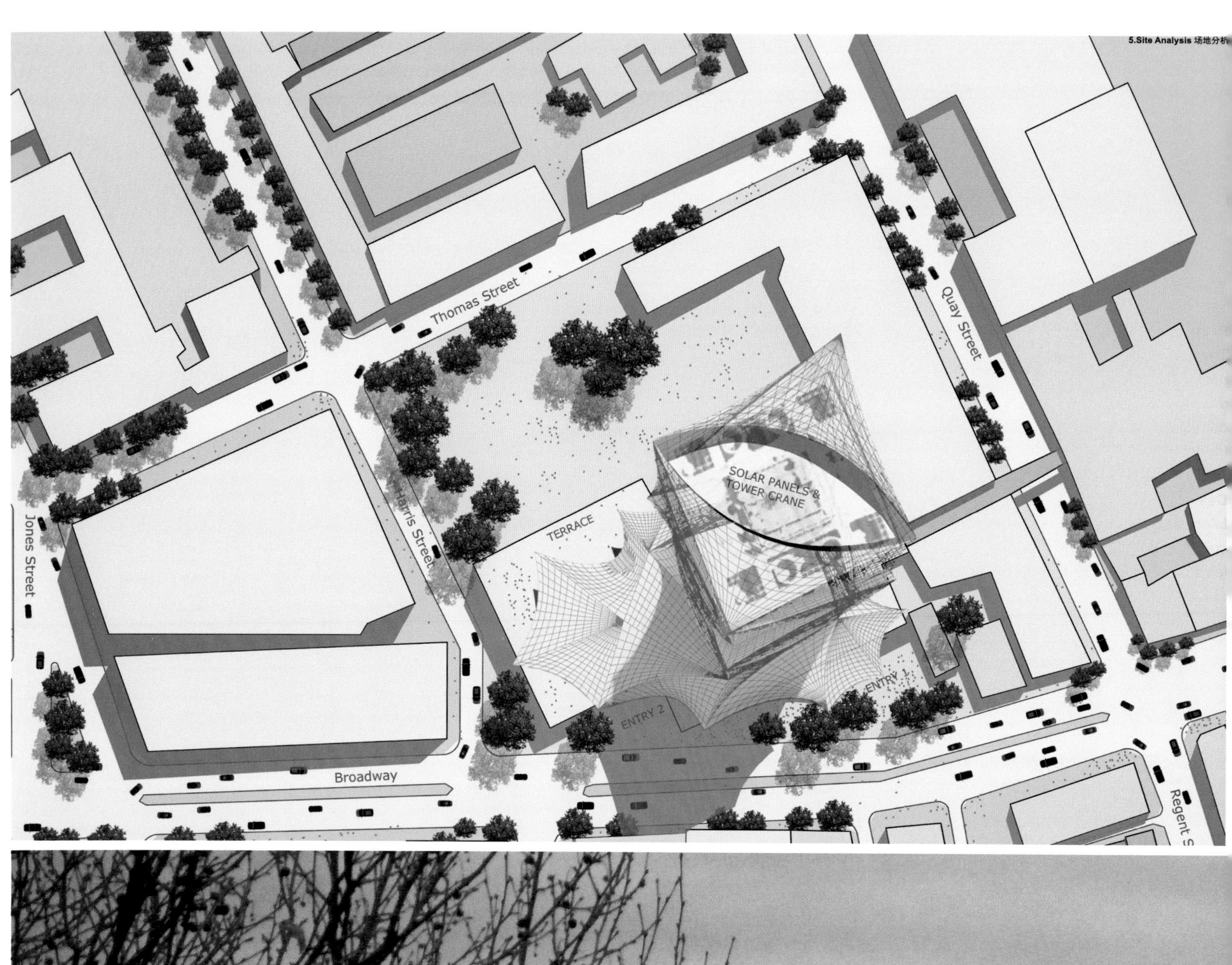
Thomas Street
Quay Street
Jones Street
Harris Street
SOLAR PANELS &
TOWER CRANE
TERRACE
ENTRY 1
ENTRY 2
Broadway
Regent S

Profile

The project is an architectural appearance renovation project for the main building of UTS. LAVA has developed a simple, cost effective and easily constructed building skin that can transform the identity, sustainability and interior comfort of an existing structure.

Design Concept

LAVA aims to build a sustainable public building by combining lightweight contemporary materials with the latest digital fabrication technologies. In the scheme, LAVA puts forward a "Re-skin" concept which plans to minimize intervention in local environment and the existing building and effectively enhance building's appearance effect and performance. This concept can also be applied in other buildings.

Design Feature

The "skin" is a translucent cocoon that can create its own "micro climate". It can generate energy with photovoltaic cells, collect rainwater, improve the distribution of natural daylight and it can use available convective energy to power the building's ventilation requirements.

A pre-existing building is wrapped with three-dimensional lightweight, high performance composite mesh textile. Surface tension allows the membrane to freely stretch around walls and roof elements. The "Re-skin" embedded with LED can be used for releasing fashion, media, communication and other information. It is regarded as an intelligent media surface and an information exchange platform.

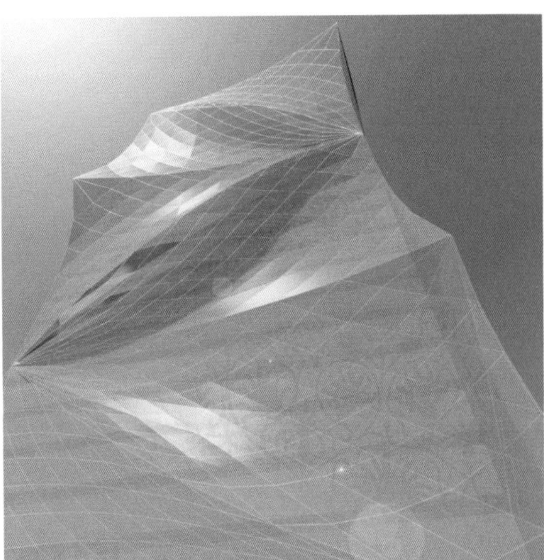

UTS

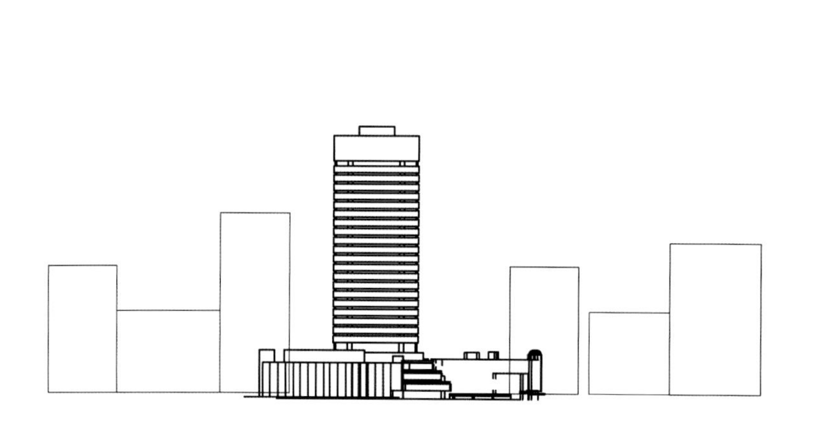

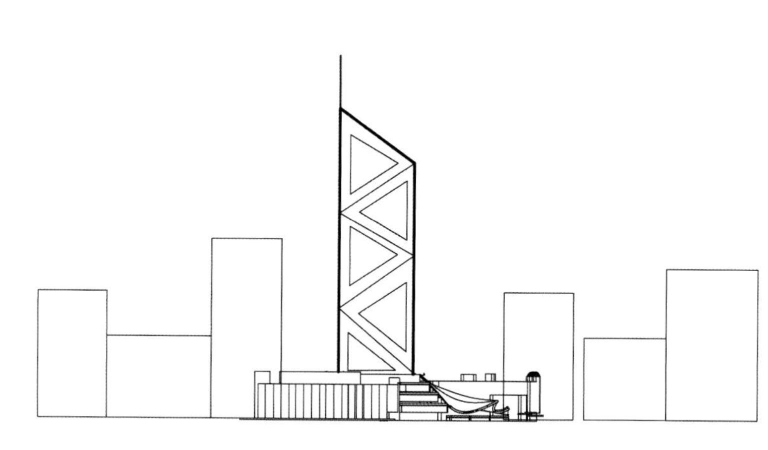

UTS

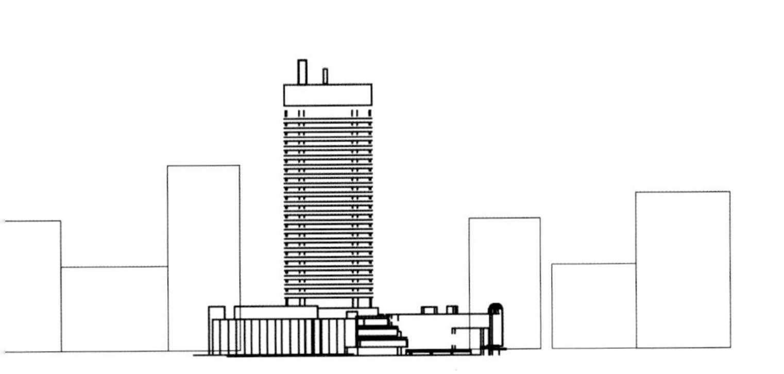

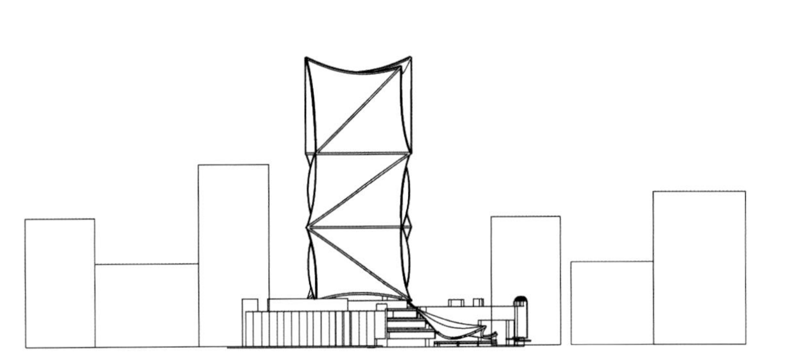

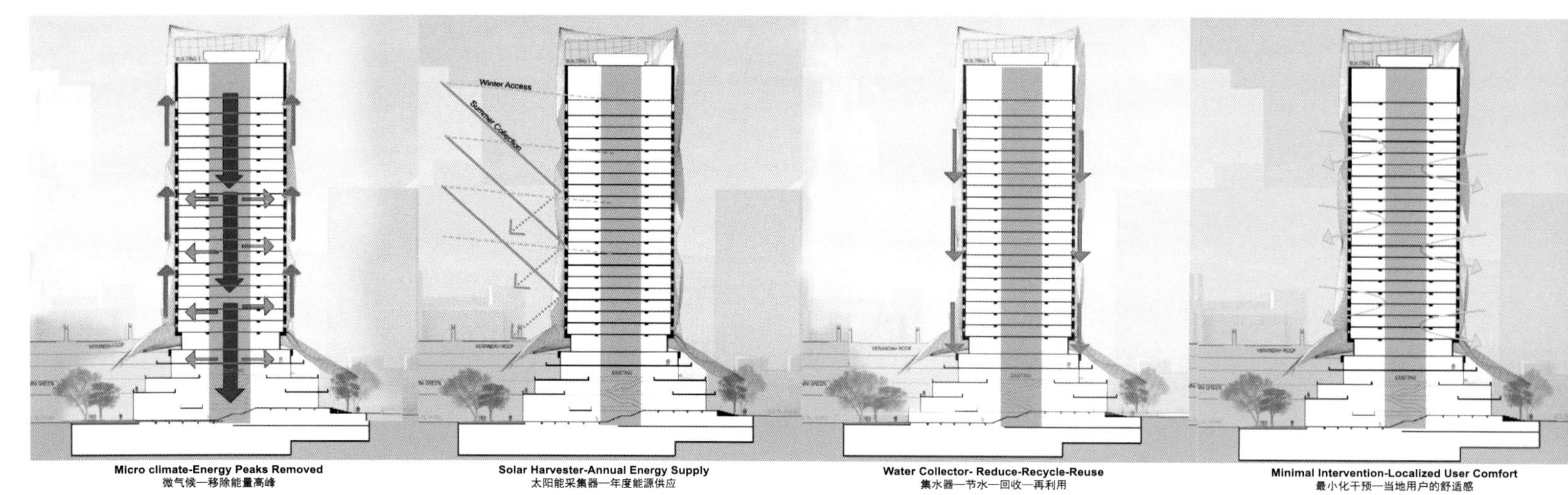

Micro climate-Energy Peaks Removed
微气候—移除能量高峰

Solar Harvester-Annual Energy Supply
太阳能采集器—年度能源供应

Water Collector- Reduce-Recycle-Reuse
集水器—节水—回收—再利用

Minimal Intervention-Localized User Comfort
最小化干预—当地用户的舒适感

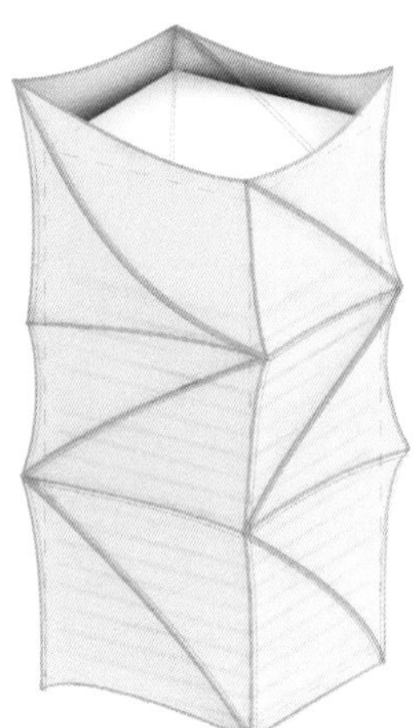

Skin 表皮

Structure 结构

Element 元素	Number of elements 元素数量	Position 位置	Bend radius m 弯曲半径 m	Diameter profile mm 侧面直径 mm	
	1	horizontal 水平	18	400	
	2	horizontal 水平	28	400	
	1	horizontal 水平	60	400	
	2	diagonal 对角线	80	600	
	2	horizontal 水平	132	400	
	1	diagonal 对角线	99	600	
	4	vertical 垂直	72	200	
	4	horizontal 水平	67	400	

Element 元素	Number of elements 元素数量	Position 位置	Bend radius m 弯曲半径 m	Diameter profile mm 侧面直径 mm	
	3	vertical 垂直	131	200	
	4	diagonal 对角线	92	600	
	4	diagonal 对角线	96	600	
	4	horizontal 水平	99	400	
	1	horizontal 水平	132	400	
	1	diagonal 对角线	99	600	
	2	diagonal 对角线	45	600	
	4	diagonal 对角线	40	400	

Bend radia
弯曲弧度

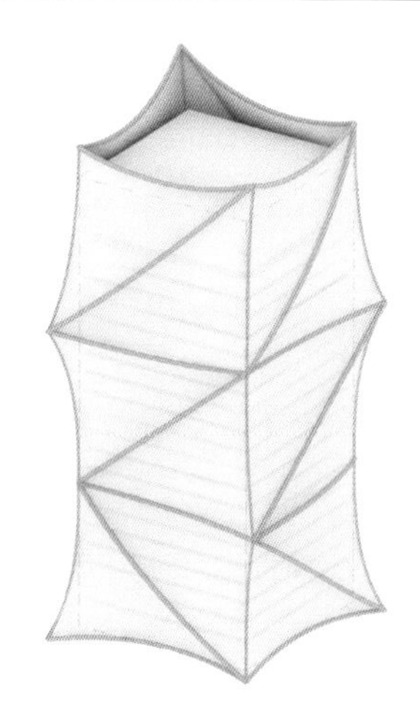

Skin 表皮

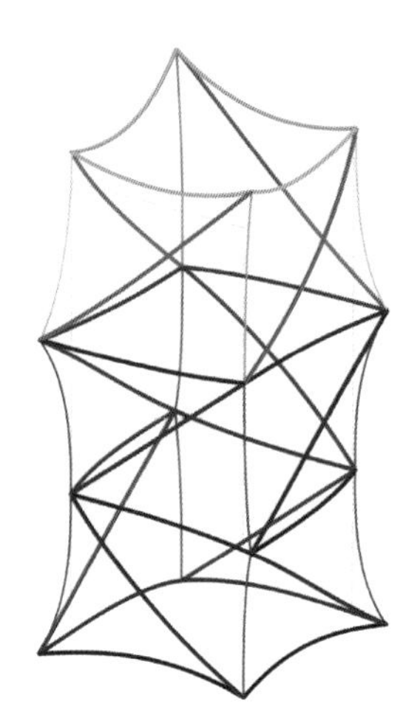

Structure 结构

Element 元素	Number of elements 元素数量	Position 位置	Bend radius m 弯曲半径 m	Diameter profile mm 侧面直径 mm	
	2	vertical 垂直	75	200	
	2	vertical 垂直	92	200	
	8	vertical 垂直	60	200	
	4	horizontal 水平	65	400	
	8	horizontal 水平	132	400	
	8	diagonal 对角线	99	600	
	4	diagonal 对角线	469	600	
	4	horizontal 水平	46	400	

Profile versions 侧面类型

Diagonal 对角线
600 mm

Horizontal 水平
400 mm

Vertical 垂直
200 mm

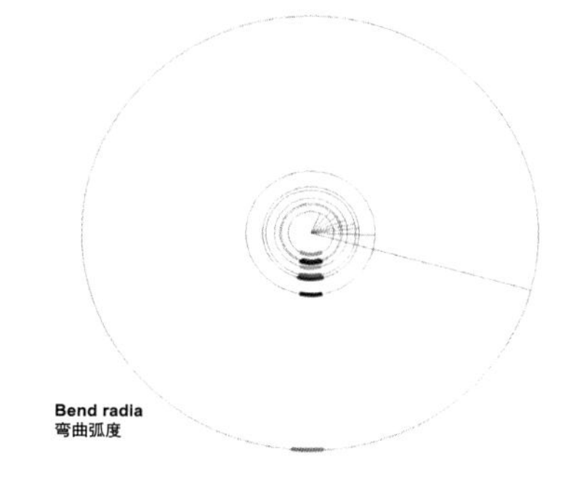

Bend radia
弯曲弧度

THOMAS ST
BROADWAY
10
2
1
4
8

SOLAR PENALS
太阳能电池板
BUILDING 1
大楼 1
BUILDING 10 大楼 10
VERANDAH ROOF
阳台屋顶
AGL RL 14,000
地面以上 RL14, 000
Harris Street
哈里斯街
BUILDING 4
大楼 4
BUILDING 6
大楼 6
Quay Street
码头街
SCALE 比例 DRAWING 图
1:1000 at A3 A3 1:1,000
SECTION EAST-WEST 东西剖面

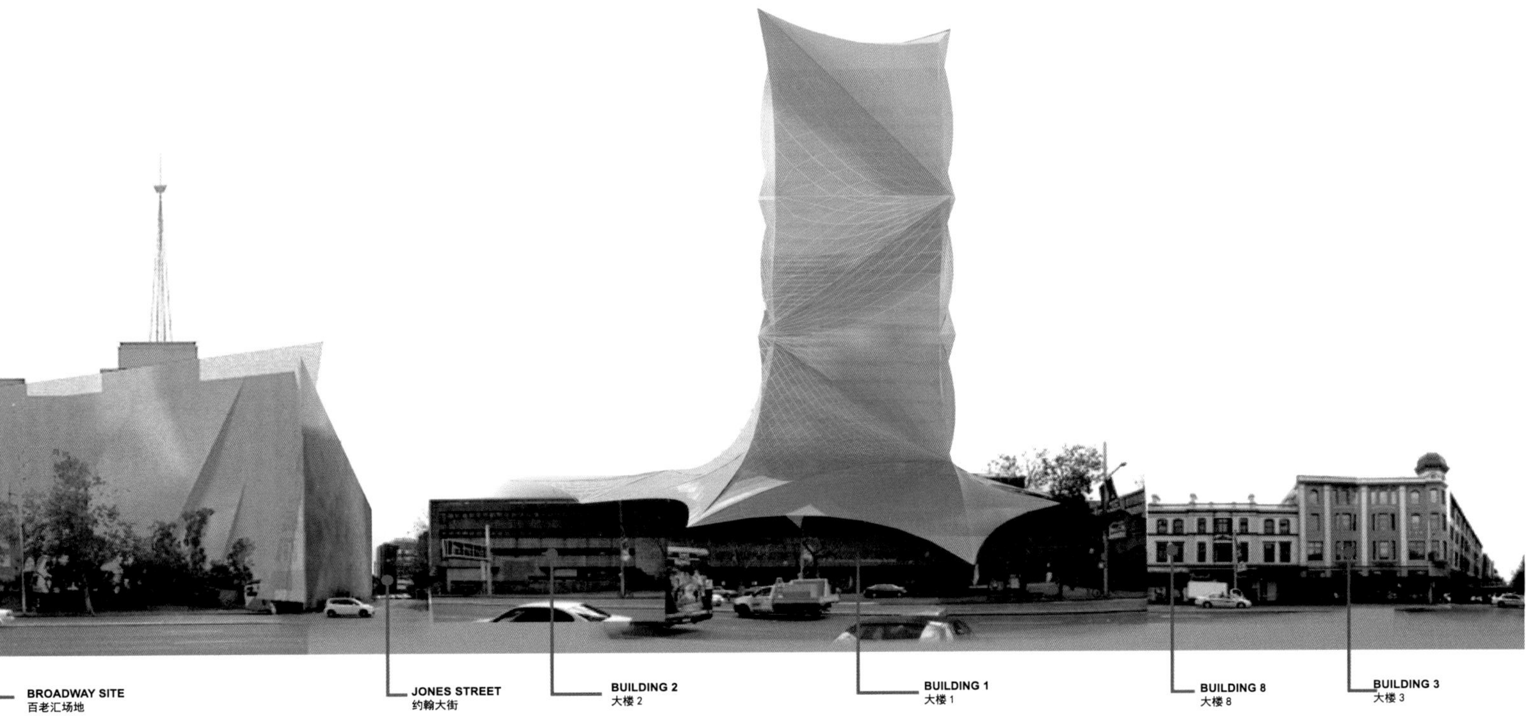

BROADWAY SITE
百老汇场地
JONES STREET
约翰大街
BUILDING 2
大楼 2
BUILDING 1
大楼 1
BUILDING 8
大楼 8
BUILDING 3
大楼 3

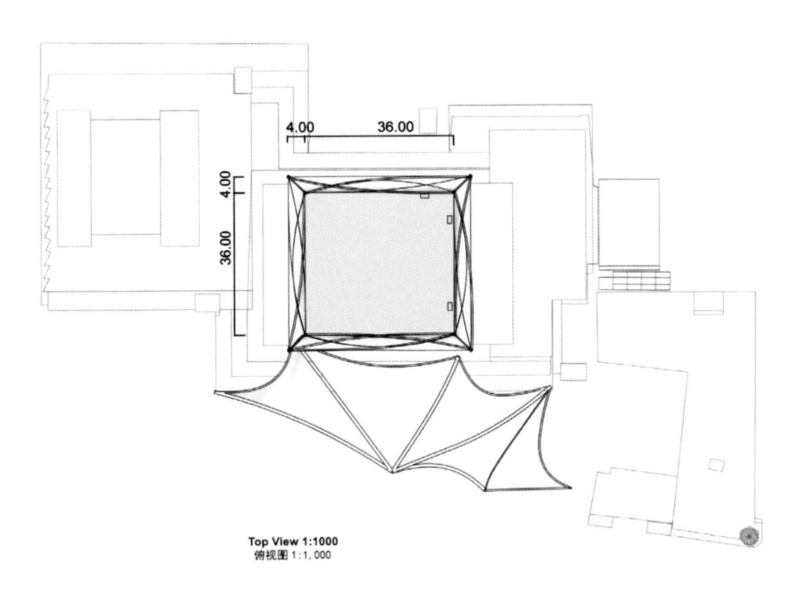

Top View 1:1000
俯视图 1:1, 000

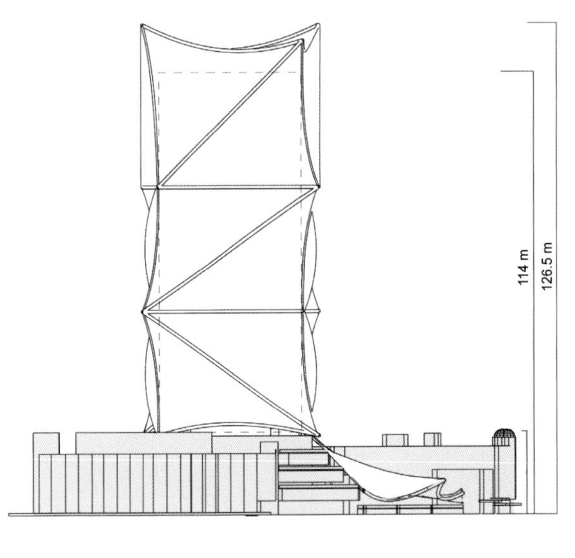

Elevation 1:1000
立面 1：1, 000

波哥大哥伦比亚国立大学新博士楼

New Doctorate's Building, National University of Colombia

设计单位：斯蒂文·霍尔建筑师事务所
开发单位：哥伦比亚国立大学
项目地址：哥伦比亚波哥大
项目面积：6 503 m²
设计团队：Steven Holl　Garrick Ambrose
Chris McVoy　Scott Fredricks
Johanna Muszbek　Dimitra Tsachrelia

Designed by: Steven Holl Architects
Client: National University of Colombia, Bogota
Location: Bogota, Colombia
Building Area: 6,503 m²
Design Team: Steven Holl, Garrick Ambrose, Chris McVoy, Scott Fredricks, Johanna Muszbek, Dimitra Tsachrelia

项目概况

这栋由斯蒂文·霍尔建筑师事务所为哥伦比亚国立大学设计的棱角分明的白色建筑构建在 20 世纪 30 年代由设计师利奥波德·洛特建立的建筑基础之上。

设计构思

洛特设计的总平面以绿地为核心，采用古典的轴对称形式和有层次的同心圆形式组织空间布局。20 世纪 70 年代新增的建筑阻隔了原始规划中内部绿地之间的相互联系，导致了很多封闭空间的出现，因此，该方案旨在打开这些封闭的空间，对绿地空间进行重新定义，使最初的总体规划重现活力。

建筑设计

这座纯净的白色建筑占地 6 400 平方米，由多个部分组成，其中包括：一个可容纳 600 人观演的音乐厅；一个悬挑的餐厅，餐厅设计了各式各样的露台，站在露台上可将不远处的山峦美景尽收眼底；还有一个截面“倒置”的人工水榭以及一个倒映池，倒映池中的水来源于雨水和再生水。

设计采用了可持续的水系统，该系统由许多位于建筑物屋顶的太阳能电池供电，这些太阳能电池同时也提供 15% 的设施用电。建筑材料将使用钢筋混凝土、当地出产的木材以及波哥大石材。

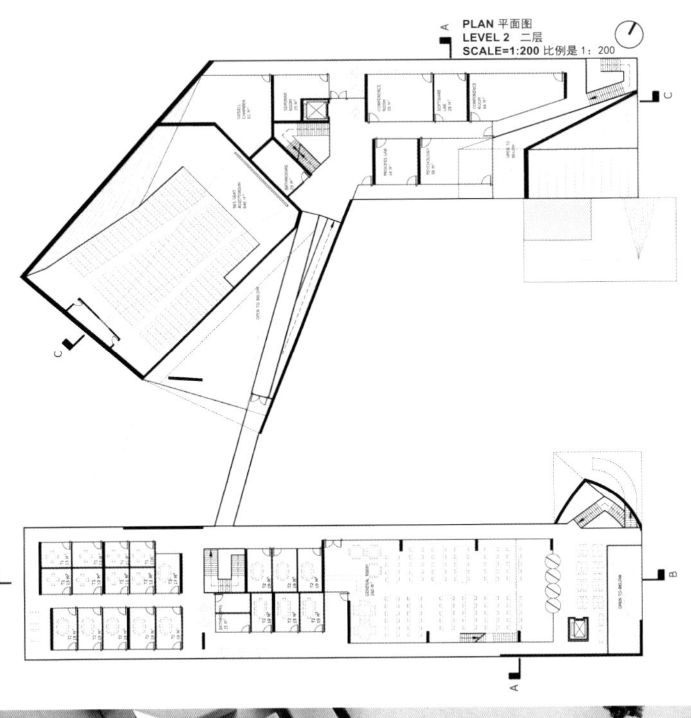

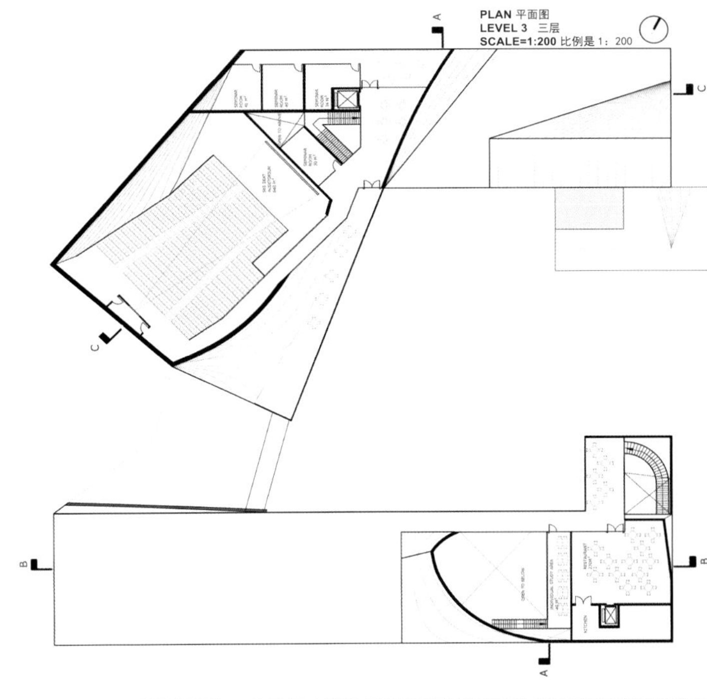

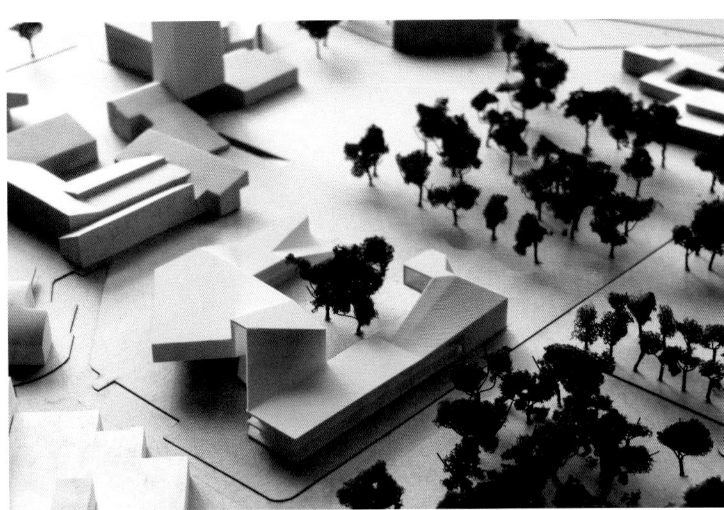

Profile

The importance of the National University of Colombia campus began with its enlightened master plan by the architect Leopold Rother in the 1930s. With its green center, classical axiality, and layered concentricity, the campus contains some wonderful examples of architecture.

Design Concept

The design takes campus green space as the core of master plan. It adopts classical axisymmetric form and concentric circles spatial layout. While buildings of the 1970s clogged the inner green of the original master plan, this new building aims to turn that closure inside out, re-establishing green space definitions and re-energizing the original master plan.

Architectural Design

The 6,400 square meters white building comprises of several parts including a 600-seat auditorium, a cantilevered restaurant with roof terraces and mountain views, a waterside pavilion with inverted section and a reflection pond. Water of this reflection pond originates from rain water and recycled water.

The water recycle system is driven by solar photovoltaic cells on the roof. These solar photovoltaic cells also provide 15% of the electrical power for the new structure. High-performance reinforced concrete joins local woods and Bogota stone in a palette of material resonance and ecological innovation.

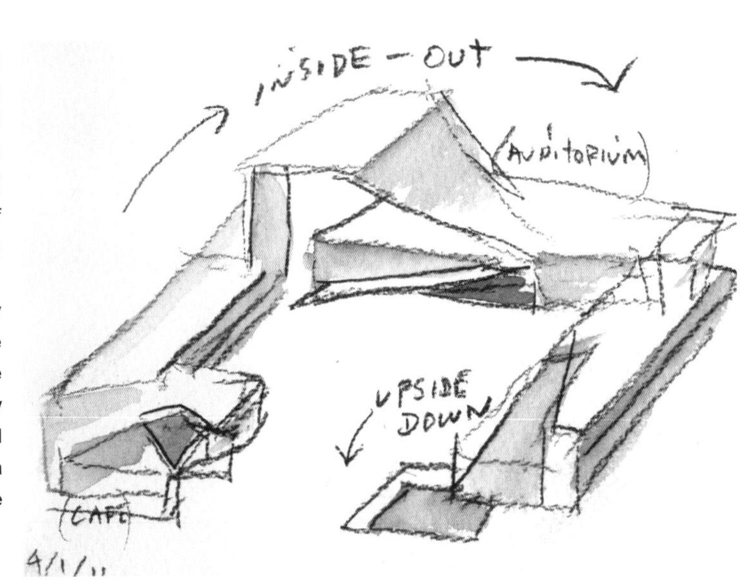

AIR FRANCE

交通建筑 Transportation Building

比利时布鲁塞尔机场连接大楼

Brussels Airport Connector, Brussels, Belgium

设计单位：UNStudio
开发商：布鲁塞尔机场公司 NV
项目地址：比利时布鲁塞尔
项目面积：23 182 ㎡
设计团队：Ben van Berkel　Gerard Loozekoot
Wesley Lanckriet　Joerg Petri
Maud van Hees　Milena Stopic
Perrine Planché　Deepak Jawahar
Hans Kooij　Benjamin Moore

Designed by: UNStudio
Client: The Brussels Airport Company NV
Location: Brussels, Belgium
Area: 23,182 m²
Design Team: Ben van Berkel, Gerard Loozekoot, Wesley Lanckriet, Joerg Petri, Maud van Hees, Milena Stopic, Perrine Planché, Deepak Jawahar, Hans Kooij, Benjamin Moore

项目概况

布鲁塞尔机场连接大楼的设计侧重于提供能与现有的机场建筑相连的、高效而灵活的基础设施，该大楼的建设也体现了布鲁塞尔机场立志成为未来欧洲交通枢纽的雄心。

建筑设计

设计通过连接大楼的建设将航站楼和 A 码头紧密结合起来，既确保了三栋大楼之间的功能衔接，同时也彰显了连接大楼自身的个性和特征。在客流量的组织与疏散等方面，连接大楼的设计既保证了流线的安全便捷性，同时也使操作流程高效而快捷。

大型零售区是项目设计的主要部分，双层的商业区为商业活动提供了大量商机。大楼内设有快速免税购物区，中央广场则涵盖了双层咖啡馆和高档精品店。中央广场旁边的高架驾驶舱成为了连接大楼的最高点，在这里，人们可以尽情观看广场和购物区内的情景以及欣赏楼外的景致。

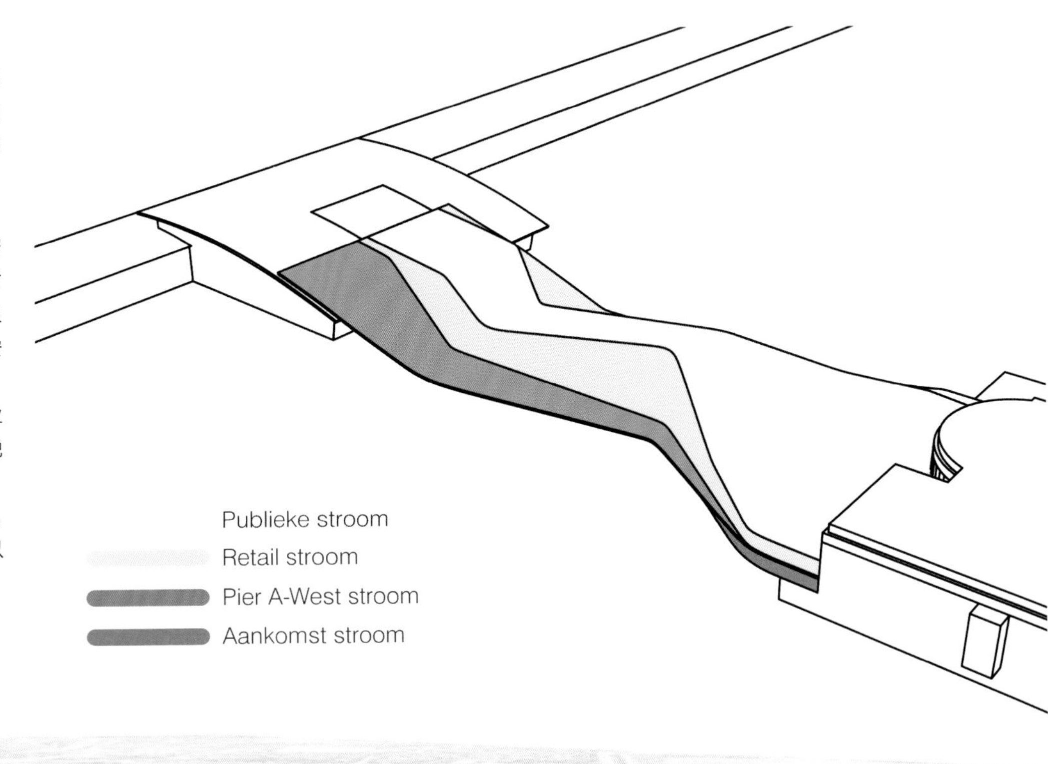

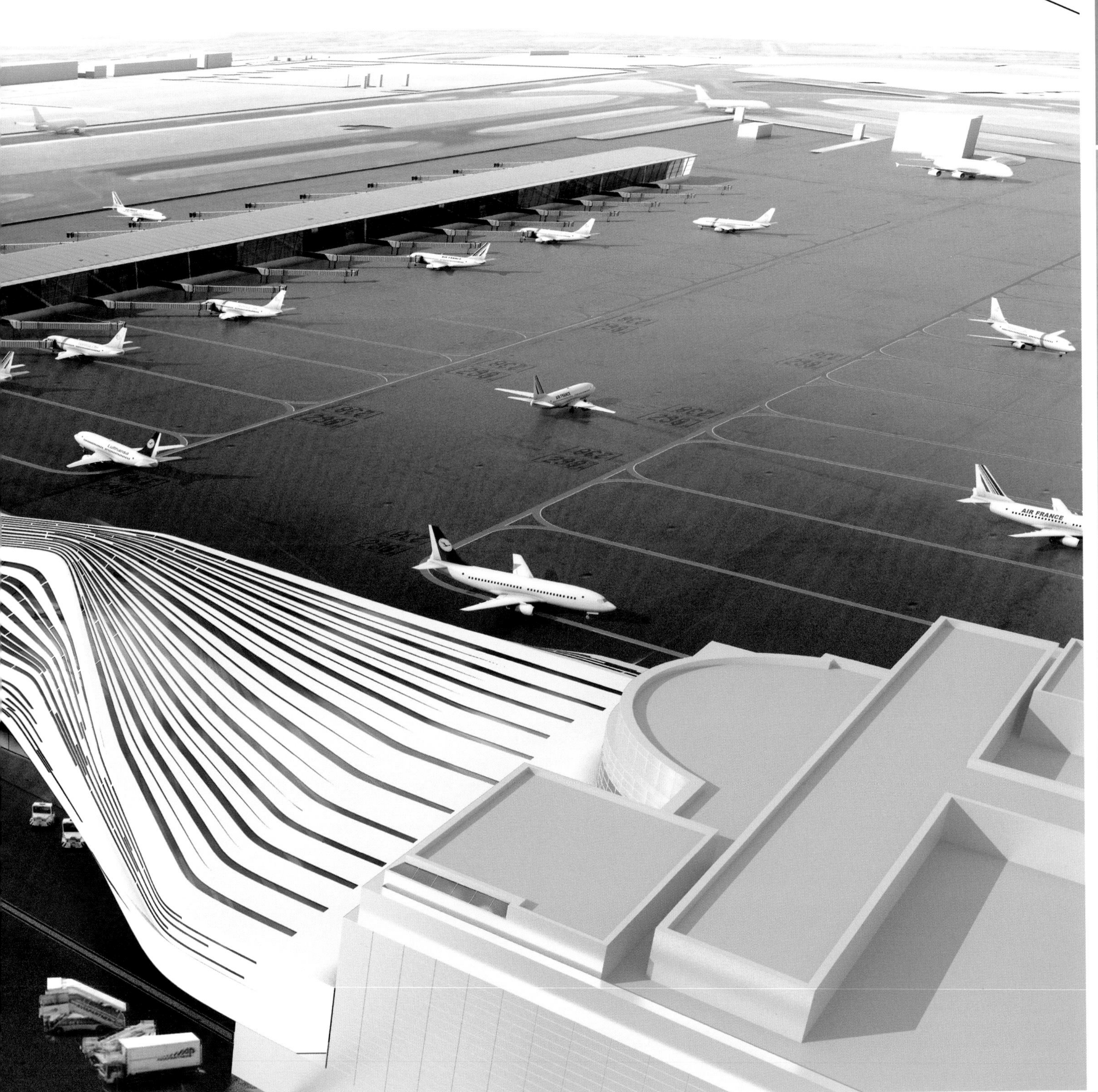

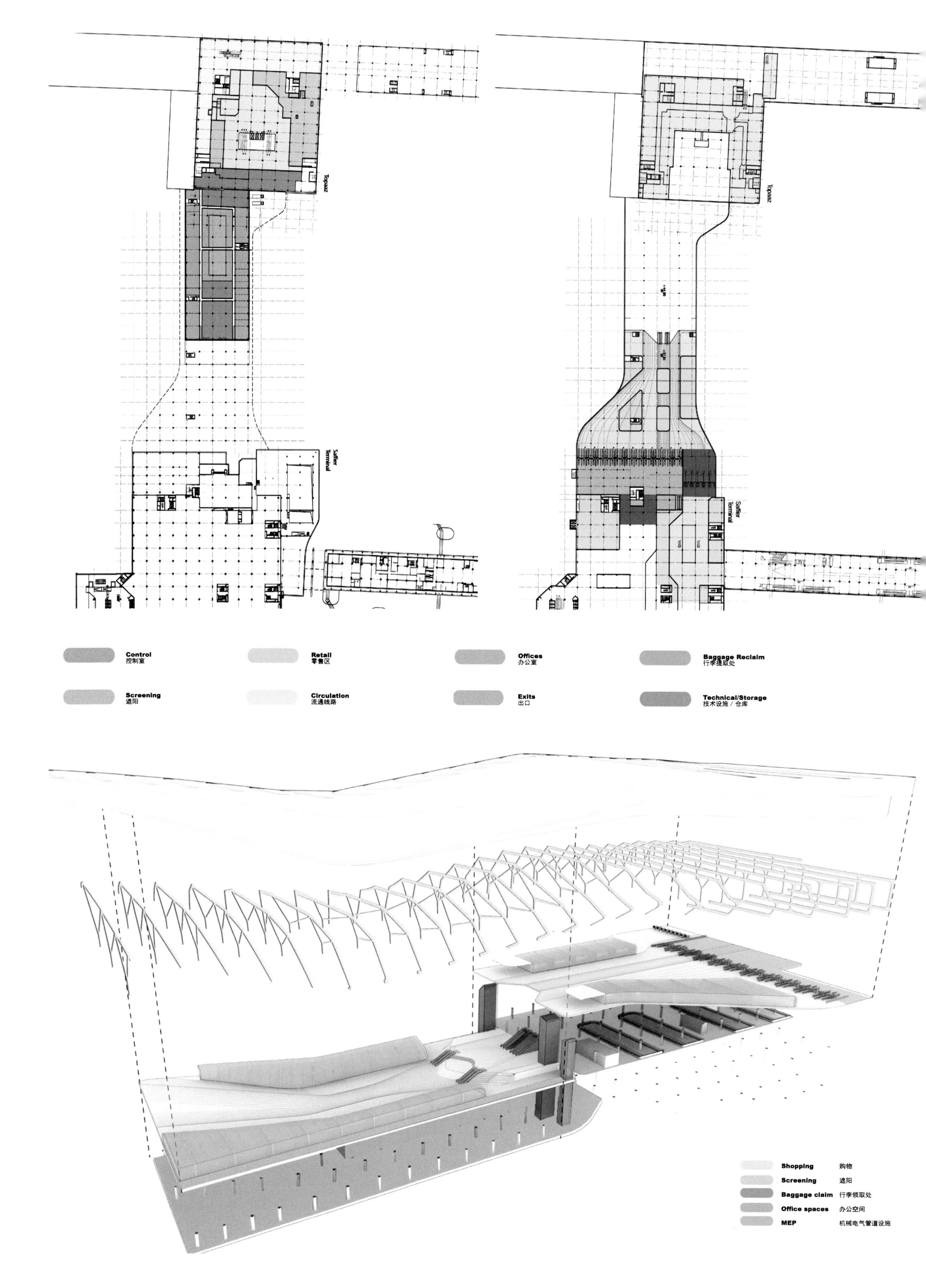
Topaaz
Saffier Terminal
Topaaz
Saffier Terminal
Control
控制室
Retail
零售区
Offices
办公室
Baggage Reclaim
行李提取处
Screening
遮阳
Circulation
流通线路
Exits
出口
Technical/Storage
技术设施／仓库
Shopping
购物
Screening
遮阳
Baggage claim
行李领取处
Office spaces
办公空间
MEP
机械电气管道设施

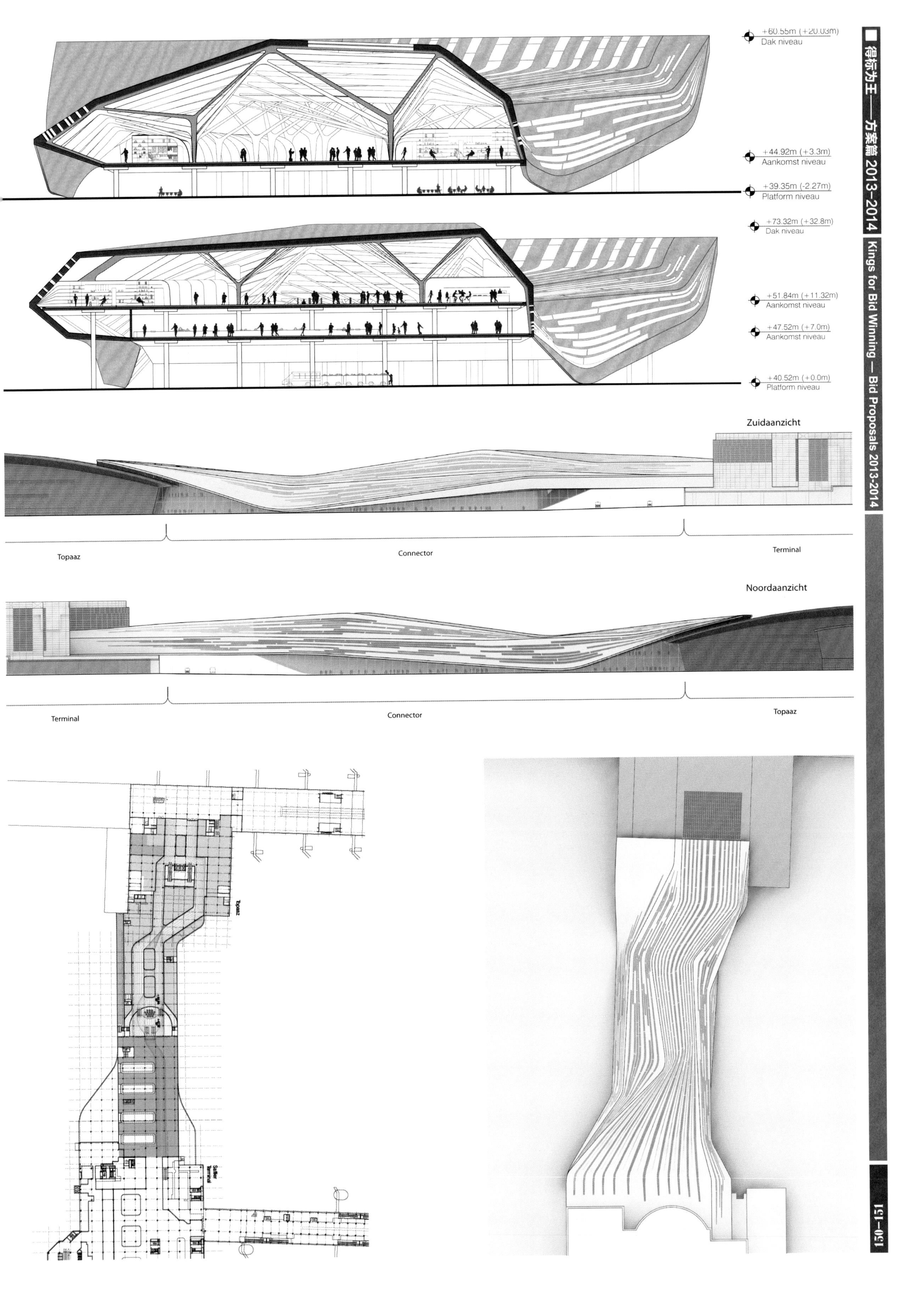
+60.55m (+20.03m)
Dak niveau
+44.92m (+3.3m)
Aankomst niveau
+39.35m (-2.27m)
Platform niveau
+73.32m (+32.8m)
Dak niveau
+51.84m (+11.32m)
Aankomst niveau
+47.52m (+7.0m)
Aankomst niveau
+40.52m (+0.0m)
Platform niveau
Zuidaanzicht
Topaaz
Connector
Terminal
Noordaanzicht
Terminal
Connector
Topaaz

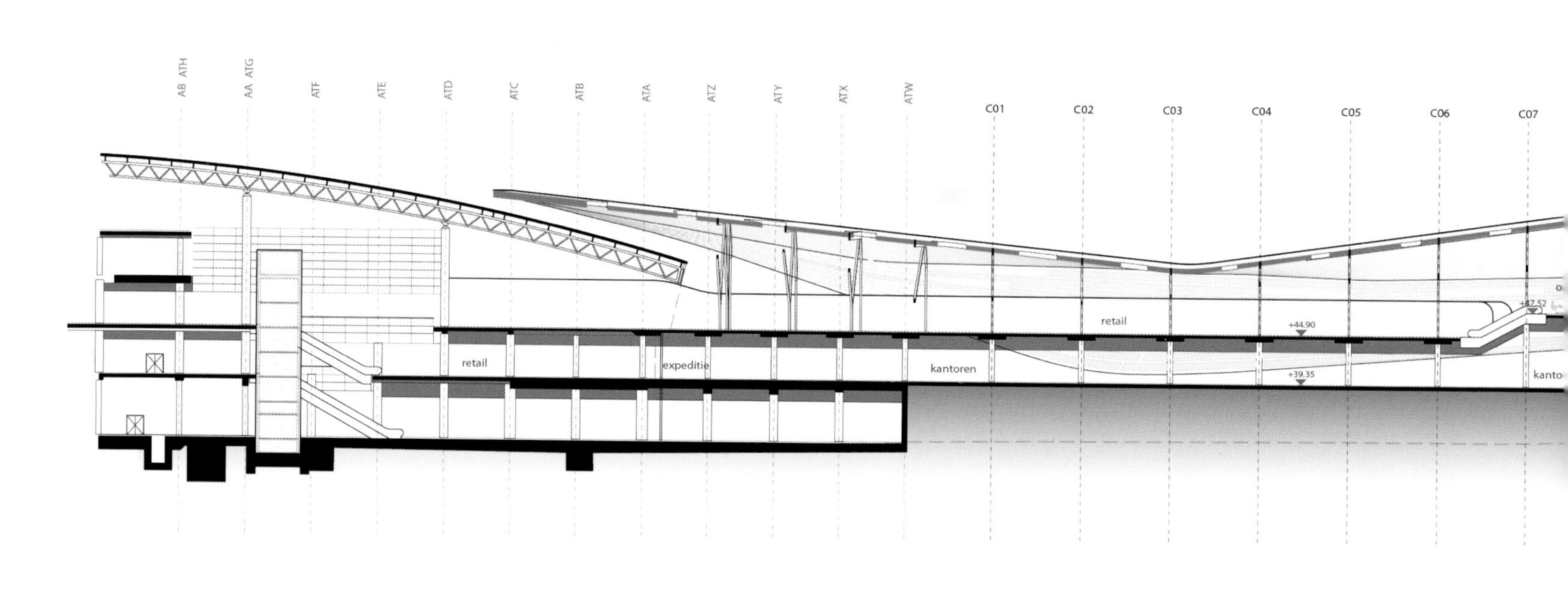
AB ATH
AA ATG
ATF
ATE
ATD
ATC
ATB
ATA
ATZ
ATY
ATX
ATW
C01
C02
C03
C04
C05
C06
C07
retail
+44,90
retail
expeditie
kantoren
+39,35

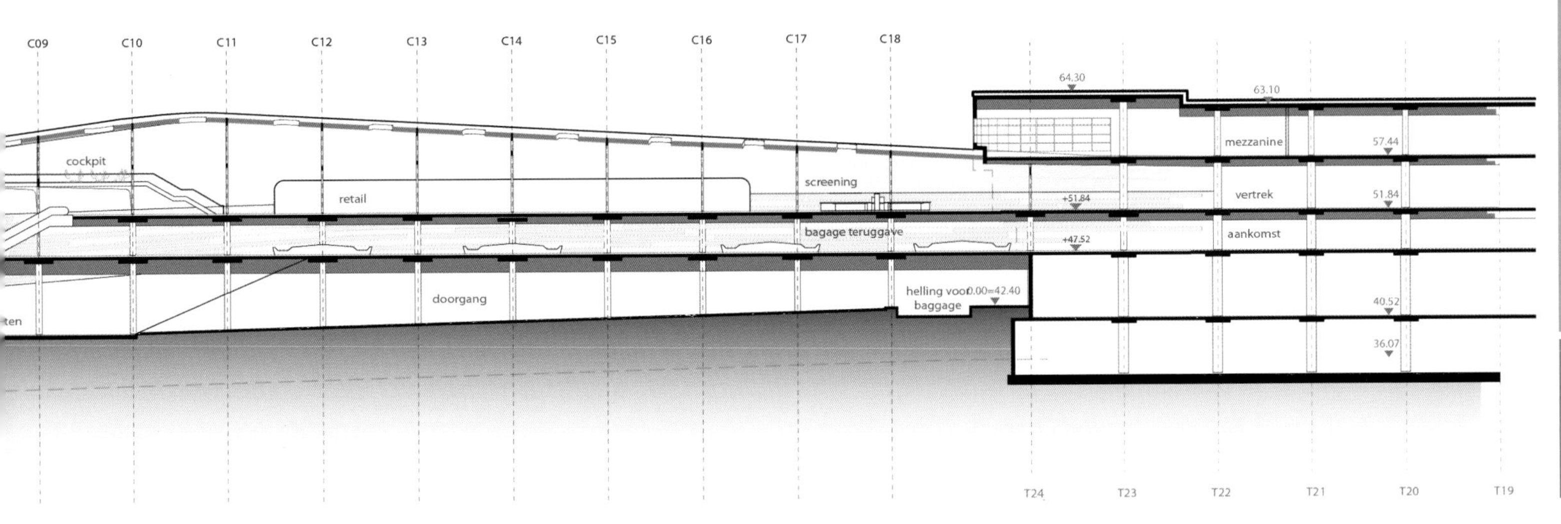
C09
C10
C11
C12
C13
C14
C15
C16
C17
C18
64.30
63.10
cockpit
mezzanine
57.44
screening
retail
vertrek
+51.84
51.84
bagage teruggave
aankomst
+47.52
helling voor baggage
0.00=42.40
doorgang
ten
40.52
36.07
T24
T23
T22
T21
T20
T19

TAX FREE

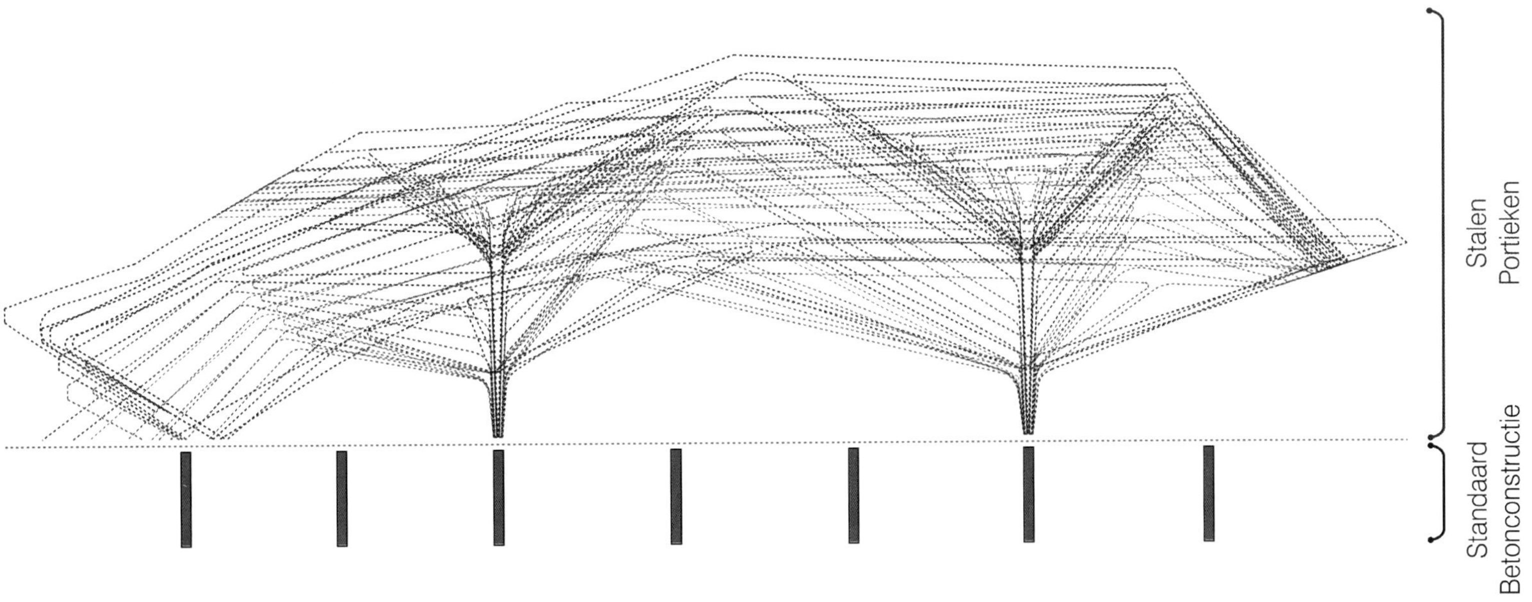

Stalen Portieken

Standaard Betonconstructie

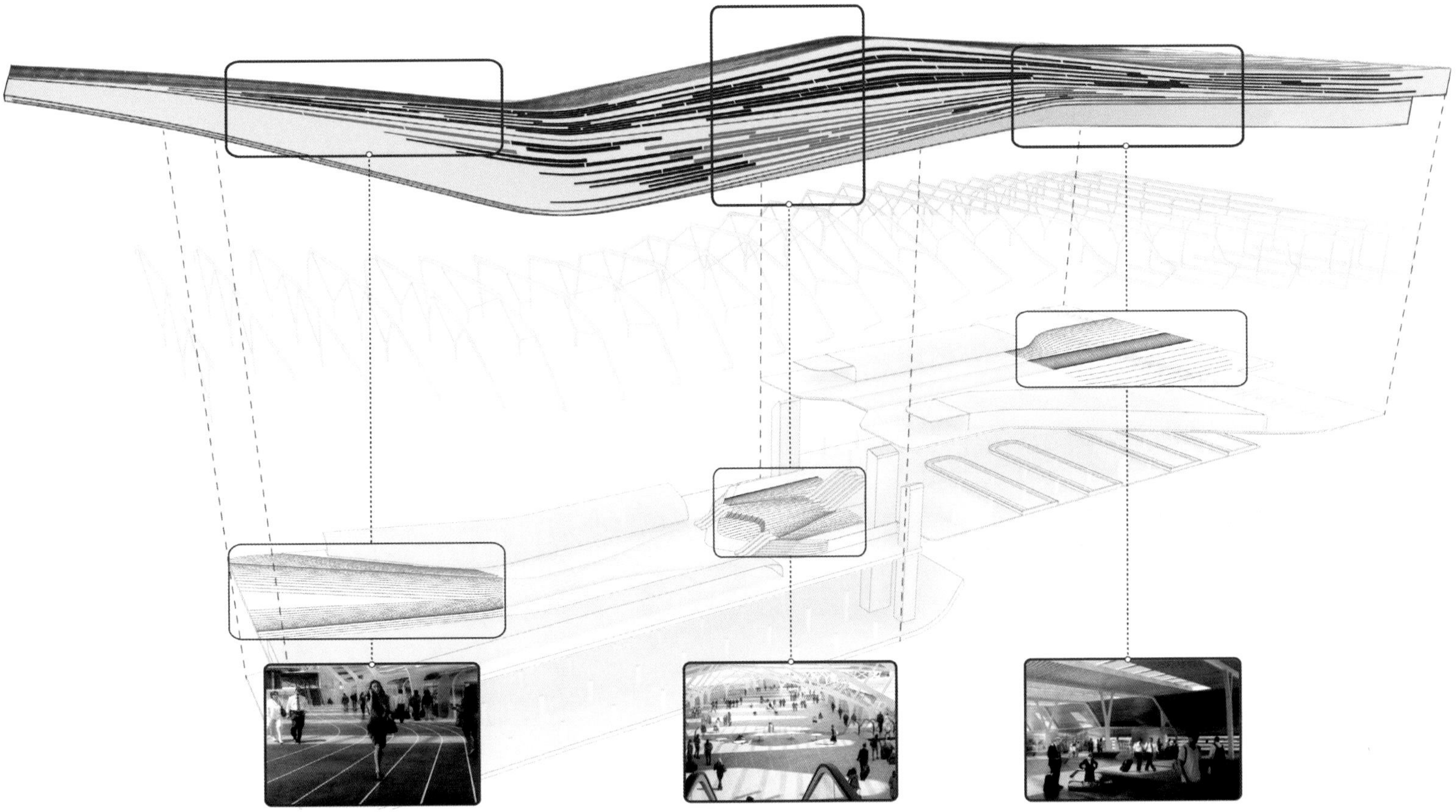

1.Shopping area 购物区

2.Cockpit area 座舱区

3.Screening area 遮阳区

Profile

The design for the Brussels Airport Connector Building focuses on providing a highly efficient, flexible infrastructural element that connects to and negotiates the existing airport architecture. The Connector is part of the future ambition of the Brussels Airport to create a European hub in Brussels.

Architectural Design

The design approach ensures cohesive functioning of the three buildings, with the Connector Building establishing a seamless connection between the two contrasting identities of the terminal and Pier A, whilst simultaneously creating its own new identity. The Connector incorporates logistical efficiency in handling passenger flows, security and operational processes.

Opportunities for commercial activity are maximized with the newly expanded, double-storey commercial area of the Connector, creating a fully-integrated retail zone as the main programmatic spine. The post-screening area with fast duty-free shopping and a horizontal organization gradually morphs into a double-height central plaza housing the main café and high-end boutique stores. Surrounding the central plaza is the elevated Cockpit - the highest point of the Connector with a light cafe boasting the best views of the bustling plaza and shopping area below, as well as the fleeting fields of airplanes outside.

拉脱维亚里加国际机场

Riga Airport

设计单位：LAVA

项目地址：拉脱维亚里加

Designed by: LAVA

Location: Riga, Latvia

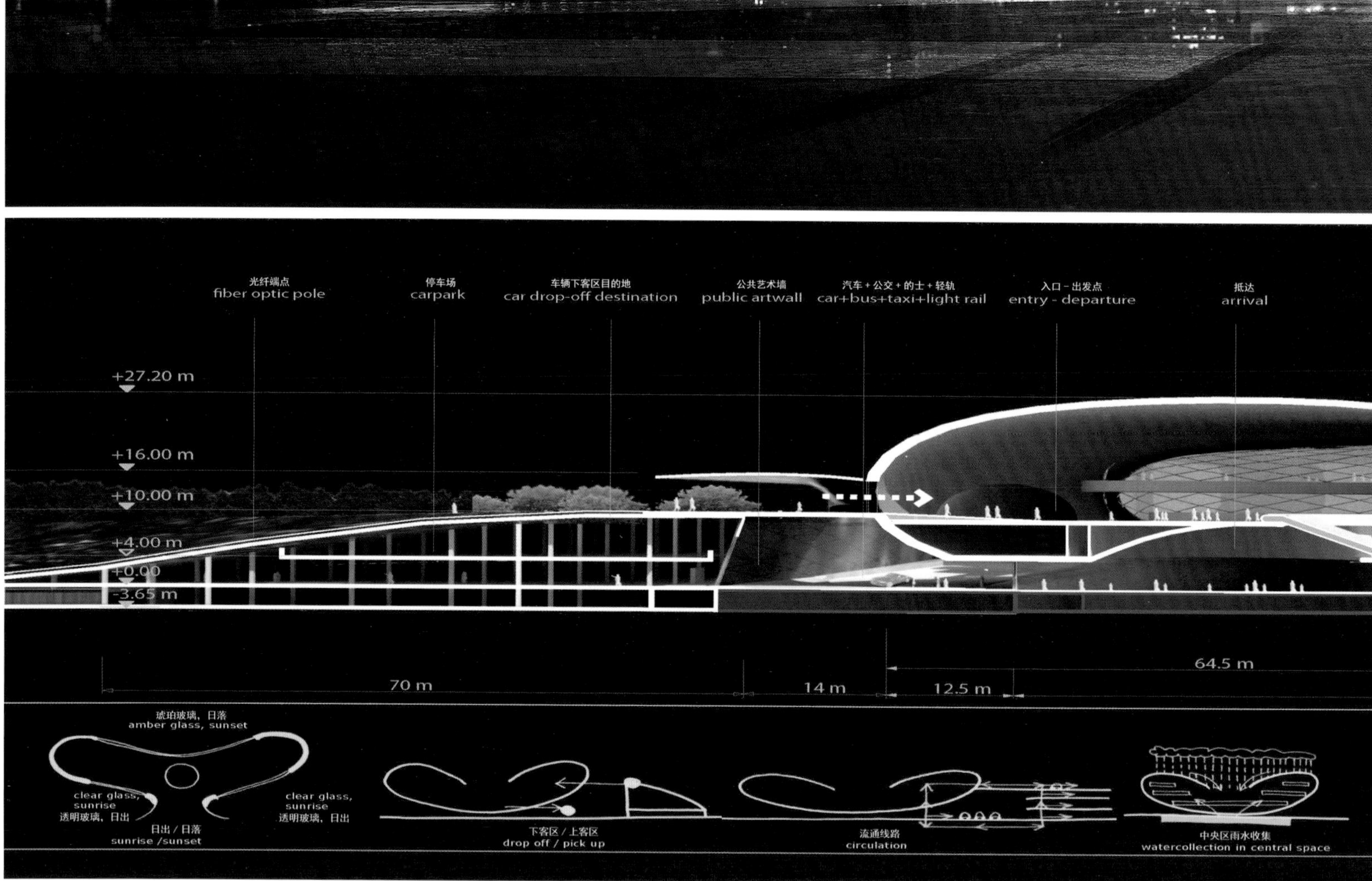

项目概况

LAVA 设计的新拉脱维亚航空公司航站楼将拉脱维亚的美景与未来航空旅游相结合，以重新燃起人们对于空中旅行的热情和活力。

设计理念

设计师从当地的景观意象，如松树、橡树、菩提树、琥珀以及瓢虫获取灵感，同时融入了对当今人们普遍关注的全球性问题的见解，从而提出了这个仿生、环保、可持续性的建筑方案。

建筑设计

航站楼卵形的建筑形态源自于拉脱维亚地区的瓢虫，当站在地面上观看这一建筑时，建筑似是从景观中升起来一样漂浮着。它闪着光，琥珀色的外观映衬着周边的景观。

这一瓢虫形态的建筑有两条履带，恍若延伸至停机坪的手臂。主入口位于两个 30 米高的琥珀色卵形窗体的侧面，两侧种植着苍翠的橡树和菩提树。这些树木连同位于中央大厅的老橡树，使建筑流露出拉脱维亚独有的景观气息。

可持续性设计

项目采用了主动和被动的可持续性设计手法，如地热系统、水资源循环回收利用、灵活的太阳能屋顶、被动式设计原理等。可持续性不仅仅体现在建筑本身，还表现在如何连接和整合各区间。设计在布局各功能区间时，重点考虑了建筑的运行效率和可持续性。

Profile

LAVA's design for a new airBaltic terminal integrates Latvia's provincial beauty with the future of air travel. It tries to reignite people's enthusiasm and passion for air travel.

Design Concept

The design is inspired by regional imagery including pines, oaks, linden trees, amber and ladybugs and infused with global concerns, which therefore contributes to this bionic, environment friendly, sustainable project.

Architectural Design

The terminal's ovoid form recalls the Latvian ladybug. It glows, its amber shades reflecting the local landscape and history. From the land, the building appears to float and rise above the landscape.

The iconic ladybug form has two caterpillar-like arms stretching out onto the tarmac. The main entrance is flanked by two, thirty meters high, amber colored ovoid windows, facing verdant stands of oaks and linden trees. These forests, together with an ancient oak in the central hall, give the sense and smell of the Latvian landscape.

Sustainable Design

The project has applied active and passive sustainable design techniques: geothermal system, water collection and circulation for reuse, flexible solar roof, passive design principles etc. Sustainability is not just in and of the building. It is also how the building connects and integrates. On the layout of the design, designers have paid special attention to functional and operational efficiency.

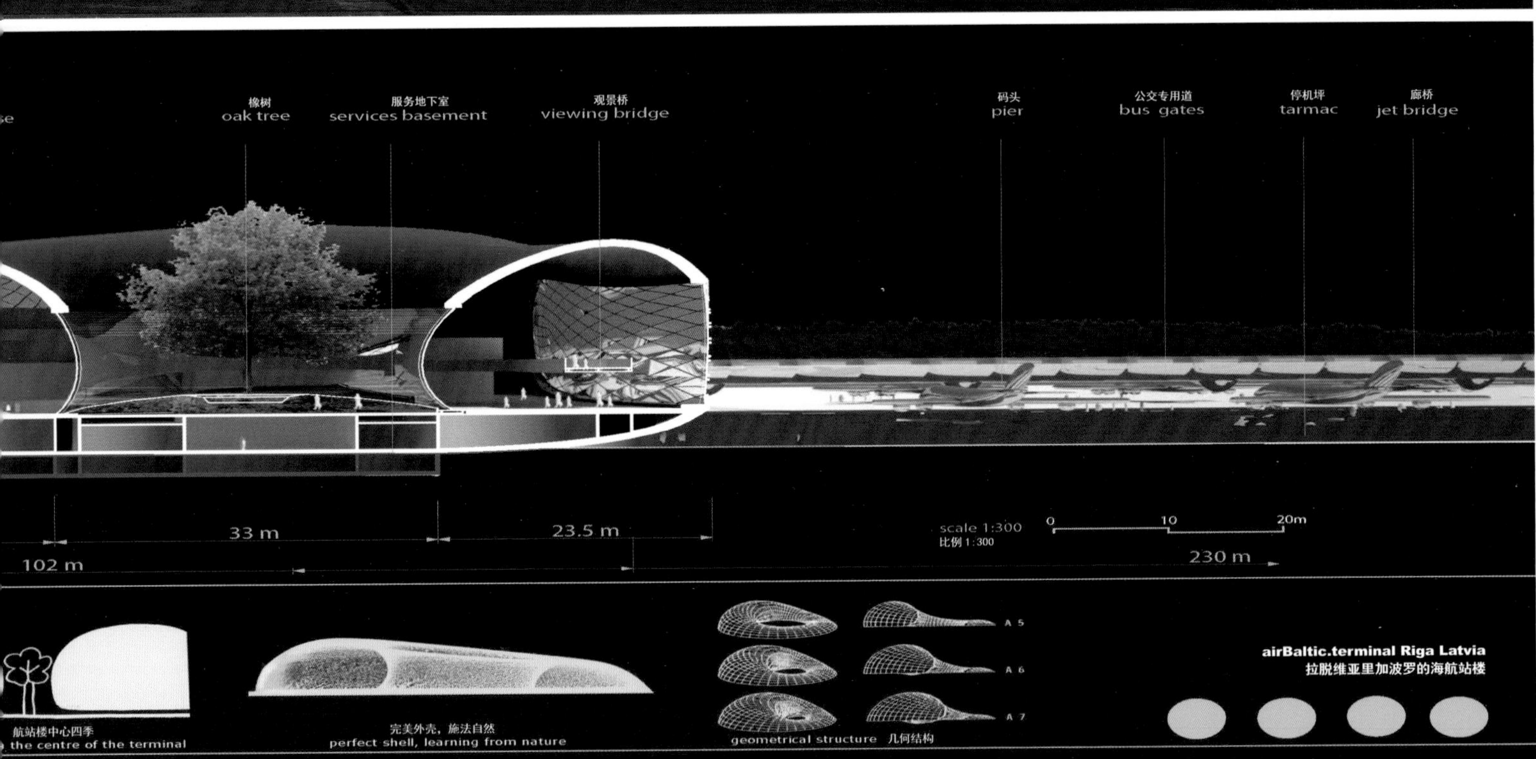

airBaltic.com
Immigration
Shops
Shops
Shops
Shops
Offices
Offices
Security
LEVEL + 05 Concourse
airBaltic
check in
shop
wine
dine
play
explore
shortcut.
where s my gate
go to gate:
白鹡鸰，国鸟
balta cielava. the national bird
architecture in Riga
斑点瓢虫，国家的标志
the 2 dot ladybug, national symbol
准备飞翔，功能分层
ready to fly, layering of functions
里加机场应用的多媒体设施
multimedia device, the Riga airport applicat

Baggage Handling System
Baggage Claim
Custom
Arrival Hall
LEVEL +-00 Arrivals
ing the seasons in the terminal
参考航空学
reference to Aviation
a field of daisies, national flower
氧气，一片松树林
oxygen, a forest of pine trees
airBaltic.terminal Riga Latvia
拉脱维亚里加波罗的海航站楼

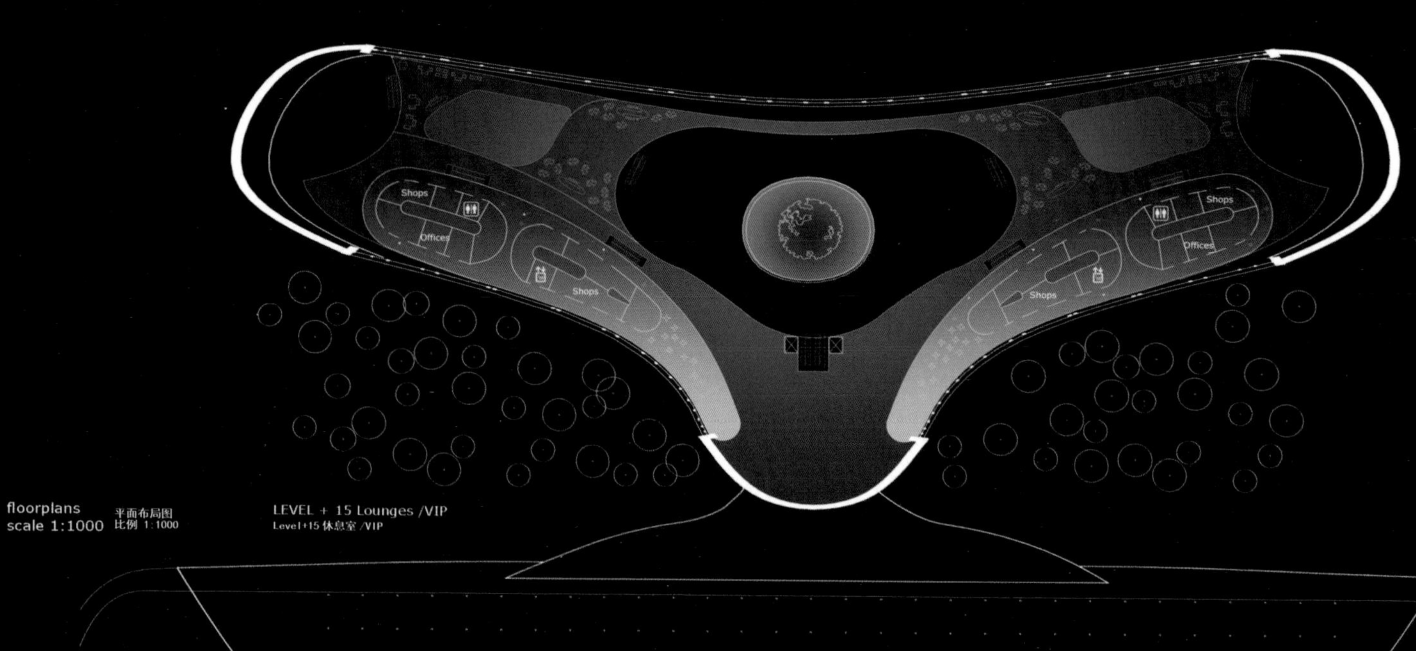

floorplans
scale 1:1000
平面布局图
比例 1:1000

LEVEL + 15 Lounges /VIP
Level+15 休息室 /VIP

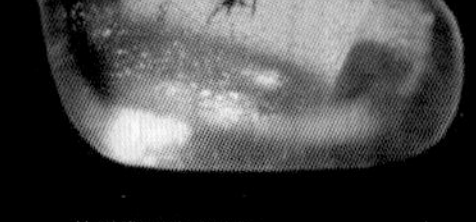

拉脱维亚的琥珀块
a piece of latvian amber

拉脱维亚树木，橡树和松树
Latvian trees, Oak Linde and Piine

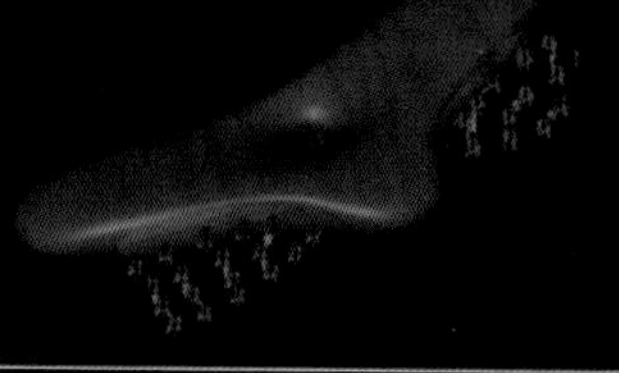

拉托维尼亚木材内层
inner skin of Latvian wood

the n

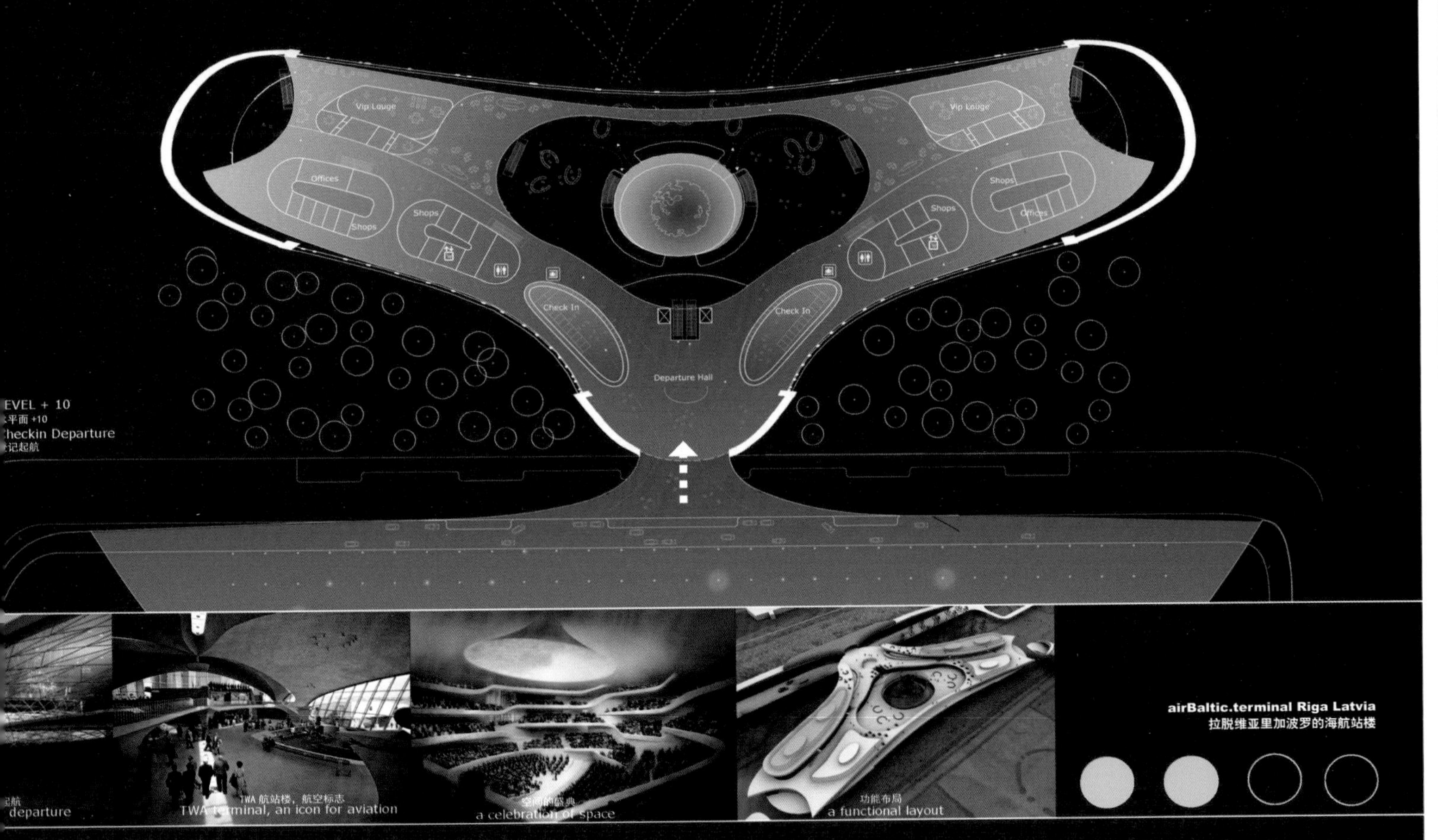

airBaltic.terminal Riga Latvia
拉脱维亚里加波罗的海航站楼

Baltic

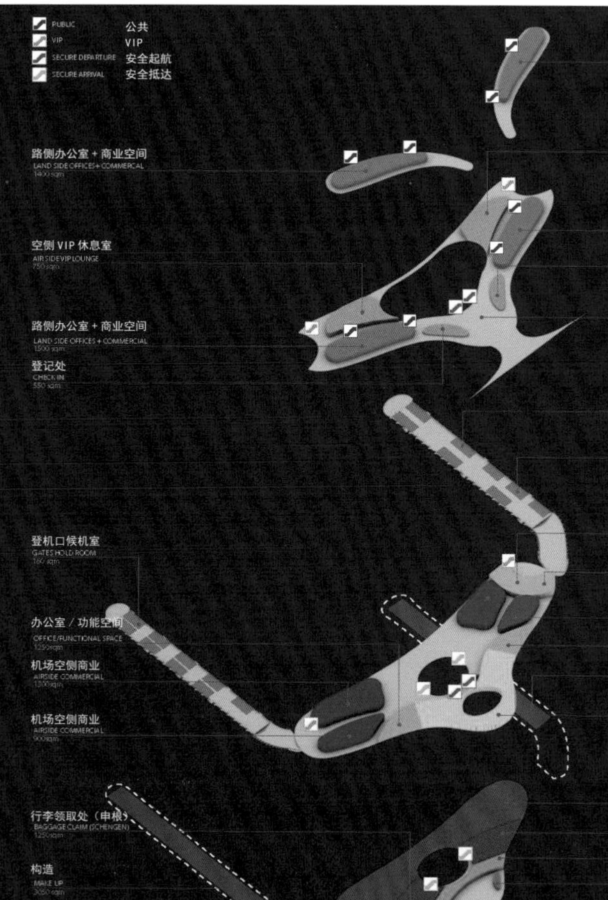
PUBLIC 公共
VIP VIP
SECURE DEPARTURE 安全起航
SECURE ARRIVAL 安全抵达
路侧办公室 + 商业空间
LAND SIDE OFFICES + COMMERCIAL
1400 sqm
空侧 VIP 休息室
AIR SIDE VIP LOUNGE
750sqm
路侧办公室 + 商业空间
LAND SIDE OFFICES + COMMERCIAL
1500 sqm
登记处
CHECK IN
550 sqm
机场候机楼
DEPARTURE HALL
2500sqm
登机口候机室
GATES HOLD ROOM
160 sqm
登机口候机室
GATES HOLD ROOM
200 sqm
出境移民 / 护照
OUT-BOUND
IMMIGRATION/ PASSPORT
400sqm
入境移民 / 护照
IN-BOUND
IMMIGRATION/ PASSPORT
350 sqm
机场空侧商业
AIRSIDE COMMERCIAL
1050 sqm
AIRSIDE COMMERCIAL
700sqm
办公室 / 功能空间
OFFICE/FUNCTIONAL SPACE
350sqm
构造
MAKE UP
2050 sqm
登机证 + 安全控制
BOARDING PASS + SECURITY
CONTROL
2800sqm
办公室 / 功能空间
OFFICE/FUNCTIONAL SPACE
1250sqm
机场空侧商业
AIRSIDE COMMERCIAL
1300sqm
机场空侧商业
AIRSIDE COMMERCIAL
900sqm
行李领取处（申根）
BAGGAGE CLAIM (SCHENGEN)
1250sqm
构造
MAKE UP
2050 sqm
行李处理系统
BAGGAGE HANDLING SYSTEM
9000sqm
行李领取处（非申根）
BAGGAGE CLAIM (NON-SCHENGEN)
1250 sqm
CUSTOM 机场海关
900 sqm
LANDSIDE COMMERCIAL 路侧商业
200 sqm
ARRIVAL HALL 抵达大厅
1320 sqm

适合大空间的宽松结构
lose fit in great space

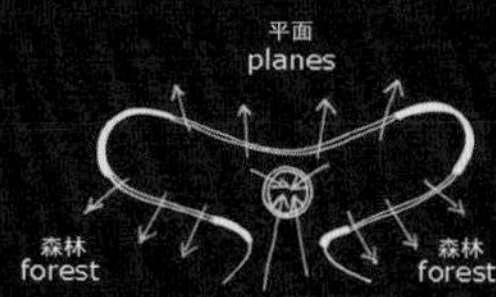
平面
planes
森林
forest
森林
forest
视线
viewlines

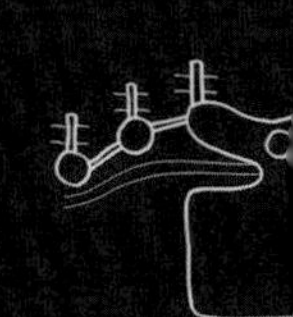
link and possible f
连接及未来的

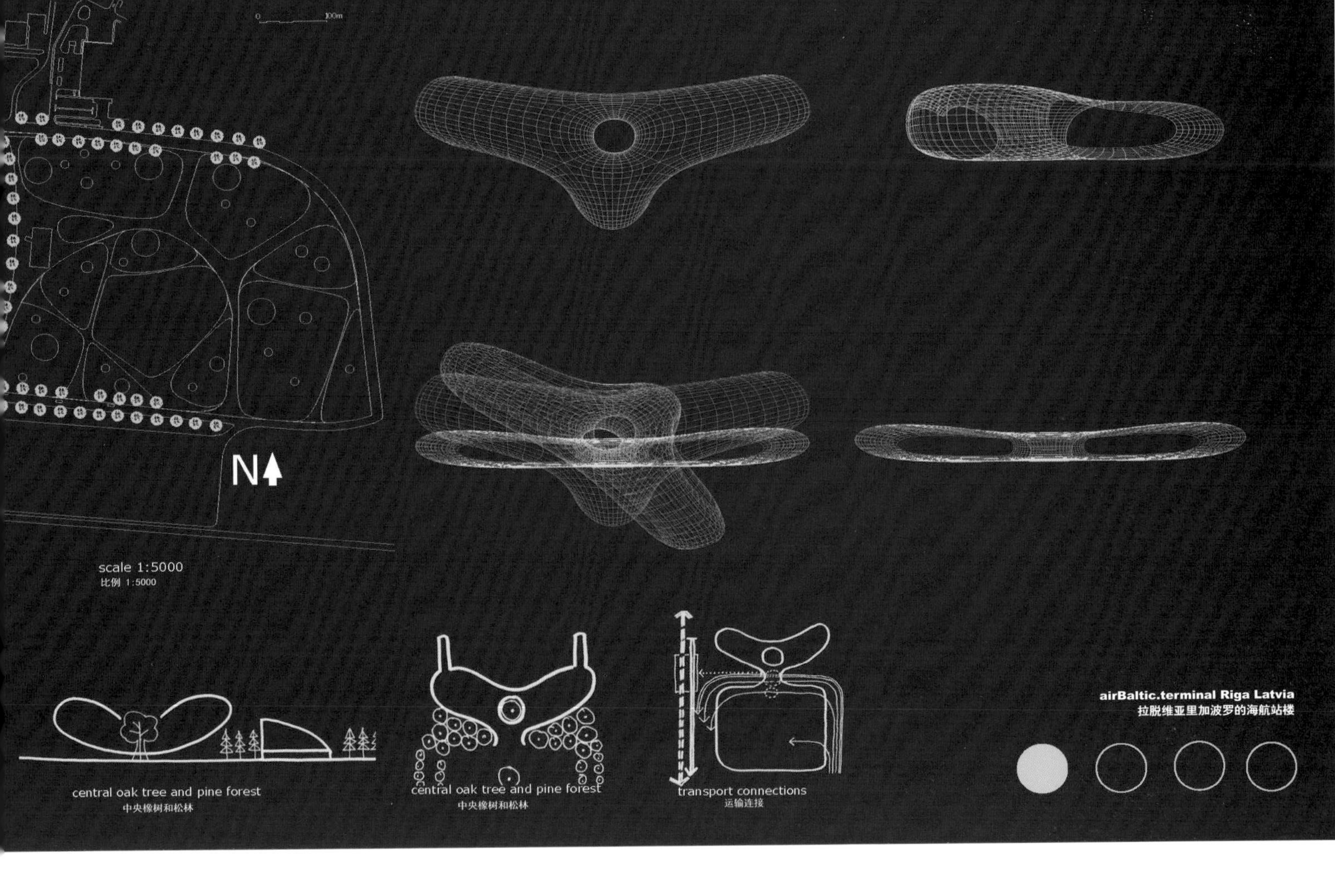

airBaltic.terminal Riga Latvia
拉脱维亚里加波罗的海航站楼

江苏常州机场

Changzhou Airport

设计单位：Alessio Patalocco
合作单位：CACC. Engineering
ADR Engineering Spa
TAU Group sas e PRVS
项目地址：中国江苏省常州市
总规划面积：52 460 m²
建筑密度：90%
绿化率：20%

Designed by: Alessio Patalocco
Collaboration: CACC. Engineering;
ADR Engineering Spa; TAU Group sas e PRVS
Location: Changzhou, Jiangsu, China
Total Planning Area: 52,460 m²
Building Density: 90%
Greening Ratio: 20%

项目概况

常州机场坐落于常州市新北区罗溪镇，北邻沪宁高速公路及长江，西靠沪宁铁路和京杭大运河，交通便捷，区位优越。由于现有机场的基础设施建设严重滞后，旅客吞吐量始终处于低位，已成为常州可持续发展的瓶颈。为维护区域市场的竞争能力，促进常州经济乃至长三角地区经济的持续快速增长，当地政府决定加快实施常州机场改扩建工程。

设计特色

建筑结构

常州机场改扩建工程的建筑主体结构形式为框架架构，局部屋面采用钢筋混凝土。为了满足建筑造型的要求，离港大厅及候机指廊屋面均采用钢屋架，形成了开阔明亮的值机大厅和候机大厅。高挑的空间搭配全景式的落地窗，使室内的乘客享有良好的视野。

低碳设计

航站楼建有大的挑檐，宛若带上了一顶“太阳帽”，具有高效的遮阳效果，极大地减少了太阳辐射进入室内，并为合理地利用阳光和自然光提供了条件。航站楼主楼二层和候机厅空侧为中空 low-e 玻璃幕墙，这种玻璃幕墙可以实现自然采光，并有效地隔音、隔热。指廊屋顶设置了一个 150 千瓦的光伏玻璃幕墙并网电站，可满足地下室及景观照明的供电需求。同时，屋面上还设置了雨水收集系统，收集后的雨水经过过滤，汇入航站楼前面的景观湖。

人性化设计

设计除了考虑本期航站楼的功能需求外，同时结合现有设施及远期扩建提出了流程设计方案。航站楼内的流程简单、直接，以尽可能地减少主要客流在航站楼内转换楼层的次数。旅客及行李的楼层转换可通过电梯及自动扶梯来完成。所有的旅客流程中，只要涉及到楼层和标高的转换、变化，均设有为残疾人服务的电梯或车道，所有提供的服务均考虑了设置残疾人专用的设施和设备。

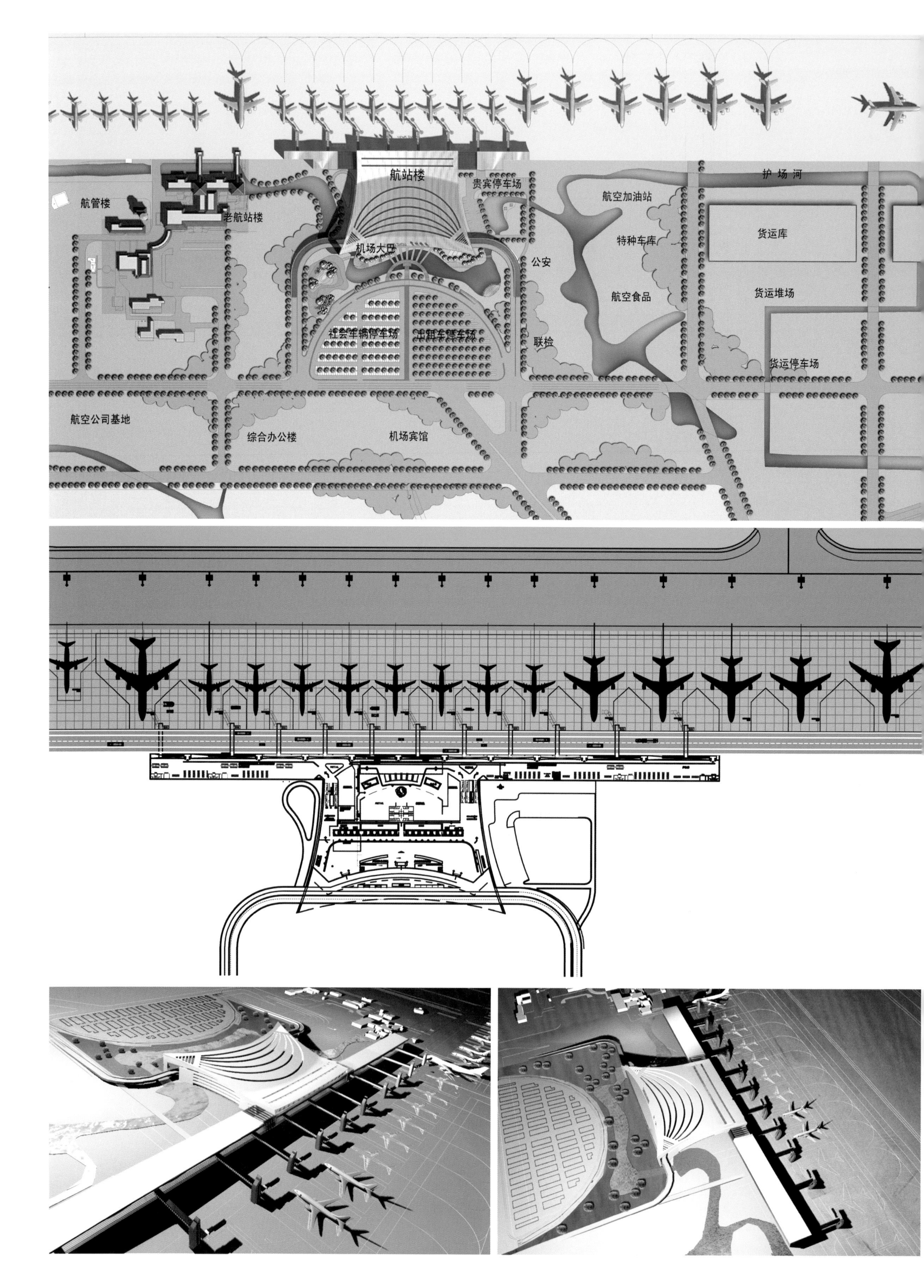
航站楼
贵宾停车场
护场河
航空加油站
航管楼
老航站楼
货运库
特种车库
机场大巴
公安
航空食品
货运堆场
社会车辆停车场
出租车停车场
联检
货运停车场
航空公司基地
综合办公楼
机场宾馆

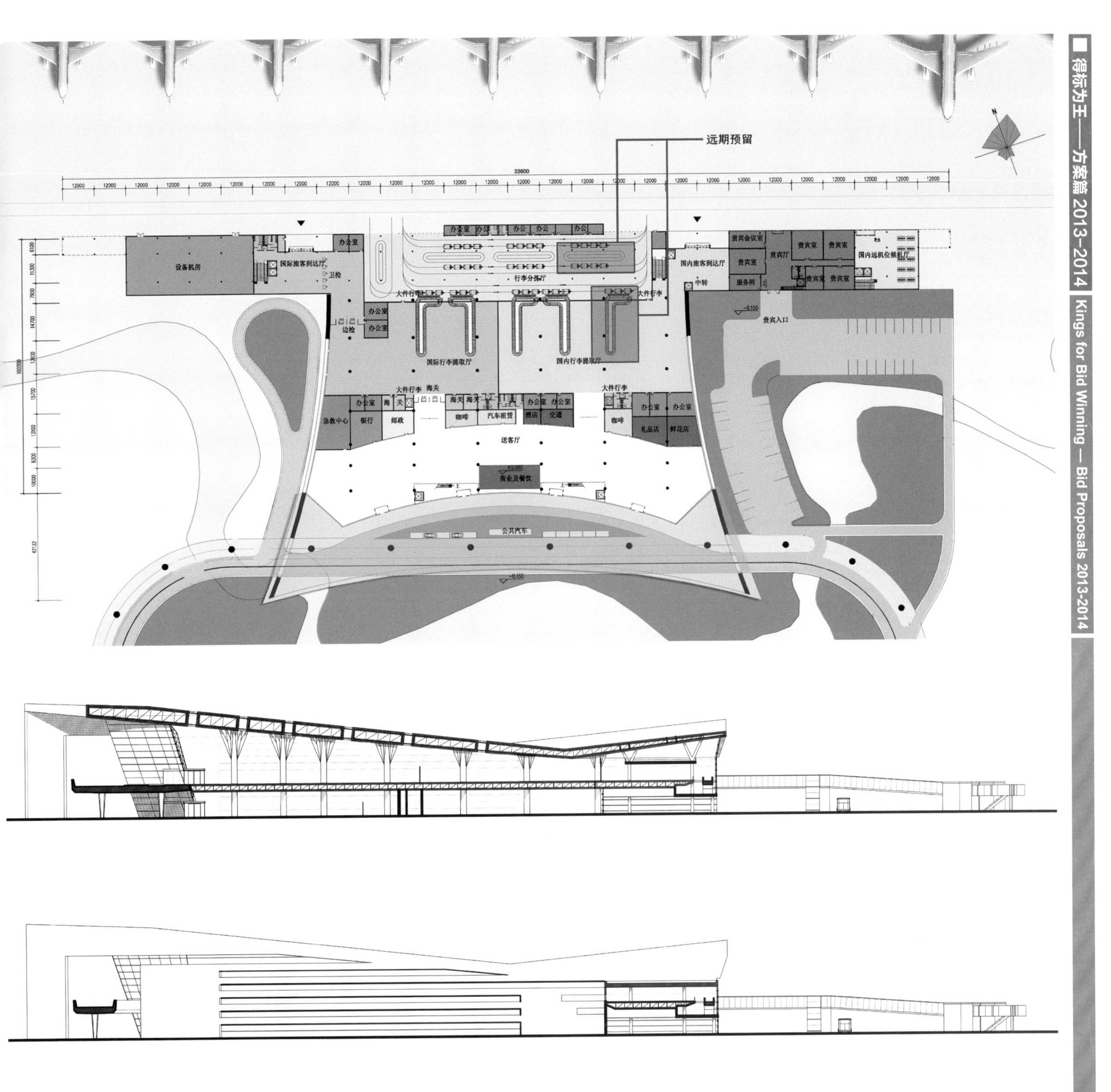
远期预留
设备机房
办公室
国际旅客到达厅
卫检
边检
行李分拣厅
大件行李
国际行李提取厅
国内行李提取厅
国内旅客到达厅
中转
贵宾会议室
贵宾室
贵宾厅
服务间
国内远机位候机厅
贵宾入口
海关
急救中心
银行
邮政
咖啡
汽车租赁
酒店
交通
礼品店
鲜花店
送客厅
商业及餐饮
公共汽车

CHANG ZHOU

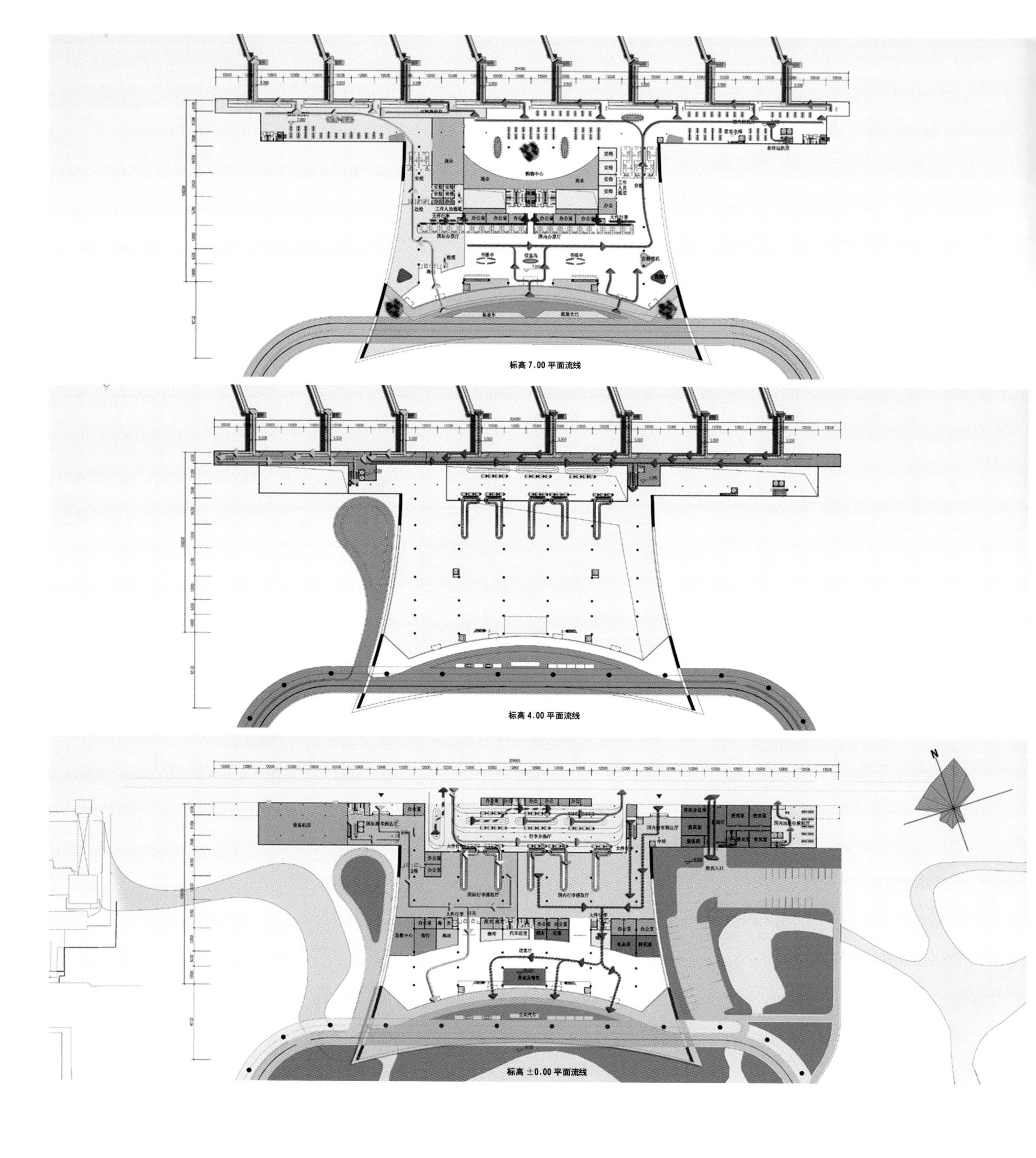

标高 7.00 平面流线

标高 4.00 平面流线

标高 ±0.00 平面流线

图例

贵宾出发　贵宾到达　国际旅客出发　国际旅客到达　国际行李出发

国际行李到达　国内旅客出发　国内旅客到达　国内行李出发　国内行李到达

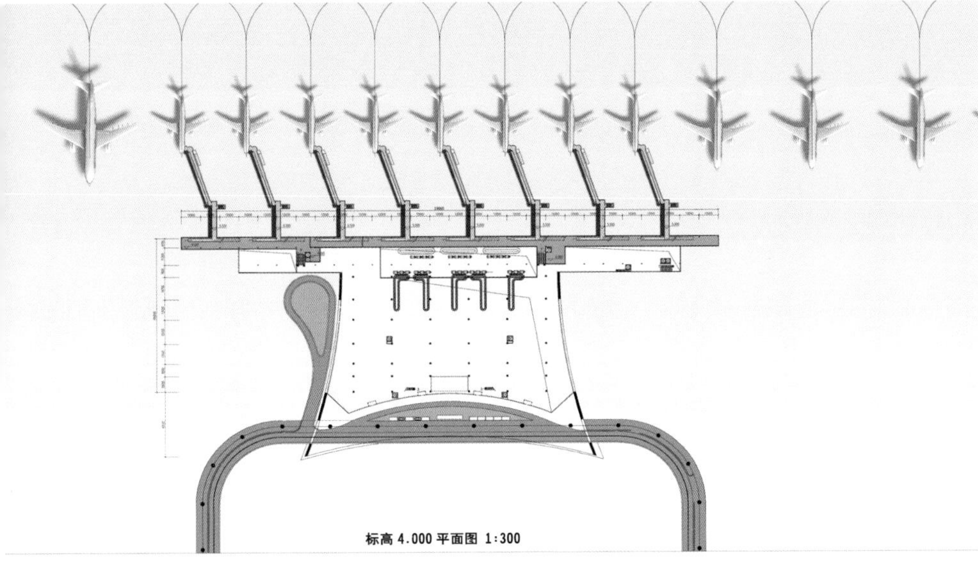

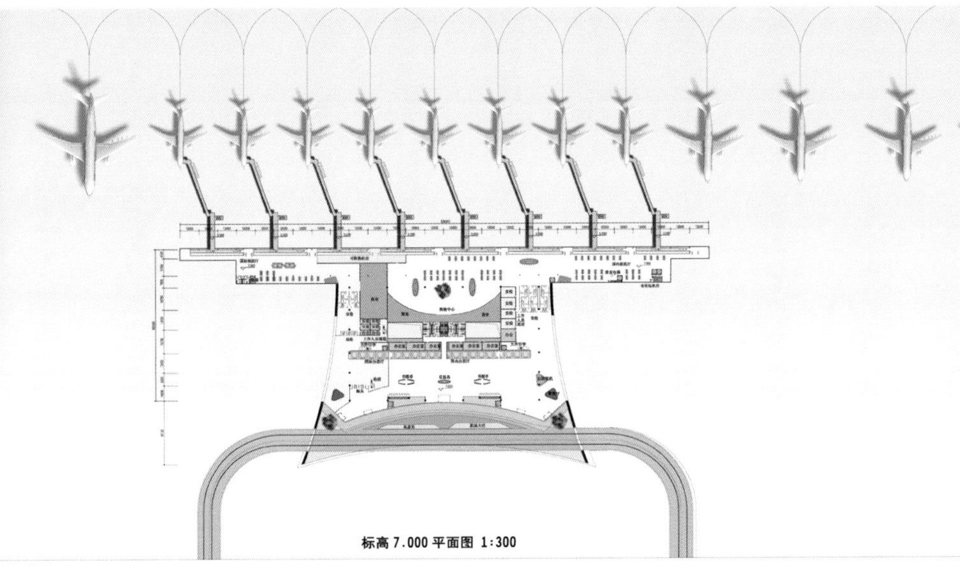

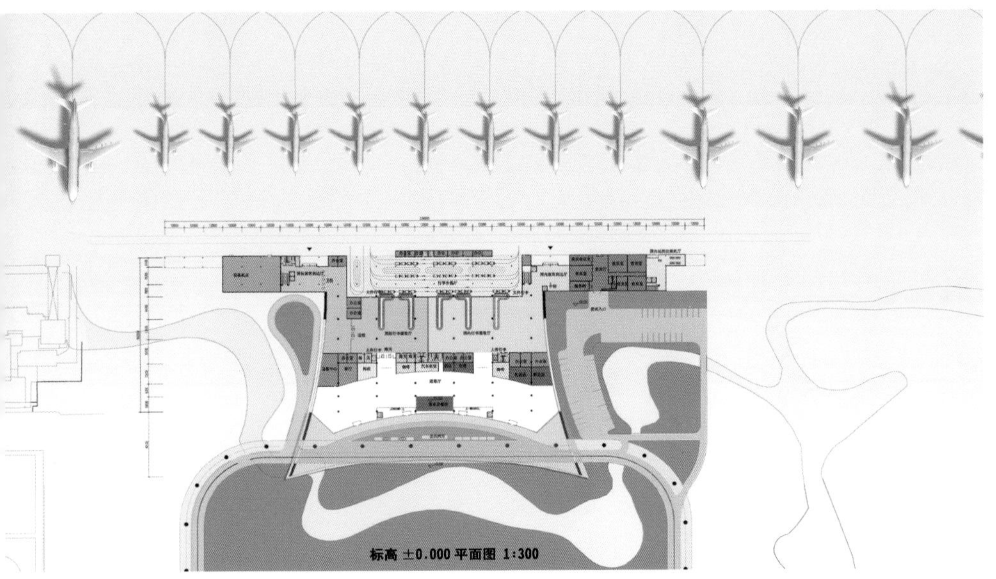

Profile

Changzhou Airport is located in Luoxi Town, Xinbei District, Changzhou City, adjoining Shanghai-Nanjing Expressway and Changjiang River to the north, close to Shanghai-Nanjing Railway and Beijing-Hangzhou Grand Canal to the west, with convenient transportation and superior location. Since infrastructure construction of the existing airport is severely lagged behind, passenger throughput remains low in Changzhou Airport, which limits the sustainable development of Changzhou. To remain market competitiveness and to promote a rapid continuous economic development of Changzhou, even Yangtze River Delta Area, Changzhou Municipal Government decides to accelerate reconstruction and extension of Changzhou Airport.

Design Feature

Architectural Structure

Main building structure of Changzhou Airport expansion is framework structure. Building roof adopts reinforced concrete locally. To satisfy specific requirements of architectural style, departure hall and waiting corridor shall all use steel roof trusses which form open bright check-in hall and waiting hall. High space with panoramic French windows offers fine views for passengers indoors.

Low Carbon Design

Large cornice of the Terminal Building looks like a "Sun Hat" which greatly decreases internal solar radiation and preserves conditions for proper usage of sunlight. The second floor of main Terminal Building and the airside of waiting hall have hollow low-e glass facade. Such glass façade can achieve natural lighting while at the same time effectively insulate sounds and heats. A 150 KW photovoltaic glass façade grid-connected power station set on the roof of waiting corridor supplies electricity for underground and landscape lighting. In the meantime, a rainwater collection system on the roof filters rainwater which inflows to landscape lake in front of the Terminal Building.

Humanized Design

Except functional requirements, the design has put forward process design scheme based on existing facilities and future expansion. Simple and direct circulation in Terminal Building minimizes the need to move between floors in the Terminal Building. Escalators and lifts bring passengers and their baggage to different floors. Every process which involves going upstairs or downstairs has arranged barrier-free elevator or lane. Also, all services provided there have set barrier-free facilities and equipments.

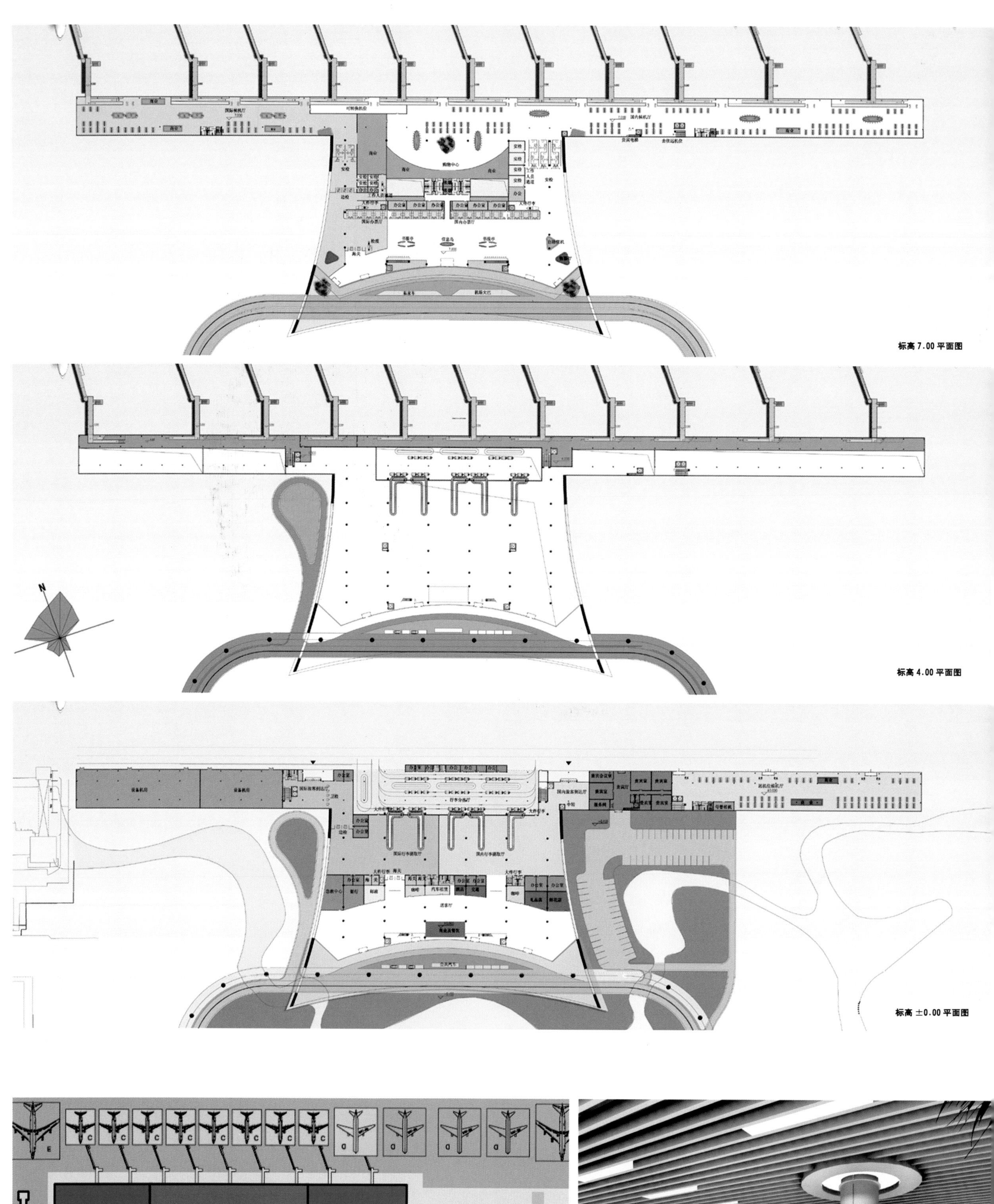

标高 7.00 平面图

标高 4.00 平面图

标高 ±0.00 平面图

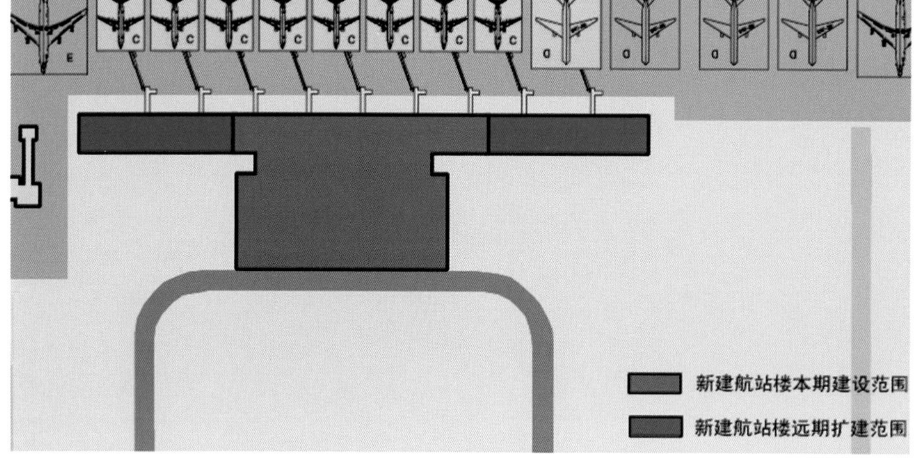

标高 7.00 平面图

标高 4.00 平面图

标高 ±0.00 平面图

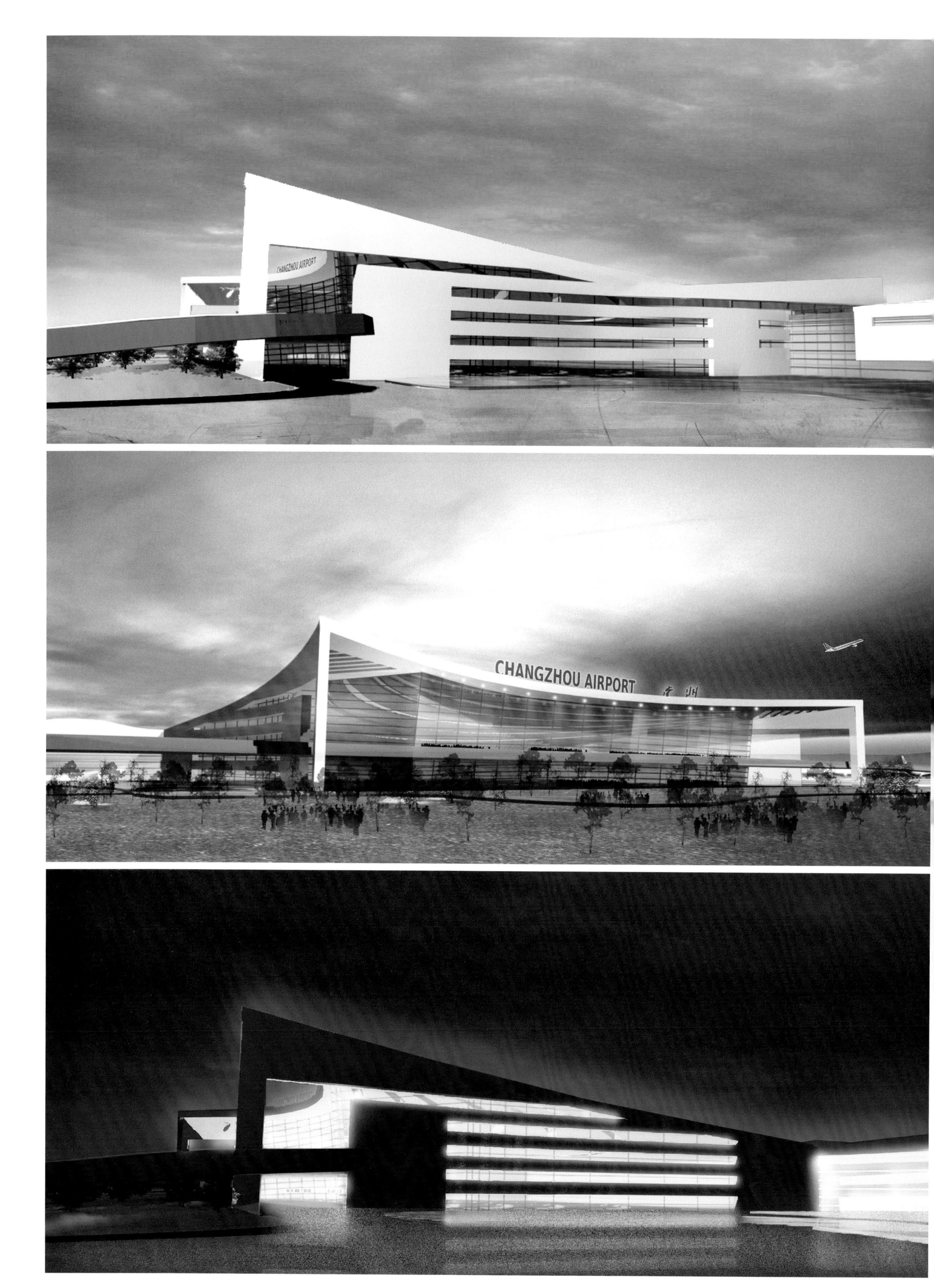
CHANGZHOU AIRPORT
CHANGZHOU AIRPORT

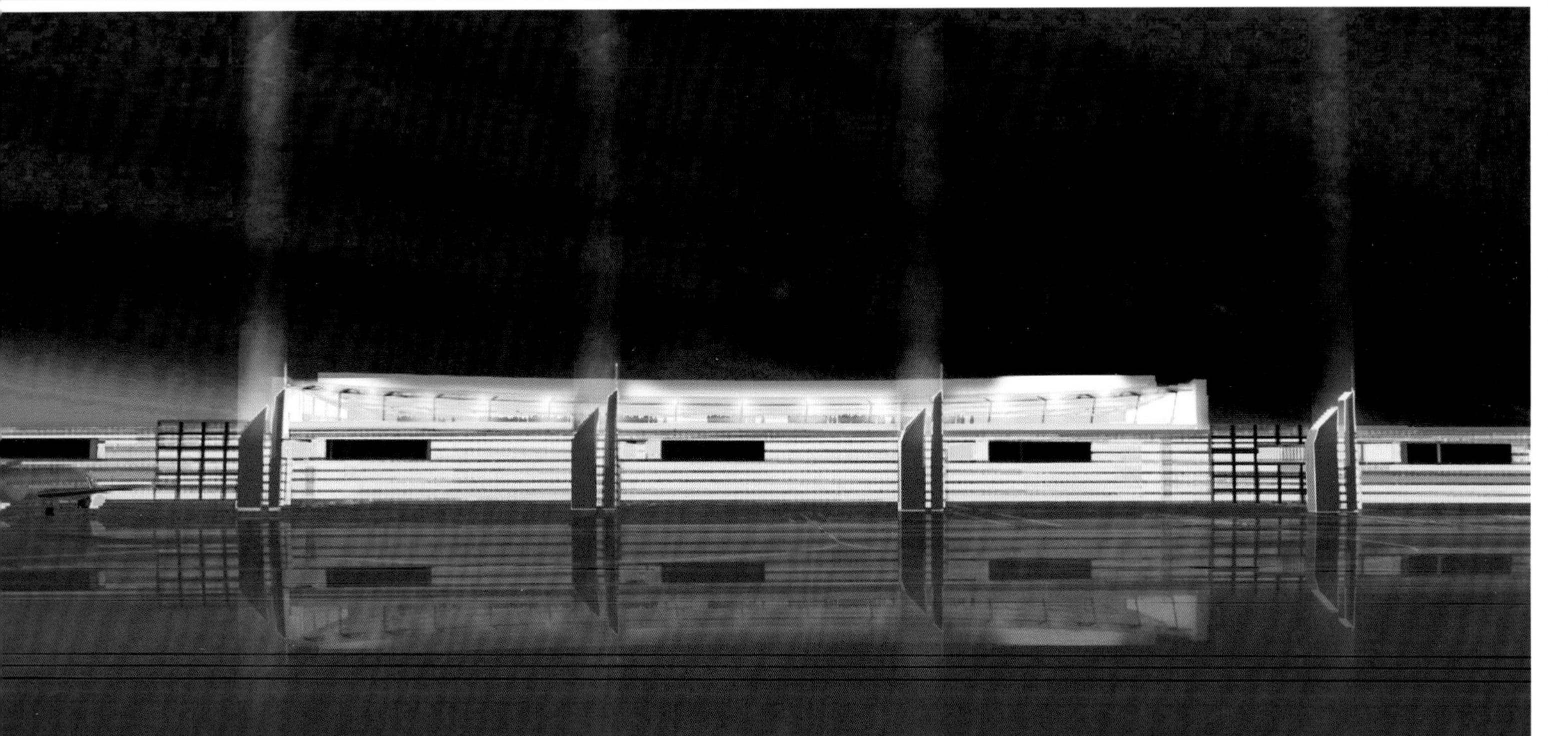

台湾高雄港游轮码头

Kaoshiung Port Cruise Terminal

设计单位：RTA-Office 建筑事务所
　　　　　Santiago Parramón
开发单位：台湾高雄交通和通讯部
项目地址：中国台湾省高雄市
建筑面积：53 448 m²
设计团队：Santiago Parramón　Simona Assiero Brá
　　　　　Mariana Rapela　Miguel Vilacha
　　　　　Luisa Garcia　Eduardo Vacotto
摄影：RTA-Office 建筑事务所

Designed by: RTA-Office; Santiago Parramón
Client: Ministry of Transportation and Communication, Kaoshiung, Taiwan
Location: Kaoshiung, Taiwan, China
Built-up Area: 53,448 m²
Design Team: Santiago Parramón, Simona Assiero Brá, Mariana Rapela, Miguel Vilacha, Luisa Garcia, Eduardo Vacotto
Photography: RTA-Office

项目概况

项目旨在重现消失在视线中的海洋，还原静默的沉思，并以此为至关重要的因素，使之成为发现自我和了解他人的媒介。

设计理念

设计旨在“了解此地”，并通过对基地的了解，在不具备建造 P&CSC 的地形的高雄，提供可建造 P&CSC 的方案。这一方案并非要吸收这座城市已形成的肌理，而是将其从建筑结构中分离，并以自然顺序寻找其自身的体系，建立景观理念。

设计特色

方案中提出的地形和建筑物的母线更多地体现在体积上，而非线性上——空间创造与外形创造相对立，可以欣赏到整个大型建筑物的外观，便于理解其作为地标的意义。而在建筑内部，则给人一种空间的晕眩感，增加游客的核心体验。

建筑物的表皮是项目的重要元素，其构成可产生一种光学效应，从而扩大或缩小建筑的视野。当处于较远的位置时，可将其看作一个单一的物体；而处于较近的位置时，则会将其看成多个物体，这种观感可充实顾客的体验。另一方面，钛表面水的倒影可在建筑内部形成灯光与映射光相互映衬的效果，以吸引行人的目光，并使建筑空间达到预先设想的自然顺序。

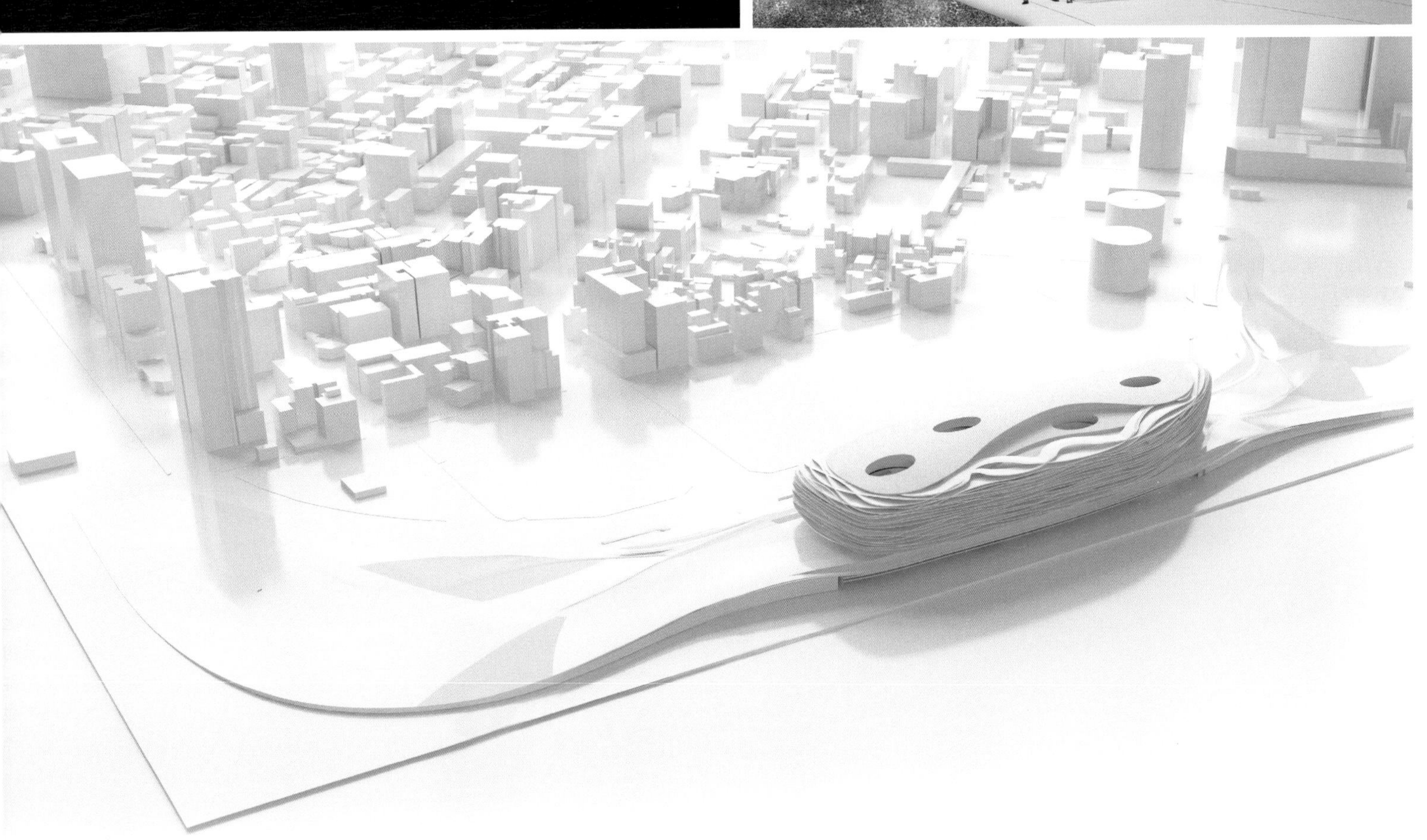

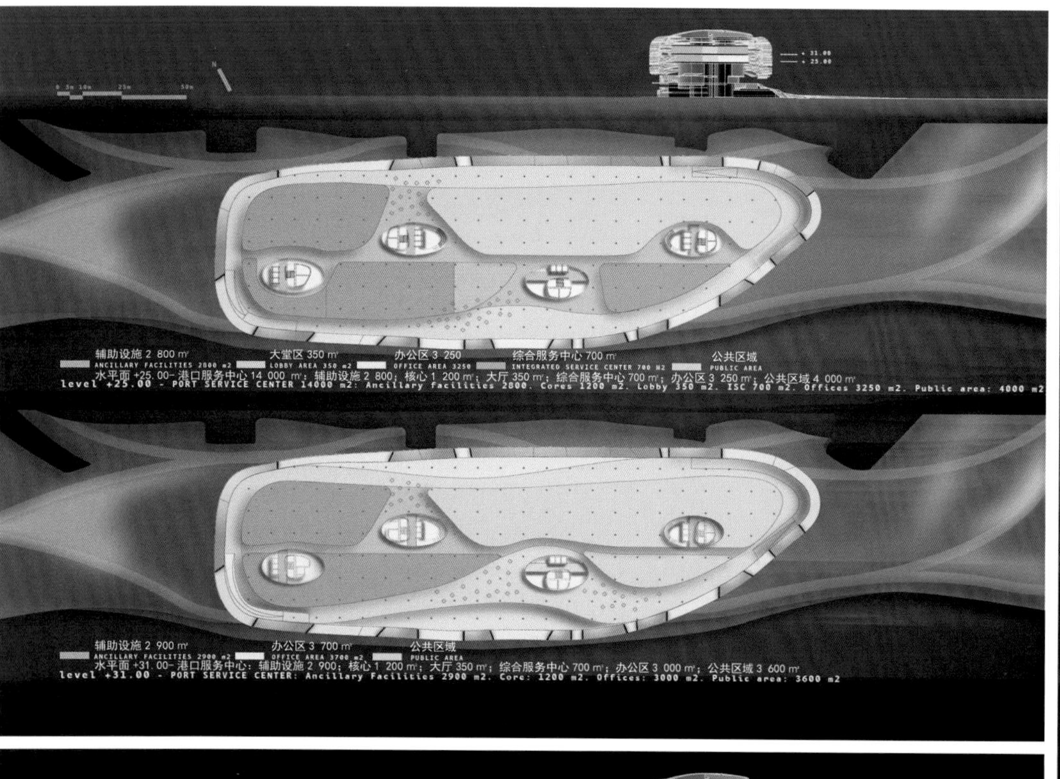
辅助设施 2 800 m²
ANCILLARY FACILITIES 2800 m2
大堂区 350 m²
LOBBY AREA 350 m2
办公区 3 250
OFFICE AREA 3250
综合服务中心 700 m²
INTEGRATED SERVICE CENTER 700 M2
公共区域
PUBLIC AREA
水平面 +25. 00- 港口服务中心 14 000 m²；辅助设施 2 800；核心 1 200 m²；大厅 350 m²；综合服务中心 700 m²；办公区 3 250 m²；公共区域 4 000 m²
level +25.00 - PORT SERVICE CENTER 14000 m2: Ancillary facilities 2800. Cores 1200 m2. Lobby 350 m2. ISC 700 m2. Offices 3250 m2. Public area: 4000 m2
辅助设施 2 900 m²
ANCILLARY FACILITIES 2900 m2
办公区 3 700 m²
OFFICE AREA 3700 m2
公共区域
PUBLIC AREA
水平面 +31. 00- 港口服务中心：辅助设施 2 900；核心 1 200 m²；大厅 350 m²；综合服务中心 700 m²；办公区 3 000 m²；公共区域 3 600 m²
level +31.00 - PORT SERVICE CENTER: Ancillary Facilities 2900 m2. Core: 1200 m2. Offices: 3000 m2. Public area: 3600 m2

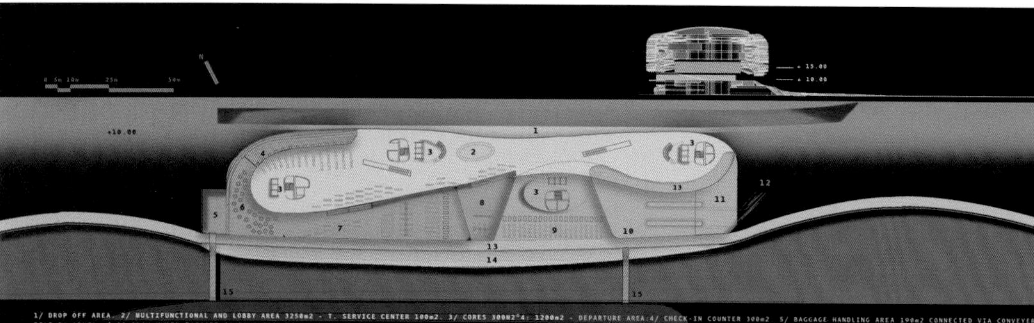
1/ DROP OFF AREA. 2/ MULTIFUNCTIONAL AND LOBBY AREA 3250m2 - T. SERVICE CENTER 100m2. 3/ CORES 300M2*4: 1200m2 - DEPARTURE AREA 4/ CHECK-IN COUNTER 300m2. 5/ BAGGAGE HANDLING AREA 190m2 CONNECTED VIA CONVEYER BELT TO LEVEL +6.60 FROM WHERE BAGGAGE IS DELIVERED TO QUAYSIDE OPERATION LINE. 6/ COMMERCIAL AREA 420m2. 7/ INTERNATIONAL DEPARTURE HALL 1000m2. 8/ DOMESTIC DEPARTURE HALL 390m2 - ARRIVALS (CONTROLLED ZONE: 9/ ARRIVAL HALL 700m2. 10/ BAGGAGE CLAIM 720m2). 11/ BAGGAGE HANDLING AREA & STORE 400m2. 12/ CONVEYER BELT TO +6.60. 13/ CUSTOM 250 m2. 14/ COVERED CATWALK: CONNECTION TO BERTHS 19, 21. 15/ MOVABLE GANGWAYS
level +10.00 - CRUISE TERMINAL 8580 m2: Lobby and multifunction area 3250 m2. Cores 1200 m2. Departure area: 2300 m2. Arrival area: 2100 m2
1. / 下客区；2/ 多功能大厅区 3 250 m² - 售票中心 100 m²；3/ 核心 300 m² *4:1200m2- 离港区；登记处 300 m²；5/190 m²行李处理区通过传送带与水平面 +6. 60 相连，从行李交付处到码头操作线的位置
6/ 商业区 420 m²；7/ 国际出发大厅 1 000 m²；8/ 国内出发大厅 390 m² - 进站（控制区：9/ 到达厅 700 m²；10/ 行李领取处 720 m²）；11/ 行李处理区 & 储藏区 400 m²；12/ 传送带到达 +6. 60；13/ 海关 250 m²
14/ 有屋顶遮盖的步行道：与 19，21 泊位相连；15/ 可移动舷梯
水平面 +10. 00- 邮轮码头 8 580 m²；大厅和多功能区 3 250 m²；核心 1 200 m²；离港区：2 300 m²；到达区 2 100 m²

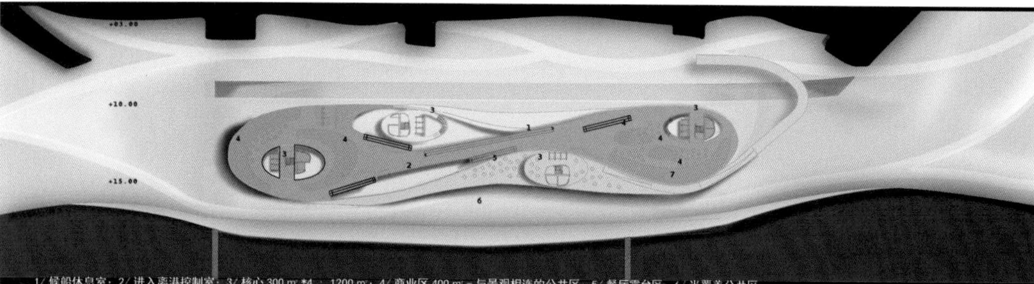
1/ 候船休息室；2/ 进入离港控制室；3/ 核心 300 m² *4 ：1200 m²；4/ 商业区 400 m² - 与景观相连的公共区；5/ 餐厅露台区；6/ 半覆盖公共区；
1/ WAITING LOUNGE AREA. 2/ ACCESS TO DEPARTURE CONTROLLED ZONE. 3/ CORES 300M2*4: 1200m2. 4/ COMMERCIAL AREA 400m2 - PUBLIC AREA CONNECTED TO LANDSCAPE. 5/ RESTAURANT TERRACE AREA. 6/ SEMI COVERED PUBLIC SPACE
7/ PUBLIC CIRCULATION: RAMP TO UPPER LEVELS. 7/ 公共流通线路：通往上层空间的坡道
level +15.00 - CRUISE TERMINAL 5200 m2: Boarding area: 3450 m2. Cores 900 m2. Commercial area 900m2.
水平面 +15. 00- 邮轮码头 5 200 m²；登船区：3450 m²；核心 900 m²；商业区 900 m²。

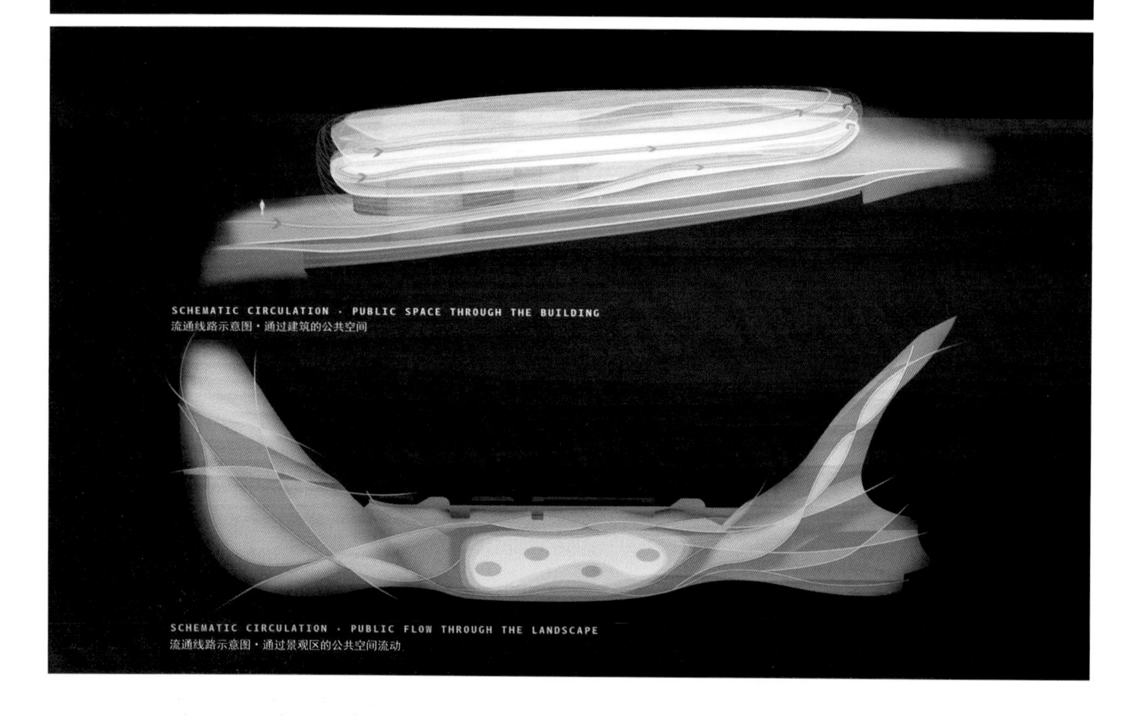
SCHEMATIC CIRCULATION · PUBLIC SPACE THROUGH THE BUILDING
流通线路示意图 • 通过建筑的公共空间
SCHEMATIC CIRCULATION · PUBLIC FLOW THROUGH THE LANDSCAPE
流通线路示意图 • 通过景观区的公共空间流动

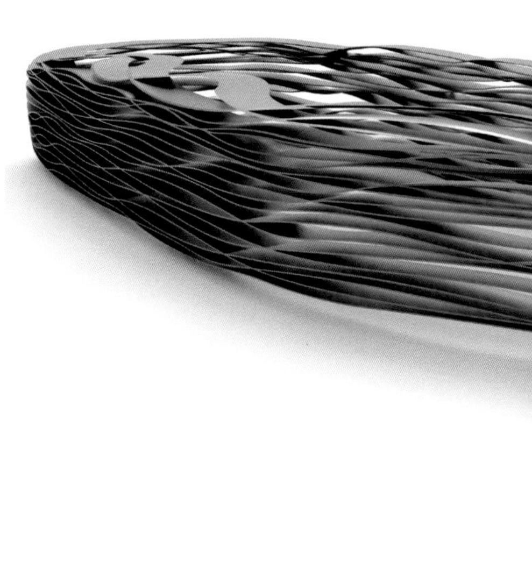

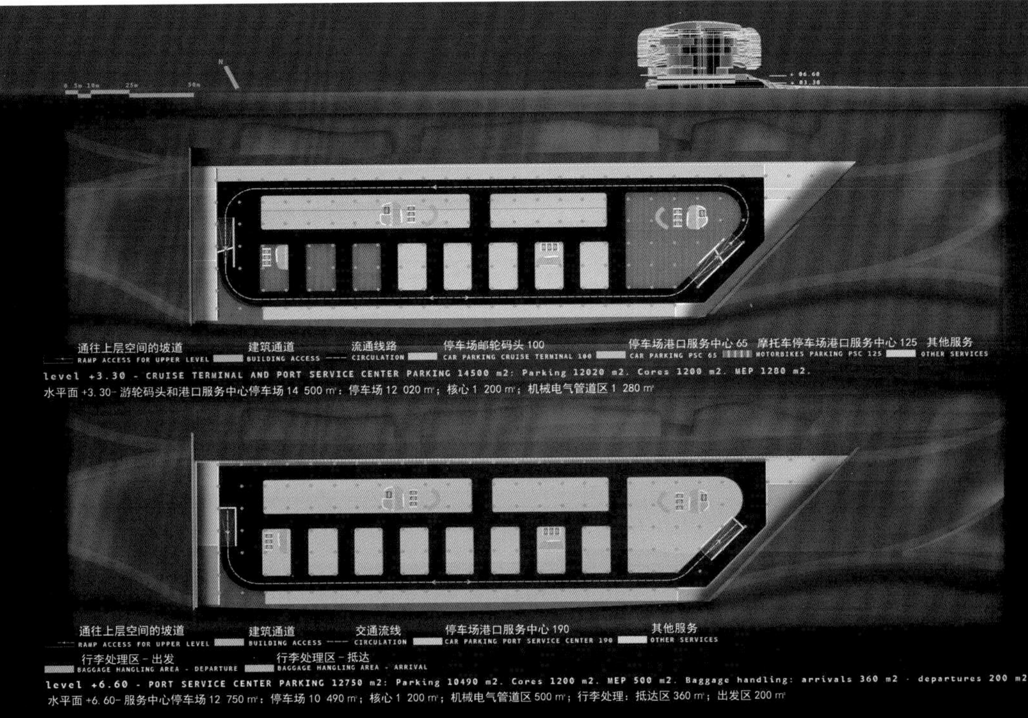

通往上层空间的坡道
RAMP ACCESS FOR UPPER LEVEL
建筑通道
BUILDING ACCESS
流通线路
CIRCULATION
停车场邮轮码头 100
CAR PARKING CRUISE TERMINAL 100
停车场港口服务中心 65
CAR PARKING PSC 65
摩托车停车场港口服务中心 125
MOTORBIKES PARKING PSC 125
其他服务
OTHER SERVICES
level +3.30 - CRUISE TERMINAL AND PORT SERVICE CENTER PARKING 14500 m2: Parking 12020 m2. Cores 1200 m2. MEP 1280 m2.
水平面 +3.30- 游轮码头和港口服务中心停车场 14 500 ㎡；停车场 12 020 ㎡；核心 1 200 ㎡；机械电气管道区 1 280 ㎡
通往上层空间的坡道
RAMP ACCESS FOR UPPER LEVEL
建筑通道
BUILDING ACCESS
交通流线
CIRCULATION
停车场港口服务中心 190
CAR PARKING PORT SERVICE CENTER 190
其他服务
OTHER SERVICES
行李处理区 - 出发
BAGGAGE HANGLING AREA - DEPARTURE
行李处理区 - 抵达
BAGGAGE HANGLING AREA - ARRIVAL
level +6.60 - PORT SERVICE CENTER PARKING 12750 m2: Parking 10490 m2. Cores 1200 m2. MEP 500 m2. Baggage handling: arrivals 360 m2 · departures 200 m2
水平面 +6.60- 服务中心停车场 12 750 ㎡；停车场 10 490 ㎡；核心 1 200 ㎡；机械电气管道区 500 ㎡；行李处理：抵达区 360 ㎡；出发区 200 ㎡

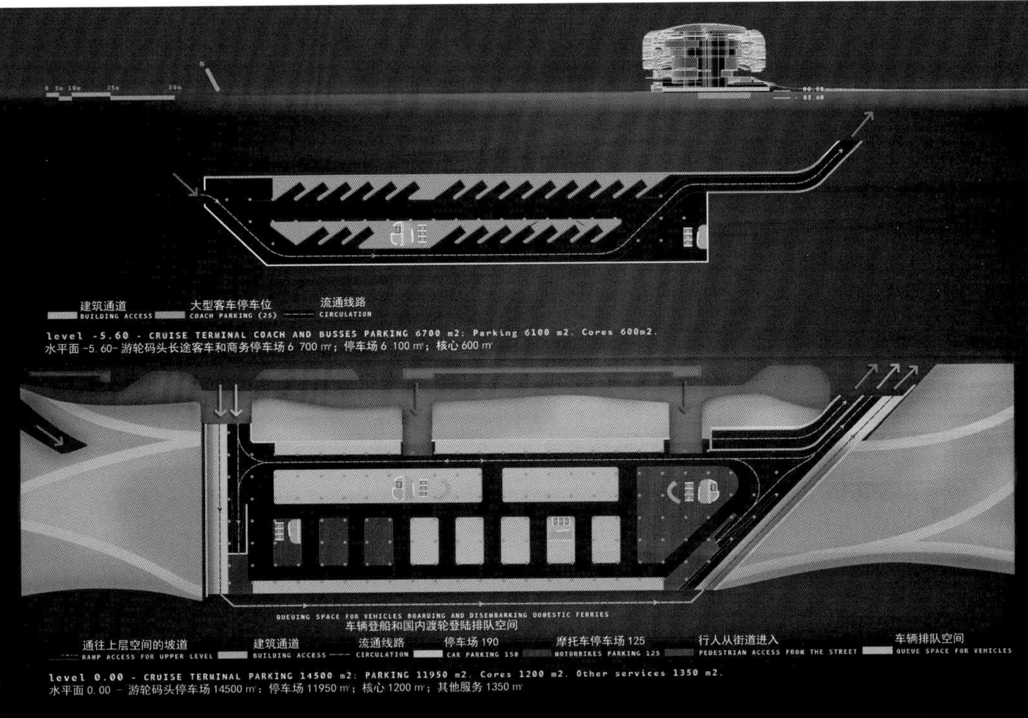

建筑通道
BUILDING ACCESS
大型客车停车位
COACH PARKING (25)
流通线路
CIRCULATION
level -5.60 - CRUISE TERMINAL COACH AND BUSSES PARKING 6700 m2: Parking 6100 m2. Cores 600m2.
水平面 -5.60- 游轮码头长途客车和商务停车场 6 700 ㎡；停车场 6 100 ㎡；核心 600 ㎡
QUEUING SPACE FOR VEHICLES BOARDING AND DISEMBARKING DOMESTIC FERRIES
车辆登船和国内渡轮登陆排队空间
通往上层空间的坡道
RAMP ACCESS FOR UPPER LEVEL
建筑通道
BUILDING ACCESS
流通线路
CIRCULATION
停车场 190
CAR PARKING 150
摩托车停车场 125
MOTORBIKES PARKING 125
行人从街道进入
PEDESTRIAN ACCESS FROM THE STREET
车辆排队空间
QUEUE SPACE FOR VEHICLES
level 0.00 - CRUISE TERMINAL PARKING 14500 m2: PARKING 11950 m2. Cores 1200 m2. Other services 1350 m2.
水平面 0.00 – 游轮码头停车场 14500 ㎡；停车场 11950 ㎡；核心 1200 ㎡；其他服务 1350 ㎡

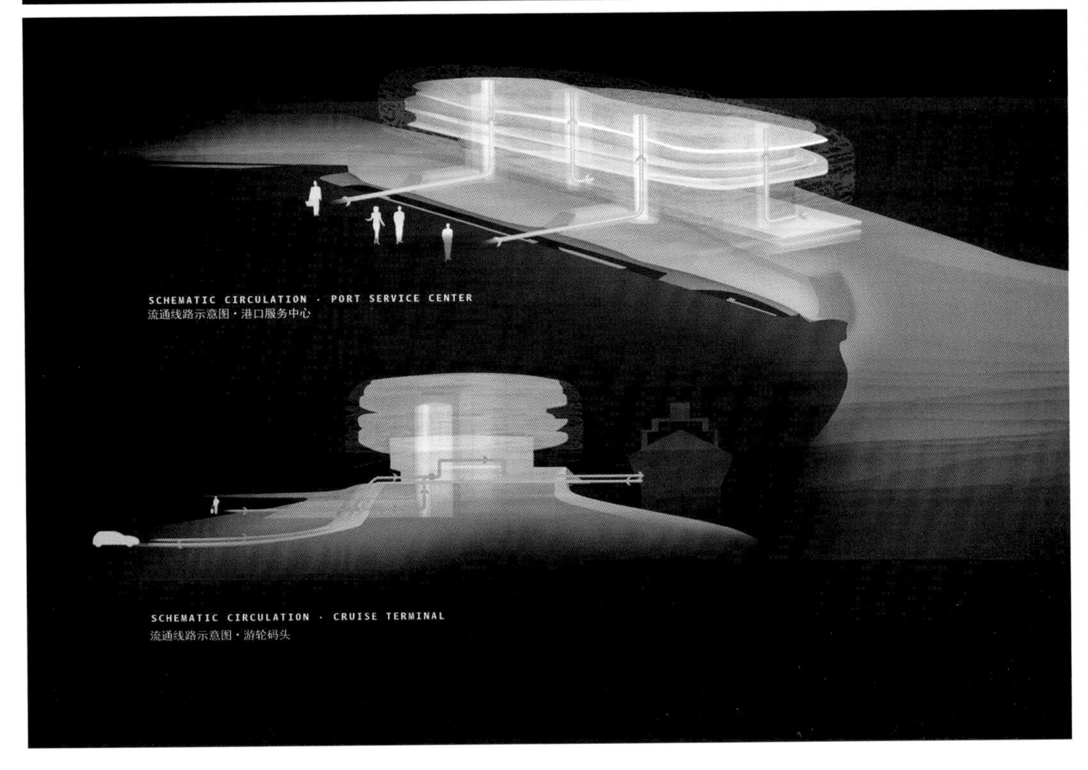

SCHEMATIC CIRCULATION · PORT SERVICE CENTER
流通线路示意图 · 港口服务中心

SCHEMATIC CIRCULATION · CRUISE TERMINAL
流通线路示意图 · 游轮码头

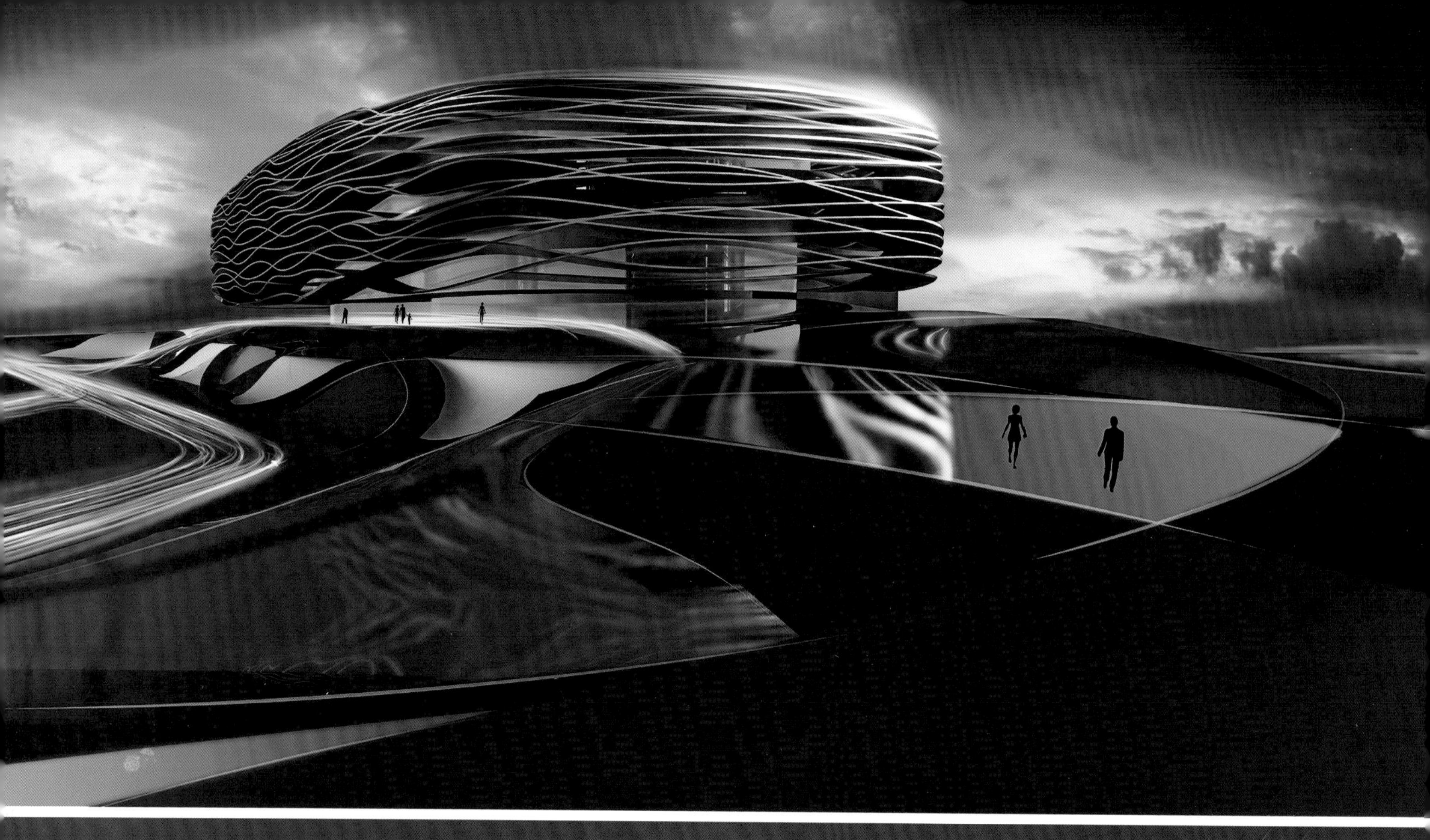

Profile

This complex seeks to recover the lost sight of the sea, and to restore the silence of contemplation as a vital and essential factor to discover ourselves and others.

Design Concept

The proposal is aimed at "Understanding the Place" and, as a result of this understanding, providing an answer to the lack of topography in the area of the city of Kaohsiung in Taiwan, where the P&CSC is to be built. This concept is not meant to be absorbed within the established magma of the city, but to emerge as a geographical element; therefore, it detaches itself from architecture to find its genealogy in the natural order to establish the idea of a landscape.

Design Feature

The generatrix of the topography and the building starts from a proposal that is more volumetric than linear – the creation of space against the creation of shapes. From the outside, we appreciate the whole magnitude of the building in order to understand it all at once as a landmark. On the inside, we find the vertigo of space.

The result is the symbiosis of a landscape that houses an object in whose core experiences are multiplied by means of spatial relationships.
The skin of the building is an essential element of the project. Its composition produces an optical effect which narrows or widens the vision of the project. This allows us to interpret it as a single object from a distance and to multiply the sight from a closer position – impressions that complement the passenger's experience. The reflection of water on a titanium surface creates a framework of lights and brightness on the inside of the building that absorbs the individual and relates the architectural space to the natural order where it was conceived.

台湾基隆游轮码头

Cruise Terminal, Keelung, Taiwan

设计单位：de Architekten Cie.
开发商：基隆港务局
项目地址：中国台湾省基隆市

Designed by: de Architekten Cie.
Client: Port Authority, Keelung
Location: Keelung, Taiwan, China

设计理念

联合办公大楼与客运/货运中心的联合，为无缝隙地整合这两个独特的项目提供了策略。设计师认为这两个相辅相成的项目在这个城市的地位，无论是在物理层面还是精神层面上，都将因两者的相互作用而得到加强。设计竭力将二者统一为一个整体，使这一对共生的建筑一起成为基隆海事活动的新标志。

建筑设计

项目场地位于港口码头，其周围 270 度视角都被水体环绕，是构建一个地标建筑的理想且独特的场所。这一场地巧妙地将市区和沿海设施连接起来，故设计选择性地将木质人行道/码头经由室内步行街与港口服务楼、游轮码头以及更远处的购物中心连接起来。

这条有顶覆盖的步行街包含了零售区、咖啡厅、酒吧以及邮轮码头设施，将 24 小时对外开放。这样，这个复合功能项目的影响力将扩展至附近的海事广场。另一方面，当码头的游轮和货船拥堵时，这一连接通道也为使用者提供了通往海事广场的路径。

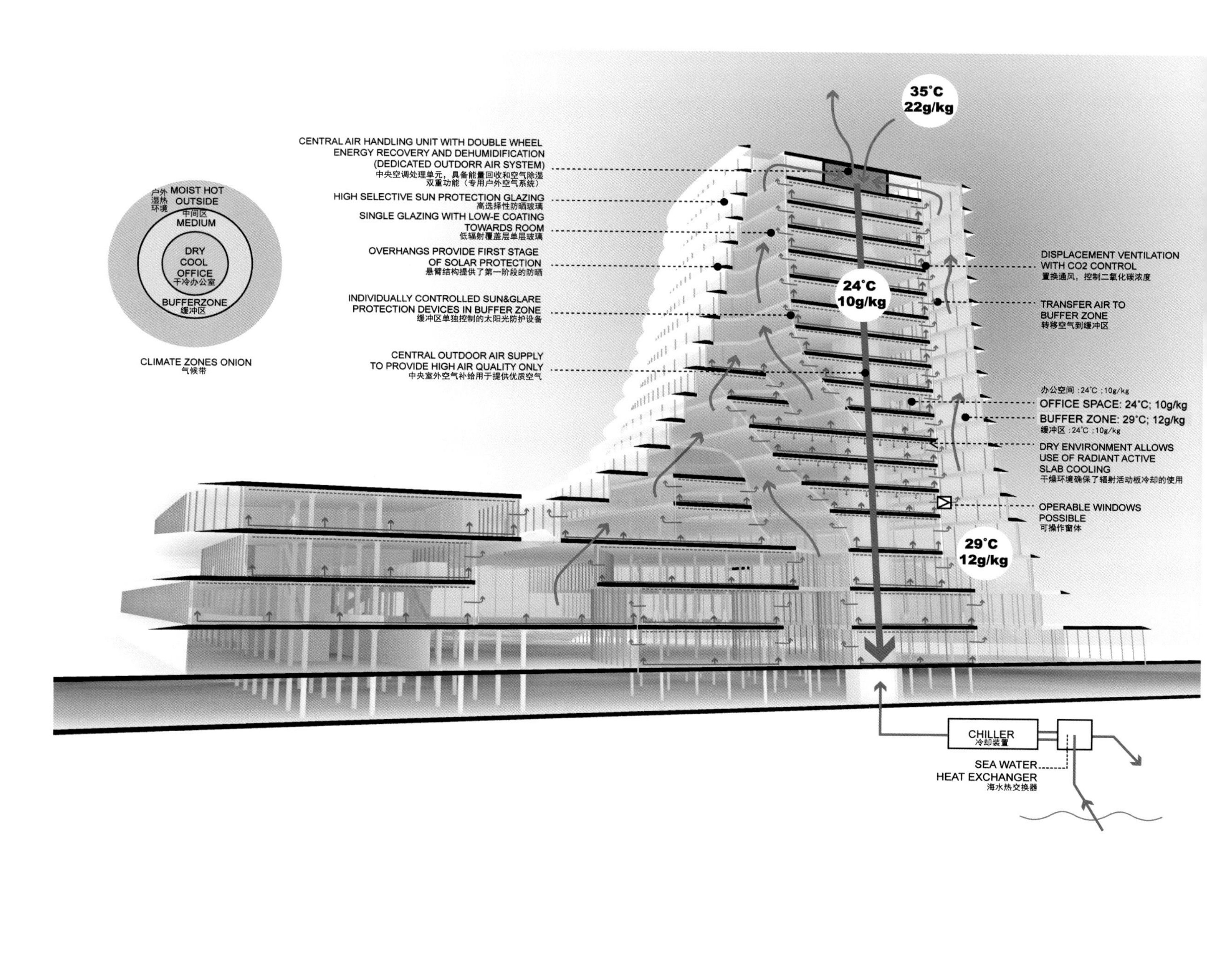

35°C
22g/kg
CENTRAL AIR HANDLING UNIT WITH DOUBLE WHEEL ENERGY RECOVERY AND DEHUMIDIFICATION (DEDICATED OUTDORR AIR SYSTEM)
中央空调处理单元，具备能量回收和空气除湿双重功能（专用户外空气系统）
HIGH SELECTIVE SUN PROTECTION GLAZING
高选择性防晒玻璃
SINGLE GLAZING WITH LOW-E COATING TOWARDS ROOM
低辐射覆盖层单层玻璃
OVERHANGS PROVIDE FIRST STAGE OF SOLAR PROTECTION
悬臂结构提供了第一阶段的防晒
INDIVIDUALLY CONTROLLED SUN&GLARE PROTECTION DEVICES IN BUFFER ZONE
缓冲区单独控制的太阳光防护设备
CENTRAL OUTDOOR AIR SUPPLY TO PROVIDE HIGH AIR QUALITY ONLY
中央室外空气补给用于提供优质空气
户外湿热环境 MOIST HOT OUTSIDE
中间区 MEDIUM
DRY COOL OFFICE 干冷办公室
BUFFERZONE 缓冲区
CLIMATE ZONES ONION
气候带
24°C
10g/kg
DISPLACEMENT VENTILATION WITH CO2 CONTROL
置换通风，控制二氧化碳浓度
TRANSFER AIR TO BUFFER ZONE
转移空气到缓冲区
办公空间：24°C ;10g/kg
OFFICE SPACE: 24°C; 10g/kg
BUFFER ZONE: 29°C; 12g/kg
缓冲区：24°C ;10g/kg
DRY ENVIRONMENT ALLOWS USE OF RADIANT ACTIVE SLAB COOLING
干燥环境确保了辐射活动板冷却的使用
OPERABLE WINDOWS POSSIBLE
可操作窗体
29°C
12g/kg
CHILLER
冷却装置
SEA WATER HEAT EXCHANGER
海水热交换器

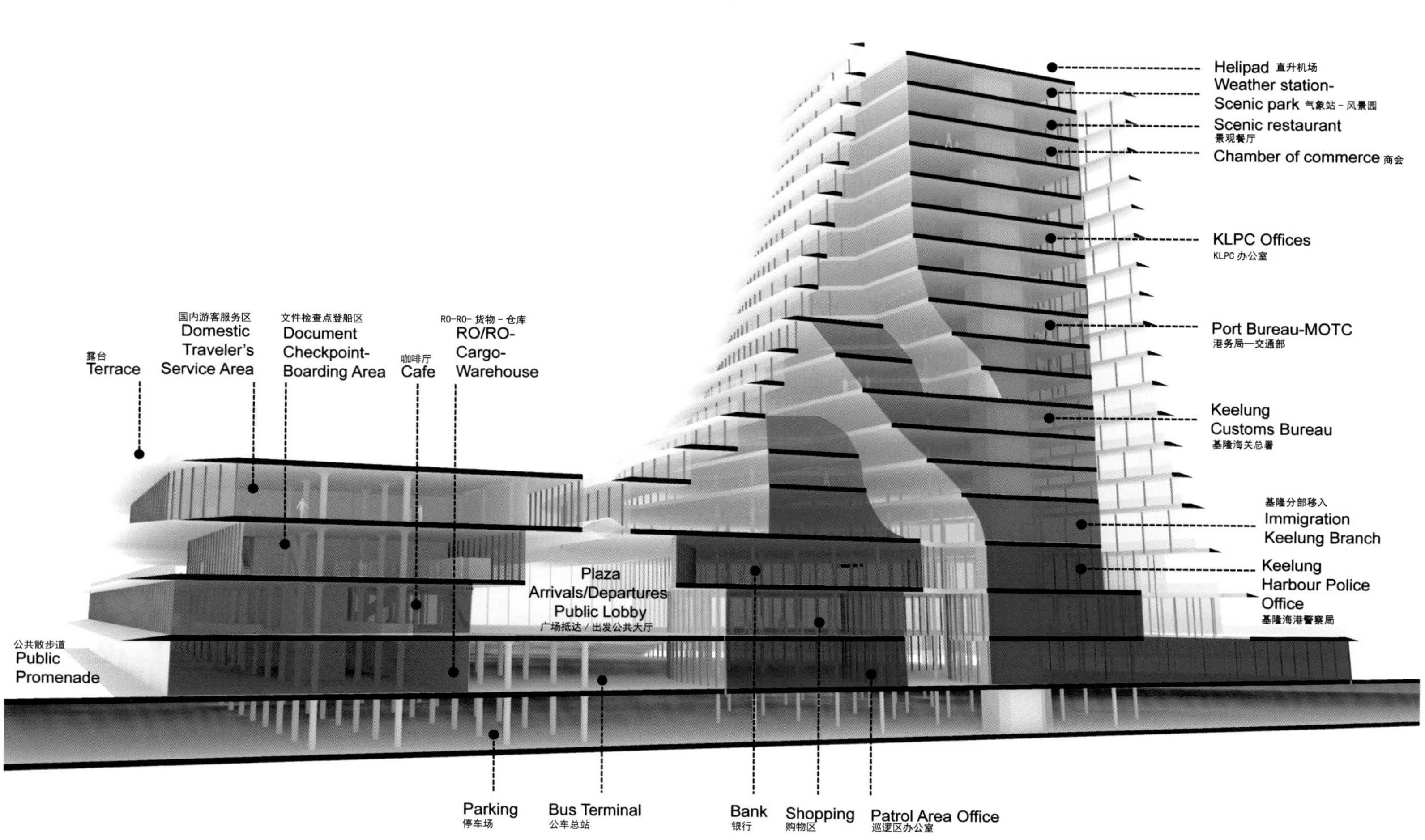

Helipad 直升机场
Weather station-Scenic park 气象站－风景园
Scenic restaurant
景观餐厅
Chamber of commerce 商会
KLPC Offices
KLPC 办公室
Port Bureau-MOTC
港务局—交通部
Keelung Customs Bureau
基隆海关总署
基隆分部移入
Immigration Keelung Branch
Keelung Harbour Police Office
基隆海港警察局
露台
Terrace
国内游客服务区
Domestic Traveler's Service Area
文件检查点登船区
Document Checkpoint-Boarding Area
咖啡厅
Cafe
RO-RO- 货物－仓库
RO/RO-Cargo-Warehouse
Plaza Arrivals/Departures Public Lobby
广场抵达／出发公共大厅
公共散步道
Public Promenade
Parking
停车场
Bus Terminal
公车总站
Bank
银行
Shopping
购物区
Patrol Area Office
巡逻区办公室

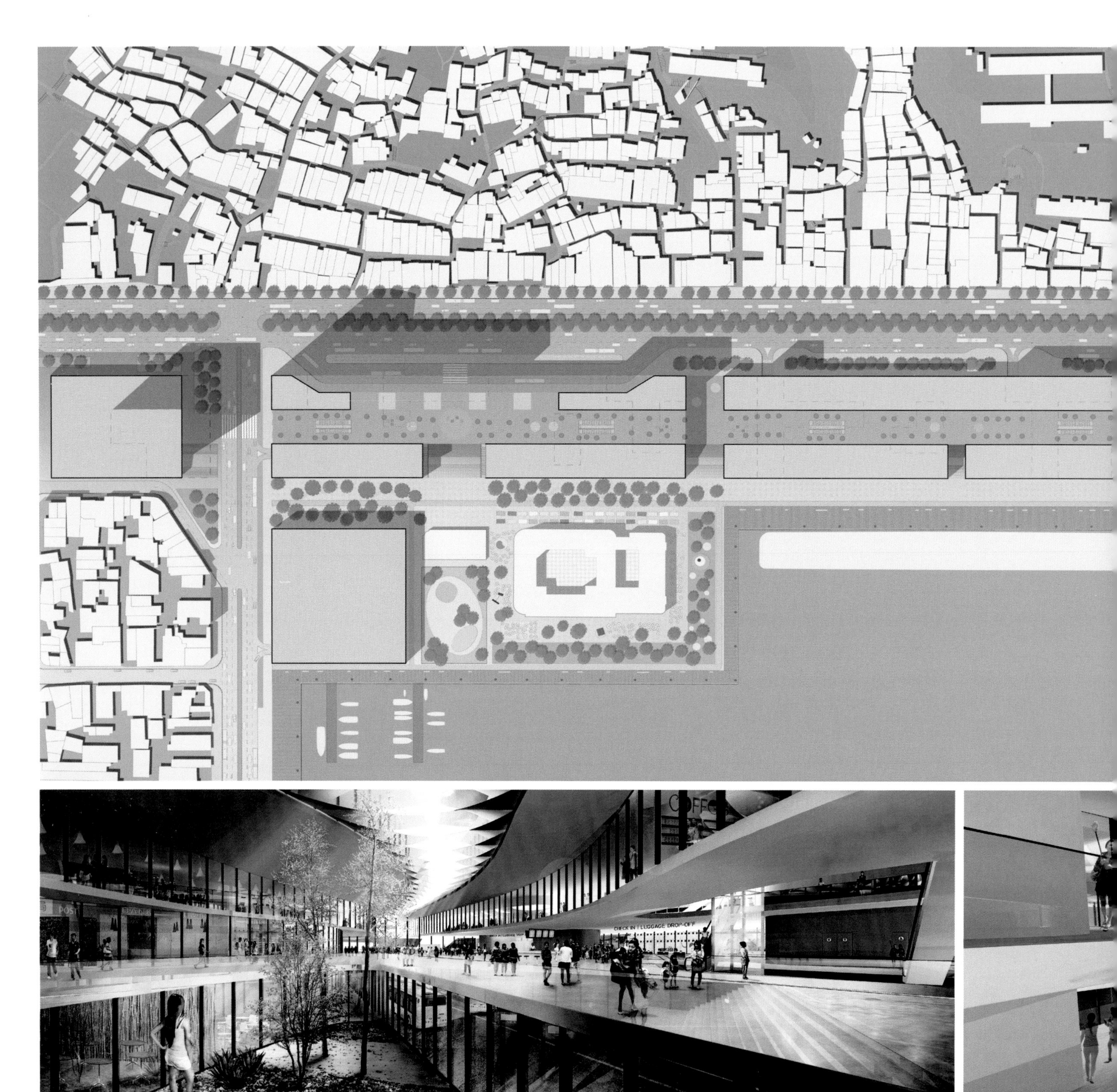
POST
CHECK IN / LUGGAGE DROP-OFF

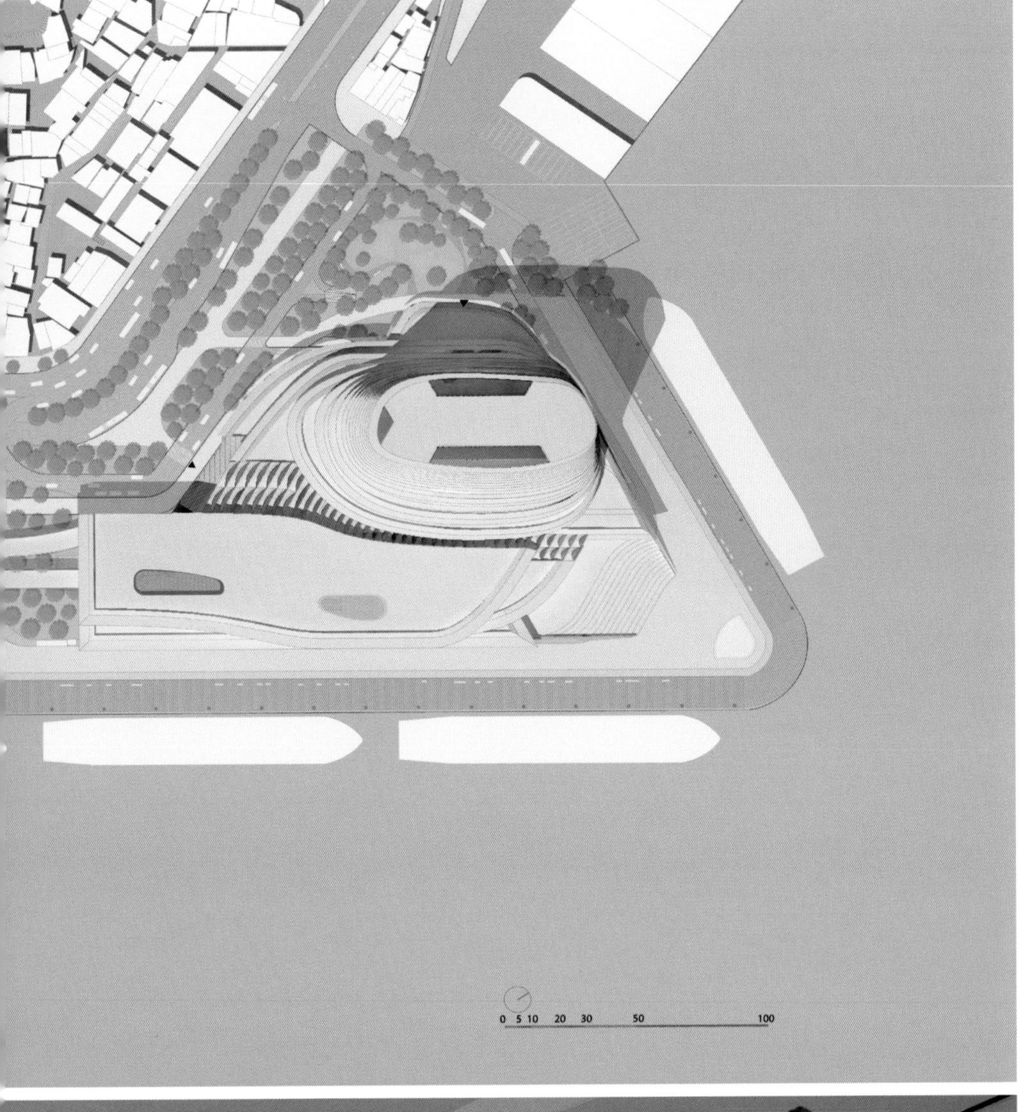

Design Concept

The combination of the Joint Office Building and Passenger & Cargo Terminal creates the ideal opportunity to explore formal strategies which seamlessly integrate the two distinct programmes. They are mutually reinforcing programmes and buildings whose individual position in the city, both physical and psychological, is strengthened through the presence of and interaction with the other. Designers have endeavored to literally and figuratively weave the two programmes together into a single entity. They have designed a symbiotic pair of buildings that combine to form a new icon of maritime activity for the city of Keelung.

Architectural Design

The incredible position of the site at the point of the harbor pier surrounded by 270 degrees of water makes it a truly unique and ideal location for a landmark building. The site is perfectly situated to connect the urban and maritime activities together. As such designers have elected to connect the boardwalk/quay through an interior promenade to the Port Services Building, Cruise Terminal and beyond to the Shopping Mall.

This covered promenade, which contains retail, cafes, bars, and cruise terminal amenities, can remain open 24 hours a day. The influence of this functional complex project shall extend all the way to the maritime gateway plaza. In addition, this connection enables users to access the maritime plaza when the pier is occupied with cruise ship or container vessel activities.

0.60m

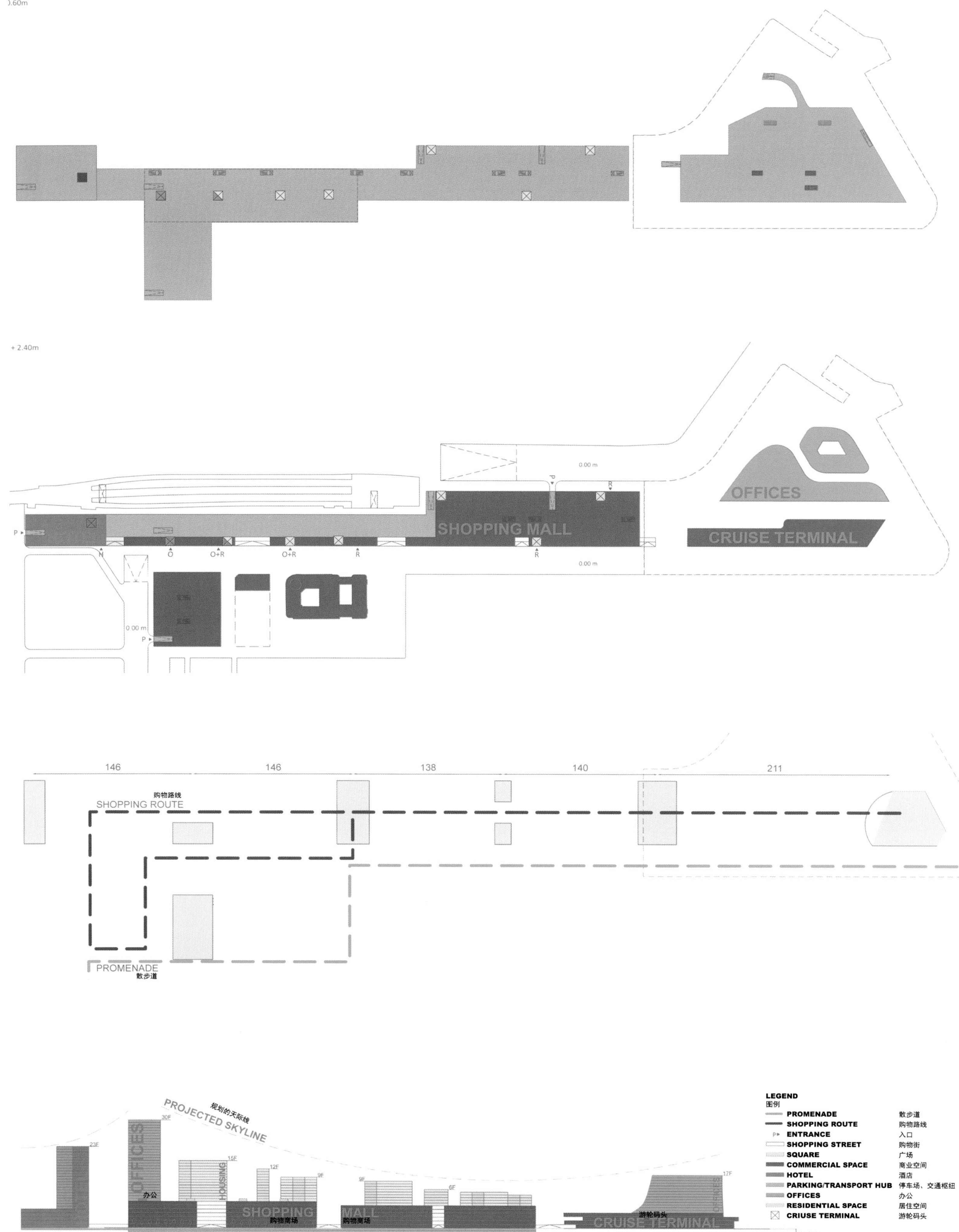
+ 2.40m
0.00 m
SHOPPING MALL
OFFICES
CRUISE TERMINAL
0.00 m
146
146
138
140
211
购物路线
SHOPPING ROUTE
PROMENADE
散步道
规划的天际线
PROJECTED SKYLINE
30F
23F
15F
12F
9F
9F
6F
17F
OFFICES
办公
HOUSING
SHOPPING
MALL
购物商场
购物商场
游轮码头
CRUISE TERMINAL
LEGEND
图例
PROMENADE 散步道
SHOPPING ROUTE 购物路线
ENTRANCE 入口
SHOPPING STREET 购物街
SQUARE 广场
COMMERCIAL SPACE 商业空间
HOTEL 酒店
PARKING/TRANSPORT HUB 停车场、交通枢纽
OFFICES 办公
RESIDENTIAL SPACE 居住空间
CRIUSE TERMINAL 游轮码头

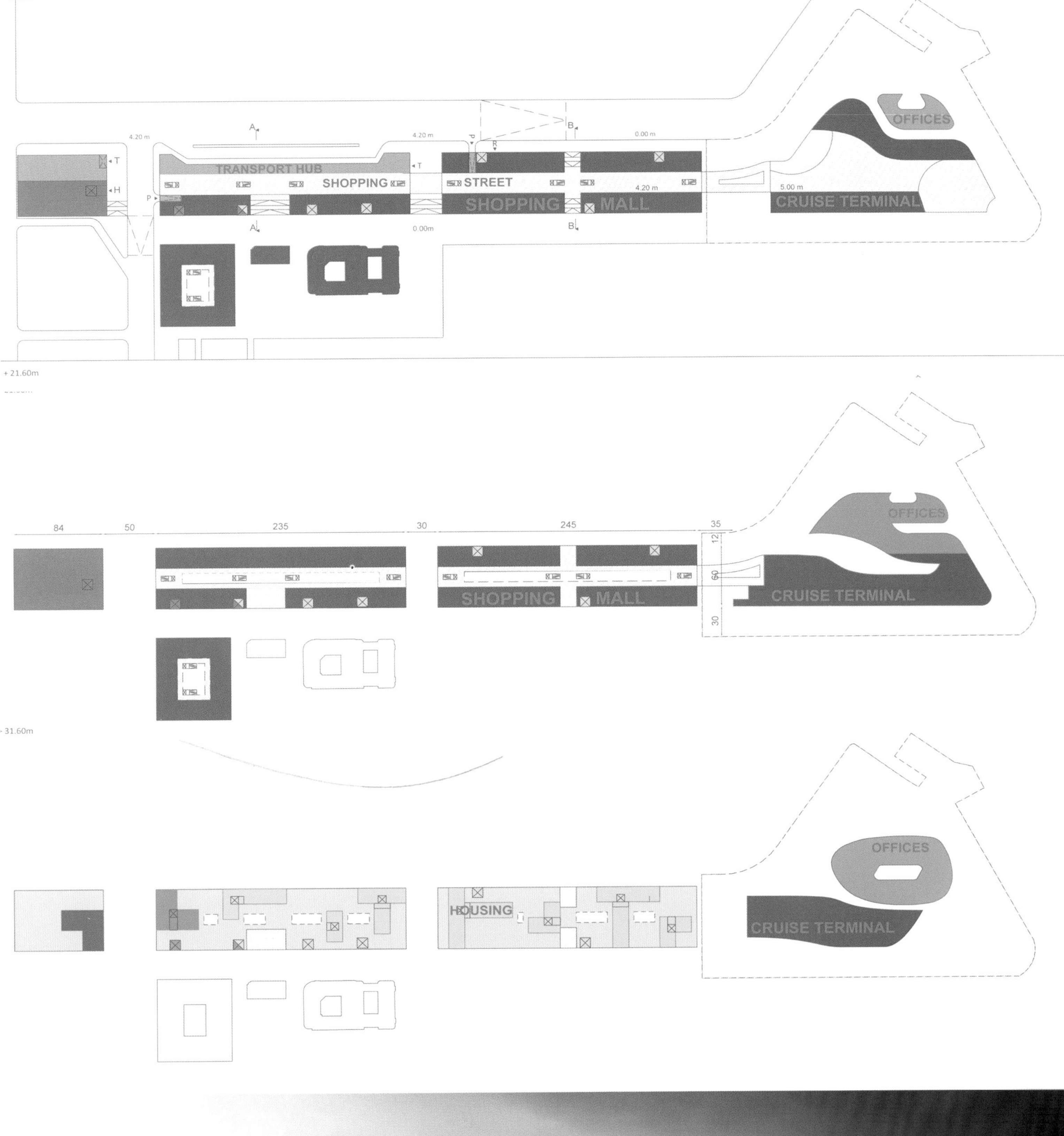
+ 6.60m
4.20 m
A
4.20 m
0.00 m
B
TRANSPORT HUB
SHOPPING
STREET
SHOPPING
MALL
OFFICES
CRUISE TERMINAL
5.00 m
4.20 m
0.00m
+ 21.60m
84
50
235
30
245
35
12
60
30
- 31.60m
HOUSING

丹麦奥尔胡斯 DSB 地区和公交总站总体规划

DSB-areas and Bus Terminal Masterplan in Aarhus

设计单位：C.F.Møller Architects
项目地址：丹麦奥尔胡斯
占地面积：80 000 ㎡

Designed by: C.F. Møller
Location: Aarhus, Denmark
Floor Area: 80,000 m²

项目概况

项目基地紧靠该区域的中心车站，位于一个活跃的、可持续发展的城市地块，这个改造项目的总体规划也是基于该地的特殊发展潜能而展开的。

规划目标

项目旨在建立一个多样化的城市空间，让更多的市民可步行或是骑自行车前往改造后的奥尔胡斯海滨区。

规划内容

项目将原有的建筑作为内部城市空间，适度地改造 Ny Banegårdsgade 大街的交通线路，并优先考虑通向海滨的线路的改造。

新公交总站的位置高于铁路广场，与车站大厅处在同一个水平面，这一项目的开发建设也有利于疏理和重组城市内部结构，完善城市内部格局。城市核心区新轻轨车站的建筑，为这一地区的发展带来了更多的活力。

拟建建筑的占地面积和体积为项目的多样化设计提供了条件，同时，也将周围密集城市街区的建筑以及邻近码头区的独立建筑体联系起来。

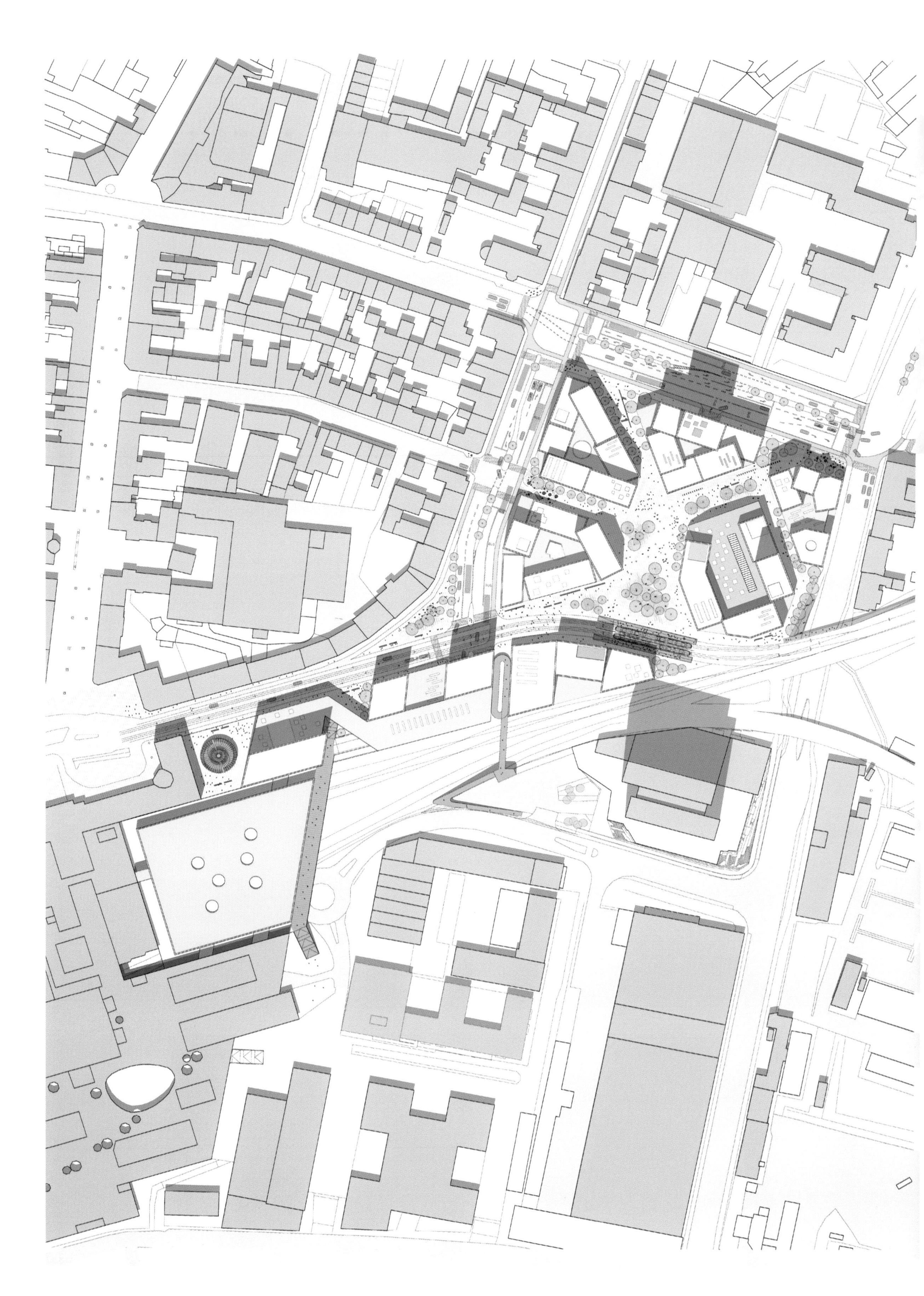

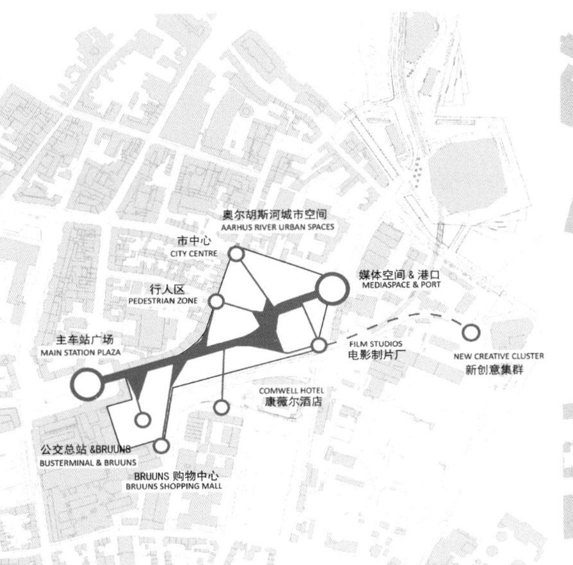
奥尔胡斯河城市空间
AARHUS RIVER URBAN SPACES
市中心
CITY CENTRE
媒体空间 & 港口
MEDIASPACE & PORT
行人区
PEDESTRIAN ZONE
主车站广场
MAIN STATION PLAZA
FILM STUDIOS
电影制片厂
NEW CREATIVE CLUSTER
新创意集群
COMWELL HOTEL
康薇尔酒店
公交总站 &BRUUNS
BUSTERMINAL & BRUUNS
BRUUNS 购物中心
BRUUNS SHOPPING MALL

D
C
B
E
A
F
G

bruun's
SONY

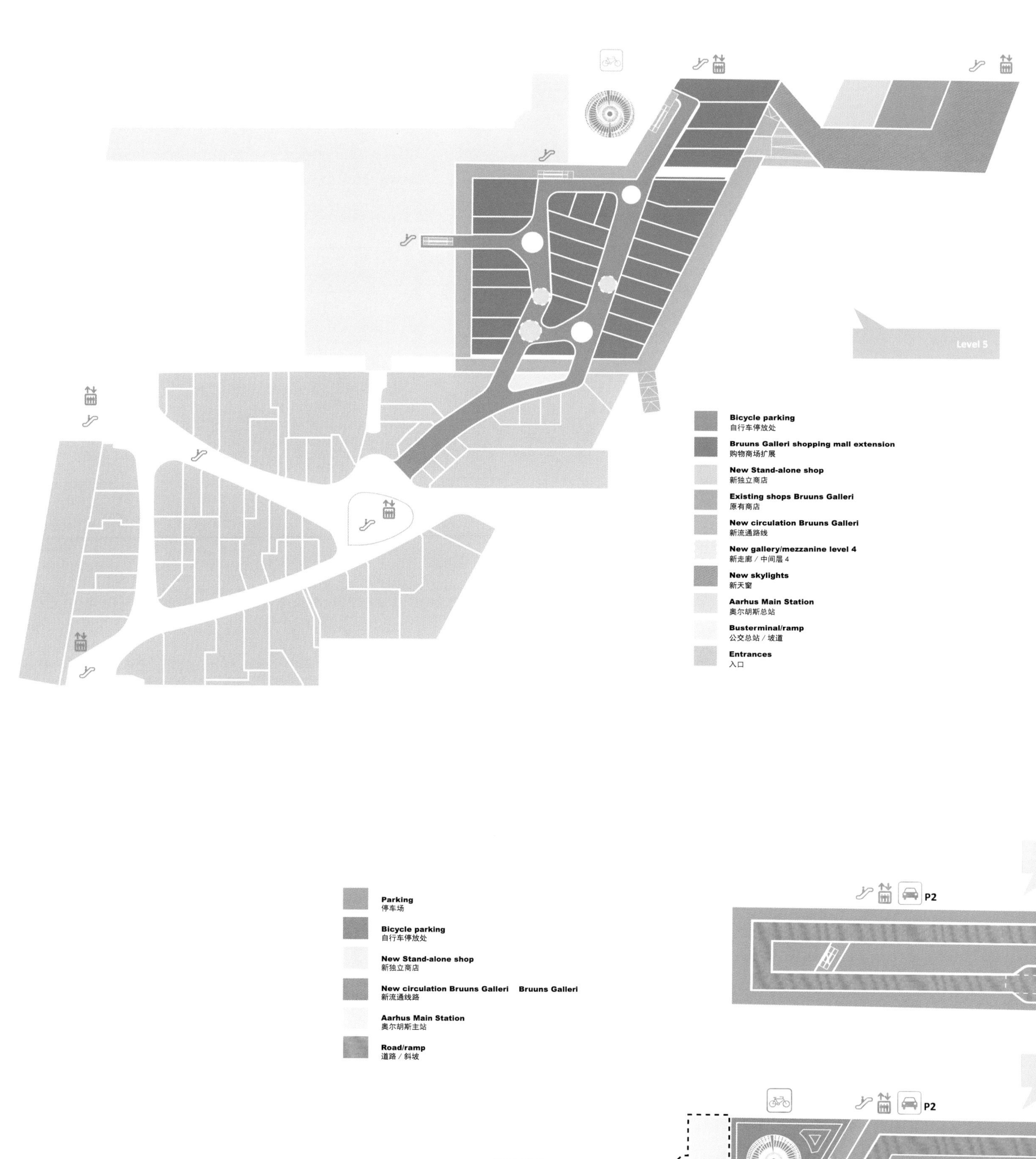
Level 5
Bicycle parking
自行车停放处
Bruuns Galleri shopping mall extension
购物商场扩展
New Stand-alone shop
新独立商店
Existing shops Bruuns Galleri
原有商店
New circulation Bruuns Galleri
新流通路线
New gallery/mezzanine level 4
新走廊 / 中间层 4
New skylights
新天窗
Aarhus Main Station
奥尔胡斯总站
Busterminal/ramp
公交总站 / 坡道
Entrances
入口

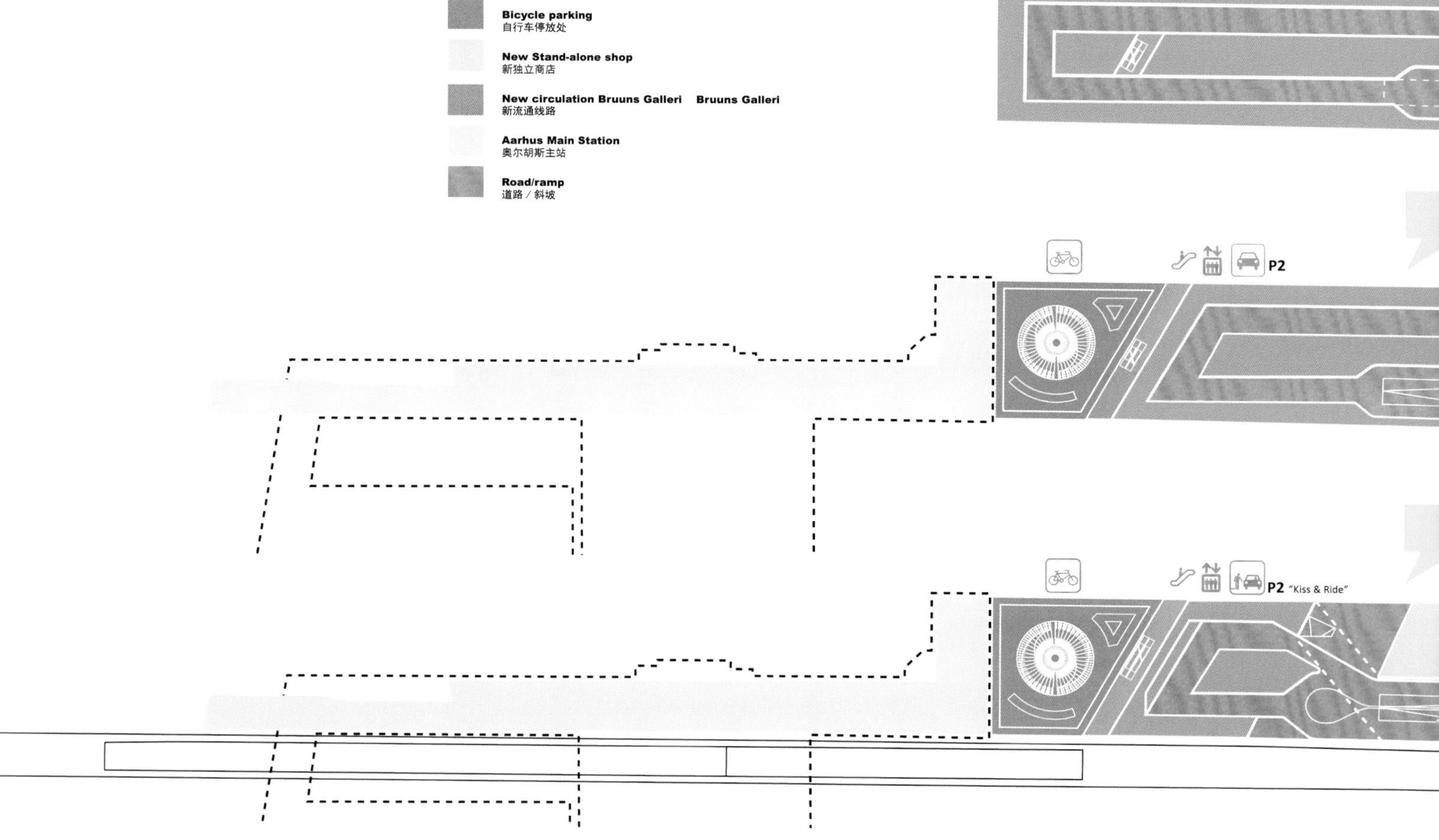
Parking
停车场
Bicycle parking
自行车停放处
New Stand-alone shop
新独立商店
New circulation Bruuns Galleri Bruuns Galleri
新流通线路
Aarhus Main Station
奥尔胡斯主站
Road/ramp
道路 / 斜坡
P2
P2
P2 "Kiss & Ride"

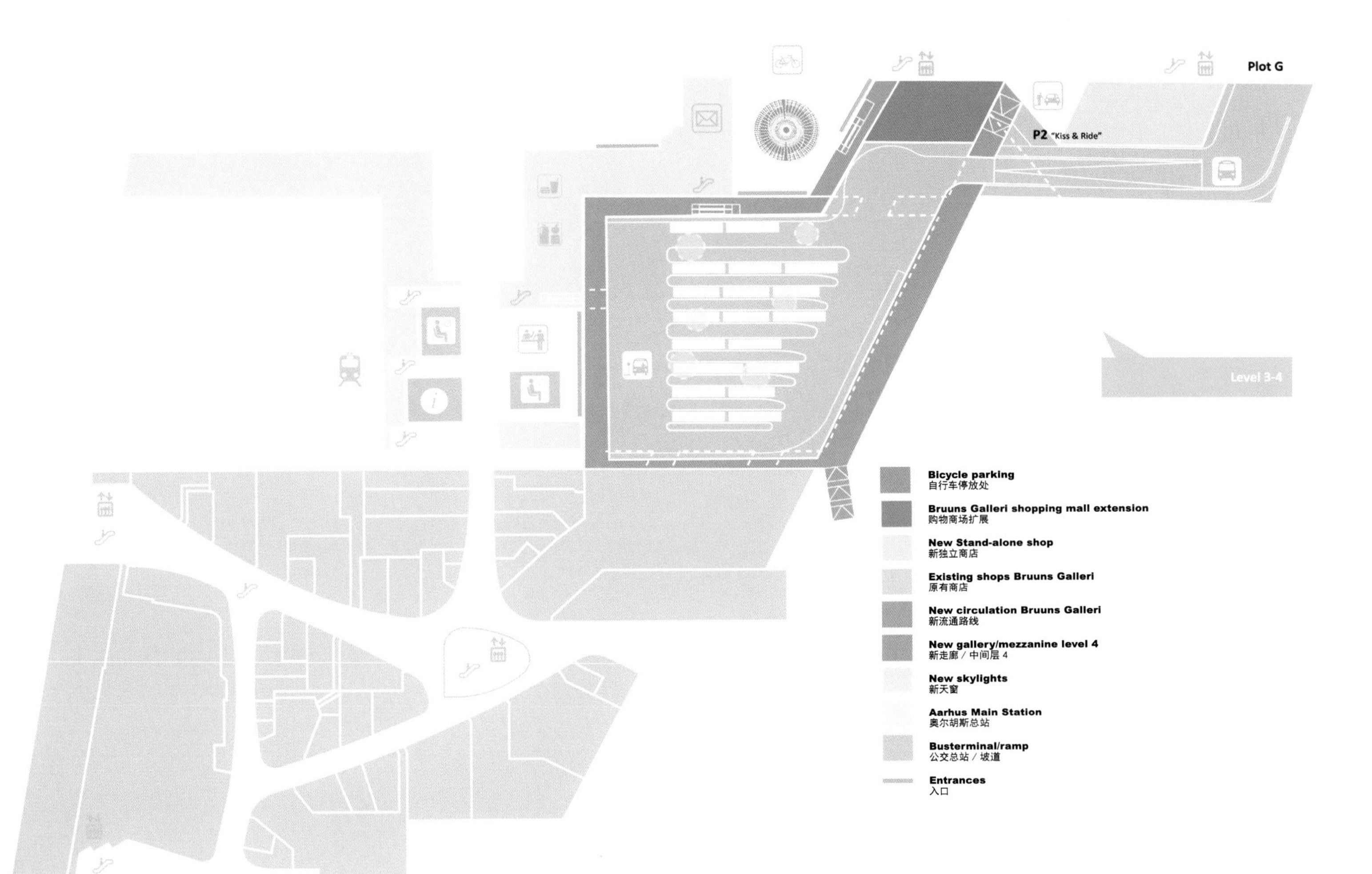

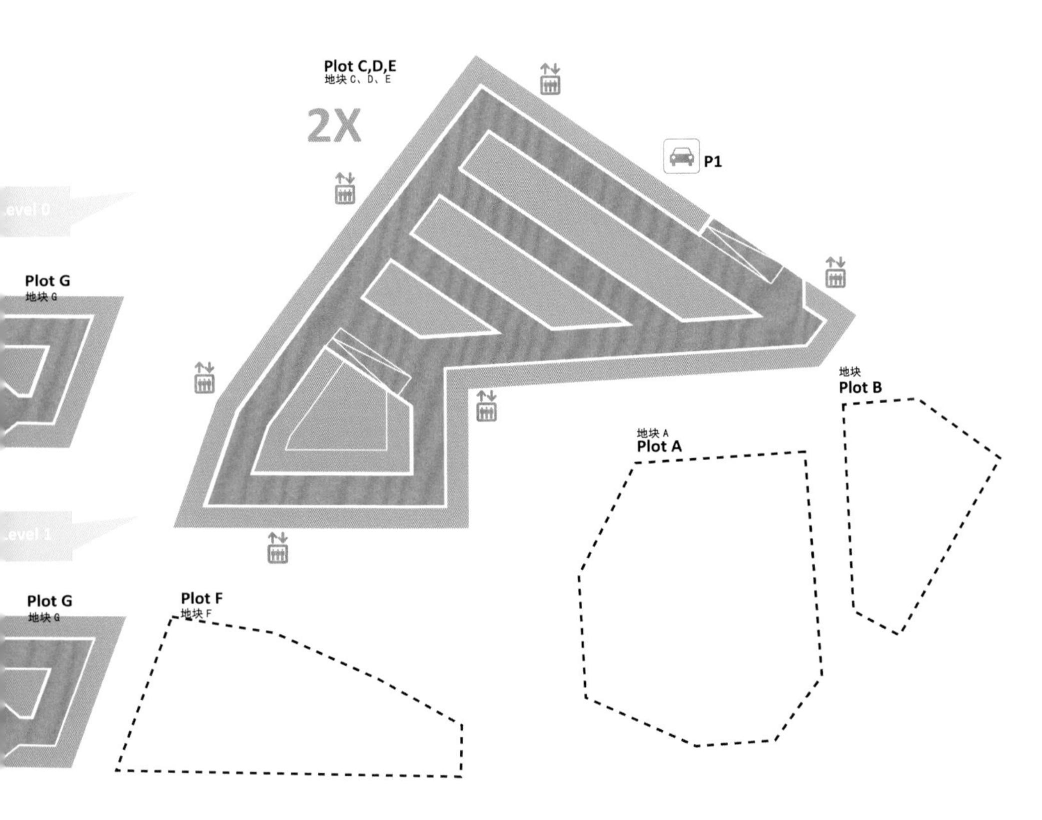

Profile

The masterplan for the transformation of the DSB-areas and the former bus terminals area in Aarhus is based on the exceptional potential of the site for a vibrant and sustainable urban development in close proximity to the central station.

Planning Goal

The aim is a diverse urban quarter that meets the need for a strong new pedestrian and bicycle connection to the newly transformed waterfront of Aarhus.

Planning Content

By re-using existing buildings as indoor urban spaces, and partially redirecting the course of the Ny Banegårdsgade Street to prioritize the direction towards the harbourfront.

A new bus terminal is located above the railyard on the level of the station hall. The development and construction of this project contribute to creating new connections and relations within the urban fabric. A new stop on the light rail in the central urban space brings even more vitality to this area.

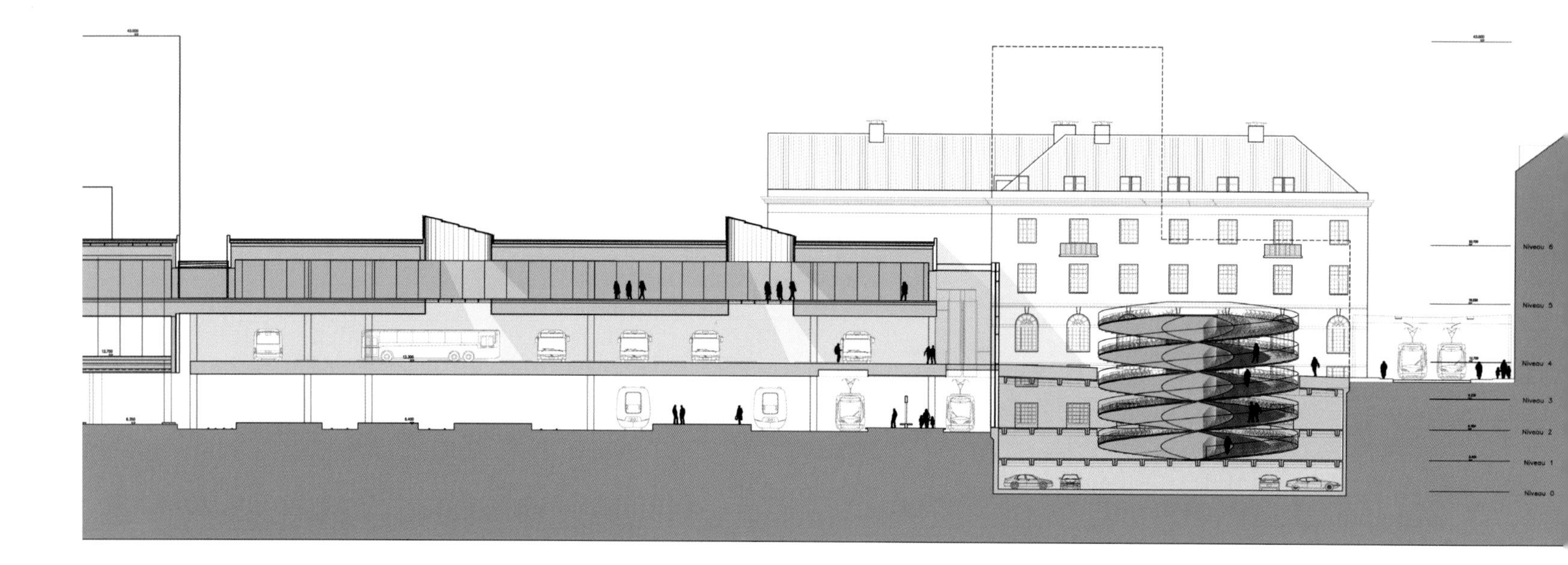

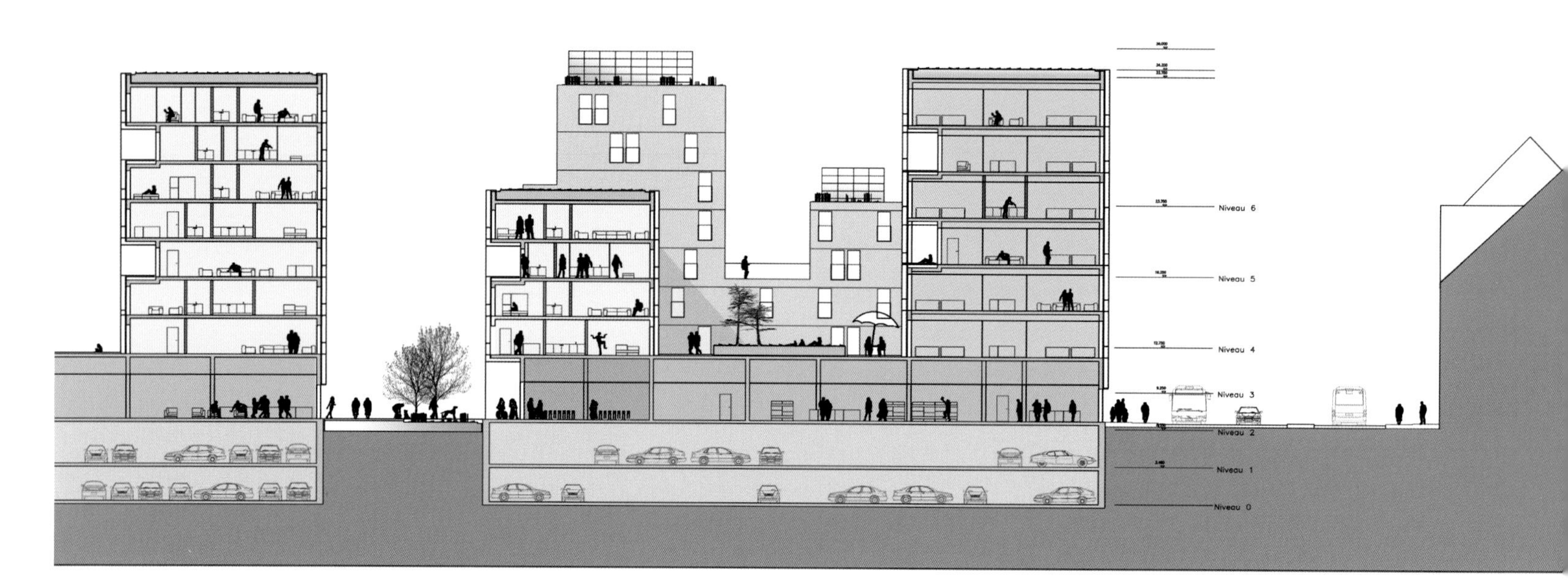

The proposed building volumes and footprints allow great freedom of interpretation and diversity, while combining the best features of, respectively, the surrounding dense city block structure and the larger free-standing volumes of the adjacent port area.

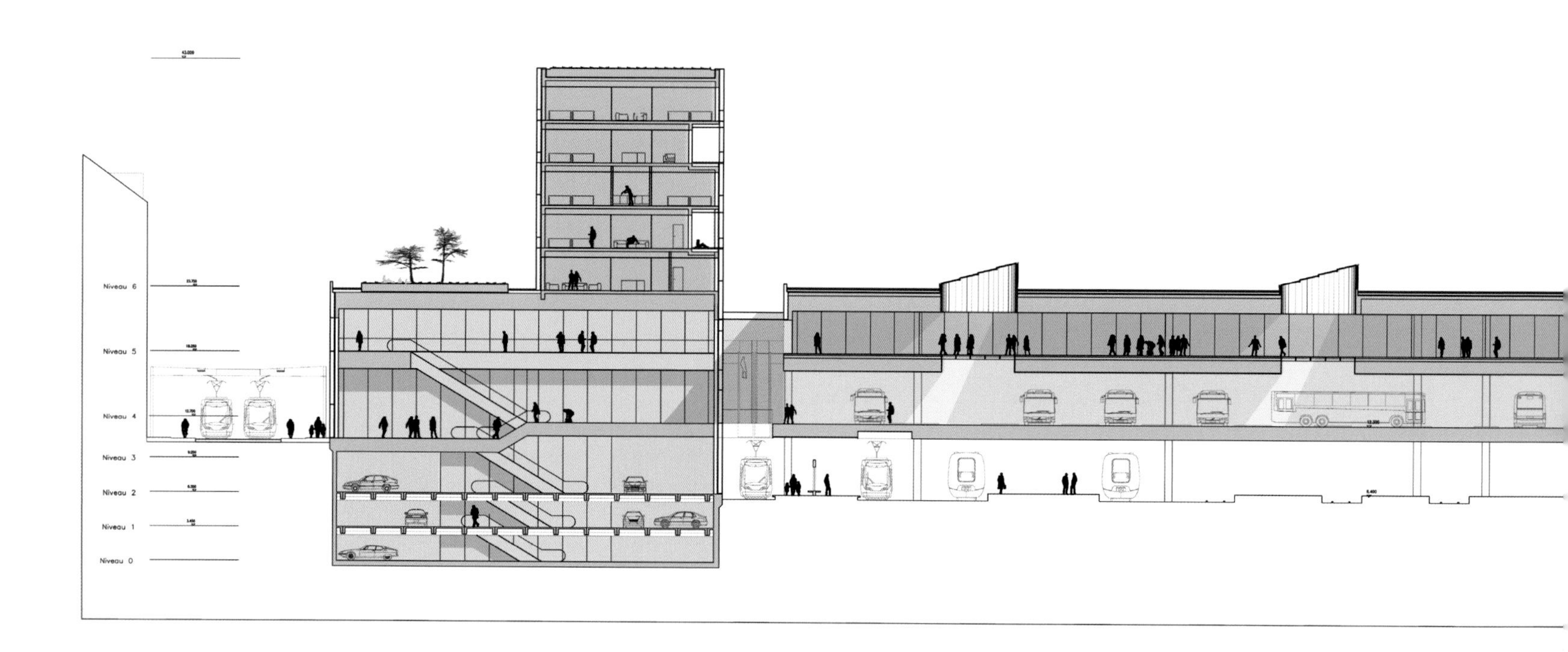

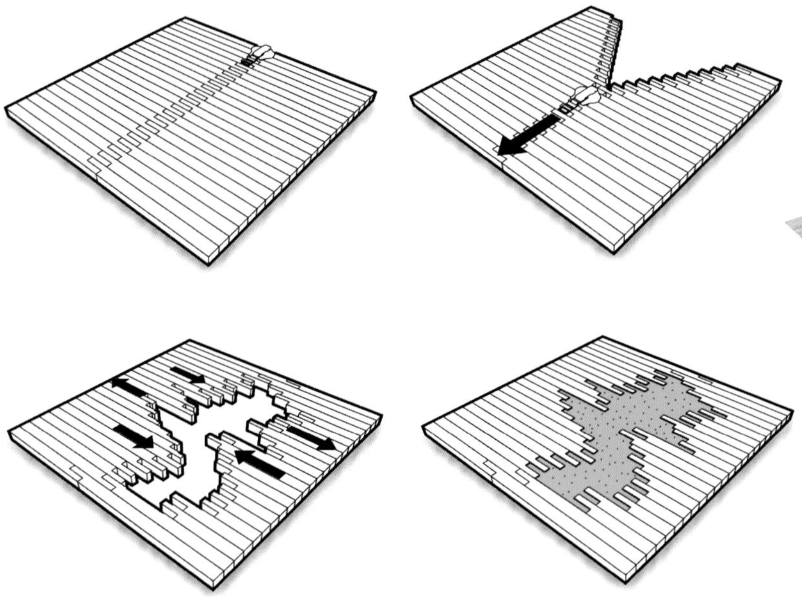

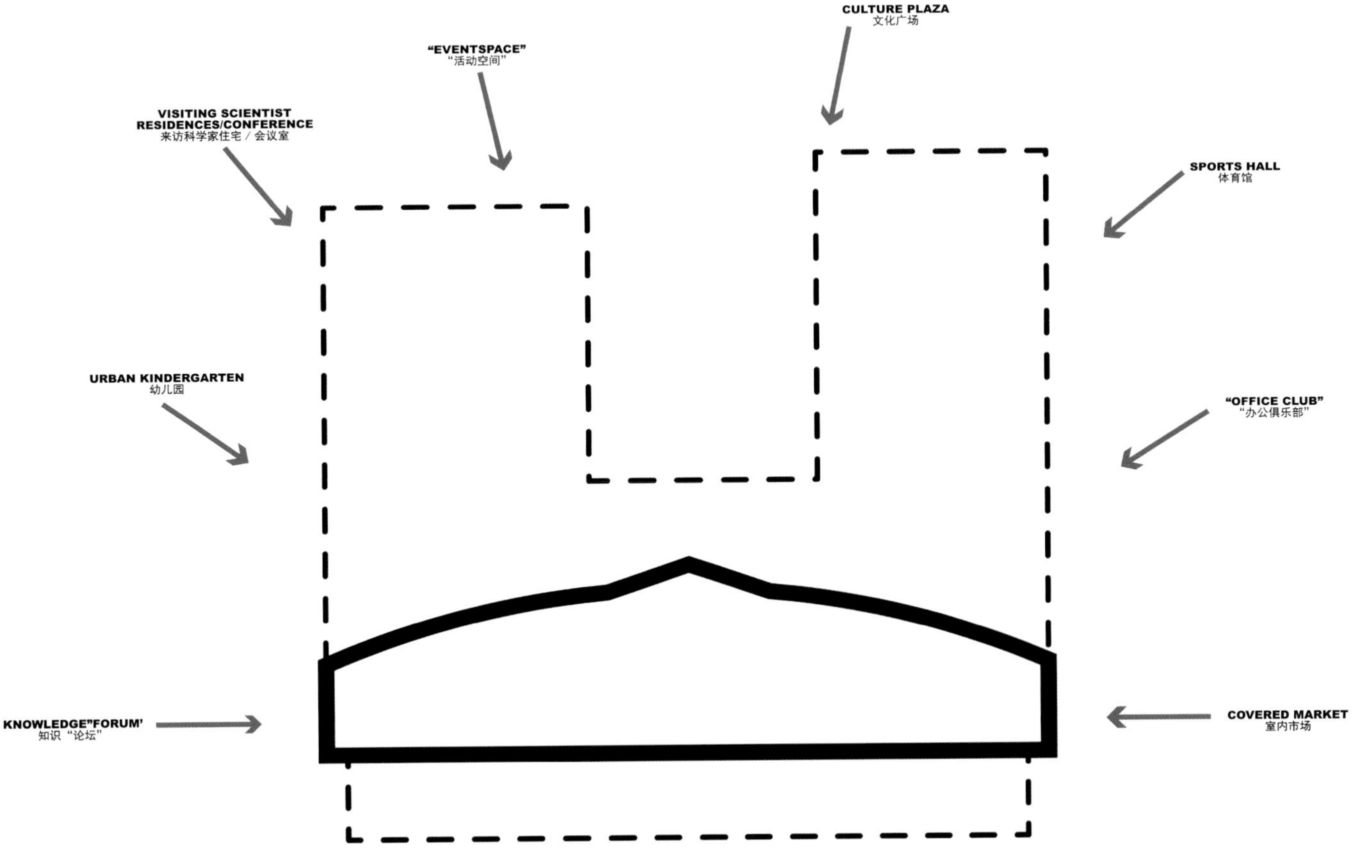

CULTURE PLAZA
文化广场

"EVENTSPACE"
"活动空间"

VISITING SCIENTIST
RESIDENCES/CONFERENCE
来访科学家住宅 / 会议室

SPORTS HALL
体育馆

URBAN KINDERGARTEN
幼儿园

"OFFICE CLUB"
"办公俱乐部"

KNOWLEDGE"FORUM'
知识"论坛"

COVERED MARKET
室内市场

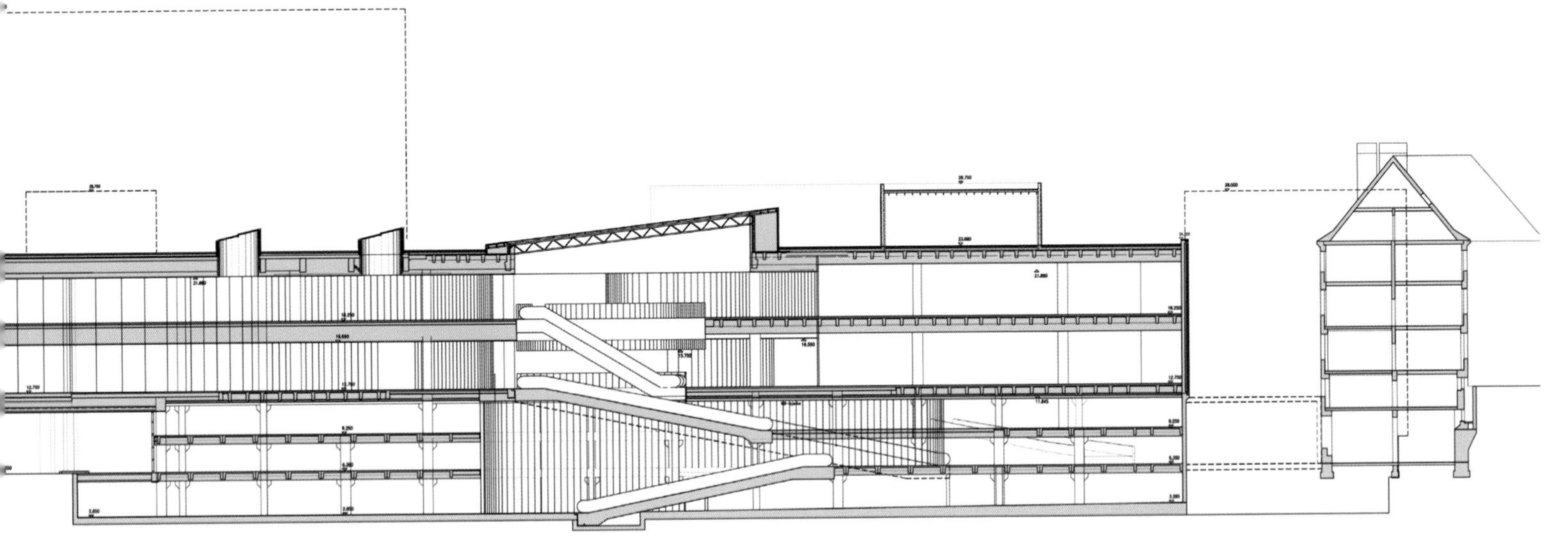

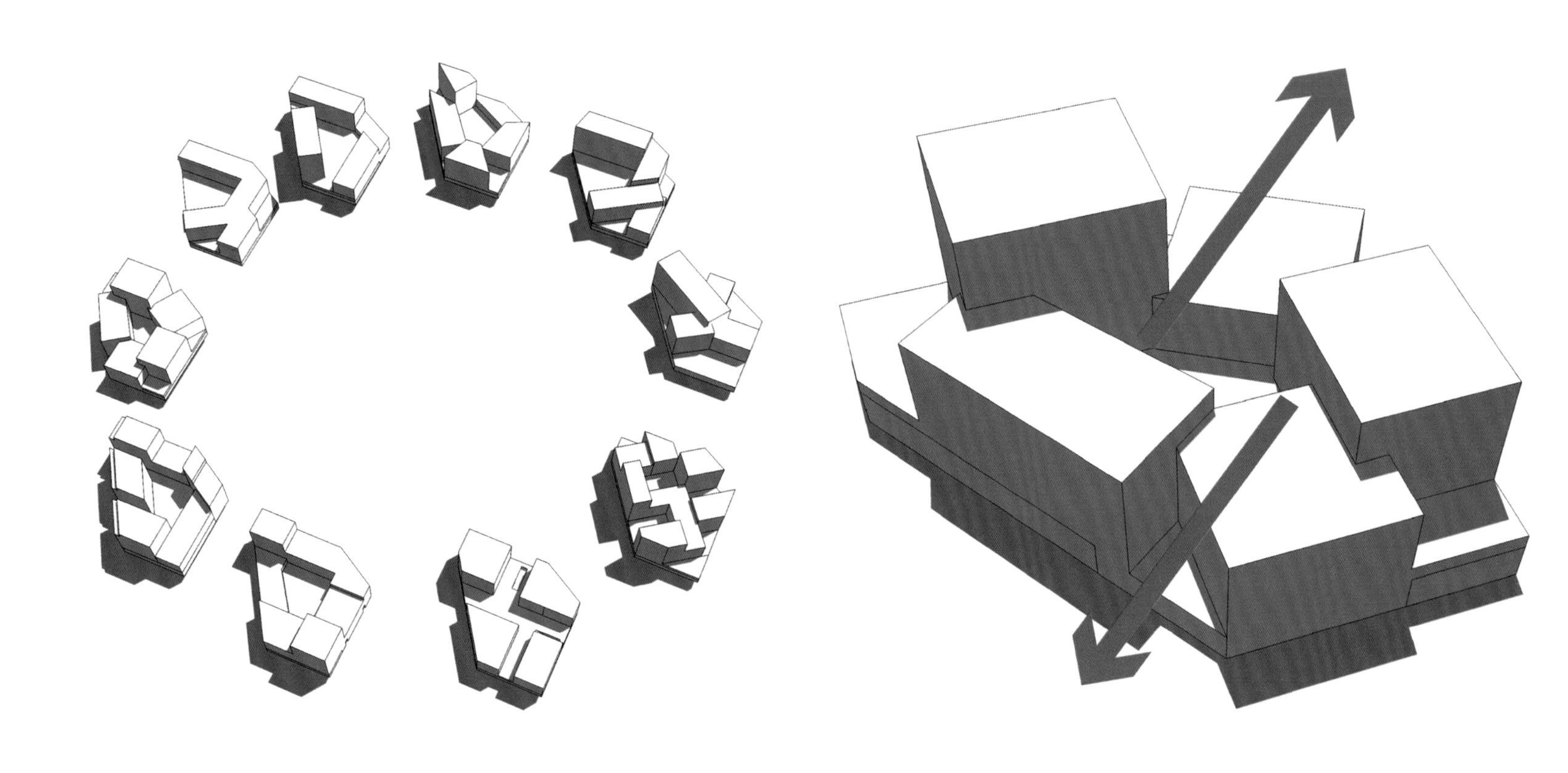

VARIATION-MULTIPLE PROPERTIES MULTIPLE ARCHITECTS MIXED-USE PROGRAMMING
变化—多重属性　多个建筑师　多用途规划

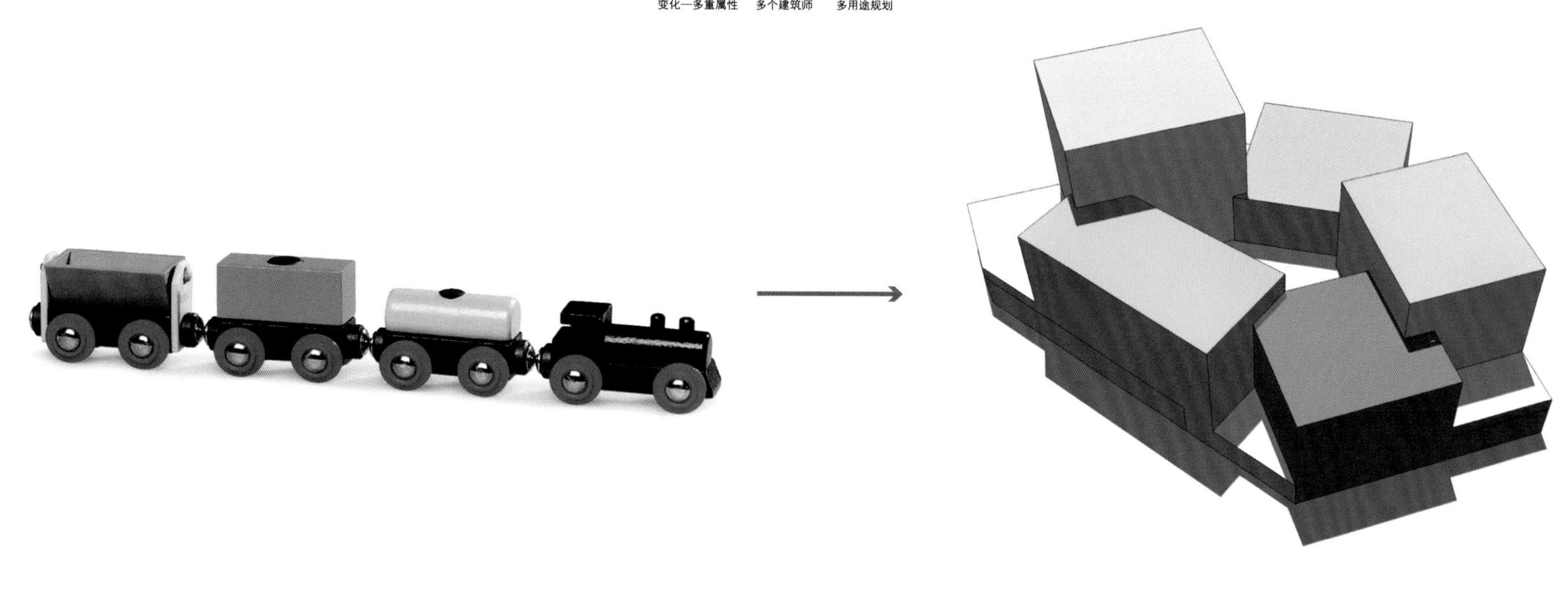

TRADITIONAL URBAN BLOCK
传统城市街区

DISPERSED URBAN BLOCK
分散型城市街区

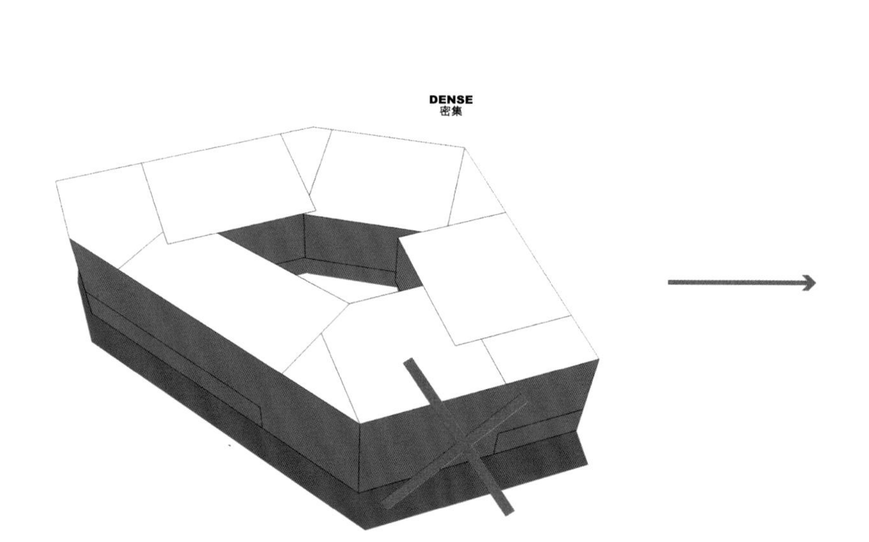

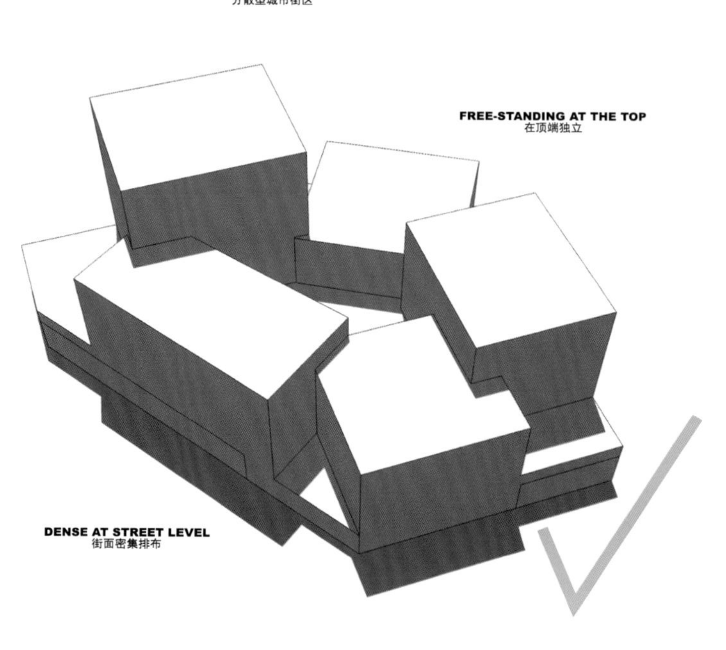

屋顶水库
ROOFTOP RESERVOIR

绿色屋顶—表面径流减少
GREEN ROOFS - LESS RUN-OFF

屋顶花园
ROOF-TOP
GARDENS

HEATING/COOLING IN BUILDINGS
建筑供热 / 制冷

RE-USE IN URBAN SPACE
城市空间的再利用

建筑再利用
RE-USE IN BUILDINGS

1.6 米表面积聚 1,6 M SURFACE BUILD-UP

WATER TREATMENT PLANT
水处理厂

城市空间下的水库
RESERVOIR UNDER URBAN SPACE

城市空间下的水库
RESERVOIR UNDER URBAN SPACE

城市空间内的明渠 OPEN CANALS IN URBAN SPACE
硬质铺面下渗透 PERCOLATION UNDER HARD SURFACING

WATER TREATMENT PLANT
水处理厂

GROUND WATER LEVEL +2,5
地下水位 +2.5

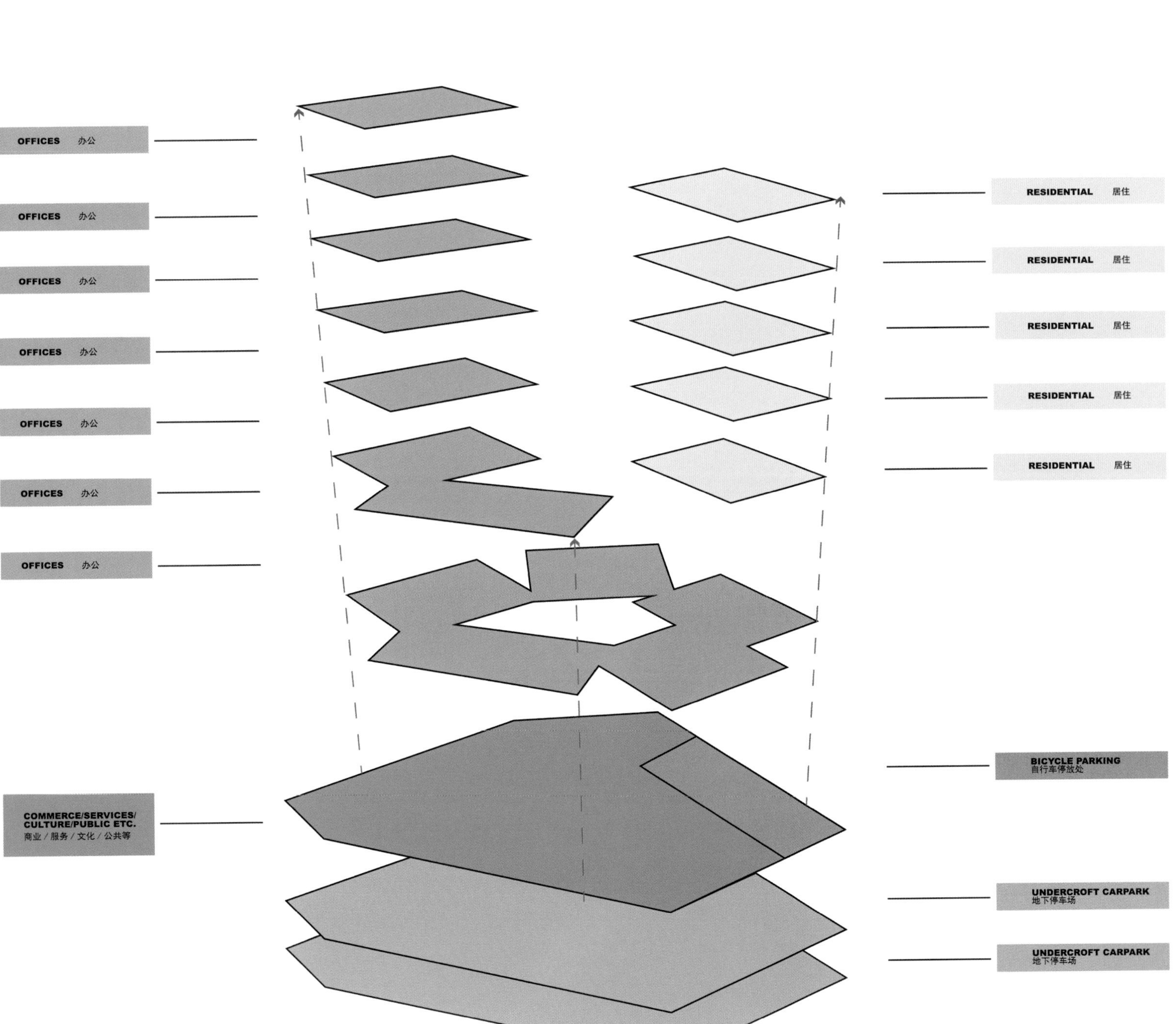

CAFE BABYLON
TUBORG

广东佛山东平新城交通枢纽中心

Foshan Dongping New City Mass Transit Center

设计单位：amphibianArc

开发商：佛山金汇海投资有限公司

项目地址：中国广东省佛山市

项目面积：600 000 m²

Designed by: amphibianArc

Client: Foshan Jinhuihai Investment Co, Ltd.

Location: Foshan, Guangdong, China

Area: 600,000 m²

项目概况

东平新城交通枢纽中心作为佛山市的交通枢纽中心之一，集地铁、轻轨、公交、的士、城巴等各种交通方式于一体，是连接东平新城与市内外交通的综合换乘中心和人车流聚散的核心平台。

建筑设计

这个 600 000 平方米的大型复合型城市中心对当地的基础设施建设起到了巨大的推动作用，有利于构建一个富有活力的动态城市。设计强调通过机械美学的运用构建一个未来主义的、非线性的、动感前卫的建筑结构。裙楼外观上超大型的 Jumbatron 显示屏以及聚能管展现了建筑强有力的速度感。

一个下沉式广场与地铁通道相连，为地铁通道提供了自然光线和清新的空气。这一广场也被一个大型的 Jumbatron 屏幕围绕，激活了这个熙熙攘攘、热闹非凡的城市。

建筑的第 1 层容纳了一个多功能大厅，它是这个交通建筑的核心，连接了位于建筑底层的地铁站、城际轻轨站、机场服务区和城市公交汽车站。围绕在这个多功能大厅周围的是零售空间，并扩展至建筑的第 3 层。

休闲空间位于裙房的顶楼，包括一个有 11 个放映厅的电影院和饮食区。3 层的地下停车场可容纳 1 000 个停车位。项目还设计了 6 栋塔楼，包括 3 栋住宅楼、2 栋办公楼以及 1 栋酒店式公寓。它们构建在这个裙房上，围绕着一个半开放的屋顶花园分布。

Profile

Dongping New City Mass Transit Center, as one of the traffic hubs of Foshan, is a comprehensive transfer center connecting Dongping New City and traffic lines inside & outside of the city as well as a vital platform accumulating pedestrians and vehicles.

Architectural Design

This 600,000 square meters gigantic mixed use city center plays an important role in the city infrastructure. It provokes a desire to contribute to a dynamic cityscape. The design addresses an image in a futuristic, non-linear, and dynamic avant-garde architecture form with mechanical aesthetics. There is a strong intention to express the sense of speed through the energy that is converged by the tubes and releases through the giant projected Jumbatron screens at the podium.

A sunken plaza is made to penetrate the underground metro passages with sunlight and fresh air. It also embraces a huge Jumbatron screen which activates the bustling and hustling cityscape.

A multi-functional hall at the center of the first floor is the hub of the transportation that connects the metro stations, the intercity train station, the airport service and the city bus terminal which locates on the ground floor. Spaces surrounding the multi-functional hall are the retail stores that grow up to the 3rd floor.

An entertainment area, including an eleven-screen cinema and food courts are in the top floor of the podium. Three-storey underground parking space is offering more than 1,000 parking spaces. Six towers, including three residential buildings, two offices and one service apartment building, sitting on the podium enclose a semi-open roof garden.

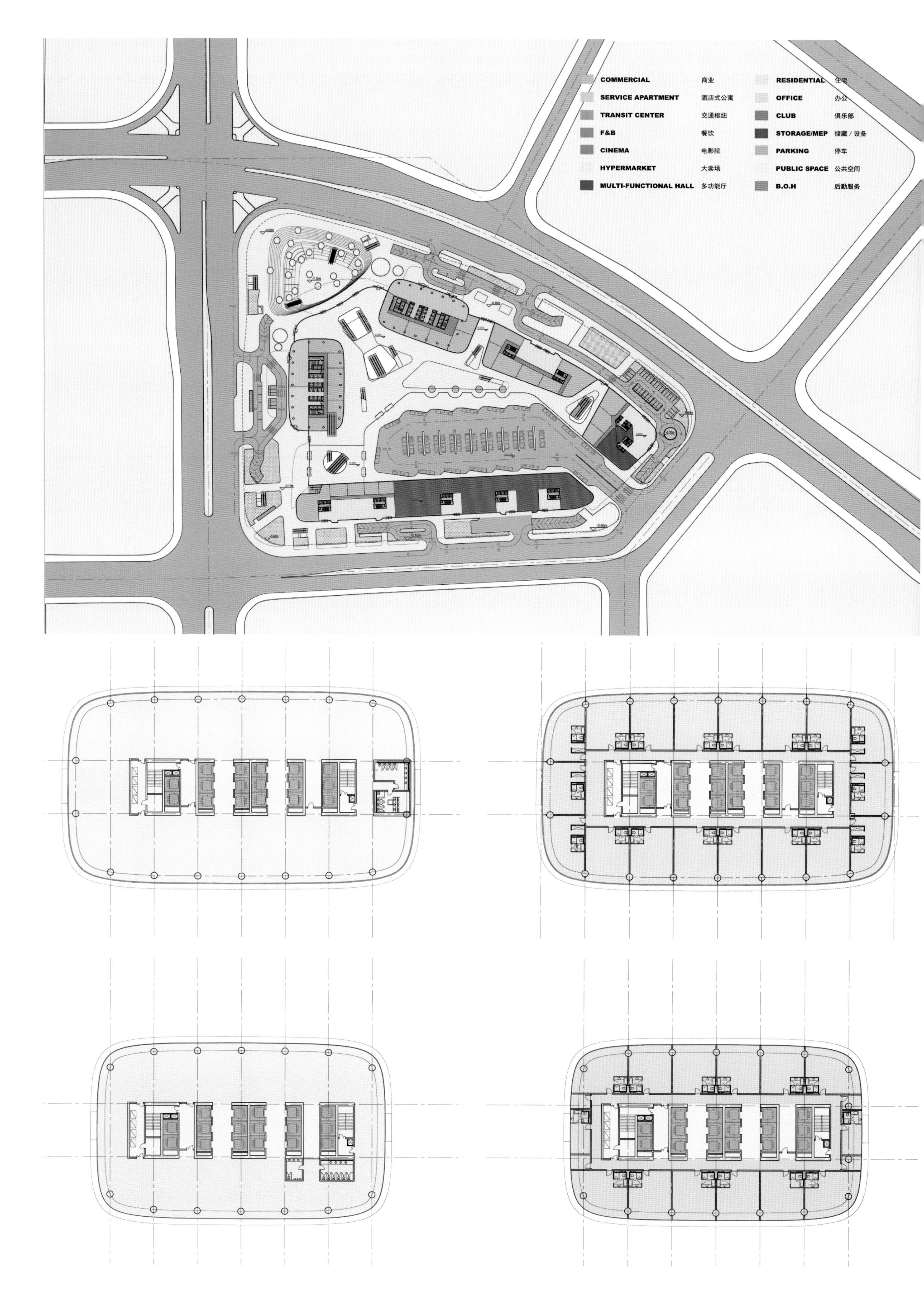
COMMERCIAL 商业
SERVICE APARTMENT 酒店式公寓
TRANSIT CENTER 交通枢纽
F&B 餐饮
CINEMA 电影院
HYPERMARKET 大卖场
MULTI-FUNCTIONAL HALL 多功能厅
RESIDENTIAL 住宅
OFFICE 办公
CLUB 俱乐部
STORAGE/MEP 储藏 / 设备
PARKING 停车
PUBLIC SPACE 公共空间
B.O.H 后勤服务

COMMERCIAL	商业	RESIDENTIAL	住宅
SERVICE APARTMENT	酒店式公寓	OFFICE	办公
TRANSIT CENTER	交通枢纽	CLUB	俱乐部
F&B	餐饮	STORAGE/MEP	储藏 / 设备
CINEMA	电影院	PARKING	停车
HYPERMARKET	大卖场	PUBLIC SPACE	公共空间
MULTI-FUNCTIONAL HALL	多功能厅	B.O.H	后勤服务

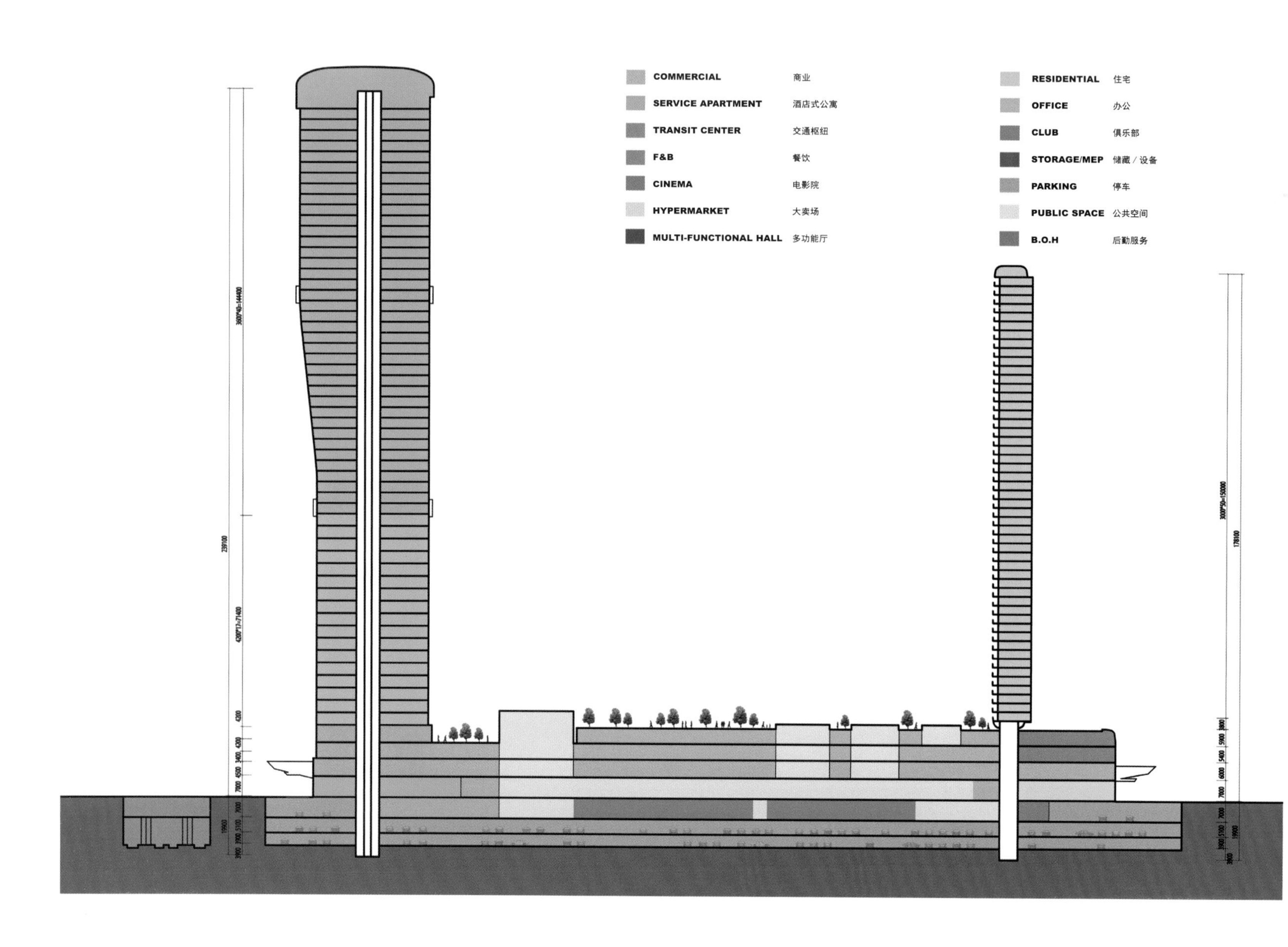
COMMERCIAL 商业
SERVICE APARTMENT 酒店式公寓
TRANSIT CENTER 交通枢纽
F&B 餐饮
CINEMA 电影院
HYPERMARKET 大卖场
MULTI-FUNCTIONAL HALL 多功能厅
RESIDENTIAL 住宅
OFFICE 办公
CLUB 俱乐部
STORAGE/MEP 储藏 / 设备
PARKING 停车
PUBLIC SPACE 公共空间
B.O.H 后勤服务

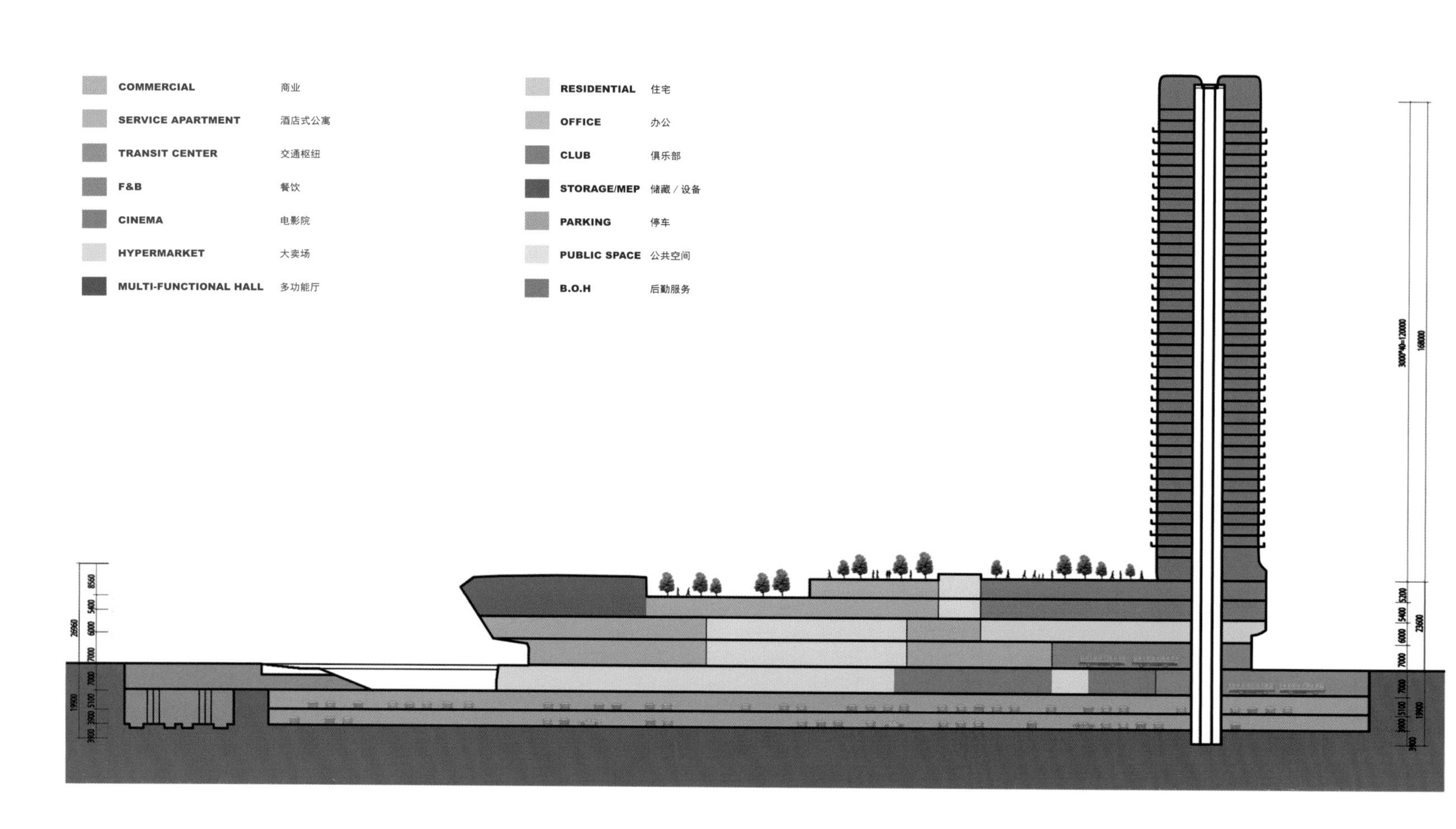
COMMERCIAL 商业
SERVICE APARTMENT 酒店式公寓
TRANSIT CENTER 交通枢纽
F&B 餐饮
CINEMA 电影院
HYPERMARKET 大卖场
MULTI-FUNCTIONAL HALL 多功能厅
RESIDENTIAL 住宅
OFFICE 办公
CLUB 俱乐部
STORAGE/MEP 储藏 / 设备
PARKING 停车
PUBLIC SPACE 公共空间
B.O.H 后勤服务

台湾基隆港新客运大楼

New Keelung Harbor Terminal Building

设计单位：Synthesis Design + Architecture

开发商：基隆港

台湾基隆港国际港务有限公司

项目地址：中国台湾省基隆市

项目面积：140 000 ㎡

Designed by: Synthesis Design + Architecture

Client: Port of Keelung

Taiwan International Ports Corporation, Ltd.

Location: Keelung, Taiwan, China

Area: 140,000 m²

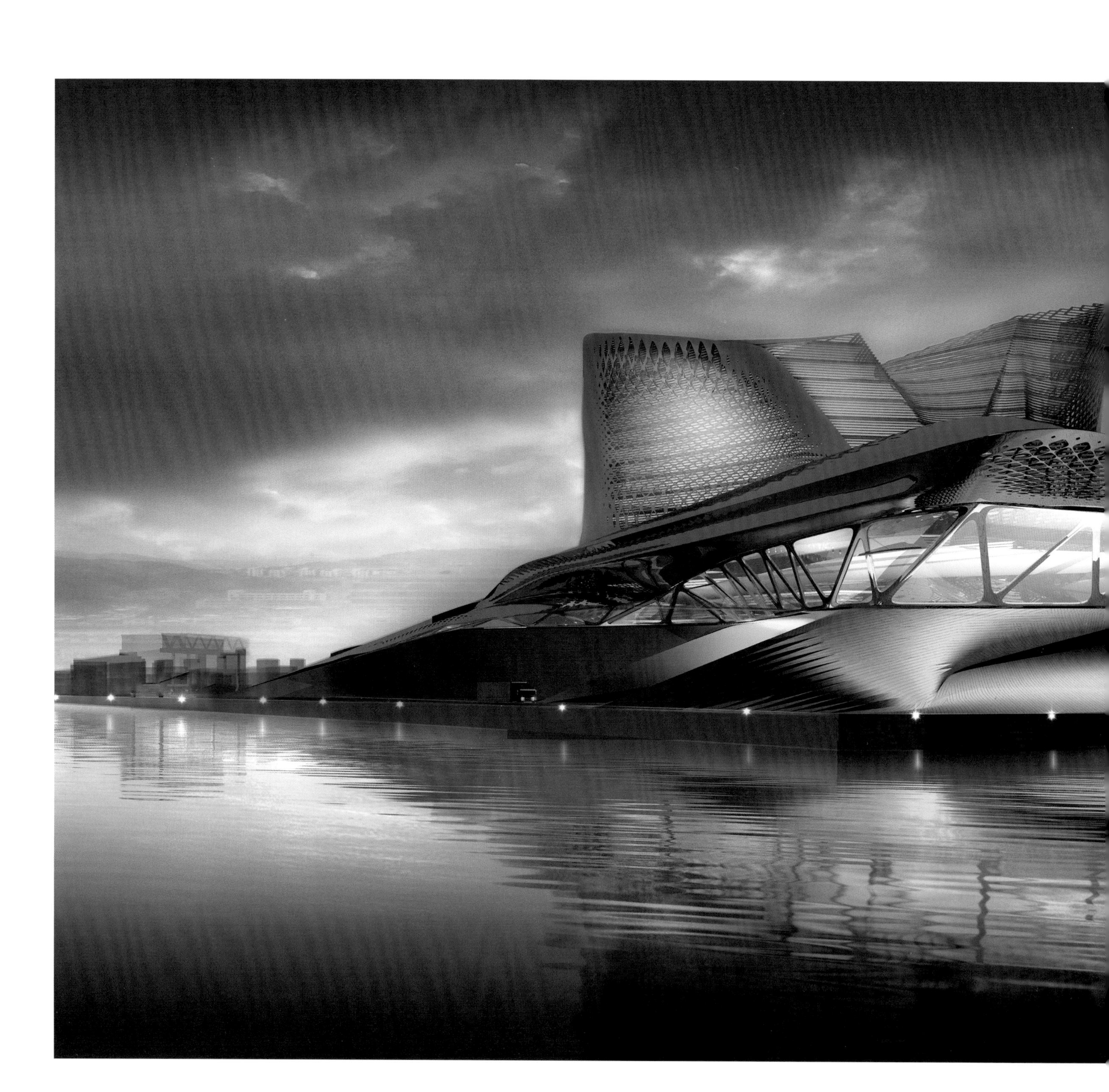

项目概况

台湾基隆港新客运大楼的设计旨在提供一个优雅、连贯、具有标志性意义的方案，使之成为基隆港一个新的标志性建筑。

建筑设计

受到“台湾鸡笼”和豪华赛艇薄壳结构的几何形态的启发，基隆港新客运大楼设计为一种由外皮演变为外骨架的动态倾斜结构，以回应项目内容和表现要求。客运大楼分为三个基本的体验群组和循环系列，设计为组群间的转换提供了衔接和过渡空间。

这个看上去很有条理的表达形式，实际上是一个“蕴含了很多信息的形式”。项目通过建筑朝向、空间布局、立面设计以及材料和结构系统的选择等综合设计元素，回应了气候、城市文脉、规划、流线以及可持续的整合。

设计综合了视觉、结构和环境因素，通过优化建筑结构和体量，进行被动的环境设计和控制，植入街区供暖 / 制冷和废物处理系统，以及利用几何形态获取风能、雨水和太阳能，来达到视觉效果、建筑功能以及环境效益的统一。

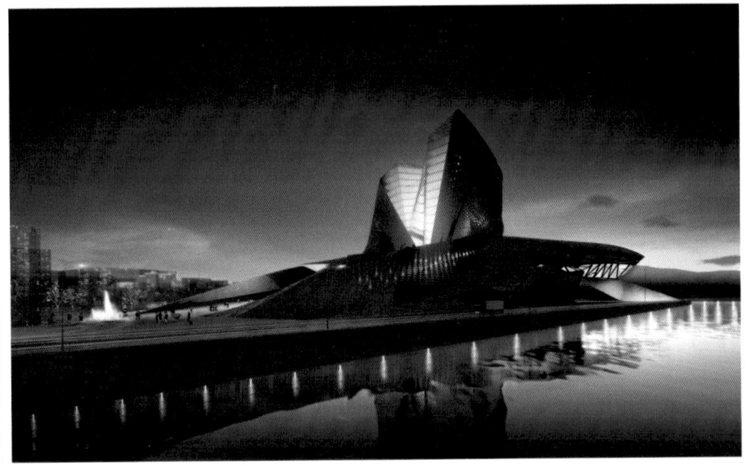

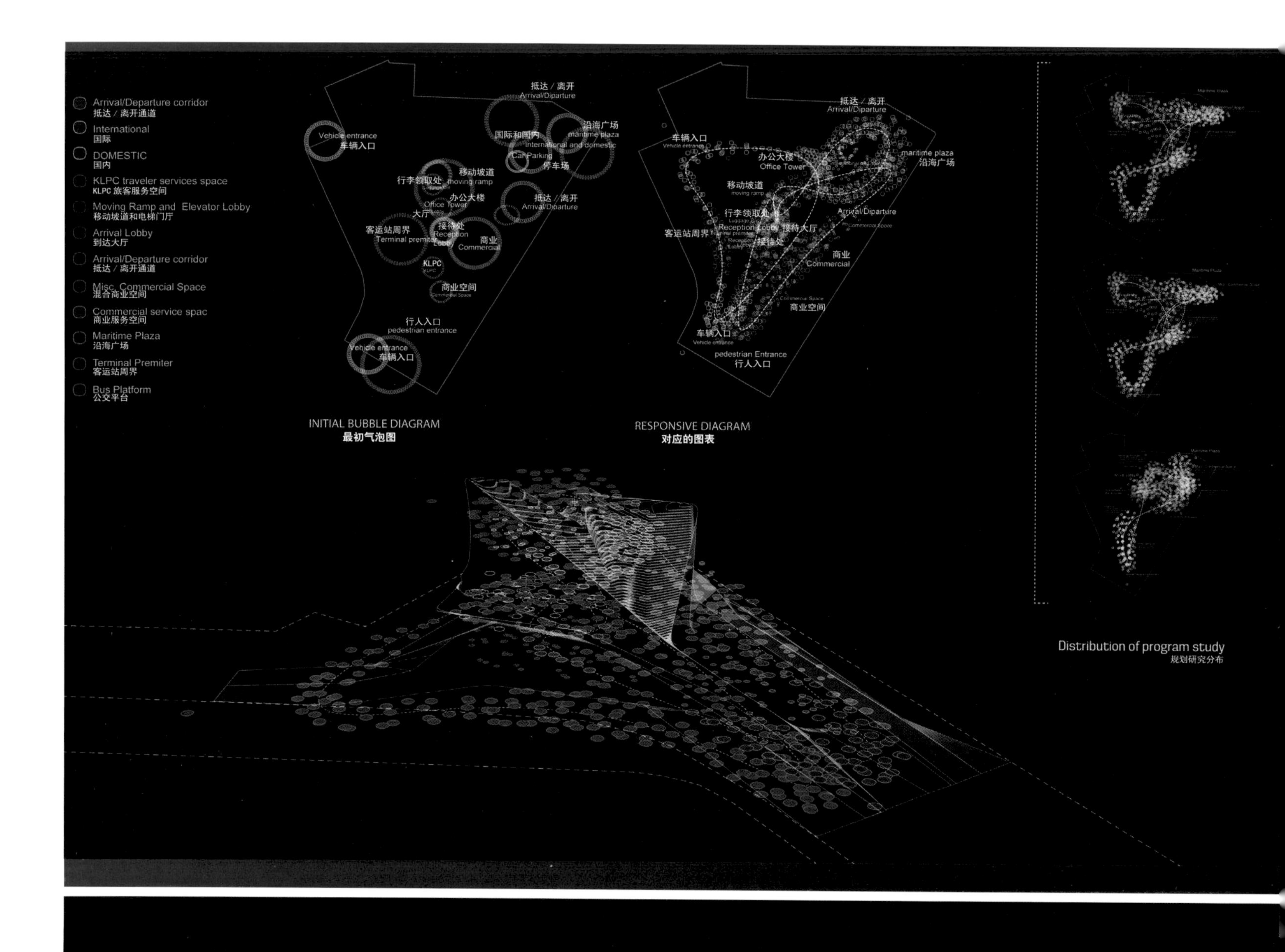

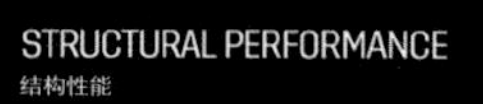

办公
OFFICE

办公
OFFICE

办公
OFFICE

STRUCTURAL CORE
结构核心

CANTILEVERED FLOOR SLABS
悬臂楼板

GLAZING SKIN
玻璃表皮

LOUVER INTERNAL SKIN
百叶窗内表皮

PREFABRICATED CARBON
FIBRE STRUCTURAL SKIN
预制碳纤维结构表皮

STRUCTURAL/CIRCULATION CORES
结构／循环核心

FACADE 2nd SKIN 立面第二层表皮
GLASS AND LOUVER FACADE 玻璃和百叶窗立面

FACADE 1ST SKIN 立面第一层表皮
CARBON FIBRE SHELL 碳化纤维外壳

中庭立面碳化
纤维束和玻璃
ATRIUM FACADE
CARBON FIBER TRUSS AND
GLAZING

COMPOSITE
CARBON FIBRE FLOOR
复合碳化纤维地板

CORE PILE FOUNDATION
核心桩基础

CONCRETE PLINTH
混凝土基座

FACADE HARVESTS
RAINWATER TO BE
RE-USED IN GREYWATER SYSTEMS
表面收集的雨水将重新用在
灰水系统中

HEAT RECOVERY
SYSTEM AT TOP
OF CORE
核心顶部
热回收系统

OUTER SKIN
PROTECTS INTERIOR FROM
EXTERNAL NOISE POLLUTION
AND DIRECT SUNLIGHT
外表皮保护室内不受
室外噪音污染和直射
阳光的影响

HEAT RISES THROUGH CORES
从核心部分上升的热气

FRESH AIR TRAVELS THROUGH
DOUBLE SKIN GAP
新鲜空气从双层表皮之间
的缝隙中流入室内

NORTH FACADE
SOLAR GAIN CONTROL
THROUGH LOUVRES
北立面通过百叶窗
控制太阳能获得

玻璃立面
GLAZED FACADE

EAST FACADE
SOLAR GAIN CONTROL
THROUGH FACADE SKIN
东立面通过外观表皮
控制太阳能获得

WEST FACADE
SOLAR GAIN CONTROL
TROUGH FACADE SKIN
西立面通过外观表皮
控制太阳能获得

SOUTH FACADE
SOLAR GAIN CONTROL
THROUGH LOUVRES
南立面通过百叶窗
控制太阳能获得

N

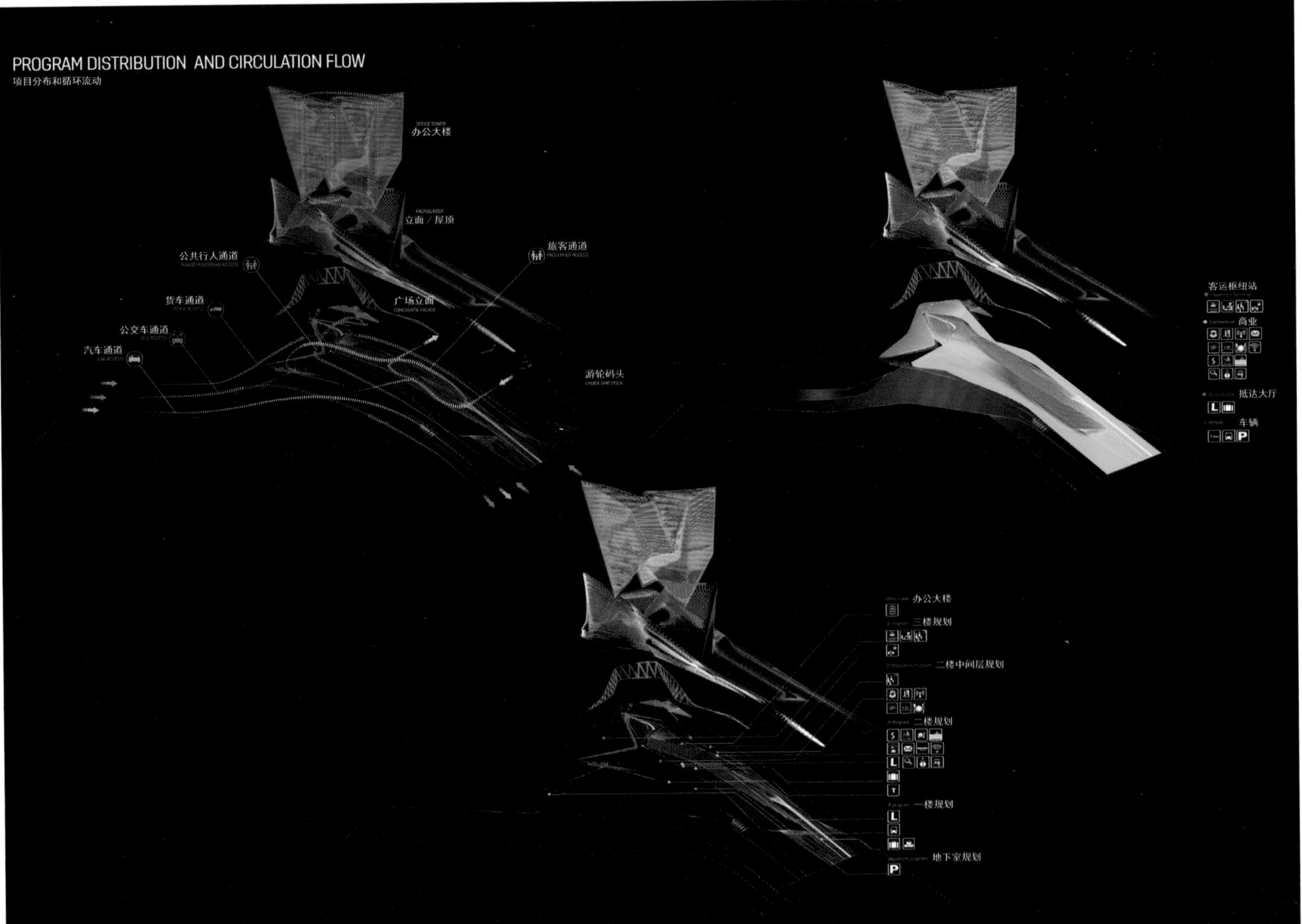

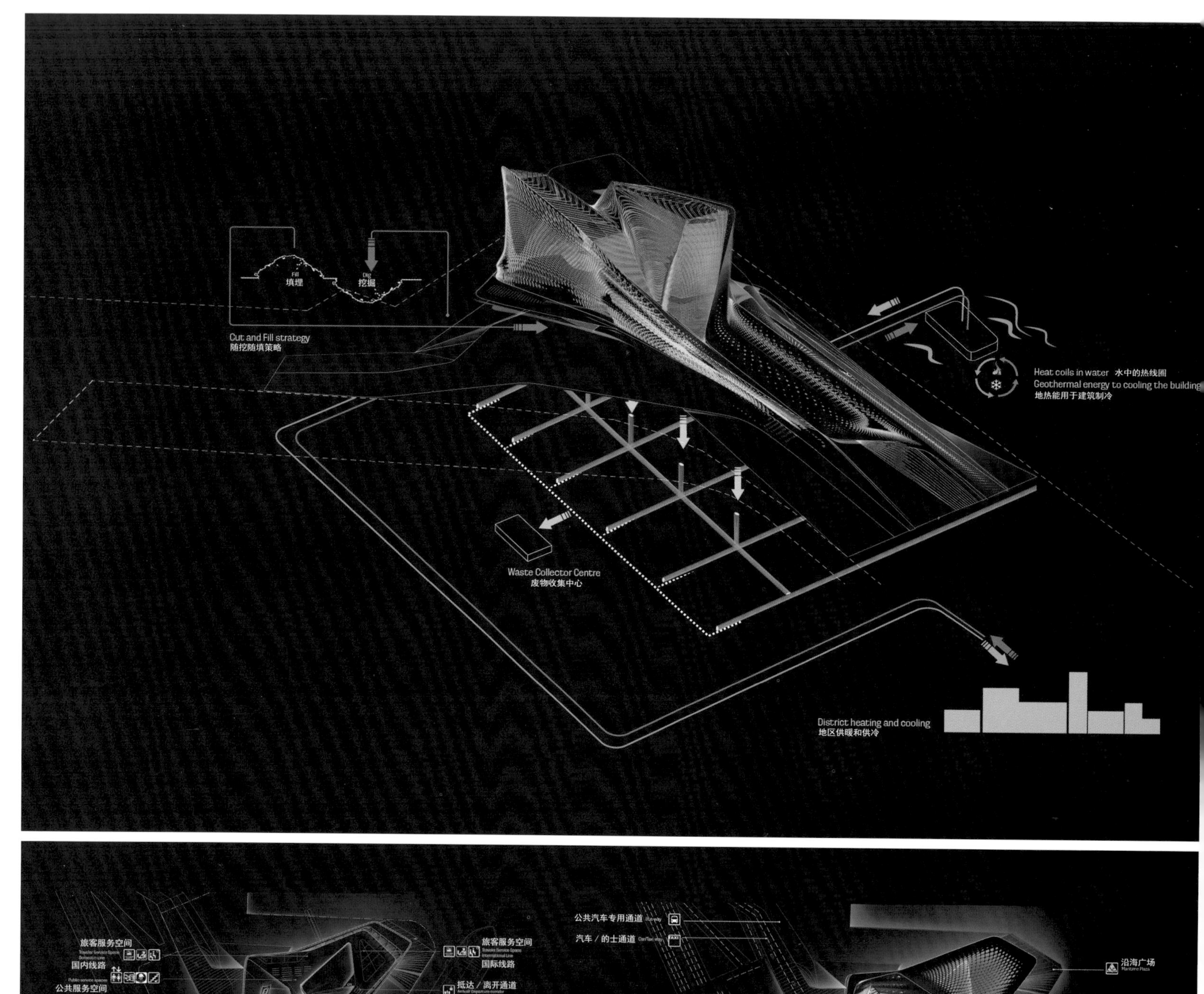

Fill
填埋
Dig
挖掘
Cut and Fill strategy
随挖随填策略
Heat coils in water 水中的热线圈
Geothermal energy to cooling the building
地热能用于建筑制冷
Waste Collector Centre
废物收集中心
District heating and cooling
地区供暖和供冷

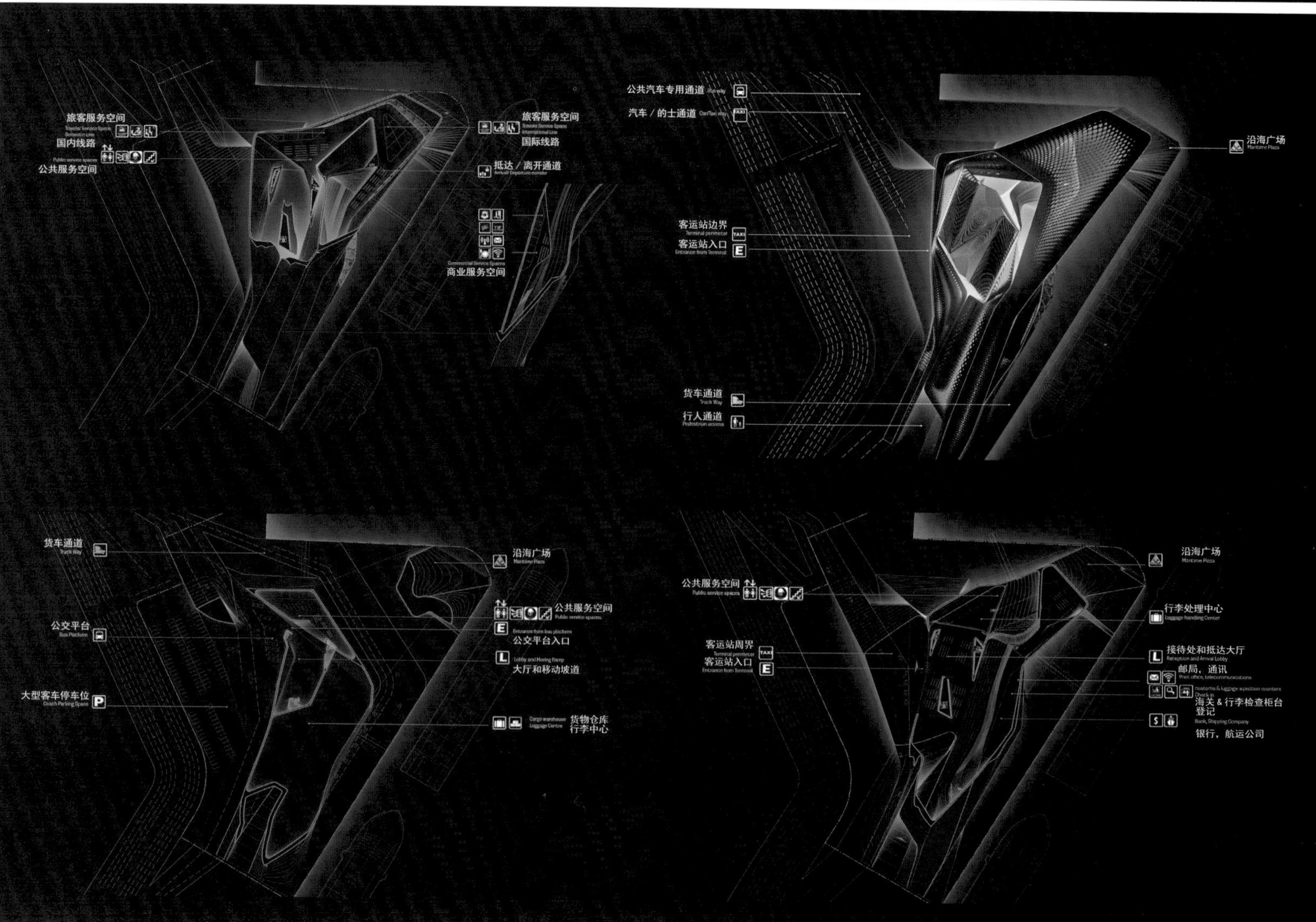

旅客服务空间
国内线路
公共服务空间
旅客服务空间
国际线路
抵达／离开通道
商业服务空间
公共汽车专用通道
汽车／的士通道
沿海广场
客运站边界
客运站入口
货车通道
Truck Way
行人通道
货车通道
Truck Way
公交平台
Bus Platform
大型客车停车位
Coach Parking Space
沿海广场
Maritime Plaza
公共服务空间
Public service spaces
Entrance from bus platform
公交平台入口
大厅和移动坡道
货物仓库
行李中心
公共服务空间
Public service spaces
沿海广场
Maritime Plaza
行李处理中心
客运站周界
客运站入口
Entrance from Terminal
接待处和抵达大厅
Reception and Arrival Lobby
邮局、通讯
海关 & 行李检查柜台
登记
银行、航运公司

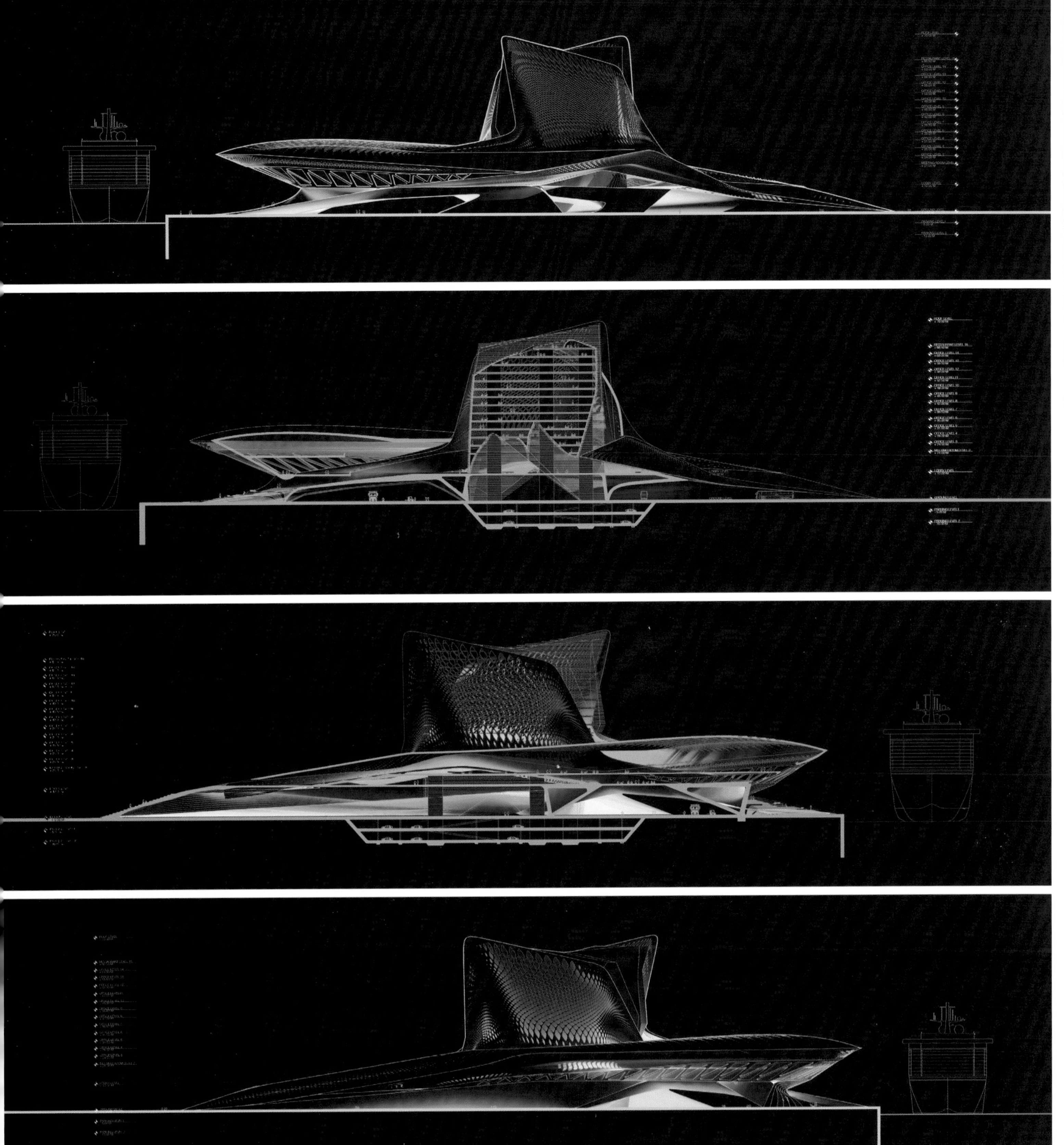

Profile

The proposal for the New Keelung Harbor terminal building focuses on providing a coherent, elegant, and iconic solution which signals the formation of a new identity for Keelung Harbor.

Architectural Design

Inspired by the geometric patterns of Taiwanese Hen Cages and the structural shells of luxury racing yachts, the building takes shape in a dynamic gradient form that transforms from exo-skin to exo-skeleton in response to programmatic content as well as performative requirements. The building program is divided into three primary experience groups and cyclical sequences. These three groups share programmatic overlaps and transitions which allow exchange between groups.

Thus, what appears to be formal expression, is actually "informed form" which responds to the integration of weather, urban context, program, circulation, and sustainability through integrated design responses that inform the building orientation, spatial layout, façade design, and choice of material and structural system.

Integrate visual, structural, and environmental performance as generative design concepts rather than additive design solutions by optimizing building orientation and massing for passive environmental design and control, harvesting wind, rain, and sun through geometric configuration, plugging in to district heating/cooling and waste disposal systems, and integrating structure and skin through geometric and material composition.

挪威奥斯陆中心车站

Oslo Central Station

设计单位：JSA-Jensen & Skodvin Arkitektkontor
项目地址：挪威奥斯陆

Designed by: JSA-Jensen & Skodvin Arkitektkontor
Location: Oslo, Norway

项目概况

奥斯陆中心车站将是挪威最重要的车站，设计旨在重塑有序、宽敞的公共空间，使其更好地连接到周围的城市肌理当中。

设计构思

出于对可持续发展的考虑，设计倾向于提高铁路运输在未来交通运输中的重要性，通过进一步扩大奥斯陆中央车站的容纳量，缓解当前交通拥堵的压力。

建筑设计

为了获得更加宽敞的空间，使其结构更加明晰，设计移除了部分不必保留的已有结构。项目的主要功能区分布在独立体量的第一、二、三层，在主体空间内形成了一个个“阶梯状岛屿”的室内空间。

车站的主要楼层是一个倾斜的平面结构，位于新兴隧道的上方，建立了与城市地铁的连接关系。这个平面结构从街面一直延伸到较高水平面的广场，跨越了南北方向的轨道。

屋顶部分是一个两层的建筑，屋顶近 60 000 平方米的楼层区内设有商业区，在距离公共交通设施非常近的范围内，为人们提供了一个极其高效的空间，大大提高了项目的商业开发价值。

Profile

Oslo Central Station is the most important station in Norway. The proposal aims to reestablish well ordered and ample public spaces that are well connected to the surrounding urban fabric.

Design Concept

For reasons of sustainability there is a desire to increase the importance of railway transport in the future, thereby increasing further the strain on the capacity of the already congested junction at Central Station.

Architectural Design

To allow for a generous space which is easy to read, existing structures not worth preserving are removed. The functions of the program are distributed in separate volumes of one, two or three levels, much like an interior landscape of terraced islands within the volume of the main space.

The main floor of the station is an inclined plane sitting on top of the emerging tunnel which constitutes the main rail connection under the city. This inclined floor runs continuously from the street level up to the higher concourse traversing the tracks in the north-south direction.

The roof is actually a 2-storey building containing areas for commercial development as much as 60,000 square meters. This allows for extremely efficient spaces with unparalleled proximity to public transport, yielding a high potential for commercial exploitation.

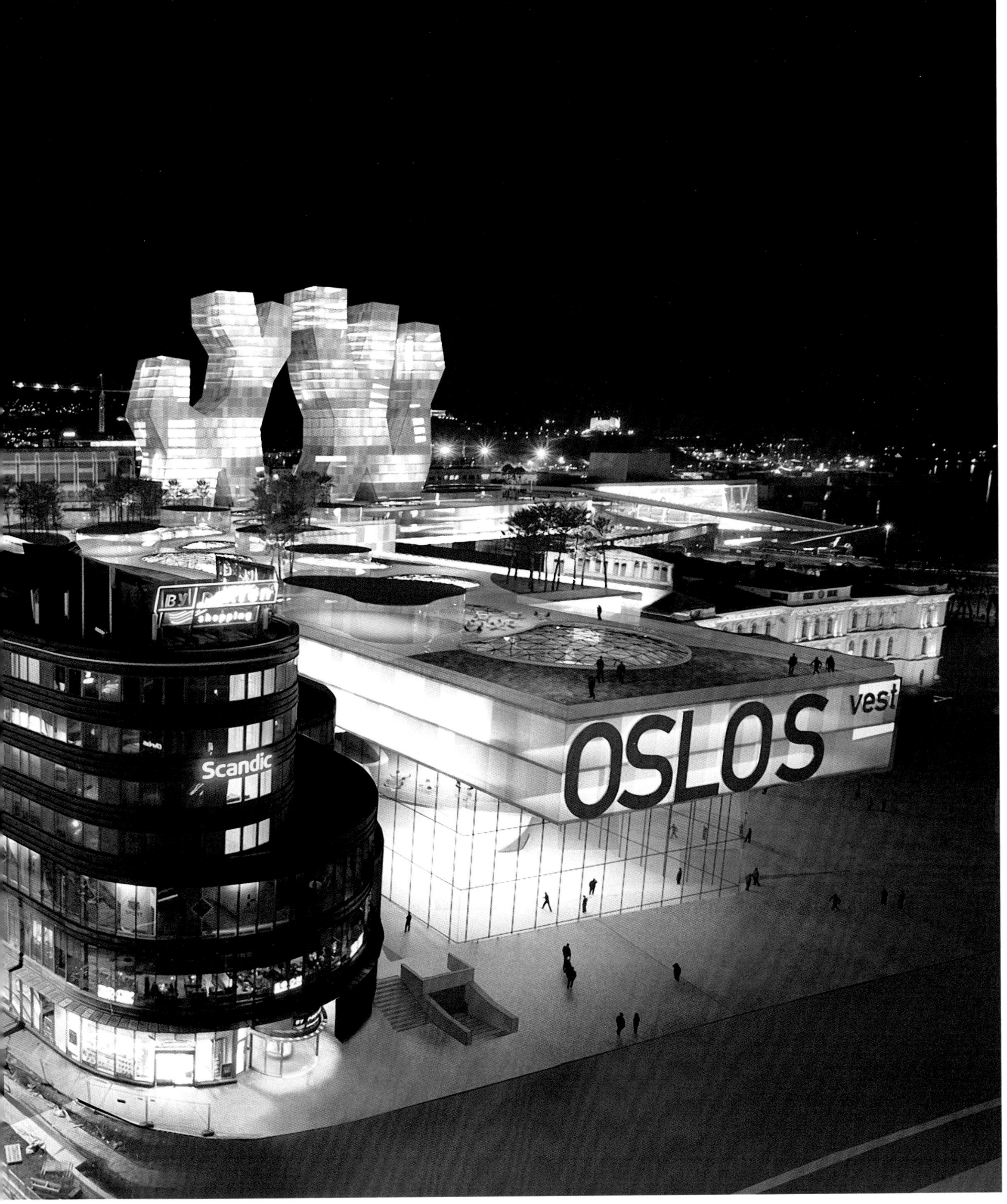
OSLOS
vest
Scandic
Shopping

BY PORTEN
BUSS

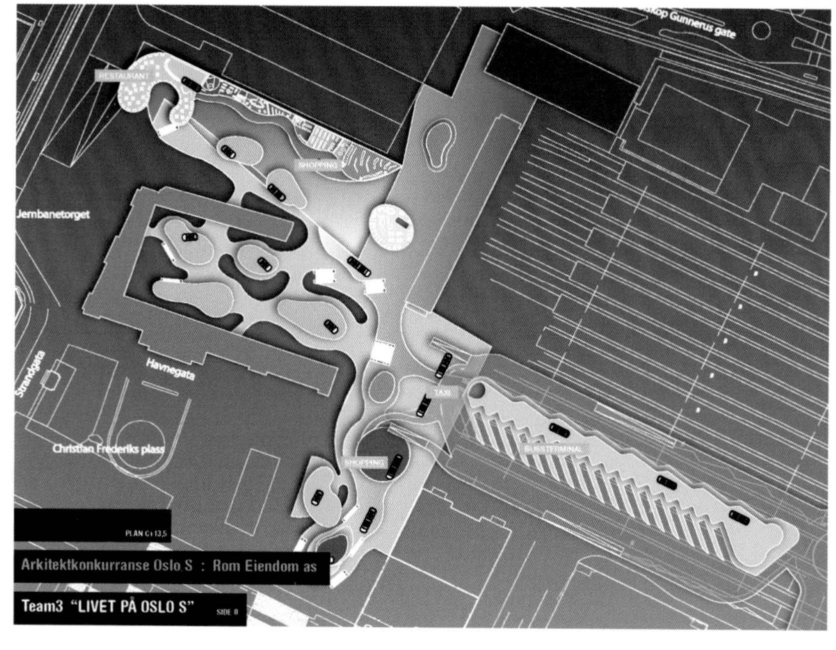
Biskop Gunnerus gate
RESTAURANT
SHOPPING
Jernbanetorget
Havnegata
Strandgata
TAXI
BUSSTERMINAL
Christian Frederiks plass
SHOPPING
PLAN C+13,5
Arkitektkonkurranse Oslo S : Rom Eiendom as
Team3 "LIVET PÅ OSLO S" SIDE 8

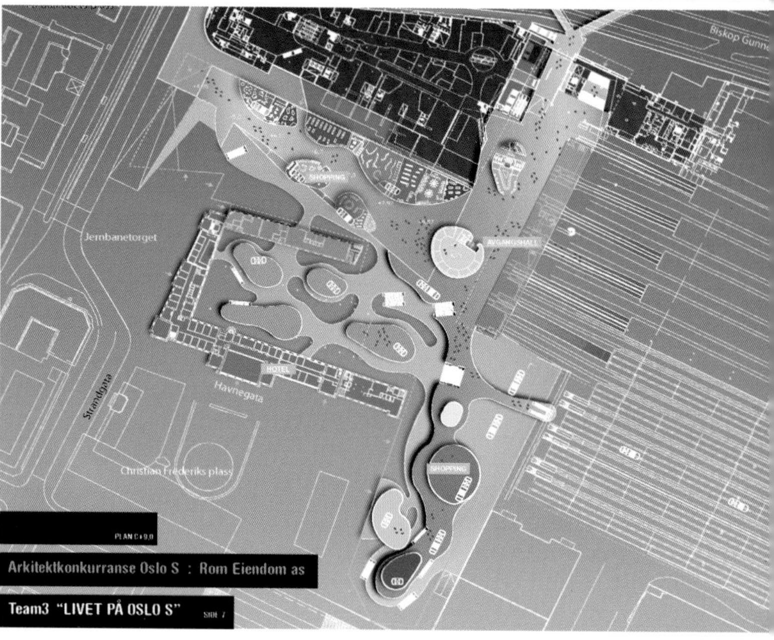
Biskop Gunne
SHOPPING
Jernbanetorget
AVGANGSHALL
HOTEL
Havnegata
Strandgata
Christian Frederiks plass
SHOPPING
PLAN C+9,0
Arkitektkonkurranse Oslo S : Rom Eiendom as
Team3 "LIVET PÅ OSLO S" SIDE 7

OSLOS
vest

OSLOS

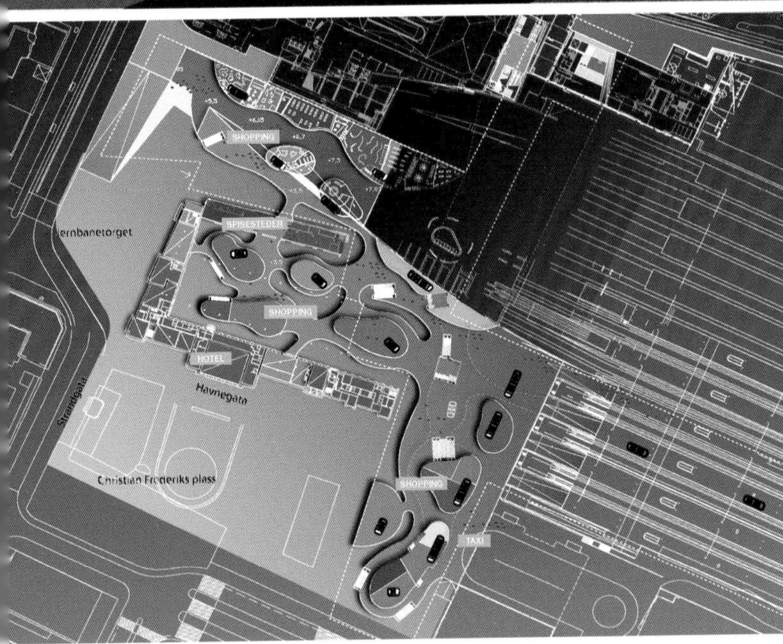
SHOPPING
Jernbanetorget
SPISESTEDER
SHOPPING
HOTEL
Havnegata
Christian Frederiks plass
SHOPPING
TAXI

Jernbanetorget
TRAFIKANTEN
SHOPPING
SALUHALL
Havnegata
Strandgata
Fred. Olsens gate
Prinsens gate
Christian Frederiks plass
SHOPPING
HOTELLOBBY
Arkitektkonkurranse Oslo S : Rom Eiendom as
Team3 "LIVET PÅ OSLO S"

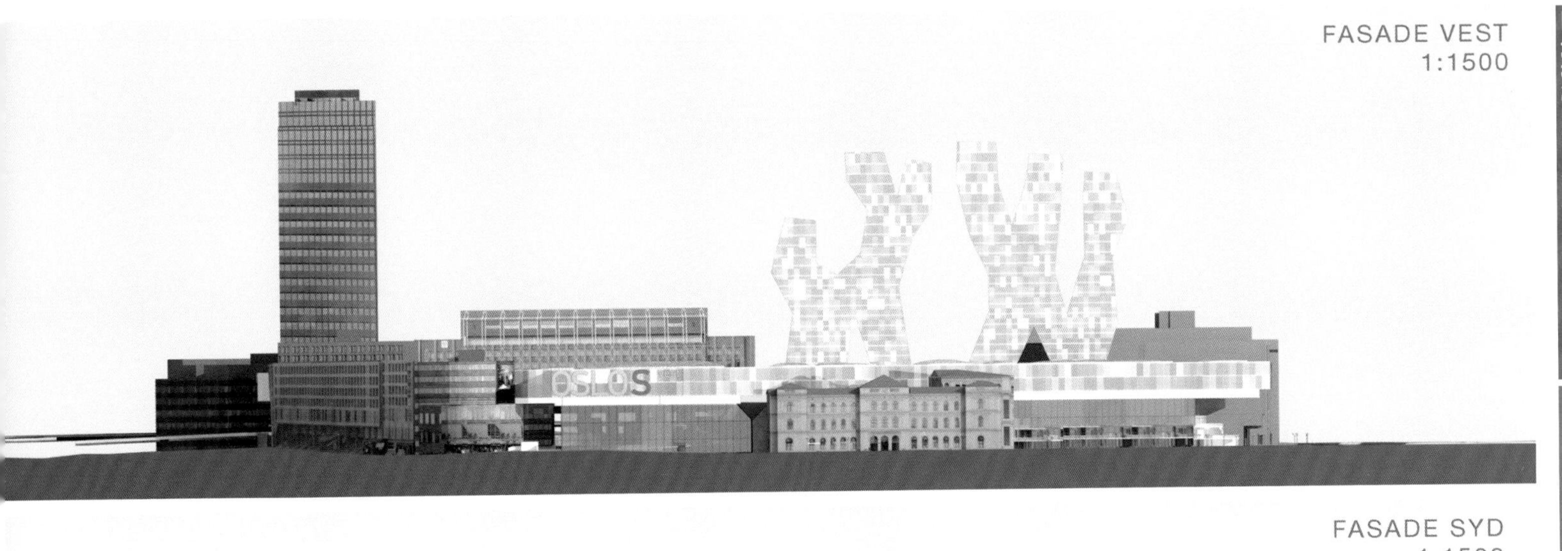

FASADE VEST
1:1500

FASADE SYD
1:1500

FASADE ØST
1:1500

FASADE NORD
1:1500

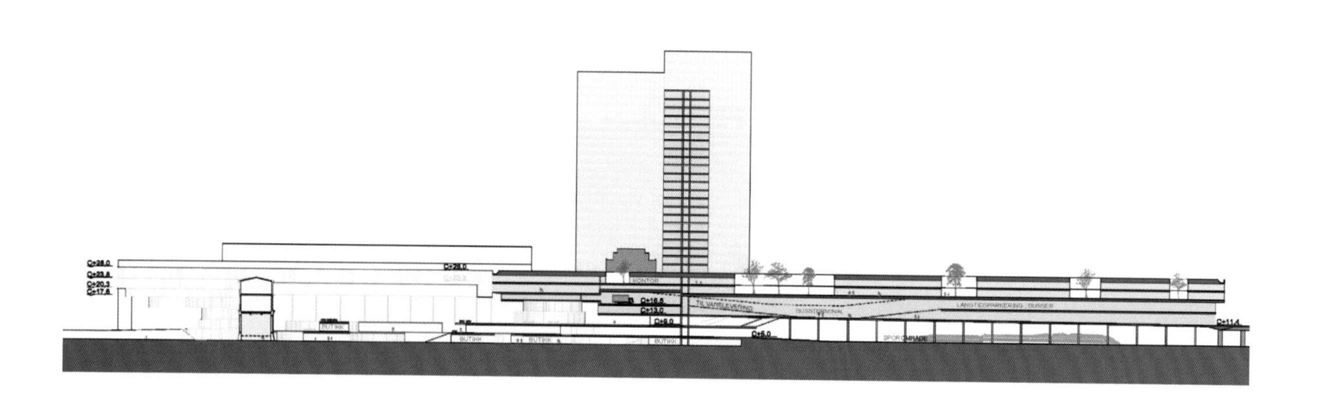

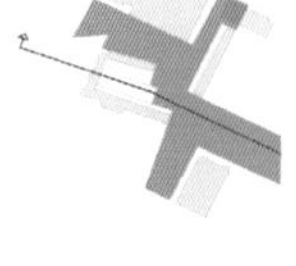

SNITT A-A
1:1500

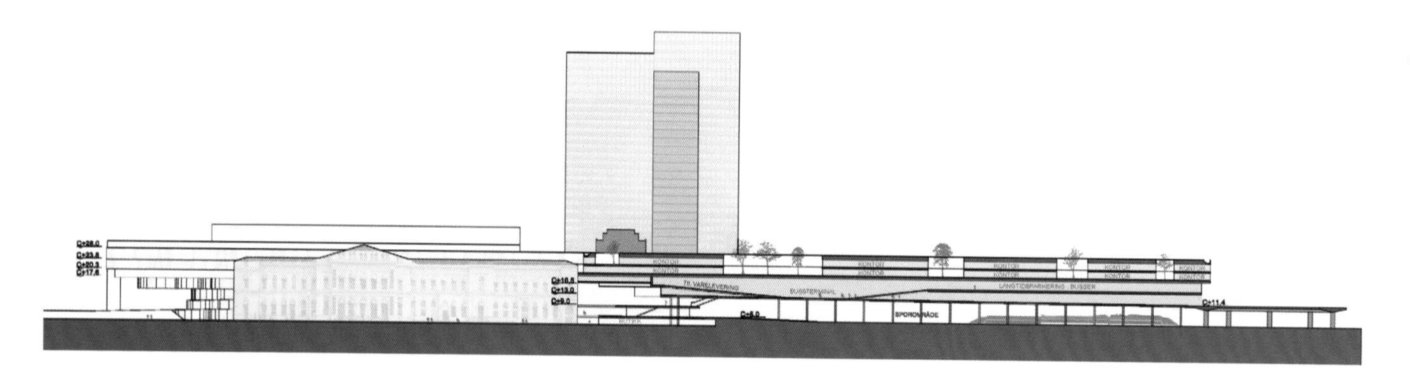

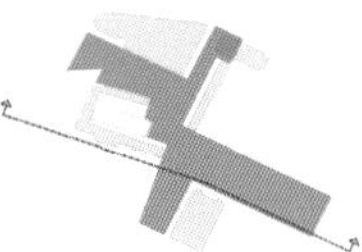

SNITT B-B
1:1500

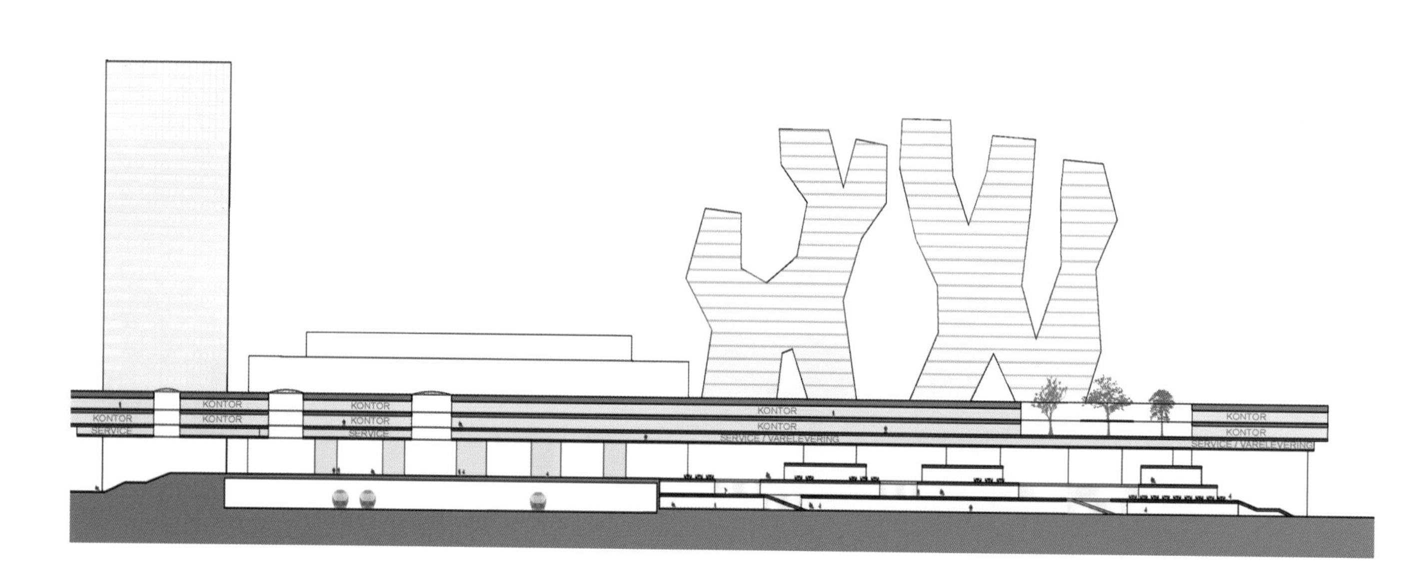

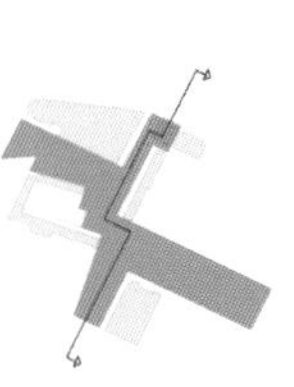

SNITT C-C
1:1000

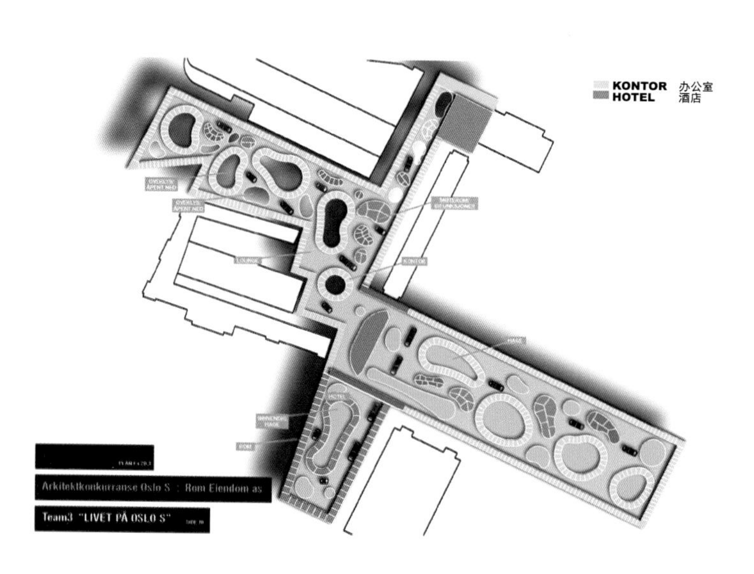

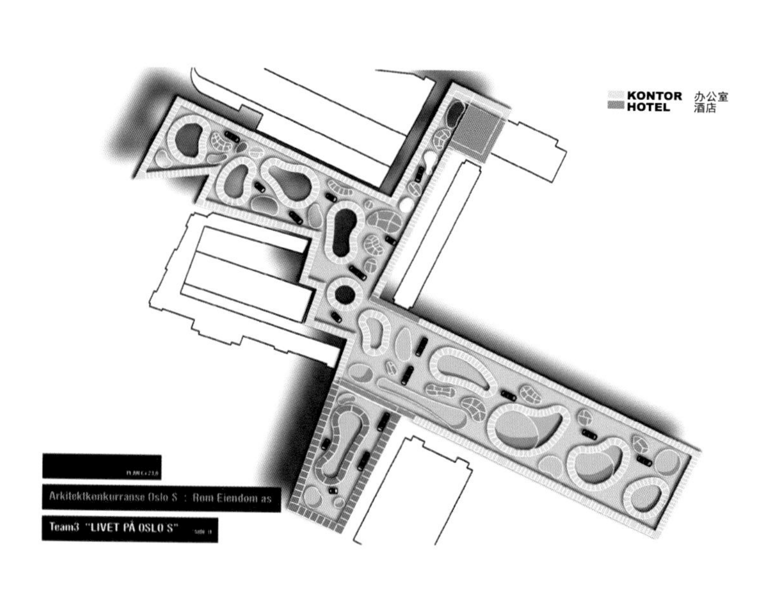

SNITT F-F
1:1000

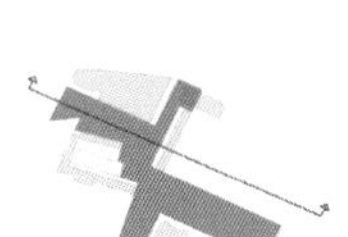

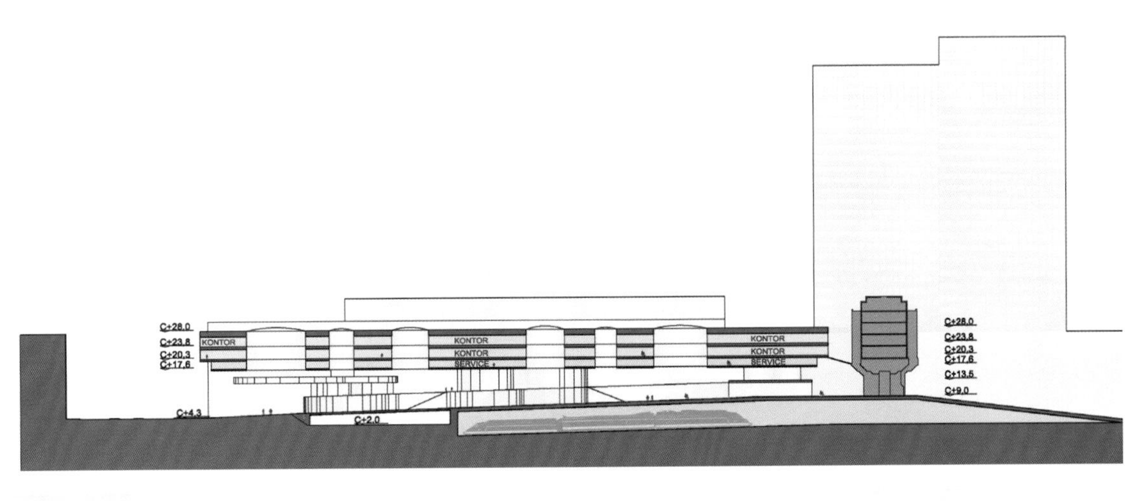

SNITT D-D
1:1000

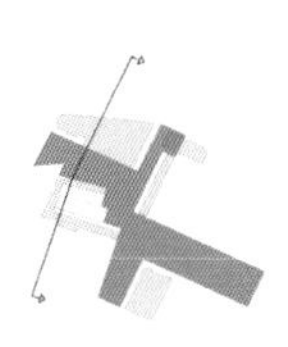

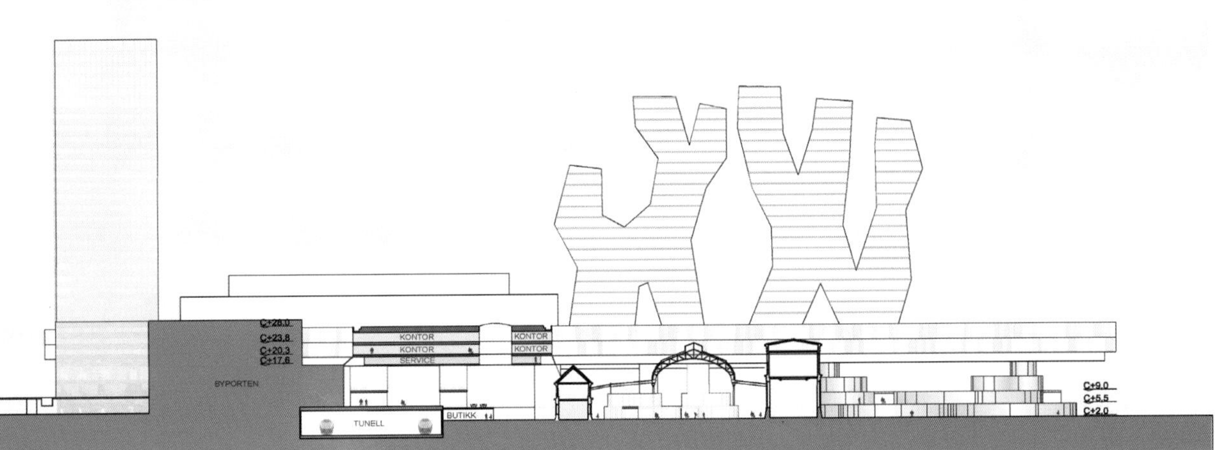

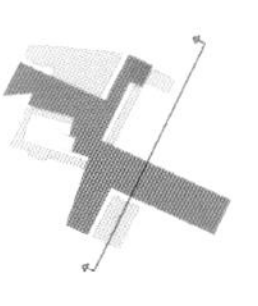

SNITT E-E
1:1000

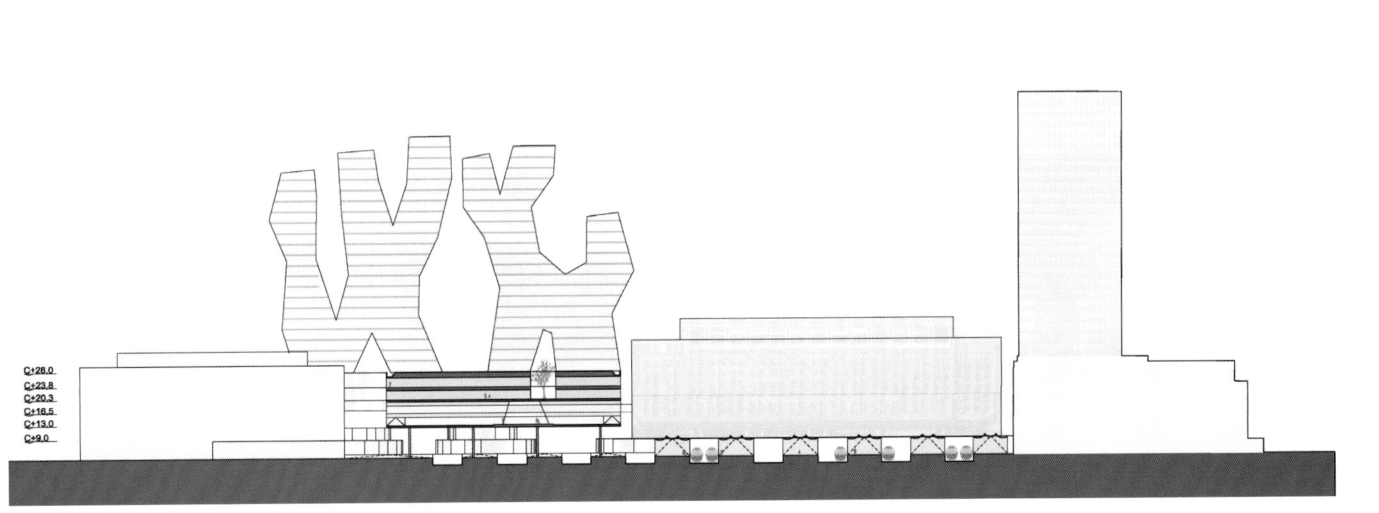

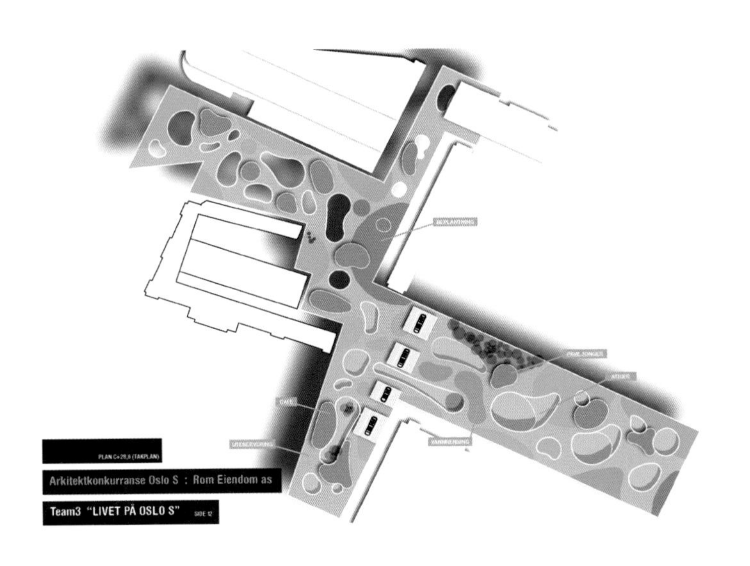

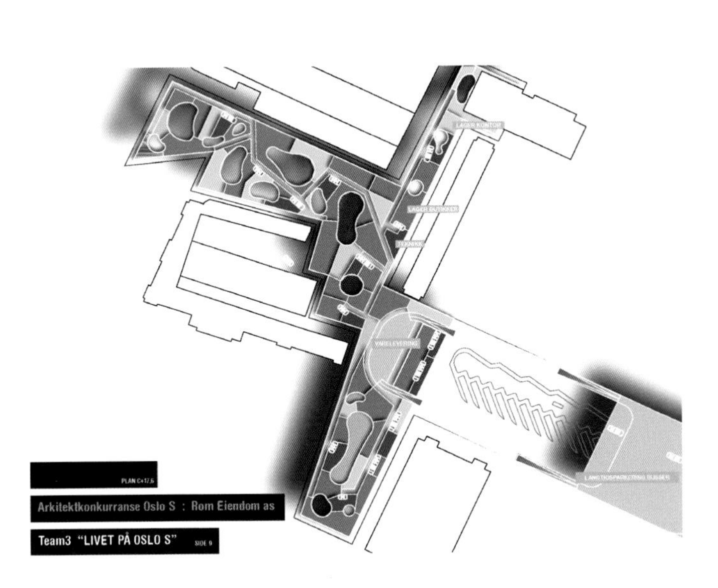

保加利亚索菲亚 20 地铁站

Metro Station 20

设计单位： M.S.B Arquitectura e Planeamento

项目地址： 保加利亚索菲亚

***Designed by:* M.S.B Arquitectura e Planeamento**

***Location:* Sofia, Bulgaria**

项目概况

项目场地位于索菲亚市中心的边缘地带，是一个非城市化的地区。设计旨在通过结构的规划和线条的梳理，构建便捷的流通空间，同时为市民提供更多的地下和地上公共设施。这些公共设施的建设，不仅有利于推动新城市开发，同时也实现了建筑功能、语汇及视觉效果的统一。

设计特色

这一地区市民的生活方式和交通习惯是这一方案的关键性因素，无论是临街面还是入口大厅或是公共平台区的设计，都重点考虑了这些因素，以实现不同层次间的交流与对话，保持建筑的整体统一性。

根据周围道路的特征和设定的街道标高，方案设计了 4 个出入口，更加方便行人通行的同时，也确保了项目建成后这一地区的舒适度。这些出入口隐藏在城市网络框架中，如雕塑般在水平方向上延伸，具有极强的视觉冲击力。

较低的层面由一个强有力的支柱网络支撑，设计保留了这些支柱，并将它们作为重要的元素来定义这一空间。设计将这些支柱结合起来，单个的支柱通过环路与其他支柱连接，共同构成一个有着多个圆顶的空间，这些圆顶结构共享等候区的平台和通往地下层的入口，不仅提高了项目的舒适度，同时也丰富了建筑的表达形式。

Profile

The site is located in a peripheral zone of the city center, a non-urbanized area. The proposal aims to create passing spaces, with the creation and delineation of structures and trees that provide the amenities for this occupation. These amenities shall facilitate the development of this new area with a philosophy of unity at a visual and functional language.

Design Feature

The way in which citizens live in this space, and walk through was a determining factor in the collection of possible solutions, both at the street level zone, as in the entrance hall, and finally to the platform zone. The sense of unity and communication between levels, in order to establish a dialogue, is an important message to retain.

According to the characteristics of the surrounding paths and assumed level of the road, designers have proposed four entrances & exits to allow more and better movement of pedestrians and ensure the necessary comfort to this intervention. These entries are embedded within the framework of the urban network, assuming a sculptural character which extends horizontally and creates strong visual impact.

The lower level is characterized by a strong network of pillars. Rather than eliminate them, the proposal sought to take advantage of these elements – as objects that define this space. The approach is to unite the pillars that glide through hoops to each other, creating a space of multiple domes that share platforms in waiting areas and entrance to the underground, creating a wealth of great comfort and formal richness.

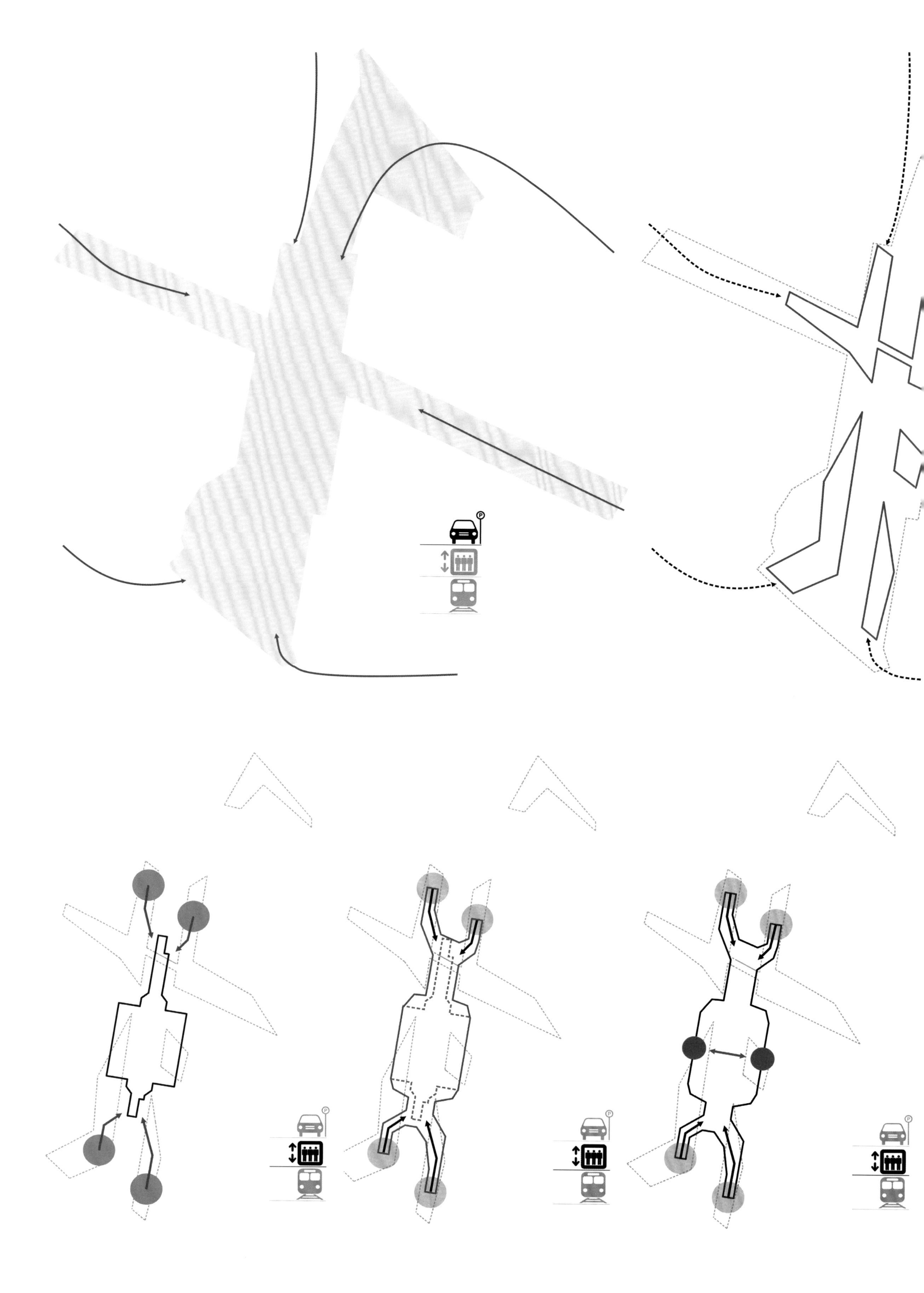

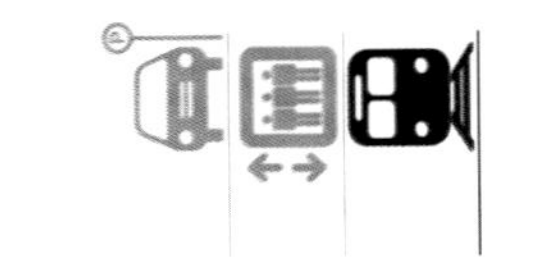

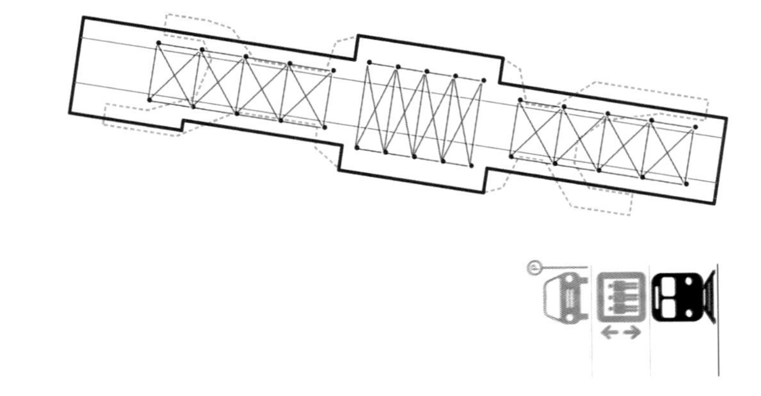

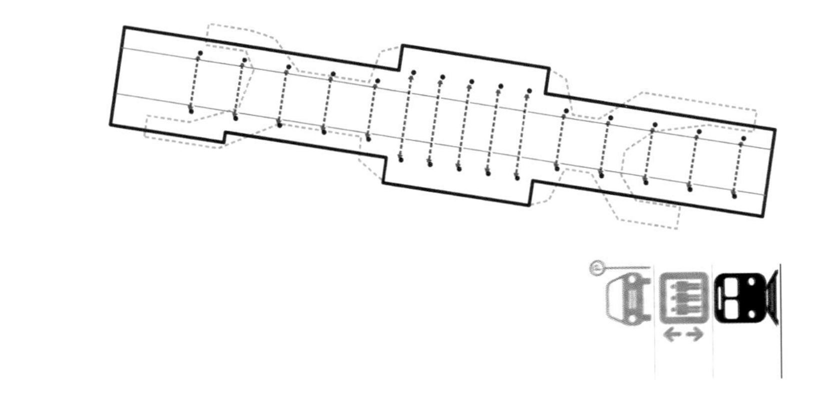

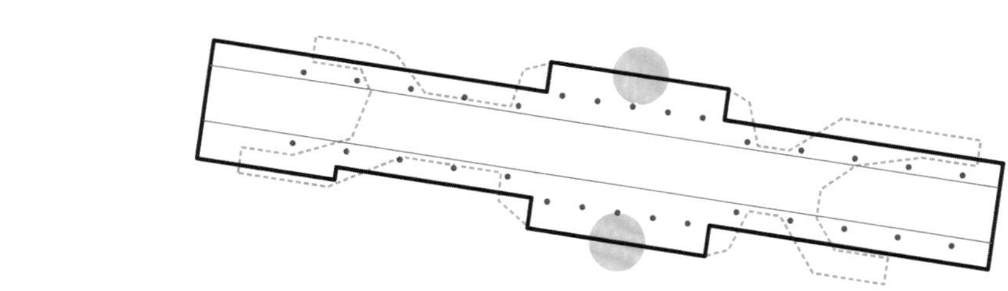

比利时沙勒罗伊消防站

Charleroi Fire Station

设计单位：SAMYN and PARTNERS， Architects & Engineers

项目地址：比利时沙勒罗伊

项目面积：19 721 ㎡

Designed by: SAMYN and PARTNERS, Architects & Engineers

Location: Charleroi, Belgium

Area: 19,721 m²

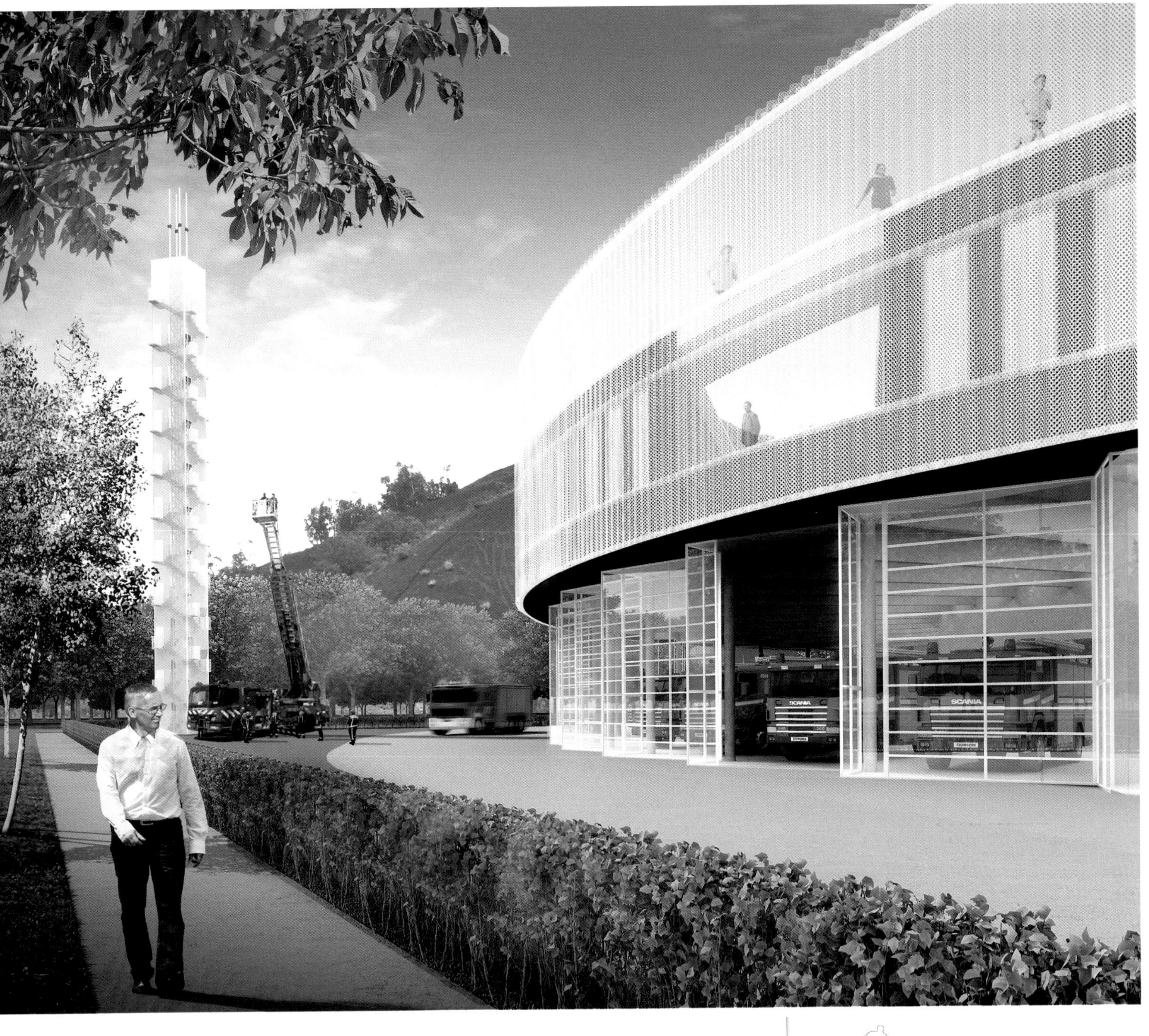

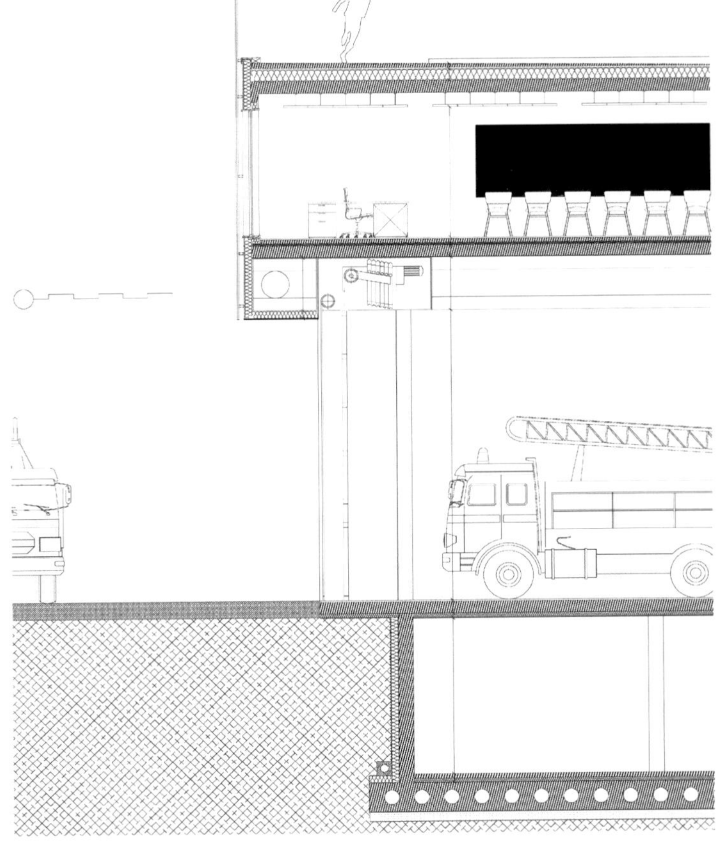

项目概况

这个简约、现代、紧凑的建筑，遵循了当地的场所特征，展现了一个标志性的、可识别的、与周围环境相融合的建筑形态。

建筑设计

项目圆形的建筑形态建立在保证应急车流快捷畅通的基础上，以确保紧急状况发生时，车辆和人员能够快速、安全地疏散。项目设计了两条循环车道：应急车辆返回车道和应急车辆离开车道，圆形的建筑形态维持了这两个车流的固有秩序，同时也可错开大厅内外的车辆调动，从而避免可能产生的混乱局面。

演习大楼位于应急道路的对面，以避免烟雾飘到邻近房屋。大楼的北侧被混凝土演习区包围，方便消防人员在此进行模拟救火活动。

所有的功能区都涵盖在这个 5 层高、直径为 90 米的圆柱体内，建筑紧凑的布局结构减少了裸露的建筑表皮面积，从而降低了能量的损失。

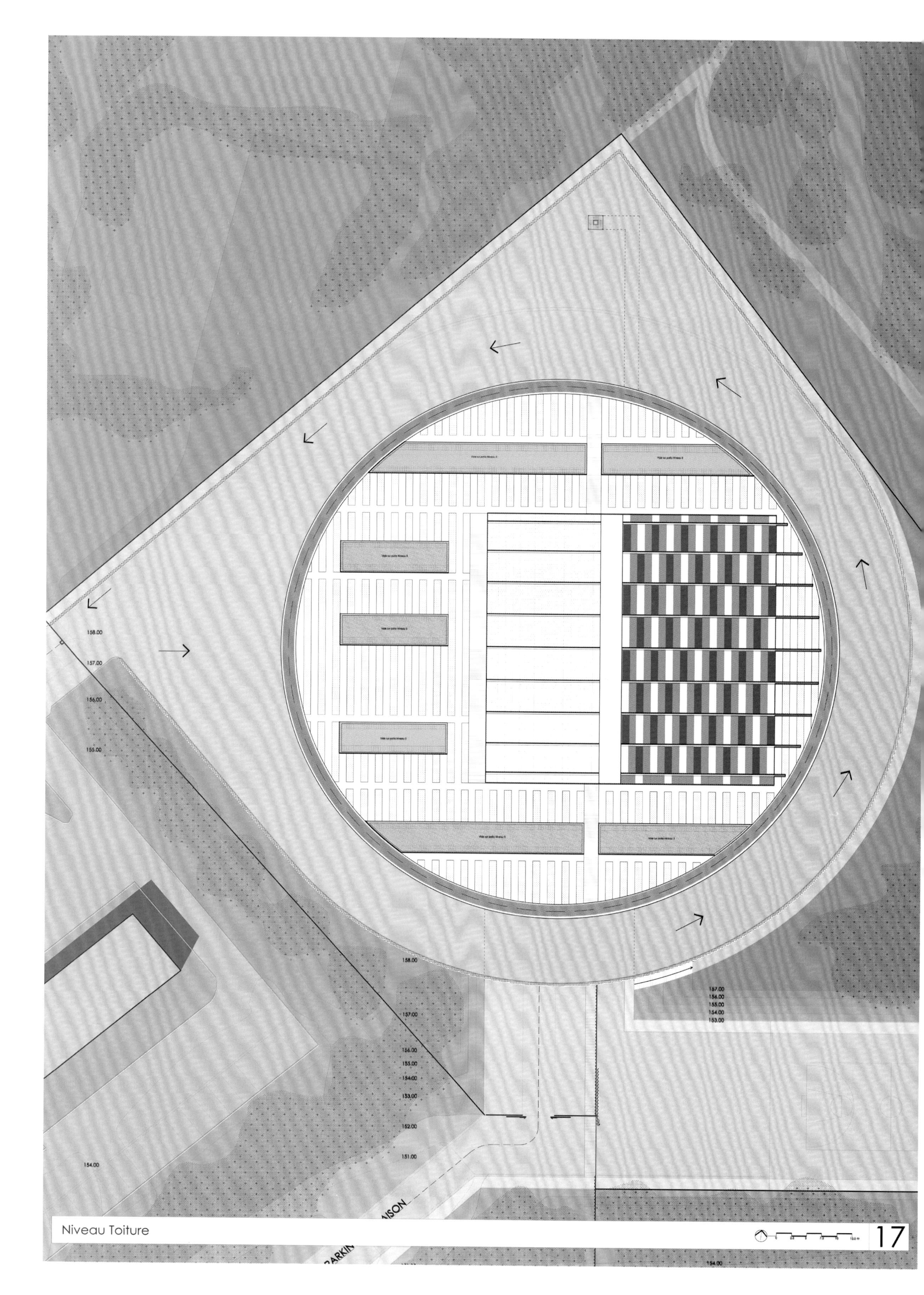

Niveau Toiture

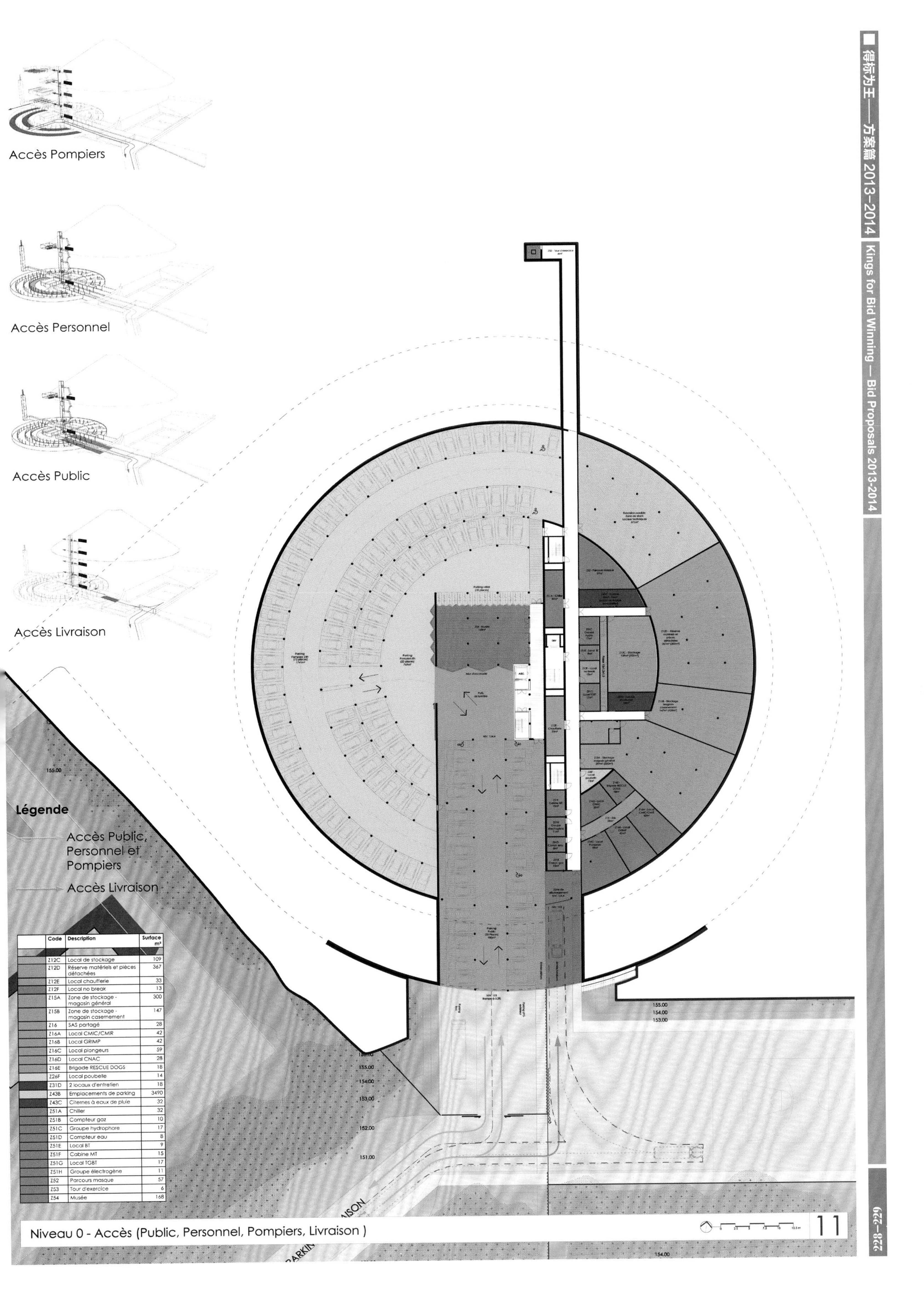

	Code	Description	Surface m²
	Z12C	Local de stockage	109
	Z12D	Réserve matériels et pièces détachées	367
	Z12E	Local chaufferie	33
	Z12F	Local no break	13
	Z15A	Zone de stockage - magasin général	300
	Z15B	Zone de stockage - magasin casemement	147
	Z16	SAS partagé	28
	Z16A	Local CMIC/CMIR	42
	Z16B	Local GRIMP	42
	Z16C	Local plongeurs	59
	Z16D	Local CNAC	28
	Z16E	Brigade RESCUE DOGS	18
	Z26F	Local poubelle	14
	Z31D	2 locaux d'entretien	18
	Z43B	Emplacements de parking	3490
	Z43C	Citernes à eaux de pluie	32
	Z51A	Chiller	32
	Z51B	Compteur gaz	10
	Z51C	Groupe hydrophore	17
	Z51D	Compteur eau	8
	Z51E	Local BT	9
	Z51F	Cabine MT	15
	Z51G	Local TGBT	17
	Z51H	Groupe électrogène	11
	Z52	Parcours masque	57
	Z53	Tour d'exercice	6
	Z54	Musée	168

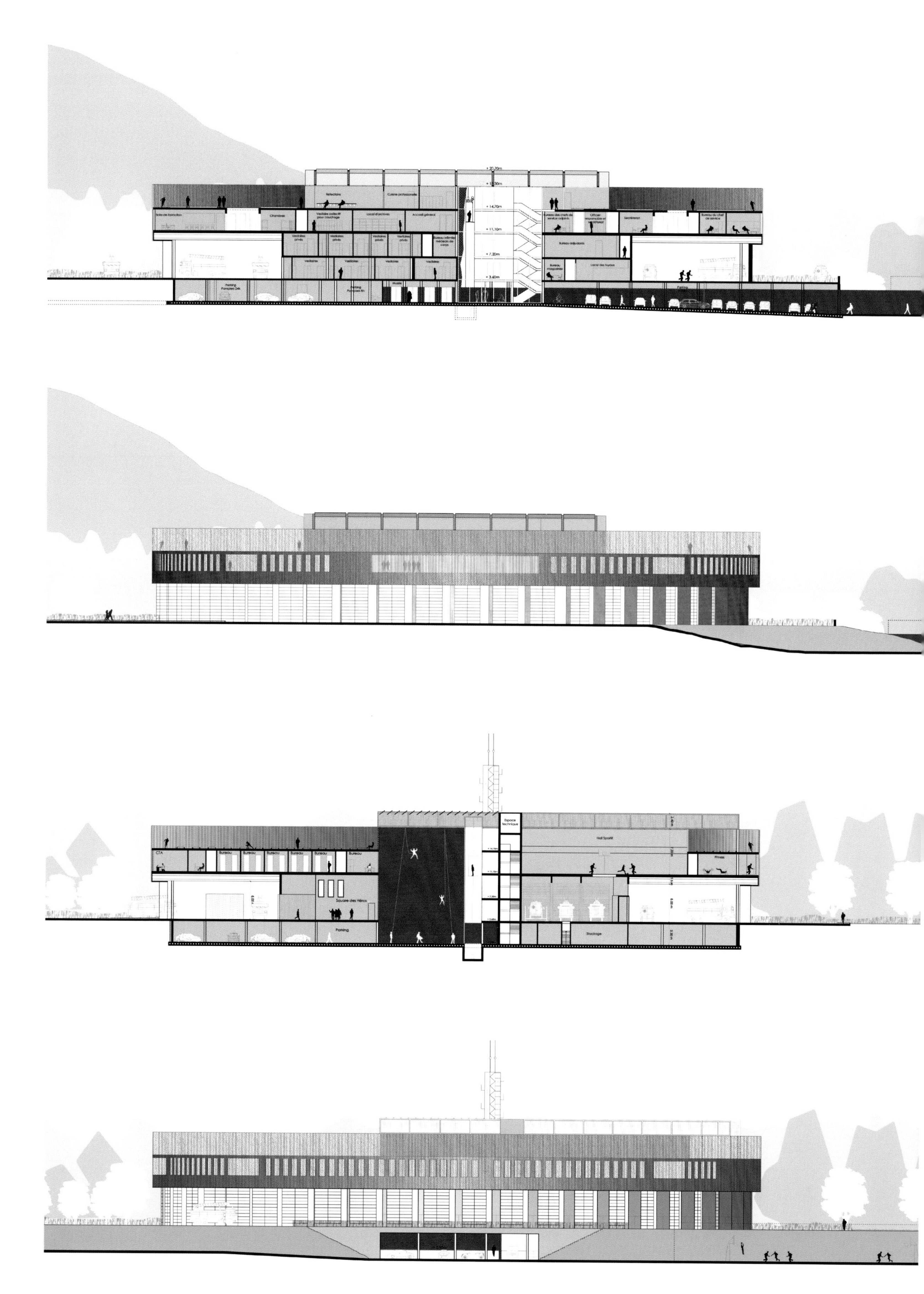
+ 20.70m
+ 14.70m
+ 11.10m
+ 7.20m
+ 3.60m
Réfectoire
Cuisine professionelle
Salle de formation
Chambres
Vestiaire collectif pour couchage
Local d'archives
Accueil général
Vestiaires privés
Bureau infirmier médecin de corps
Vestiaires
Parking Pompiers 24h
Musée
Bureau des chefs de service adjoints
Secrétariat
Bureau du chef de service
Bureau adjudants
Bureau magasinier
Local des tuyaux
Parking
CTA
Bureau
Square des Héros
Espace technique
Hall Sportif
Fitness
Stockage

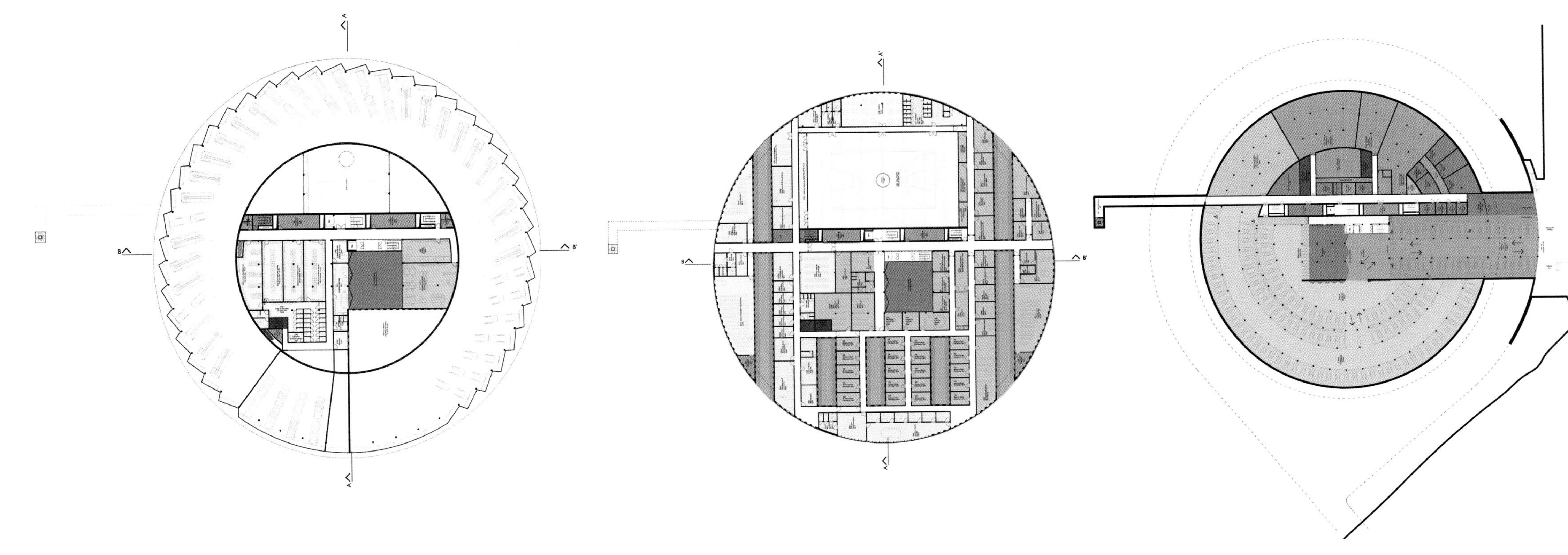

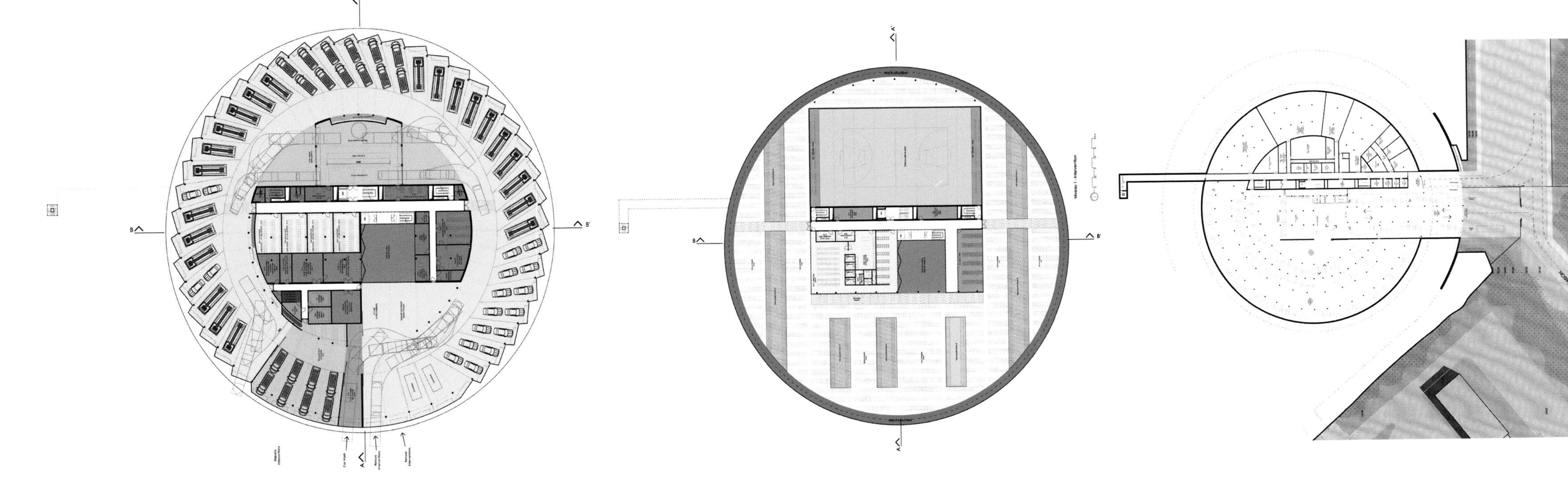
Niveau 1 - Intervention

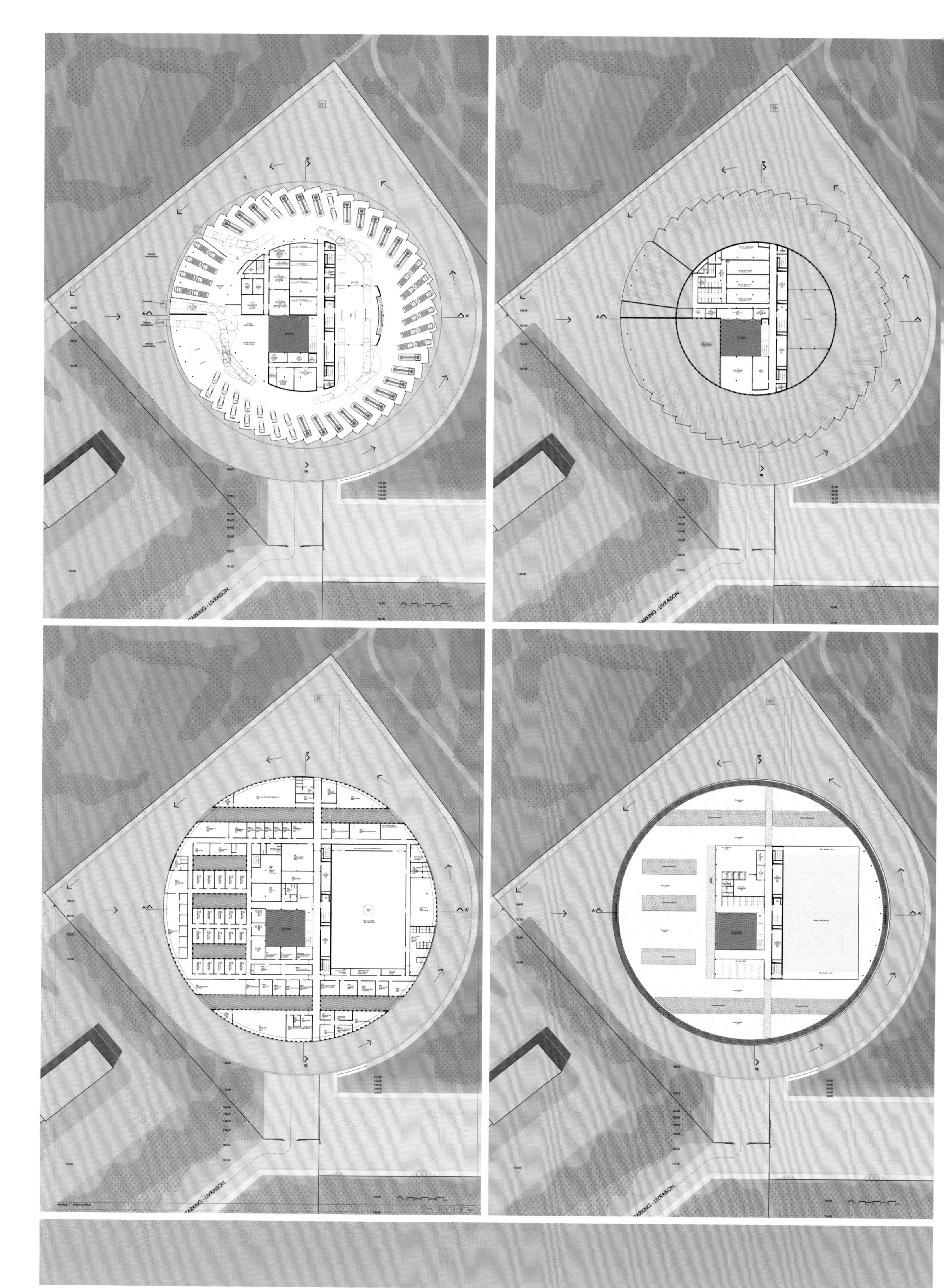

A
A'
PARKING - LIVRAISON
PARKING - LIVRAISON
PARKING - LIVRAISON
PARKING - LIVRAISON

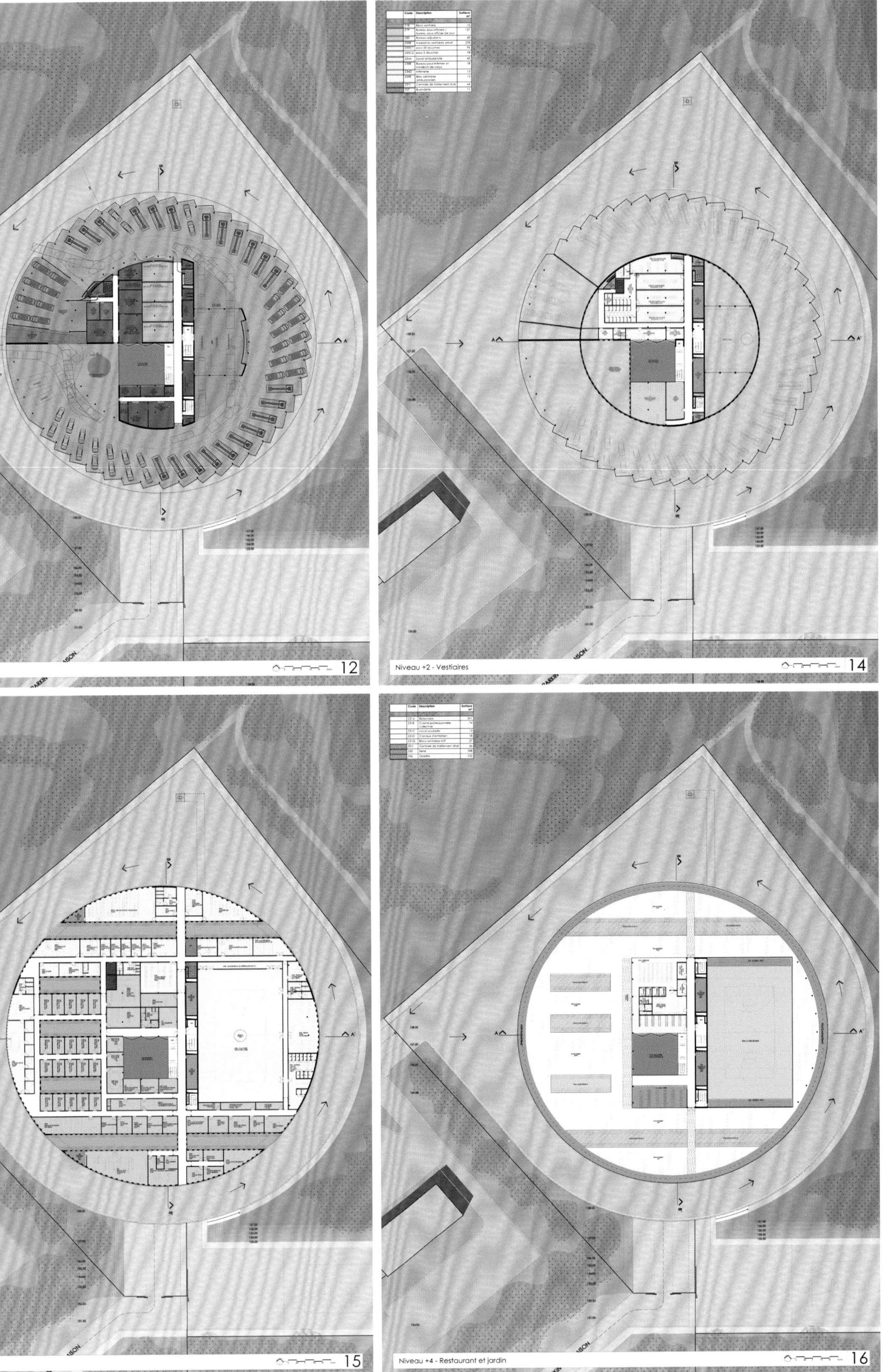
Légende
Escalier Ambulanciers
Toboggans Ambulanciers
Escalier Pompiers
Toboggans Pompiers
OUT
IN
Square des Héros
Niveau +1 - Intervention (IN, OUT)
12
Niveau +2 - Vestiaires
14
Niveau +3 - Administration et vie
15
Niveau +4 - Restaurant et jardin
16

Profile

Considering the specific characteristics of the site, the project has simple, contemporary, compact and functional architecture while offering an iconic and recognizable form, compatible with the character of the environment.

Architectural Design

The round form of the plan issued directly from the flow of emergency vehicles. It ensures a fast and safe transport of vehicles and people in case of emergency. Indeed, two circulation driveways are planned: the first, internal, for return from emergencies and the other, external, for departures. It has a round form in order to avoid any confusion between the two flows and to ensure that no manoeuvre is performed both inside and outside of the great hall.

The exercise tower is situated across the emergency road. It is surrounded by a concrete exercise area to the north, which allows trucks to easily turn around the tower and to simulate fires. The site report suggests establishing the area as a place for fire drill exercises to the north of the field, in order to avoid smoke drifting toward neighboring houses.

All functions are housed in this 90 meters diameter and 5 level high cylinder. The compactness of this volume reduces surfaces with external exposure and thus reduces energy loss.

住宅建筑

Residence

阿联酋阿布扎比雪花大厦

Snowflake Tower

设计单位：LAVA
项目地址：阿联酋阿布扎比
建筑面积：106 550 m²

Designed by: LAVA
Location: Abu Dhabi, UAE
Gross Floor Area: 106,550 m²

项目概况

迈克尔－舒马赫世界冠军大楼是由 F1 传奇赛车手迈克尔－舒马赫提议建造的，全球将建成 7 栋这样的大楼，雪花大厦就是这一建筑系列的第一座。建筑位于迪拜首都阿布扎比 ReemIsland 中央商务区，耸立在海滨一隅，与大海融为一体。

设计理念

设计旨在将品牌文化融入一个具有标志性意义的建筑之中。设计师从传统角度出发，并从雪花的几何逻辑形态和一级方程式赛车的空气动力学中获取灵感，构建了一栋集速度感、流体动力学、未来技术和天然的几何形态于一体的建筑。

建筑设计

立面上连续的表皮使建筑呈现出重复弯曲的形态，螺旋状的楼体结构凸显了流线形的建筑曲线。私人阳台位于垂直设计的狭长凹槽中，极富特色。鳞片状的反光设计使整栋大厦“活”了起来，建筑外观也随之发生变化。

大楼较低的楼层通常被认为是最难设计和最没有吸引力的区域，在本方案中，设计师将这一区域设置为码头公寓，阶梯状的设计使之恰似一艘巡洋舰的甲板，建筑的基底也得以扩展，显得独特又别致。建筑顶端的空中别墅享有 270 度的景观视角，人们可从这里看到对面 Saadyiat 岛屿上的新文化区。

Profile

Formula 1 legend Michael Schumacher presented the design for The Michael Schumacher World Champion Tower, the first in a series of seven towers planned worldwide. Snowflake Tower is located in ReemIsland CBD of Abu Dhabi, UAE, standing on the seashore and integrating with the ocean.

Design Concept

The design aims to develop a prototypical building translating brand values into an iconic architecture. The project is inspired by the geometrical order of a snowflake and the aerodynamics of a Formula 1 racing car, the tower encapsulates speed, fluid dynamics, future technology and natural patterns of organization.

Architectural Design

The facade's continuous surface enables curvature with a lot of repetition. Spiral building structure highlights streamlined building curves. The building features a facade characterized by vertical slots with private balconies. A series of reflective fins generates vertical dynamic and gives the building a constantly changing appearance.

The lower levels of the tower, traditionally the most difficult and least attractive area, were reinterpreted as a series of prestigious wharf apartments, terraced similar to that of cruise ship decks. The extended base of the building appears unique and distinctive. The top sky villas offer 270-degree views opposite the new cultural district on Saadyiat Island.

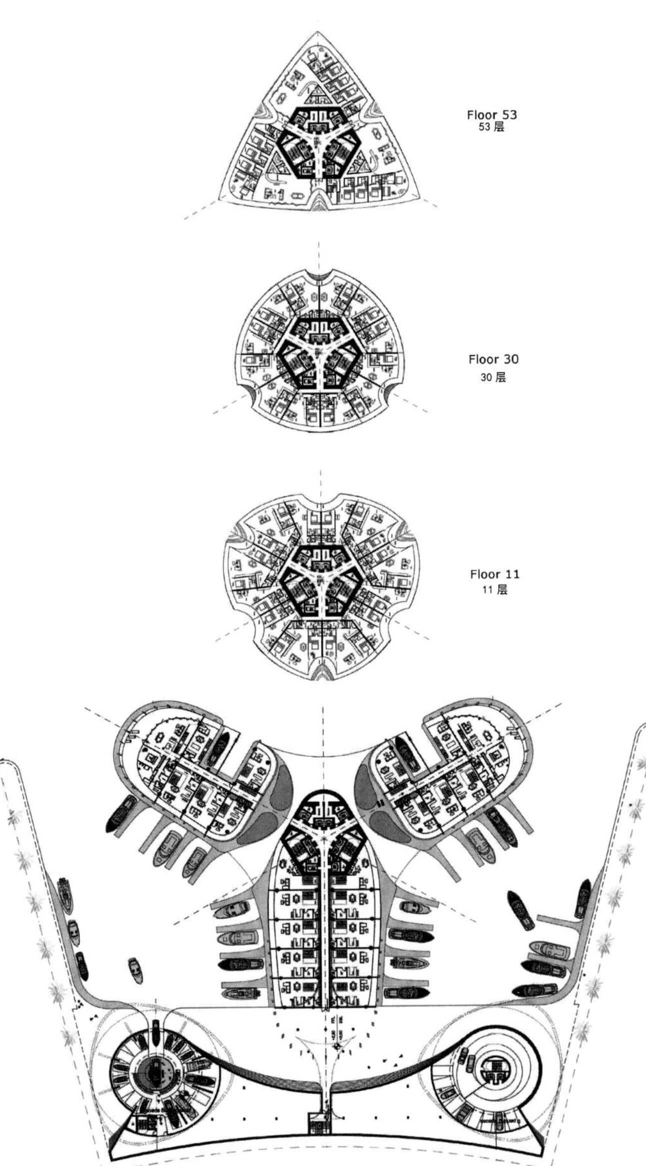
Floor 53
53 层
Floor 30
30 层
Floor 11
11 层

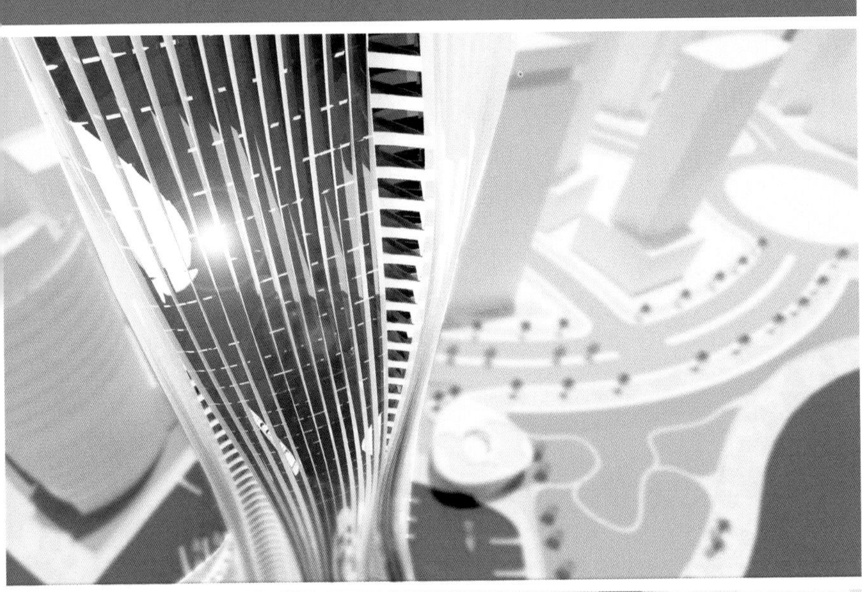

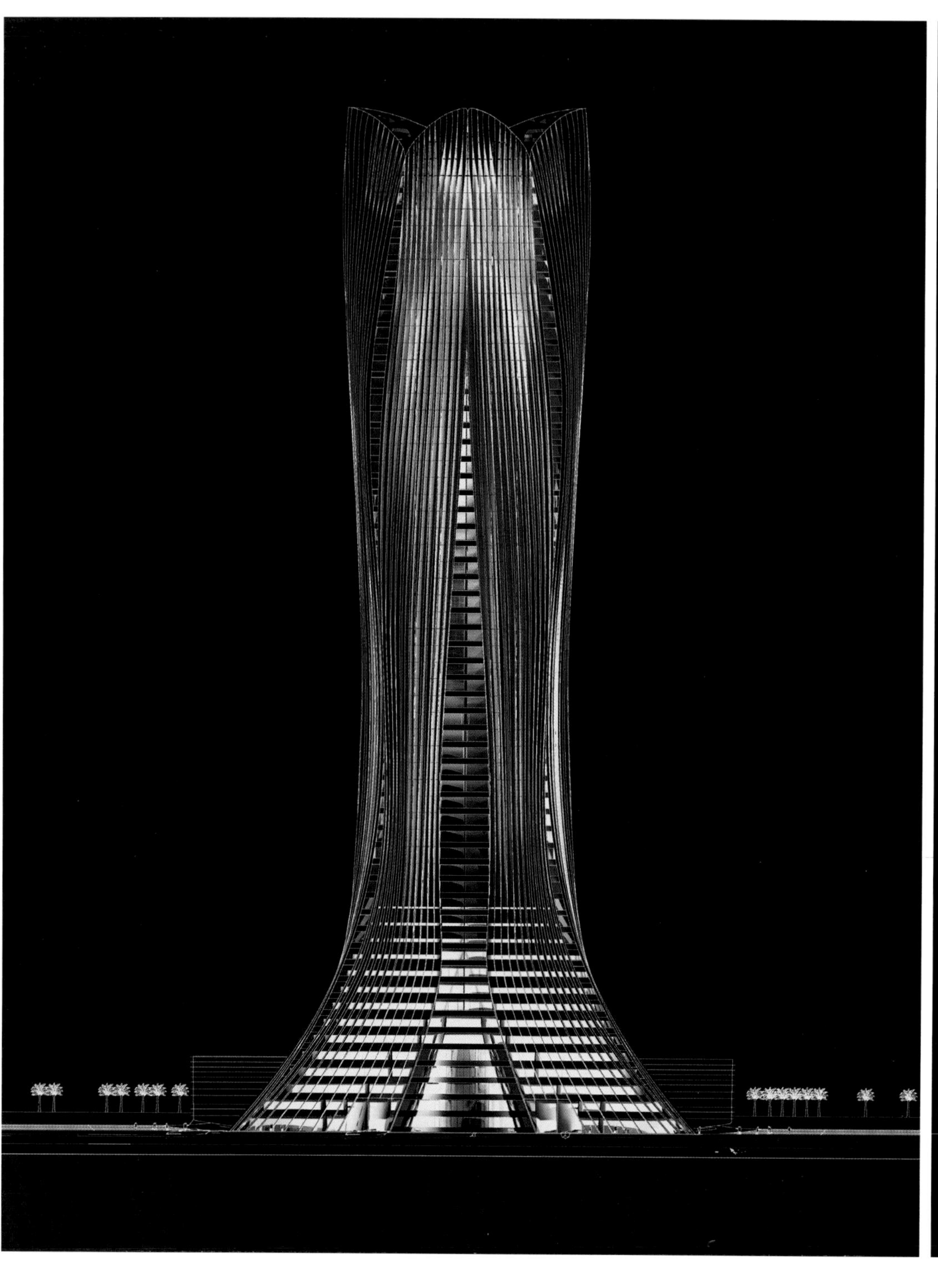

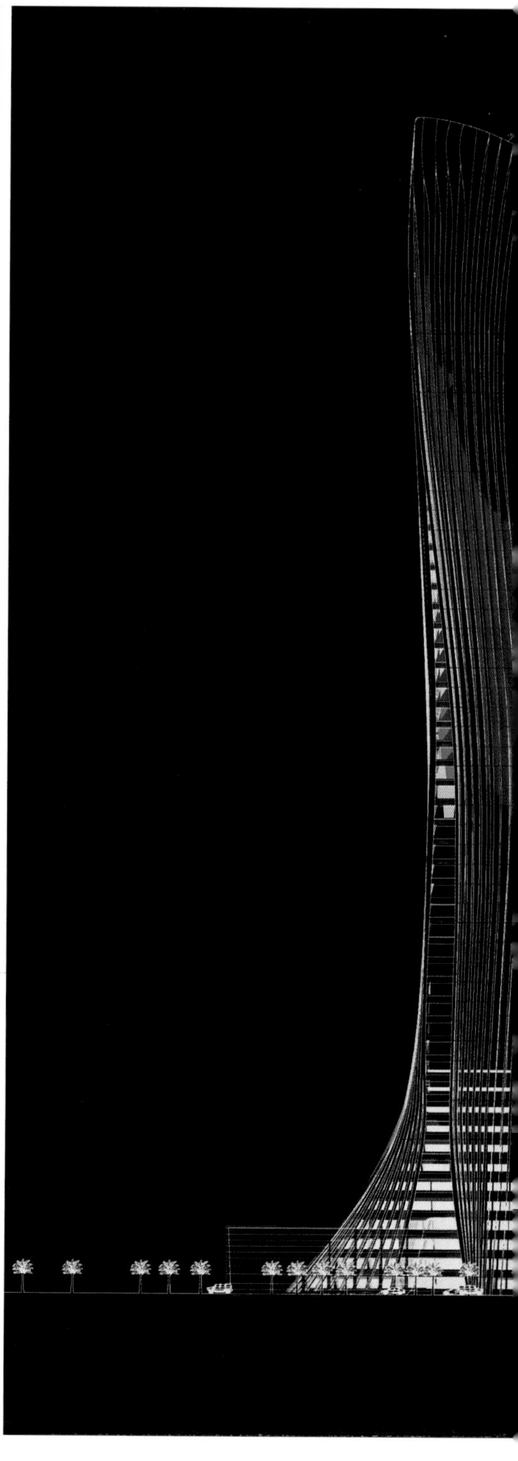

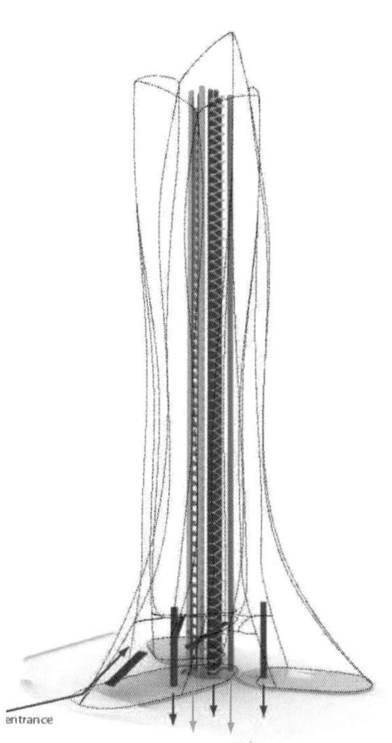

Movement Vertical 垂直移动
Cor-elevator-Staircase 核心电梯－楼梯

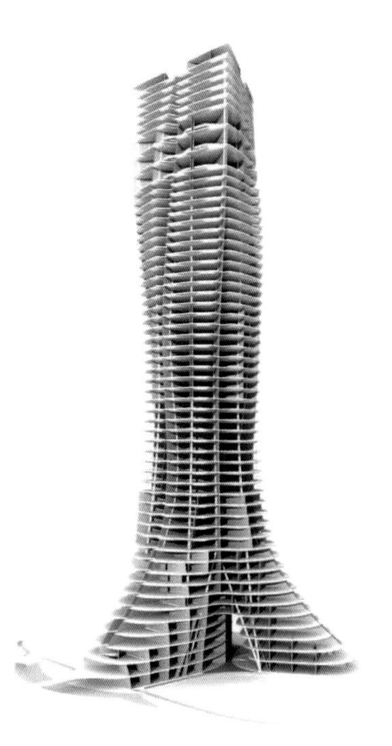

Structure Vertical 垂直结构
Columns-Core-Floorplates 柱状－核心－楼板

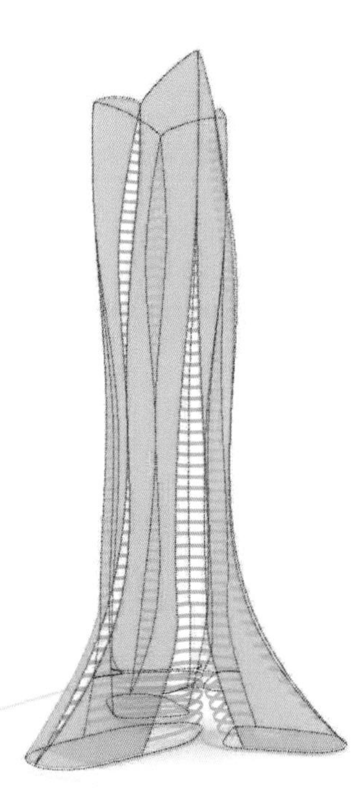

Balustrades Horizontal 水平栏杆
Balconies 阳台

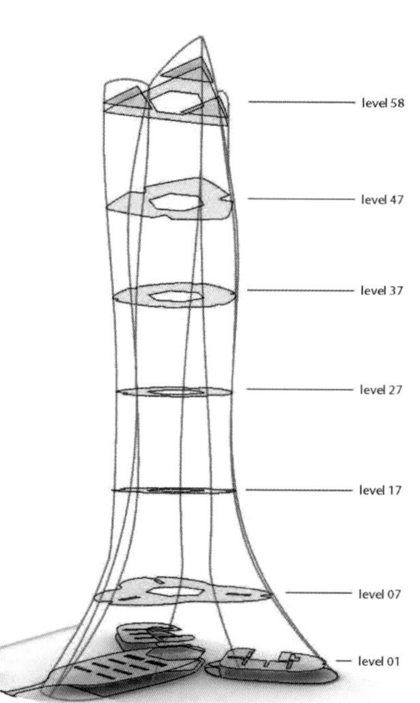

Geometry Concept 几何概念
Design Approach 设计方法

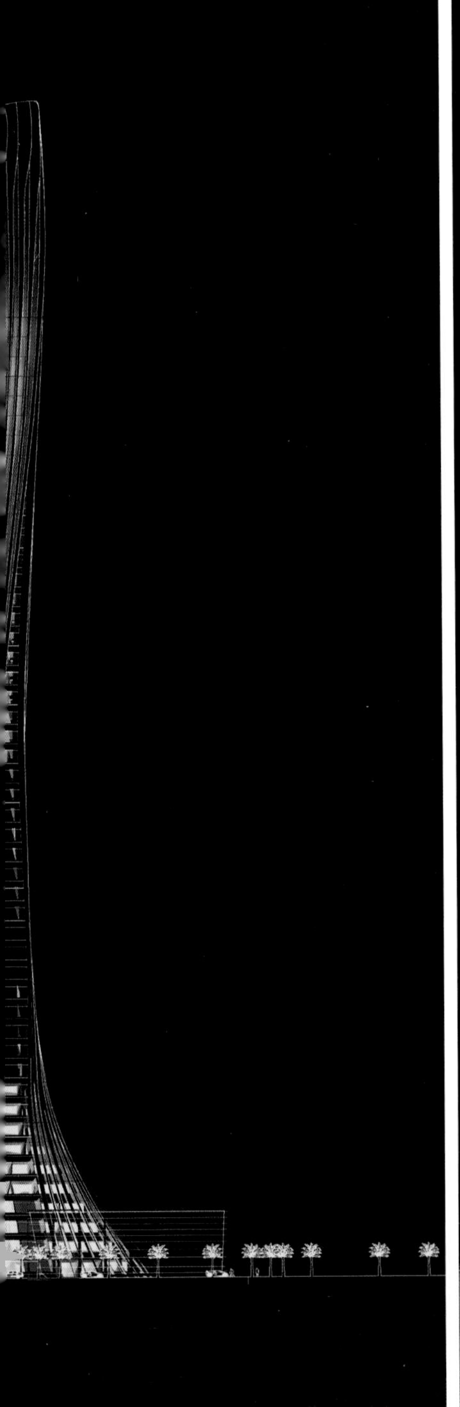

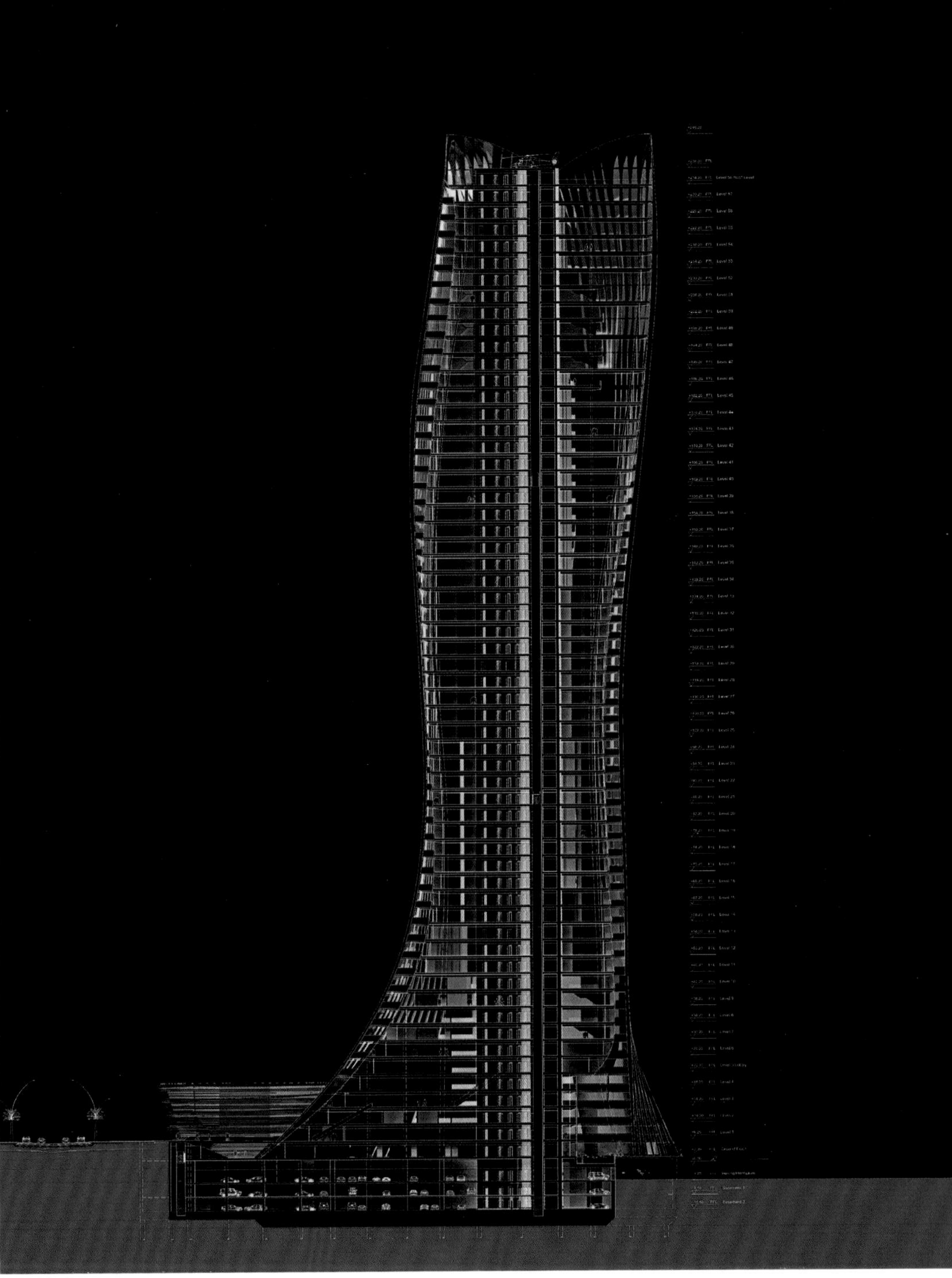

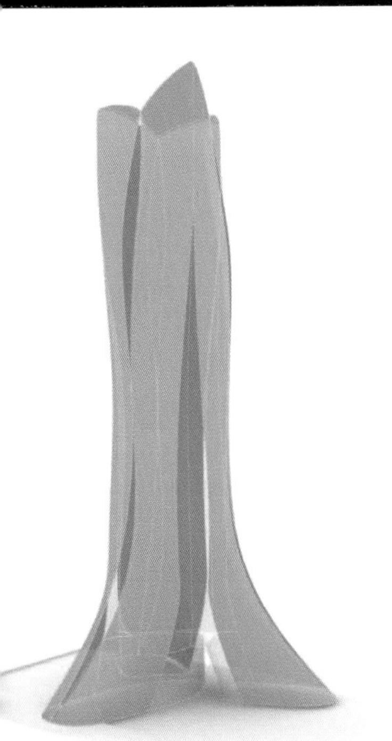

Macro Structure 宏观结构
Tower Geometry 大楼几何形态

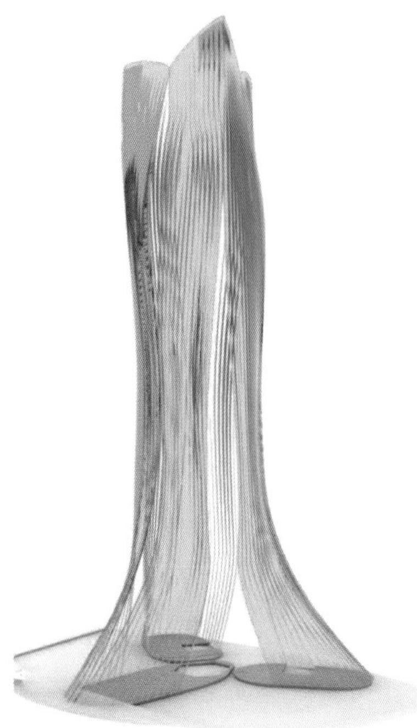

Micro Structure 微观结构
Lamellas-White & Silver 薄片－白色和银色

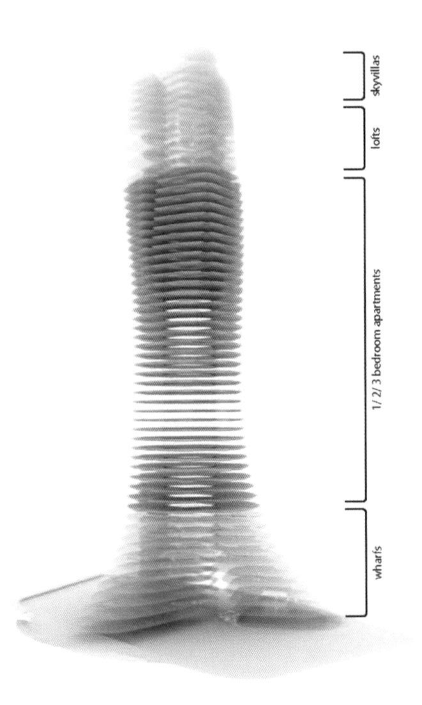

Unit Mix Apartments 单元组合公寓
Wharfs-1/2/3 Bedroom Apartments-Lofts - Skyvillas 码头－1/2/3 室公寓－阁楼－空中别墅

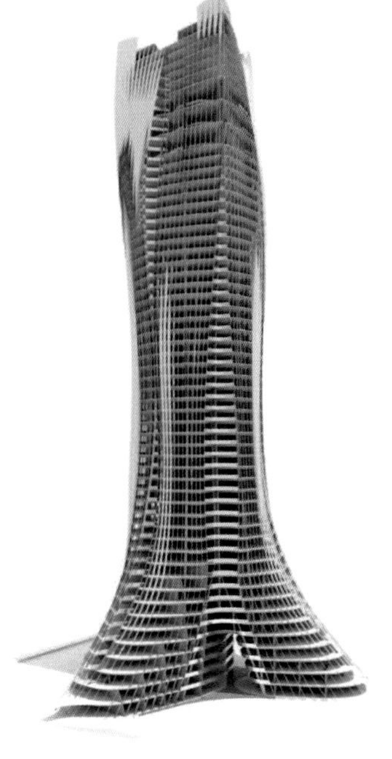

Tower Structure 大楼结构
Geometry of Tower 大楼几何形态

马来西亚布城滨水住宅楼

Putrajaya Precinct 4 Waterfront Development, Kuala Lumpur

设计单位：Studio Nicoletti 建筑事务所
　　　　　Hijjas Kasturi 建筑事务所
开发商：Putrajaya Holdings Ltd
项目地址：马来西亚布城
项目面积：280 000 m²

Designed by: Studio Nicoletti Associati;
Hijjas Kasturi Associates
Client: Putrajaya Holdings Ltd
Location: Putrajaya, Malaysia
Area: 280,000 m²

项目概况

普特拉贾亚滨水住宅楼也被称作 Precinct 4，位于马来西亚吉隆坡以南 30 千米处，是一个大型的可持续性绿色建筑。设计融合了伊斯兰元素，给人耳目一新之感。

设计特色

设计旨在探索一种可持续住宅的设计模式。设计师从大海和当地城市的独特景观中获取灵感，并融入伊斯兰的设计元素，这使建筑不仅拥有了鲜明的海洋特色，也赋予了建筑独特的异域风情。

设计采用了独特的外部骨架结构，创造了独一无二的外观。这些半透明的圆形骨架被设计师设想为宏伟的拱门，由此形成的帆船式建筑形态构成了源自伊斯兰文化的壮观穹顶，既美观又简约。

这种未来主义的骨架结构兼顾了美观性和功能性。骨架建筑群具有较高的能源效率，每座建筑的能源需求减少了 50%。此外，大楼还配备了创新的冷却系统和通风系统。

Profile

Putrajaya waterfront residential building, also called Precinct 4, is located 30 kilometers south of Kuala Lumpur, Malaysia. It is a large sustainable green building. Integrated with Islamic elements, the design presents a completely new and fresh appearance.

Design Feature

The project aims to explore a design mode of sustainable residence. Designers, inspired by the ocean and local urban landscape, integrate buildings with Islamic elements, which enable the building to possess distinctive waterfront features and unique exotic style.

Unique external framework creates unparalleled architectural appearance. These semi-transparent circular frameworks are envisaged as grand arches. The resulting sailboat-shaped architecture constitutes magnificent domes originated from Islamic culture, gorgeous and simple.

Such futurism framework structure takes aesthetics and functions into account. Architectural cluster with framework structure has relatively high energy efficiency. The energy demand of each building has decreased 50%. Besides, innovative cooling system and ventilation system are also equipped in these buildings.

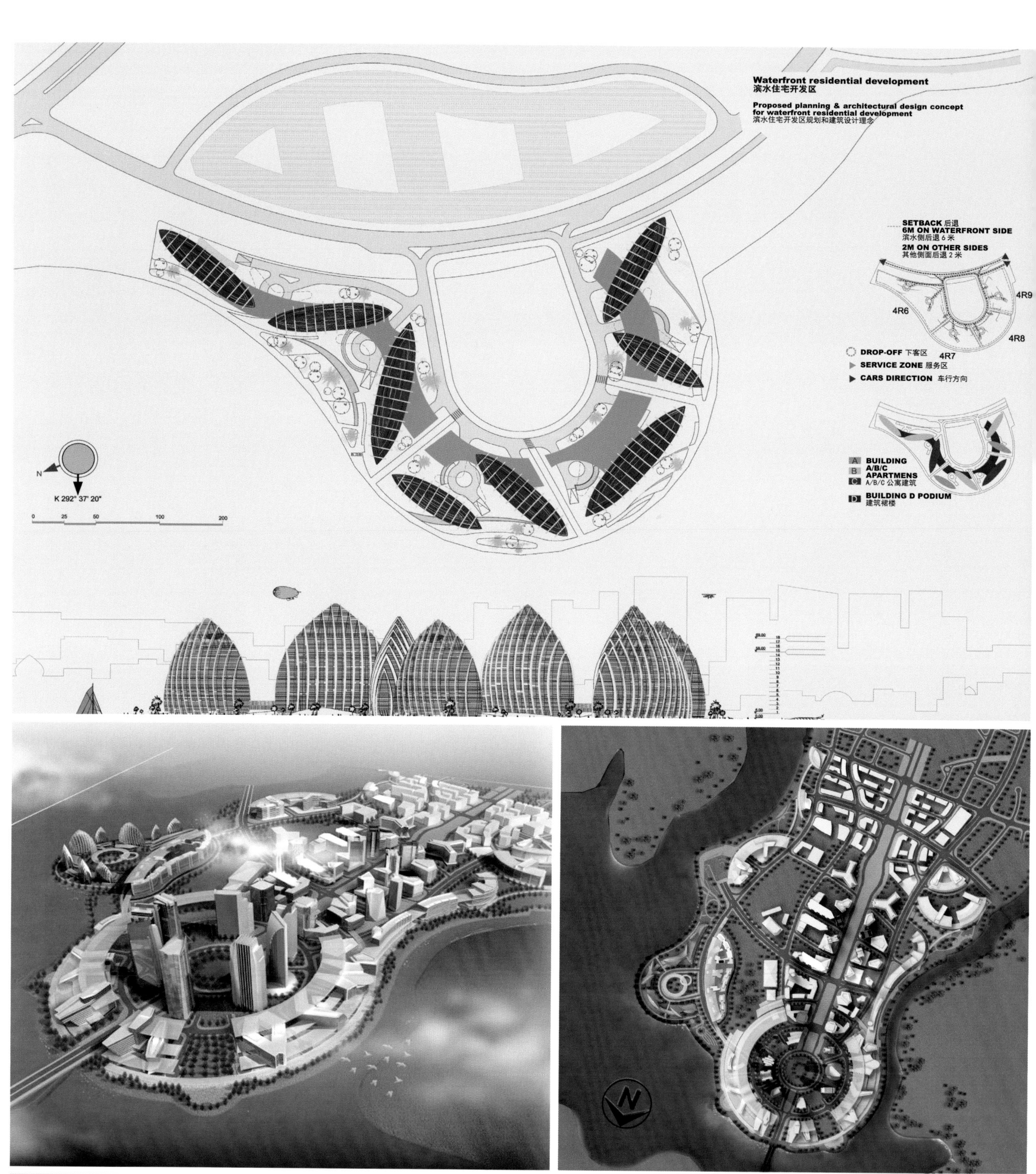
Waterfront residential development
滨水住宅开发区
Proposed planning & architectural design concept for waterfront residential development
滨水住宅开发区规划和建筑设计理念
SETBACK 后退
6M ON WATERFRONT SIDE
滨水侧后退 6 米
2M ON OTHER SIDES
其他侧面后退 2 米
4R6
4R7
4R8
4R9
DROP-OFF 下客区
SERVICE ZONE 服务区
CARS DIRECTION 车行方向
A BUILDING
B A/B/C
C APARTMENS
A/B/C 公寓建筑
D BUILDING D PODIUM
建筑裙楼
N
K 292° 37' 20"
0
25
50
100
200

88000
8000
BALCONY
MASTER BEDROOM
FAMILY AREA
LIVING
DINING
D. KITCHEN
W. KITCHEN
UTILITY
A/C & YARD
BEDROOM 1
MASTER BATH
LANAI
AV ROOM
GUEST ROOM
STUDY
VOID
CHILDREN'S PLAY AREA
INFINITY CHILDREN POOL
POOL BAR
OPEN FUNCTION AREA
LIFT LOBBY
INFINITY SWIMMING POOL

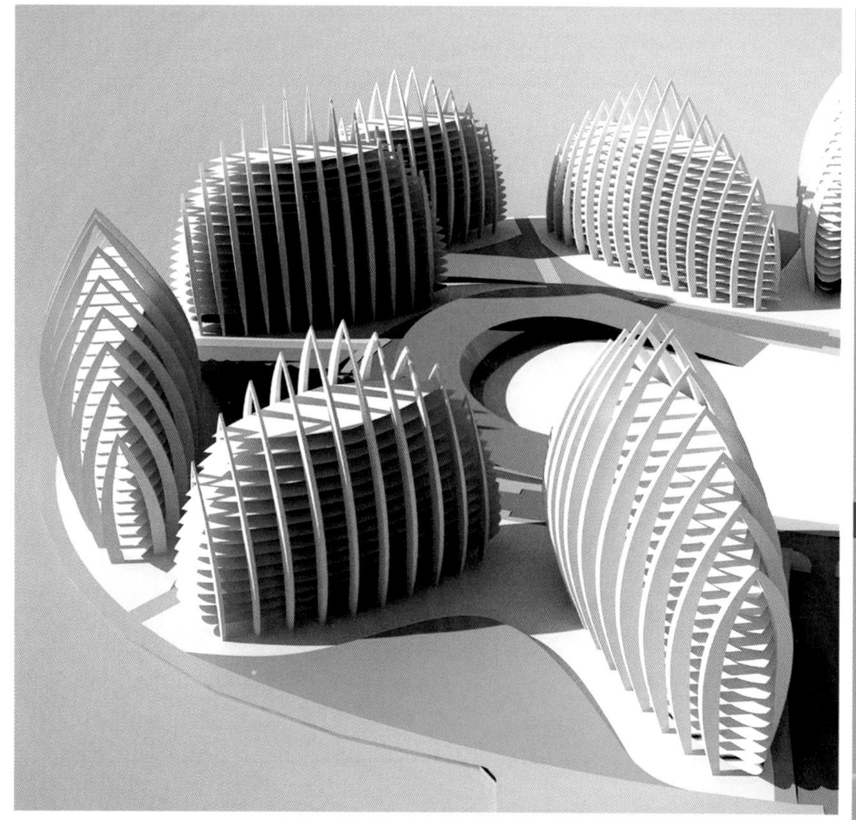

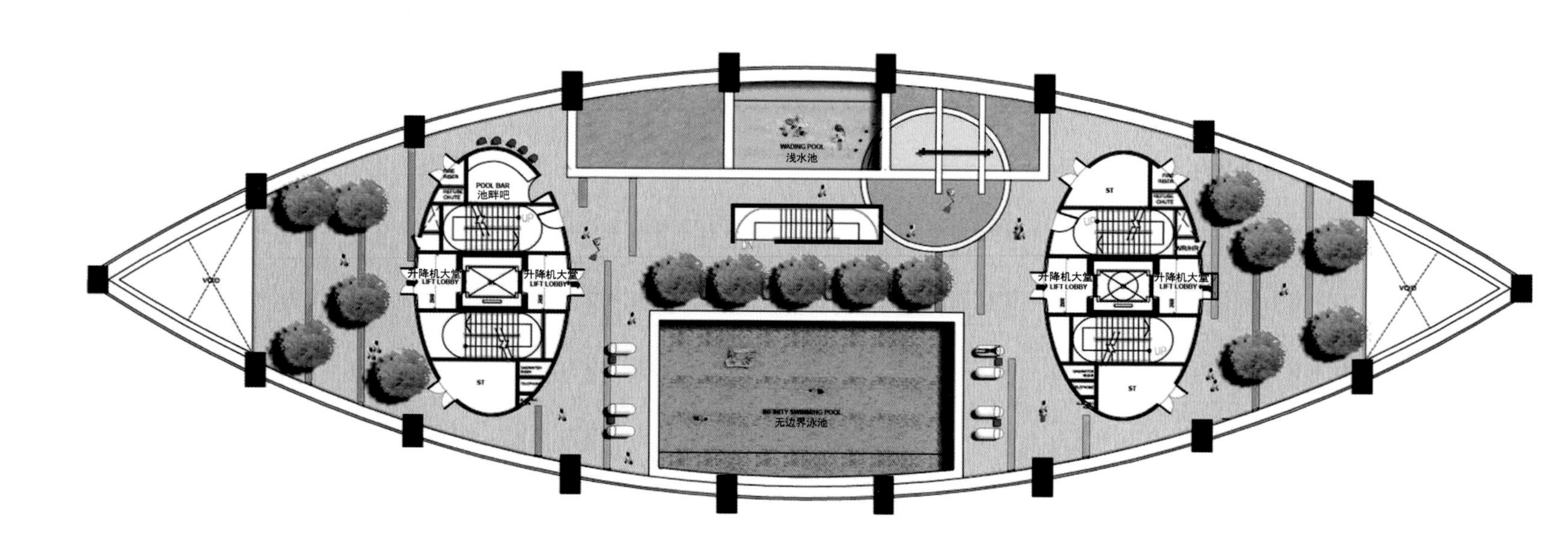
WADING POOL
浅水池
POOL BAR
池畔吧
REFUSE CHUTE
FIRE RISER
升降机大堂
LIFT LOBBY
升降机大堂
LIFT LOBBY
ST
INFINITY SWIMMING POOL
无边界泳池
ST
升降机大堂
LIFT LOBBY
升降机大堂
LIFT LOBBY
ST
ST
VOID

A/C& 庭院
A/C & YARD
ENTRANCE
入口
DINING
就餐
LIVING
生活
卧室 1
BEDROOM 1
A/C
浴室
BATH
主浴室
MASTER BATH
主卧
MASTER BEDROOM
D. KITCHEN 厨房餐间
W. KITCHEN
工作厨房
UTILITY BATH
实用浴室
BALCONY
阳台
家庭区
FAMILY AREA
BEDROOM 2
卧室 2
BATH
MASTER BATH
主浴室
MASTER BEDROOM
主卧
工作厨房
W. KITCHEN
PR
储备室
厨房餐间
D. KITCHEN
DINING
就餐
LIVING
生活
入口
ENTRANCE
A/C & YARD
A/C& 庭院
BALCONY
阳台
28288
46466
8000
8000
8000

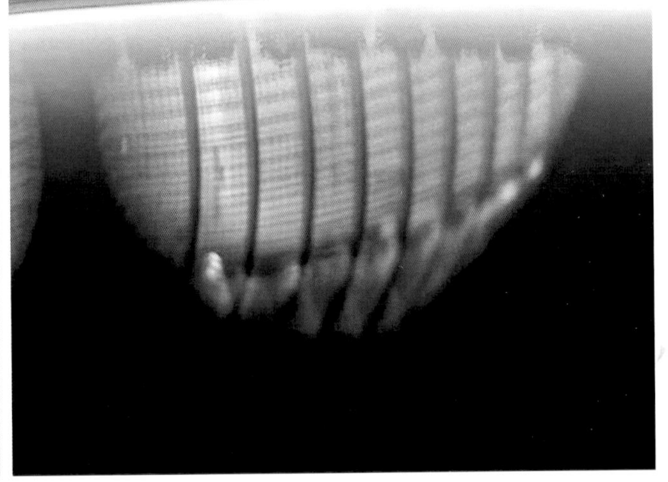

上海黑森林高级住宅三期

Black Forest Residences Phase Ⅲ

设计单位：欧博迈亚工程咨询（北京）有限公司

开发商：上海中鹰集团

项目地址：中国上海市

总用地面积：20 000 ㎡

总建筑面积：69 880 ㎡

容积率：3.5

Designed by: OBERMEYER Engineering Consulting Co., Ltd.

Developer: CEG Group

Location: Shanghai, China

Site Area: 20,000 m^2

Gross Floor Area: 69,880 m^2

Plot Ratio: 3.5

项目概况

黑森林高级住宅三期项目是上海最具创新的住宅建筑之一，设计融入了创新的技术，旨在为当地居民提供一个根植于景观之中、隐匿在繁华大都市间的，有着舒适、宜人环境的豪华住宅。

建筑设计

3栋雪白色的建筑自蓊郁的绿色景观中拔地而起，这些堆叠而成的建筑体量，每 5 个楼层就会形成些微的偏移。这种楼层平面布局方式，使建筑每 4 层或 5 层就可形成一个环绕整栋大楼的阳台，部分公寓也因此获得一个额外的空间。

楼板结构也十分新颖。设计师没有将横梁设置在楼板的下面，而是搭建在楼板的上面，这就为安装楼层冷却管道预留了空间，有利于空气自由流通。设计采用了全高的窗体结构，保证阳光可照射到建筑内部。窗户配备了高科技自动遮阳系统，既可以过滤阳光，又可以方便人们欣赏到外面的风景。

每一套公寓都配备了家庭自动化系统，这一系统由电脑系统精确操控，通过对室内湿度和温度进行设定，从而达到自定义室内气候环境的效果。同时，该系统也可以对光照条件进行程序化控制。

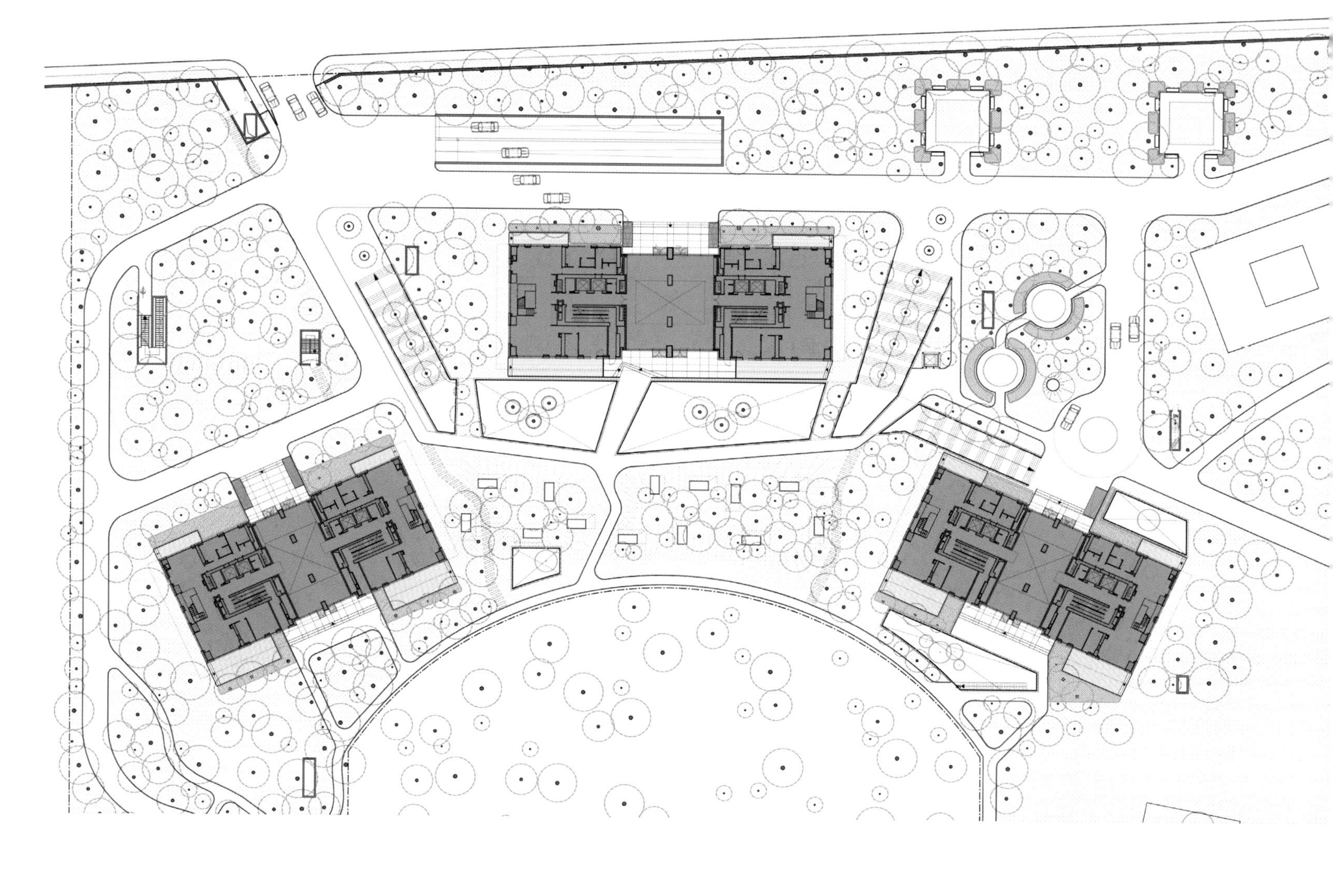

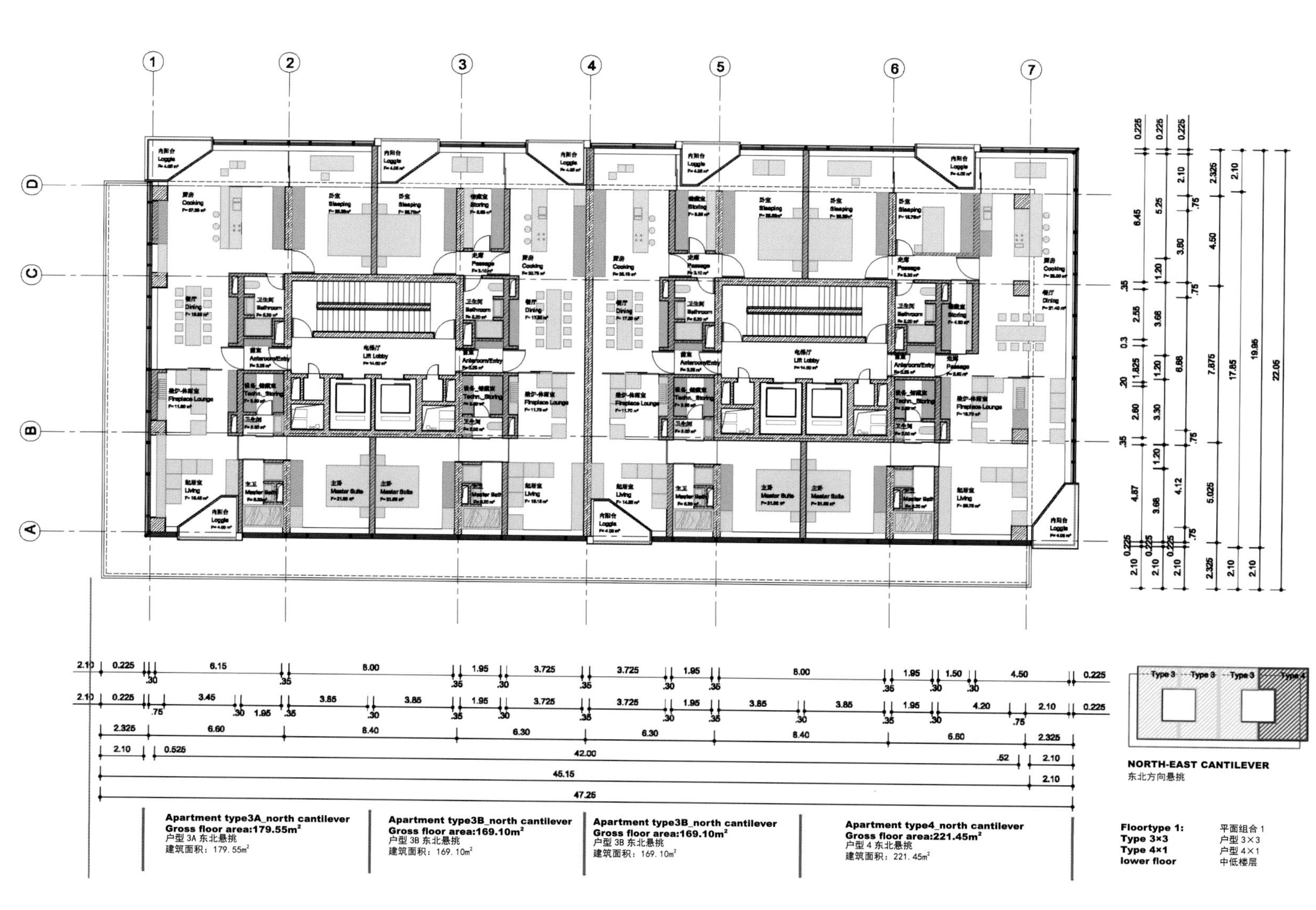
1
2
3
4
5
6
7
A
B
C
D
Type 3
Type 4
NORTH-EAST CANTILEVER
东北方向悬挑
Apartment type3A_north cantilever
Gross floor area:179.55m²
户型 3A 东北悬挑
建筑面积：179. 55m²
Apartment type3B_north cantilever
Gross floor area:169.10m²
户型 3B 东北悬挑
建筑面积：169. 10m²
Apartment type3B_north cantilever
Gross floor area:169.10m²
户型 3B 东北悬挑
建筑面积：169. 10m²
Apartment type4_north cantilever
Gross floor area:221.45m²
户型 4 东北悬挑
建筑面积：221. 45m²
Floortype 1: 平面组合 1
Type 3×3 户型 3×3
Type 4×1 户型 4×1
lower floor 中低楼层

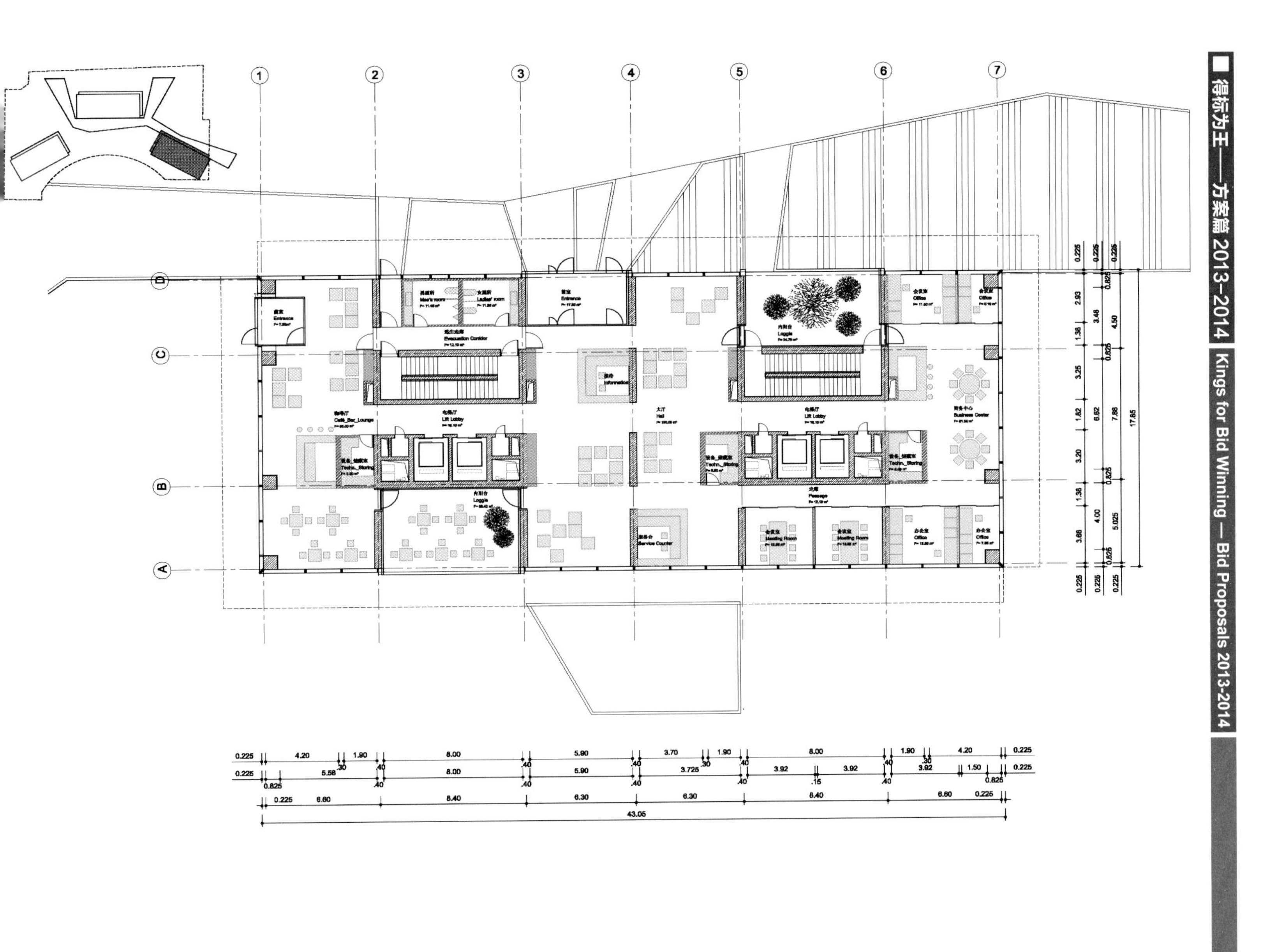
Entrance
Evacuation Corridor
Cafe_Bar_Lounge
Lift Lobby
Hall
Information
Loggia
Business Center
Passage
Service Counter
Meeting Room
Office
43.05

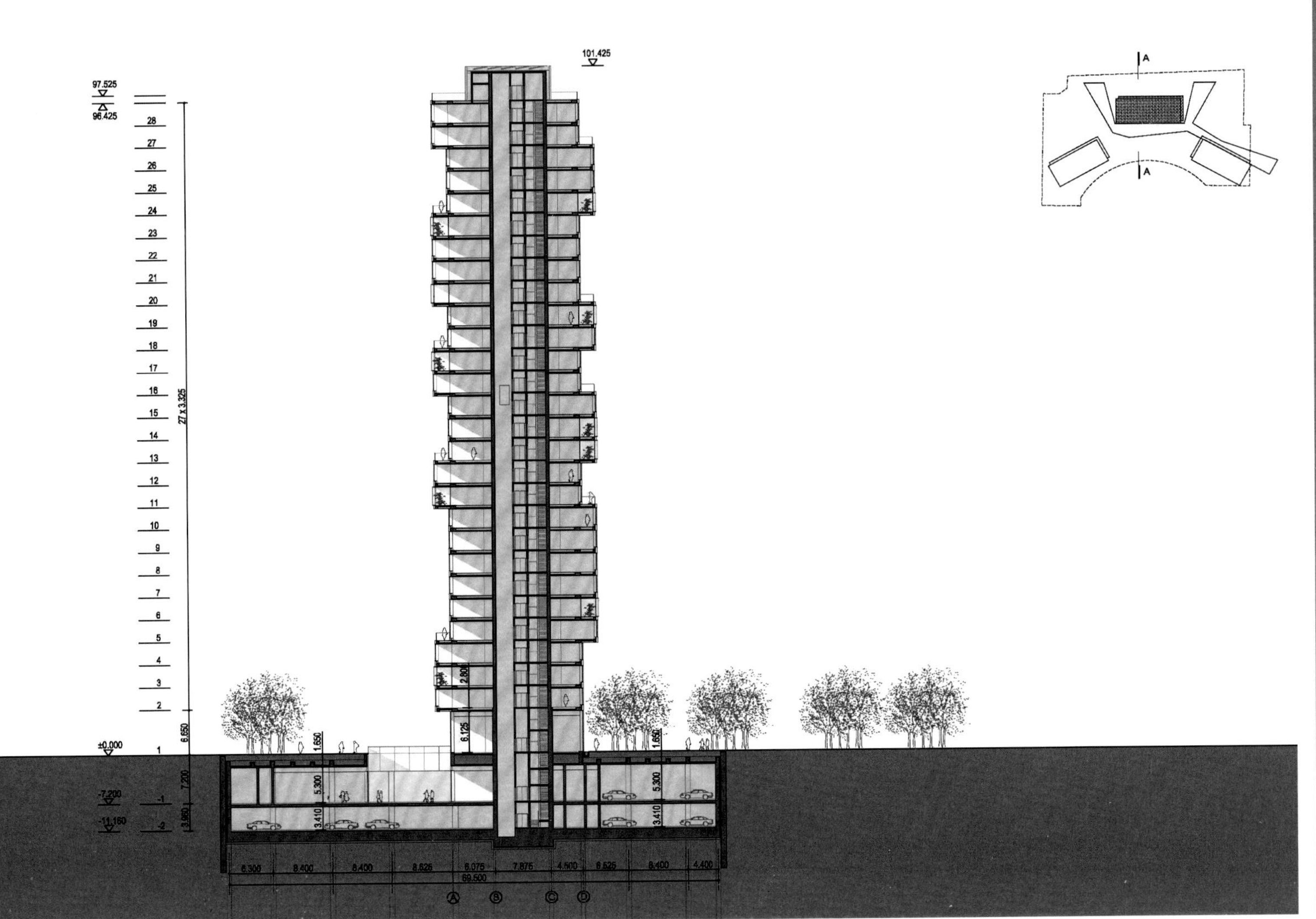
101.425
97.525
96.425
27 x 3.325
±0.000
-7.200
-11.150
69.500

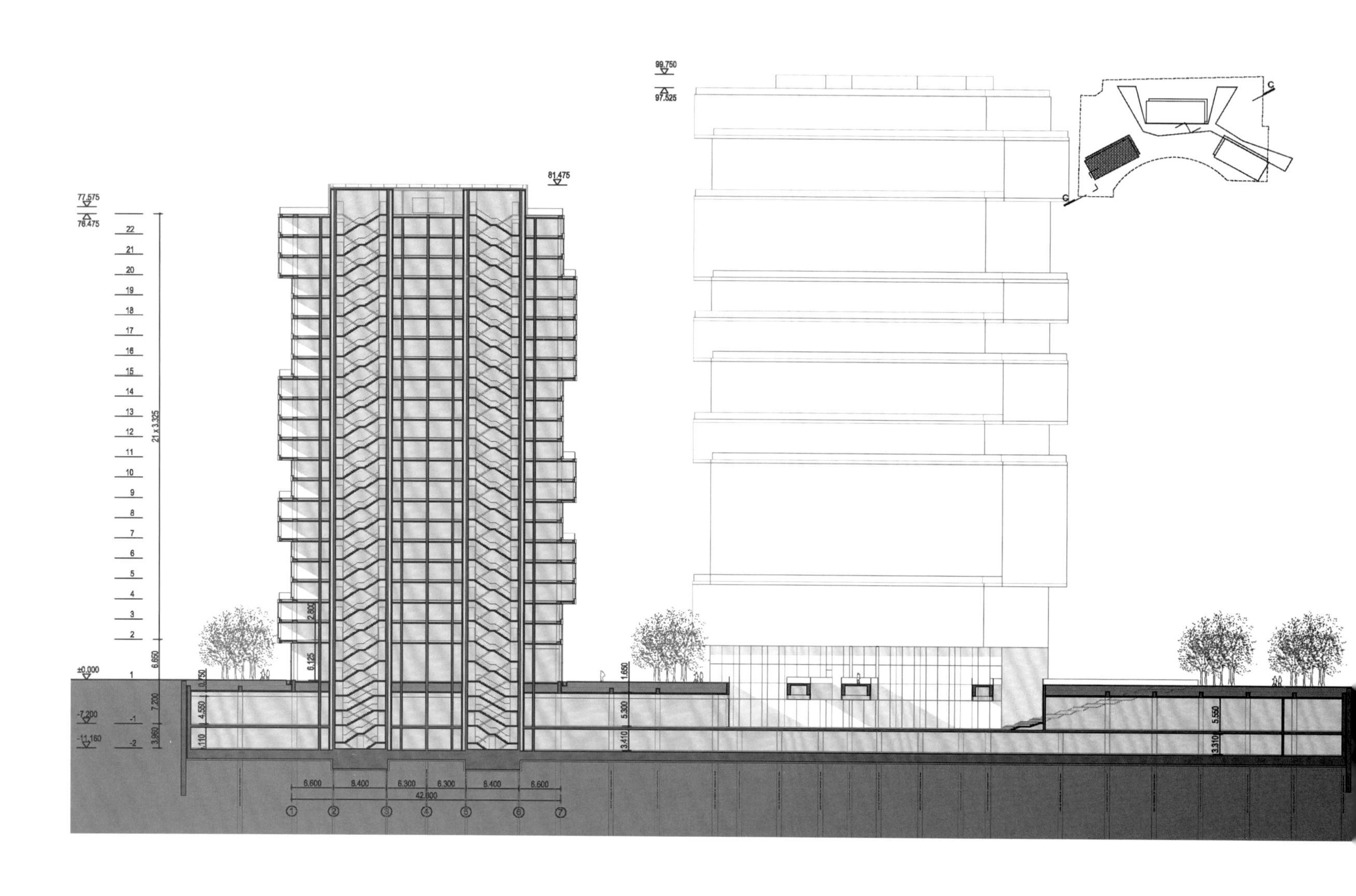

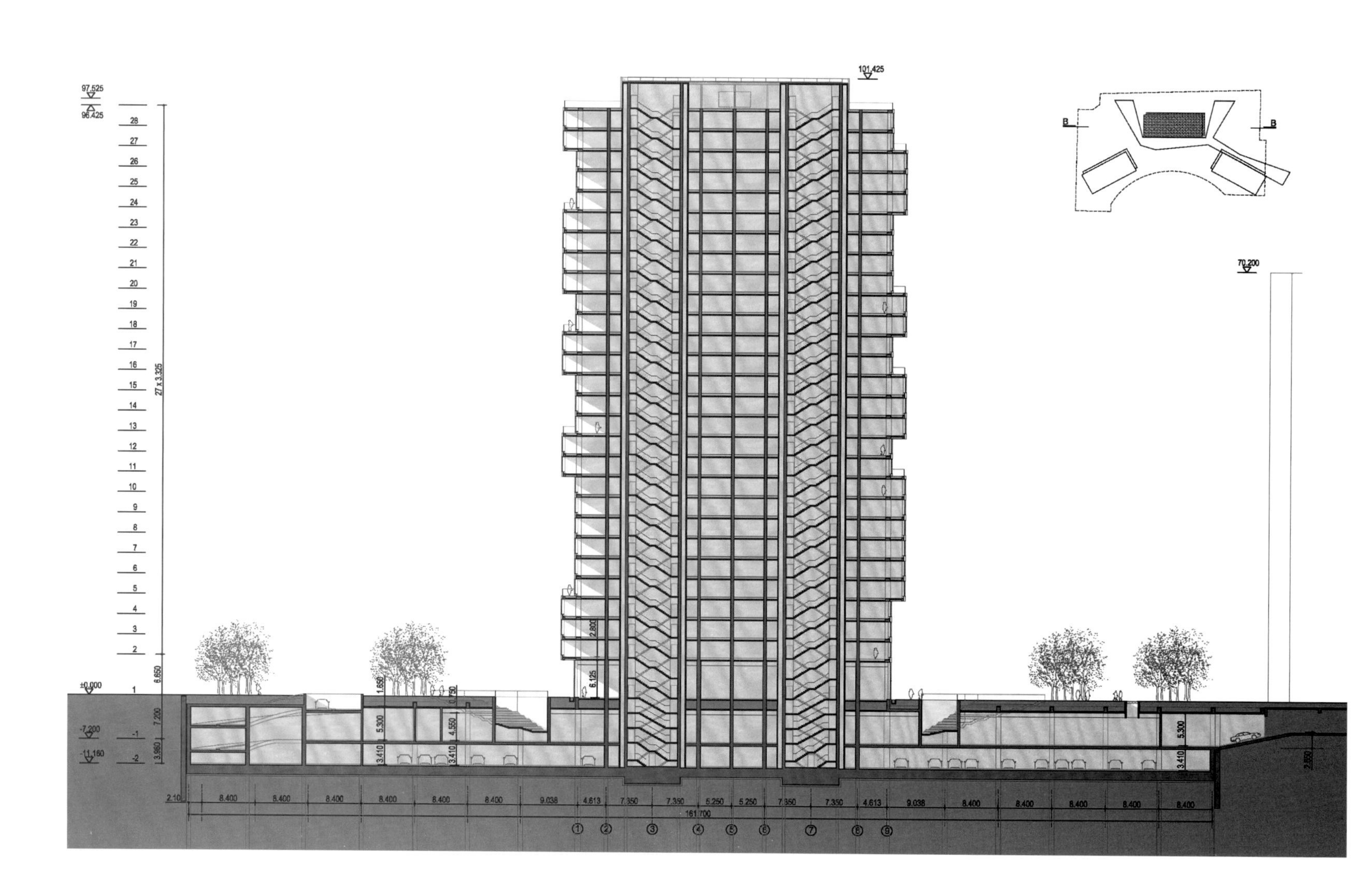

Profile

Upcoming Phase 3 of “Black Forest Residences” is one of Shanghai’s most innovative residential projects. The project intends to provide the residents a home alike: embedded in landscape it is a hideout from the busy Shanghai, with a comfortable climate and atmosphere.

Architectural Design

Three snow-white towers emerge from the lush green landscape, stacked volumes, shifted elegantly every five floors. The apartments gain an additional room and every 4 or 5 floors, balconies wrap around the whole building.

New is also the structure of the floor slabs. Instead below, the beams are above the slab. This allows space for the cooling ducts in the floors, which facilitates free air circulation. The full height windows let the daylight enter the rooms. The windows are equipped with a high-tech automatic sunshading system, which filters the light but keeps the view.

Each apartment includes a home automations system to customize the settings for the room climate. The humidity and temperature can be set and will be exactly controlled by the computer system. Light-atmosphere settings can also be programmed.

法国格勒诺布尔 138 住宅

138 Dwellings

设计单位：ECDM Architectes
开发商：法国巴黎银行
项目地址：法国格勒诺布尔
场地面积：2 518 m²
建筑面积：8 000 m²

Designed by: ECDM Architectes
Client: BNP Paribas
Location: Grenoble, France
Site Area: 2,518 m²
Building Area: 8,000 m²

项目概况

这一住宅大楼位于伊泽尔省德拉克流域附近的一个岛屿上，临近阿尔卑斯山脉的第一个山麓，与格勒诺布尔地区具有直接的视觉联系。建筑以层层叠叠的形态向阶梯式花园敞开，连续的形态给人无穷无尽之感，具有极强的表现力。

建筑设计

3 栋山形的建筑有着重叠繁复的形态，仿若多个别墅住宅堆叠而成。层层叠叠的空间在水平方向上呈不同的角度向外凸出，形成一个个开放的阳台，这些凸出墙体的阳台不仅形成了户外的开放空间，也成为这 3 栋建筑的主要外部特征。

这些阳台在不同的楼层之间是错落排布的，有时会在楼层间形成双层高度的间隔，这不仅使每套公寓都享有充足的日光，也保证了每套公寓的独立性和私密性。

游戏室、餐厅、夏季温室或空中花园等外部空间通过支架固定在建筑外立面上，赋予了阳台多种功能。这些外部空间由预制混凝土部件构成，既降低了热传递，也满足了防震的要求。

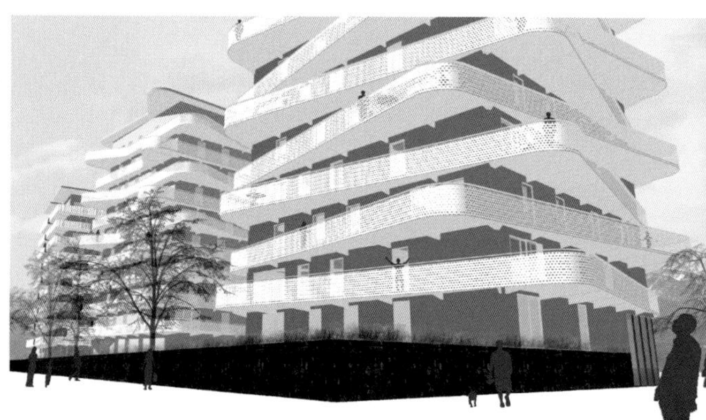

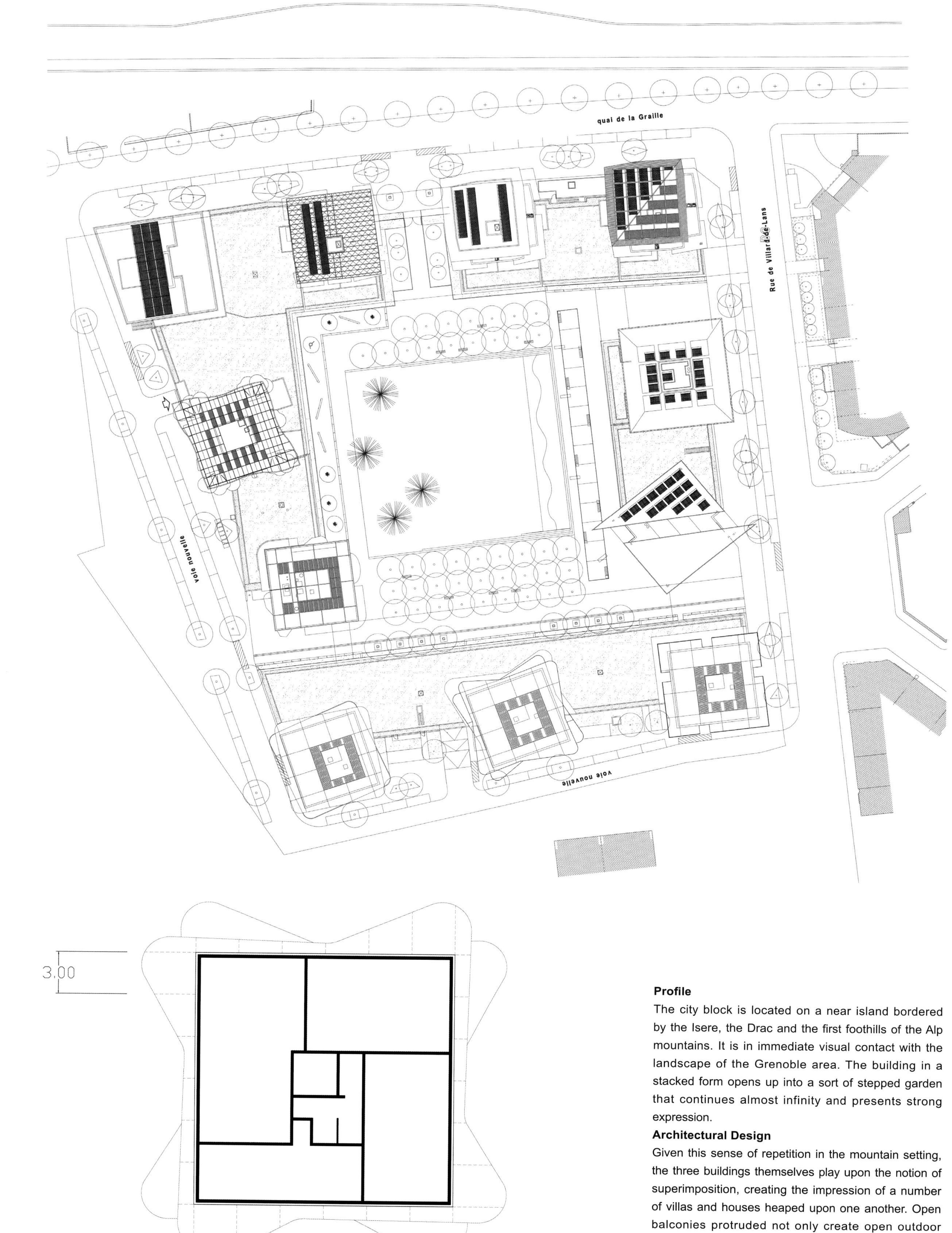

Profile

The city block is located on a near island bordered by the Isere, the Drac and the first foothills of the Alp mountains. It is in immediate visual contact with the landscape of the Grenoble area. The building in a stacked form opens up into a sort of stepped garden that continues almost infinity and presents strong expression.

Architectural Design

Given this sense of repetition in the mountain setting, the three buildings themselves play upon the notion of superimposition, creating the impression of a number of villas and houses heaped upon one another. Open balconies protruded not only create open outdoor spaces but also become a main external characteristic of these three buildings.

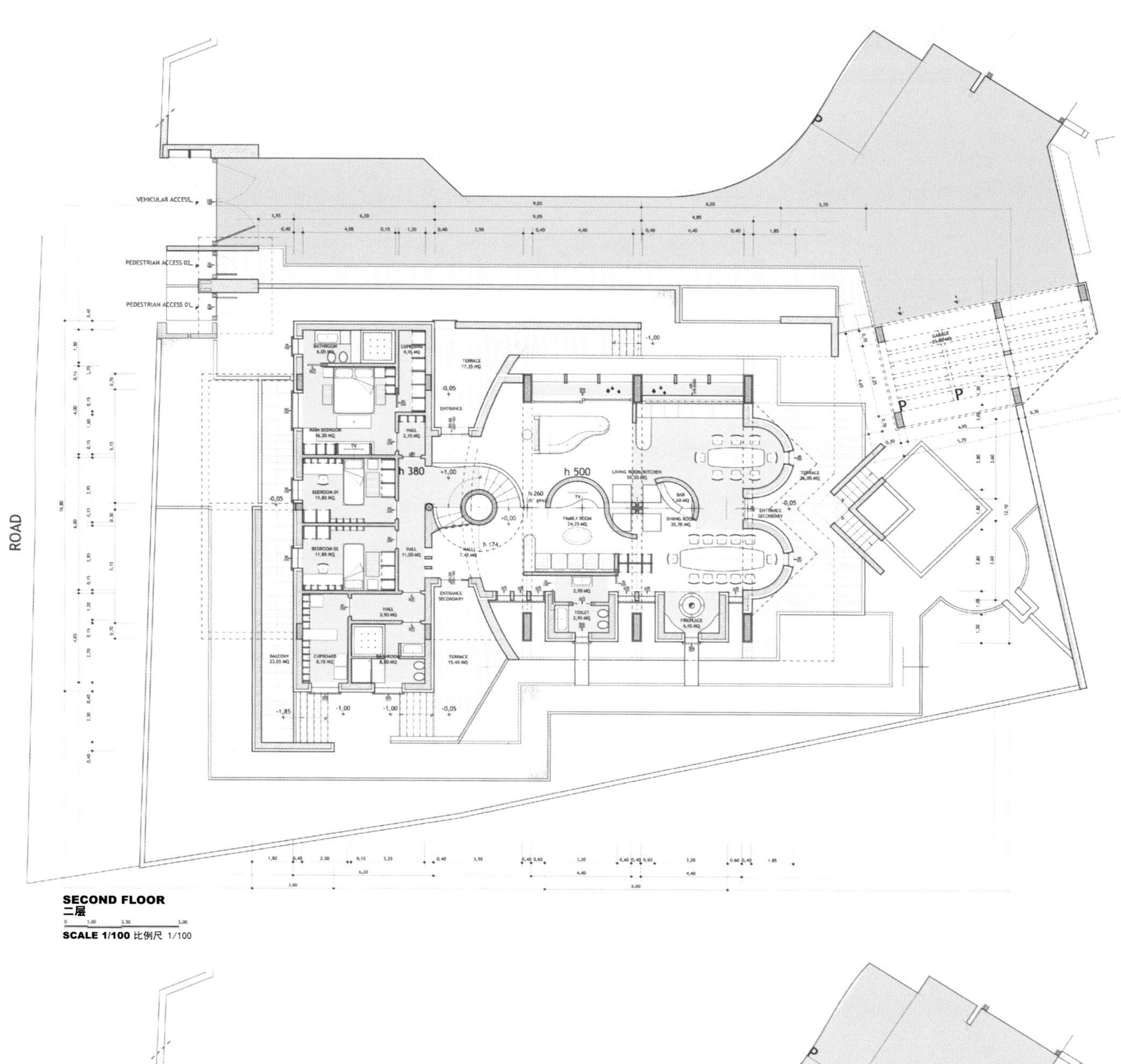

SECOND FLOOR
二层
SCALE 1/100 比例尺 1/100

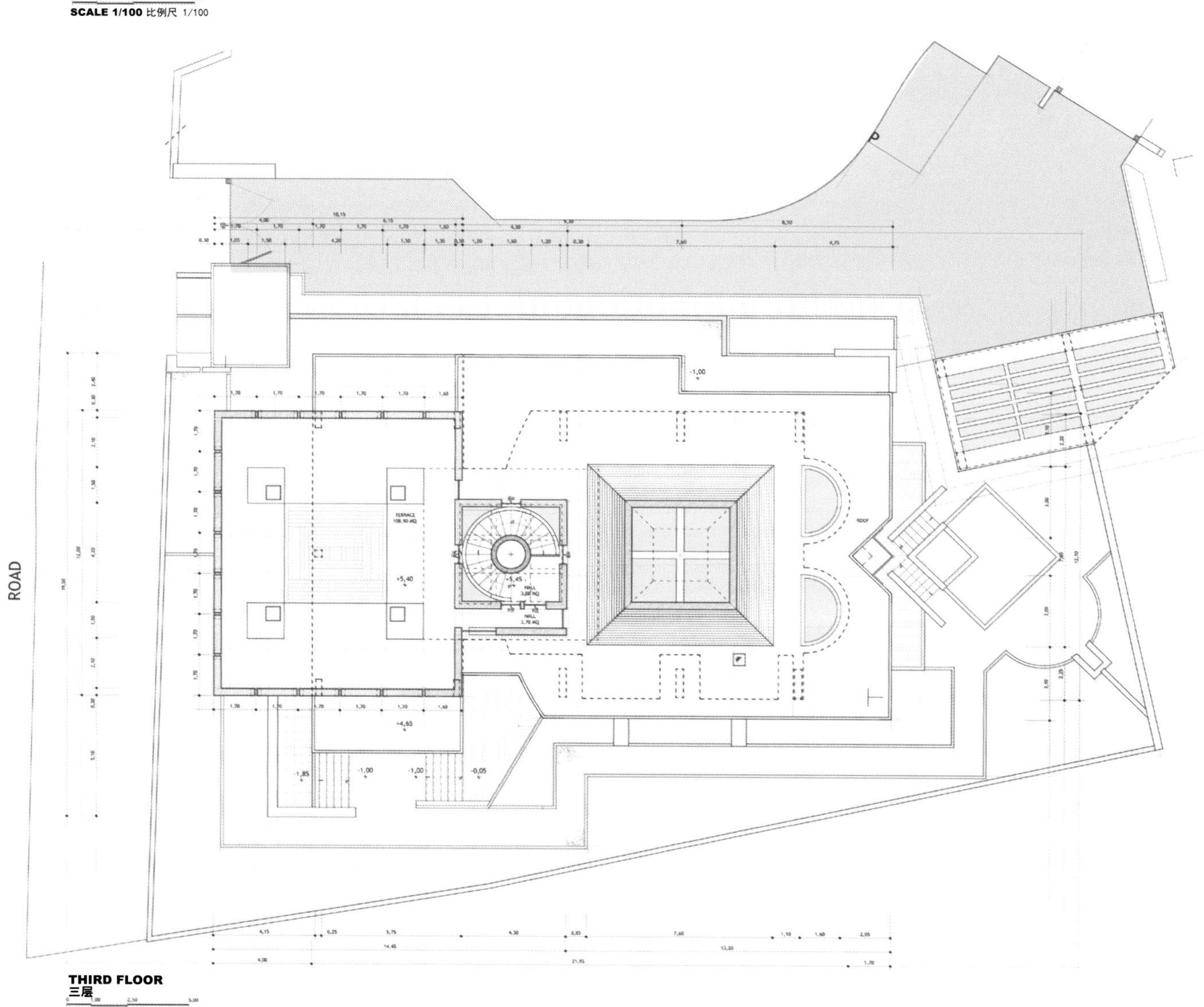

THIRD FLOOR
三层
SCALE 1/100 比例尺 1/100

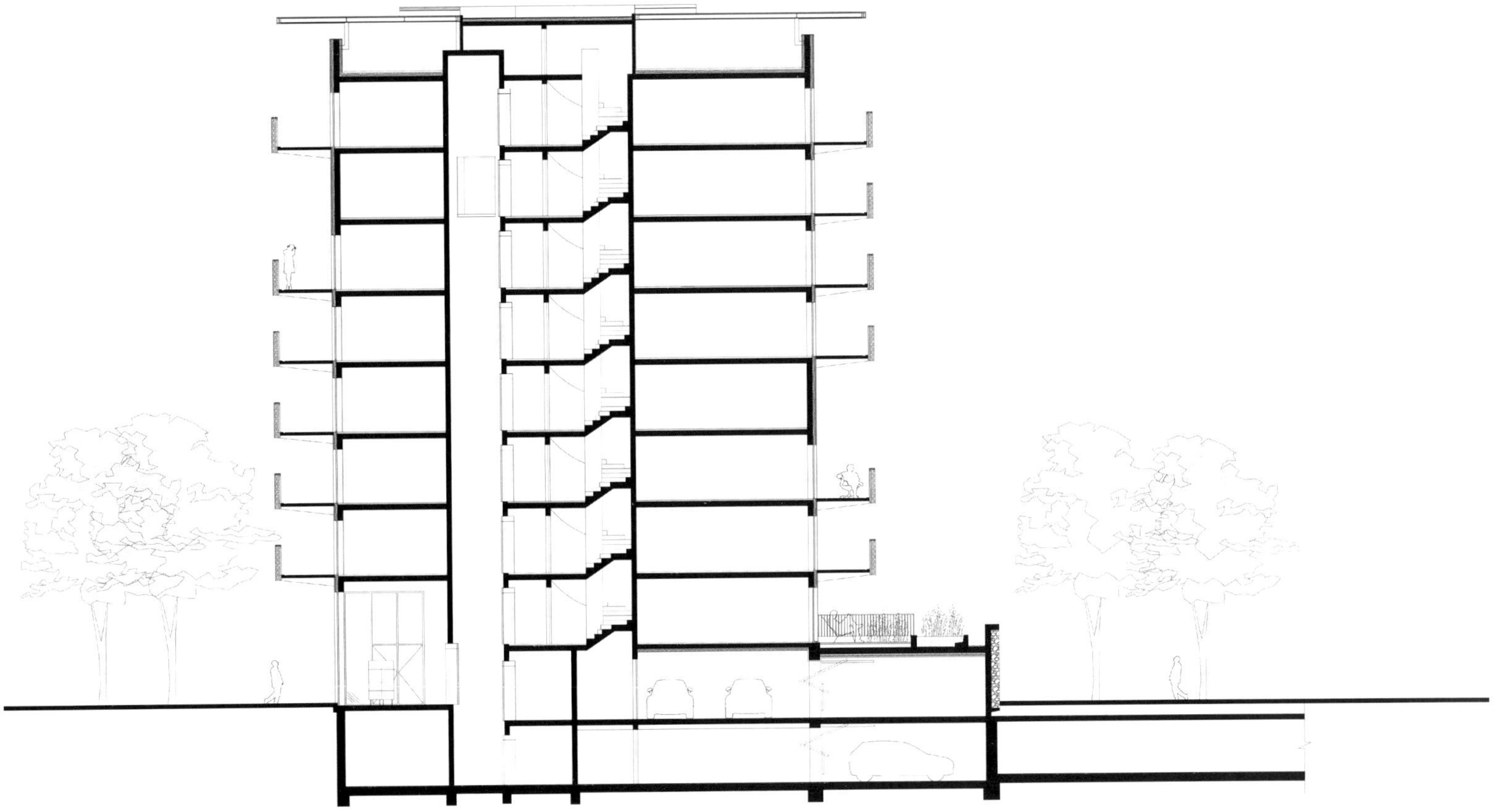

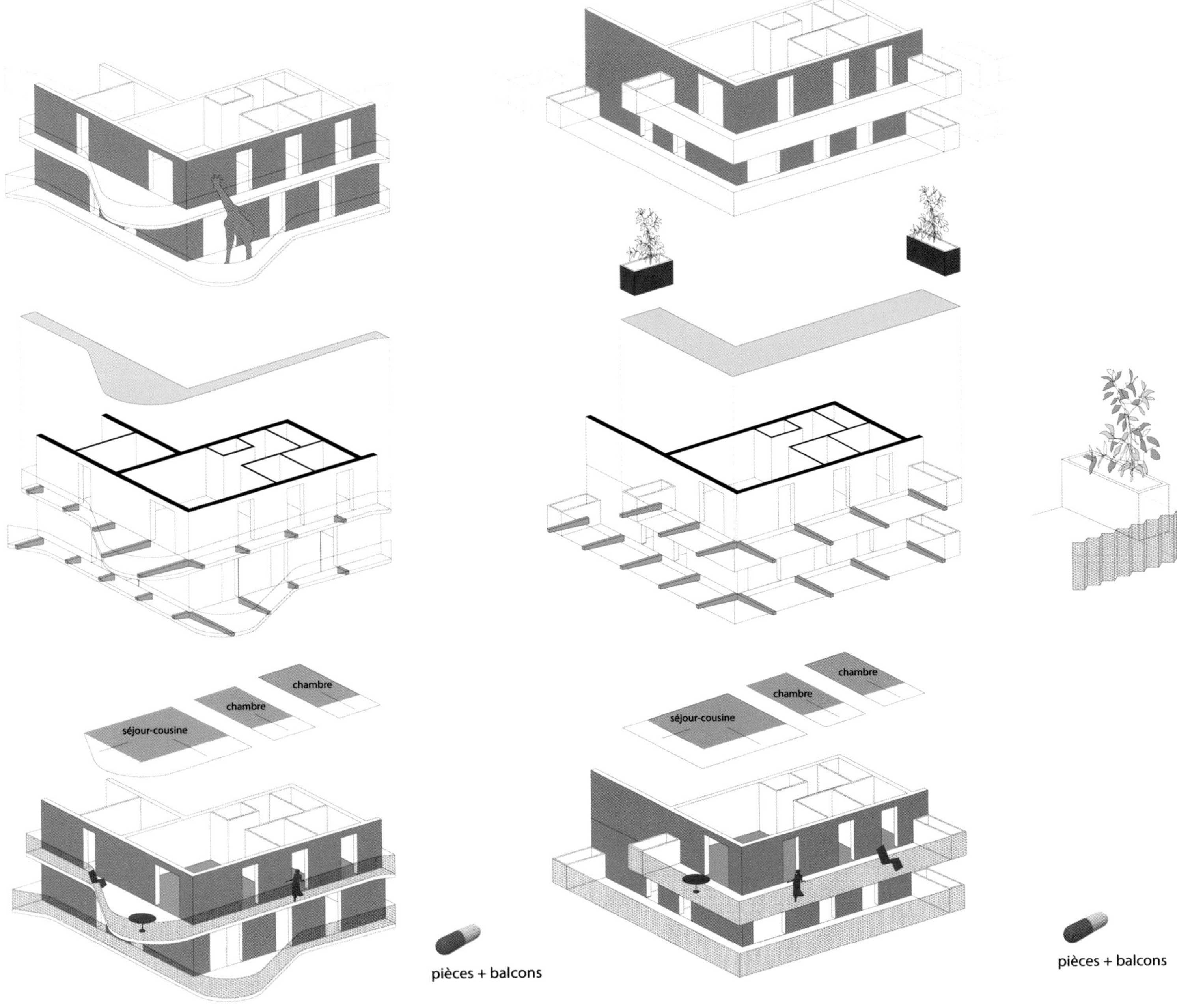

chambre
chambre
séjour-cousine
pièces + balcons

chambre
chambre
séjour-cousine
pièces + balcons

These balconies are "staggered" according to differences in floor level, sometimes leaving double height intervals. Therefore, each apartment can receive large quantities of natural light whilst enjoying independence and privacy.

Games rooms, dining areas, summer conservatories or hanging gardens, these multifunctional exterior "rooms" are fixed to the facades by means of console brackets. These external spaces composed of prefabricated concrete components are used to cut down heat transference and meet anti-seismic requirements.

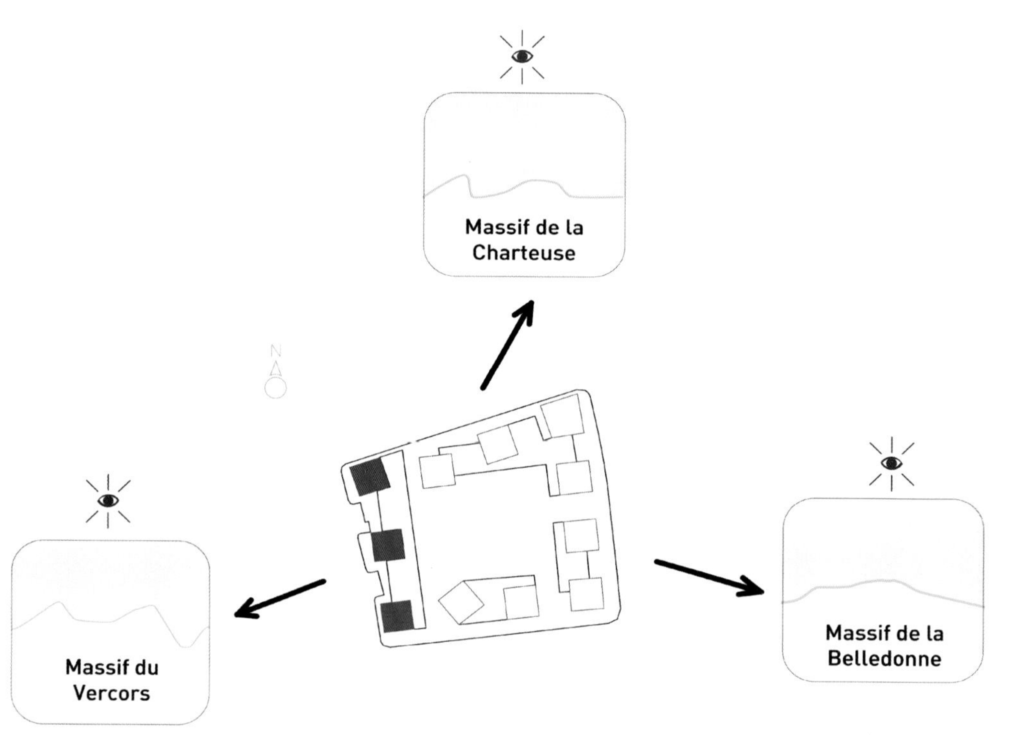

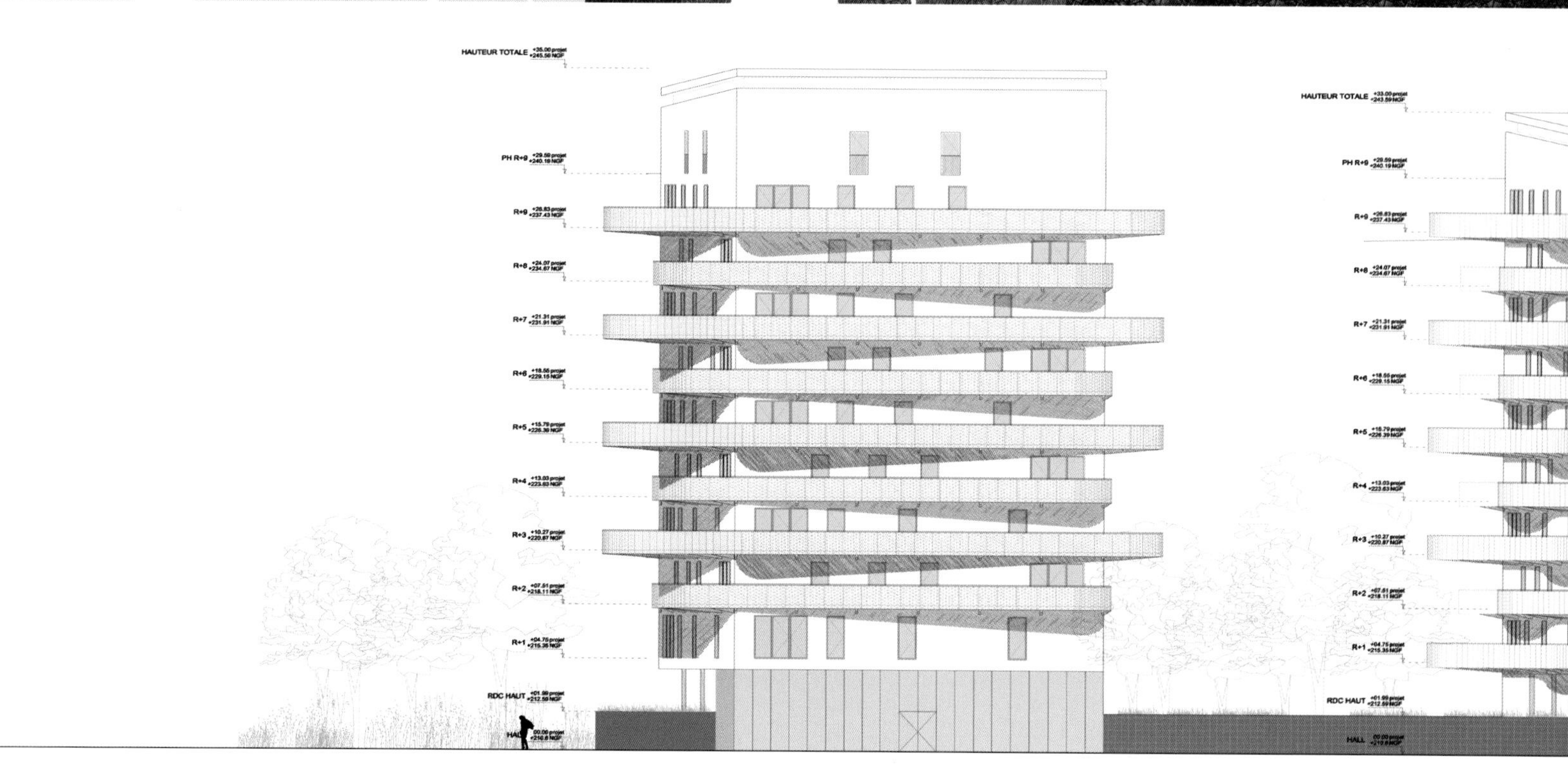

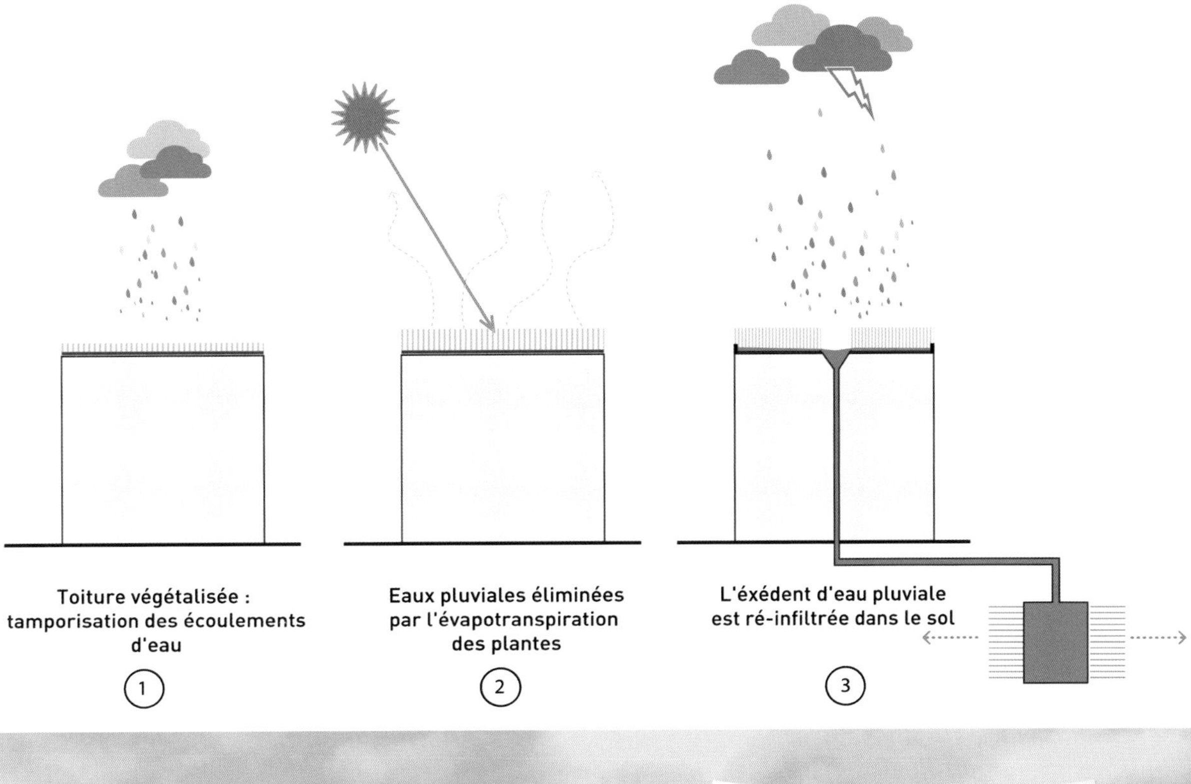
Toiture végétalisée :
tamporisation des écoulements
d'eau
1
Eaux pluviales éliminées
par l'évapotranspiration
des plantes
2
L'exédent d'eau pluviale
est ré-infiltrée dans le sol
3

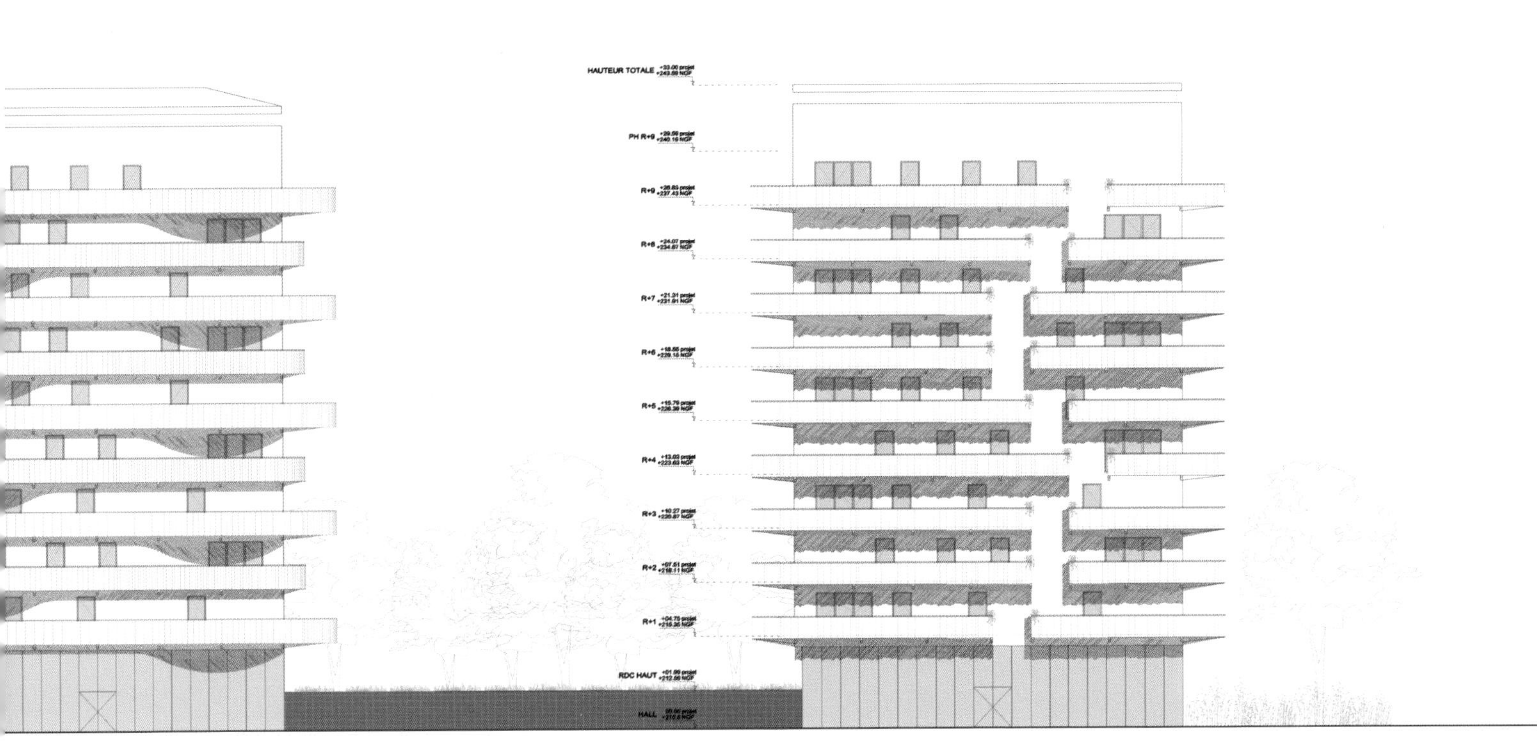
HAUTEUR TOTALE
PH R+9
R+9
R+8
R+7
R+6
R+5
R+4
R+3
R+2
R+1
RDC HAUT
HALL

澳大利亚达尔文市 Gateway 广场大楼

Gateway Plaza Building

设计单位：Bell Gabbert Associates

项目地址：澳大利亚达尔文市

Designed by: Bell Gabbert Associates

Location: Darwin, Australia

项目概况

项目位于 McMinn 街和 Daly 街桥的交界处，且与周边的其他建筑相对独立，优越的地理位置为构建一栋轮廓清晰、视野开阔、视觉效果良好的标志性建筑提供了条件。

建筑设计

这栋高 55 米的 18 层建筑包括了近 1 500 平方米的沿街商住用房，这些灵活的空间的前面是一个 300 平方米的覆盖式庭院广场，可俯瞰 McMinn 街。其后是 250 平方米的“开放式空间”，朝向旧铁路沿线的绿化公共开发区。

设计改变了玻璃扶栏和固体扶栏的比率，使塔楼的上层空间呈现出金字塔的形态，加强了建筑外立面的视觉效果。为了突出项目作为“城市门户”的形象和优势，设计将位于建筑顶部 3 层核心区的住宅单元的阳台延伸出来，形成一个拱门，赋予建筑“门户”的象征意义。

建筑的外表皮上基本由配有遮阳构件的开口和外墙组成。这些开口形成连续的阳台，包裹在建筑表层，使每个居住单元都可自然采光和自然通风，保障室内空气的畅通。阳台上的遮光墙和滑动门窗，也可起到遮阳的作用，降低了住户对空调的依赖度，有利于提高建筑的能源效率。

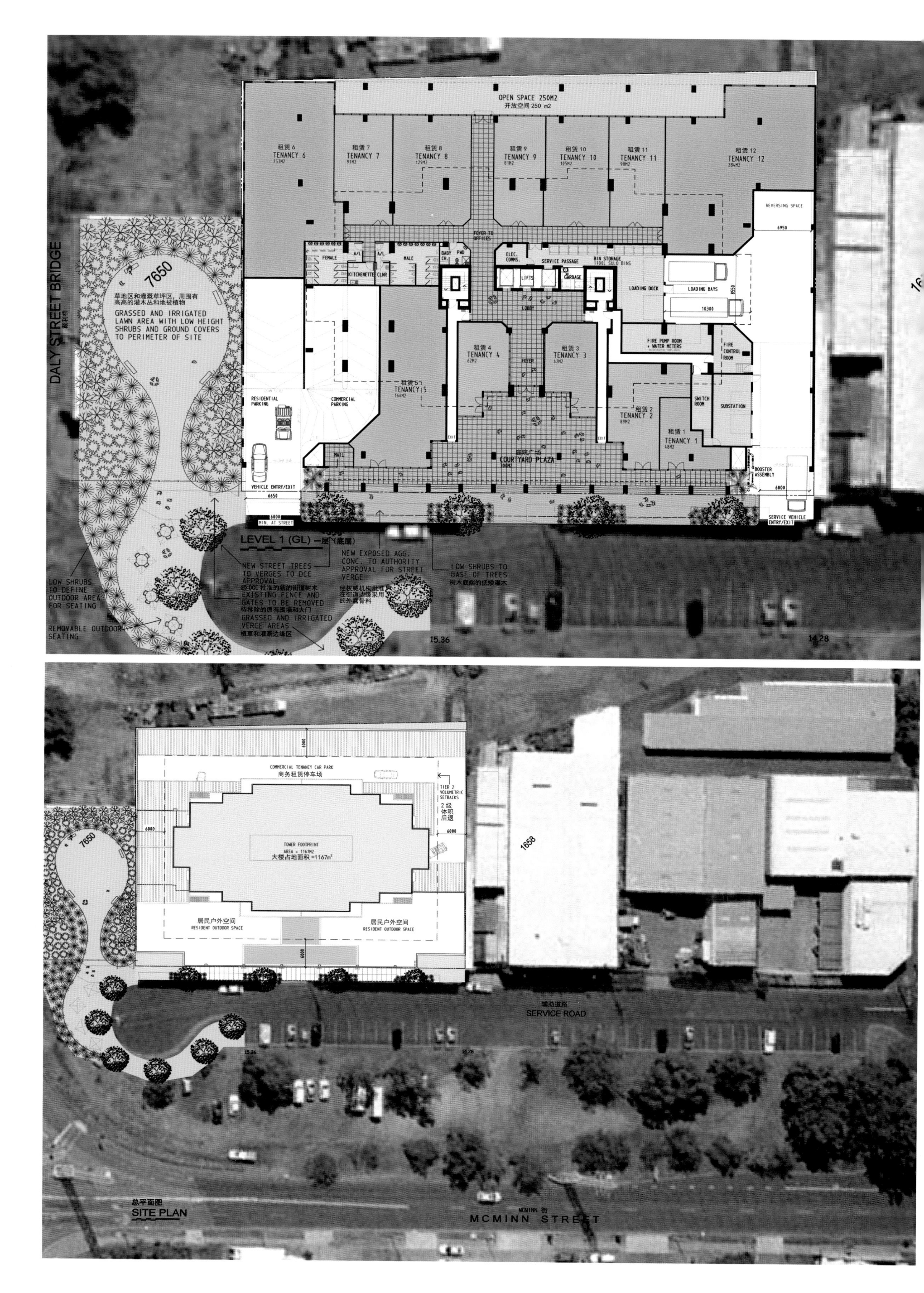

OPEN SPACE 250M2
开放空间 250 m2
租赁 6
TENANCY 6
租赁 7
TENANCY 7
租赁 8
TENANCY 8
租赁 9
TENANCY 9
租赁 10
TENANCY 10
租赁 11
TENANCY 11
租赁 12
TENANCY 12
REVERSING SPACE
FOYER TO OFFICES
ELEC. COMMS.
SERVICE PASSAGE
BIN STORAGE
LIFTS
GARBAGE
LOADING DOCK
LOADING BAYS
LOBBY
租赁 4
TENANCY 4
租赁 3
TENANCY 3
FOYER
FIRE PUMP ROOM + WATER METERS
FIRE CONTROL ROOM
租赁 5
TENANCY 5
RESIDENTIAL PARKING
COMMERCIAL PARKING
租赁 2
TENANCY 2
SWITCH ROOM
SUBSTATION
租赁 1
TENANCY 1
COURTYARD PLAZA
BOOSTER ASSEMBLY
VEHICLE ENTRY/EXIT
SERVICE VEHICLE ENTRY/EXIT
MIN. AT STREET
DALY STREET BRIDGE
7650
草地区和灌溉草坪区，周围有高高的灌木丛和地被植物
GRASSED AND IRRIGATED LAWN AREA WITH LOW HEIGHT SHRUBS AND GROUND COVERS TO PERIMETER OF SITE
LEVEL 1 (GL)
NEW STREET TREES TO VERGES TO DCC APPROVAL
经DCC批准的新的街道树木
EXISTING FENCE AND GATES TO BE REMOVED
待移除的原有围墙和大门
GRASSED AND IRRIGATED VERGE AREAS
植草和灌溉边缘区
NEW EXPOSED AGG. CONC. TO AUTHORITY APPROVAL FOR STREET VERGE
经权威机构批准在街道边缘采用的外露骨料
LOW SHRUBS TO BASE OF TREES
树木底端的低矮灌木
LOW SHRUBS TO DEFINE OUTDOOR AREA FOR SEATING
REMOVABLE OUTDOOR SEATING
15.36
14.28
COMMERCIAL TENANCY CAR PARK
商务租赁停车场
TIER 2 VOLUMETRIC SETBACKS
2级体积后退
TOWER FOOTPRINT
大楼占地面积 =1167m²
居民户外空间
RESIDENT OUTDOOR SPACE
1658
辅助道路
SERVICE ROAD
总平面图
SITE PLAN
MCMINN 街
MCMINN STREET

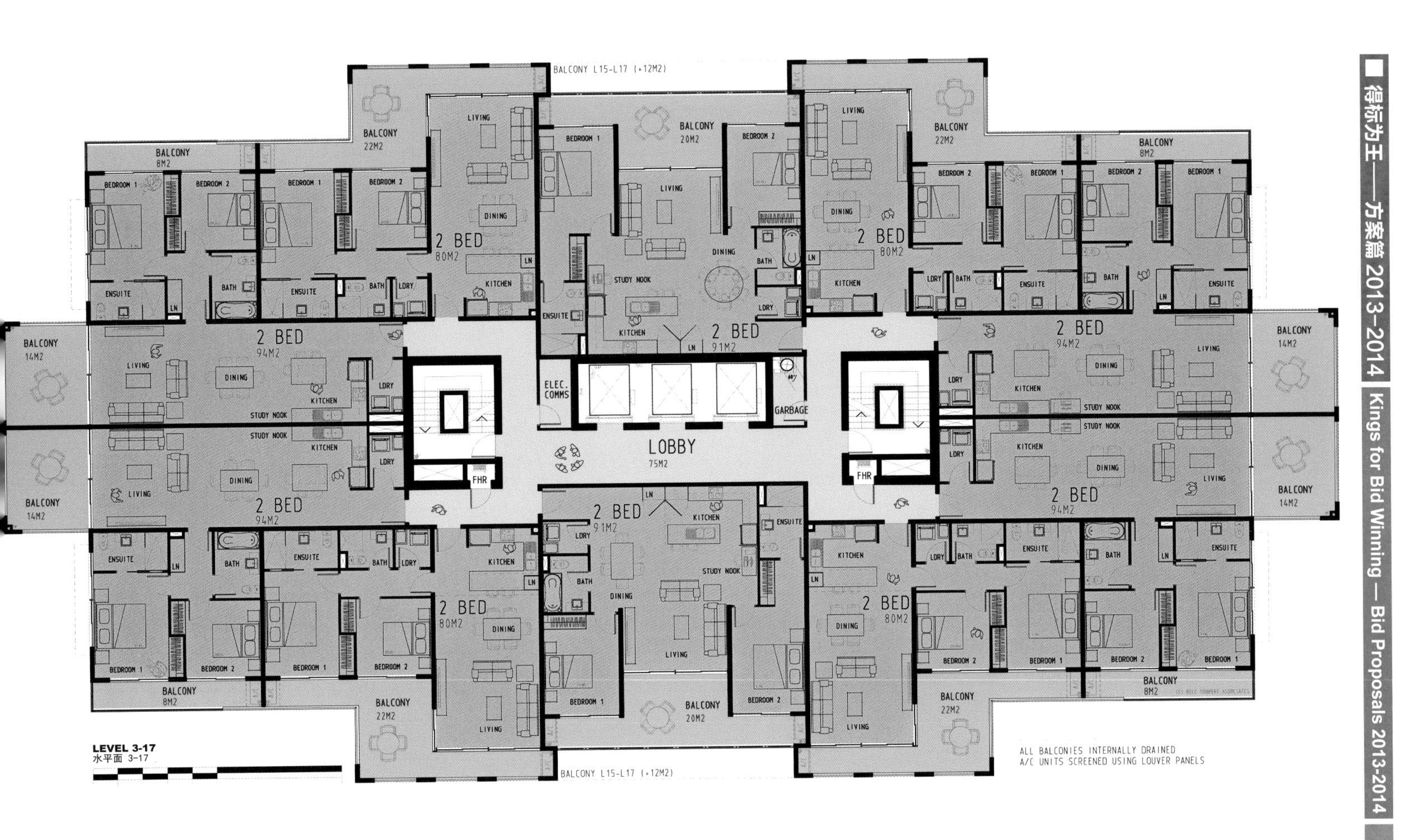

Profile

The project's strategic location on the highly visible corner intersection of McMinn Street and Daly Street Bridge offers high exposure for the building and commercial tenancies at street level. Its stepped plan form provides good articulation and visual interest whilst delivering privacy to all apartments.

Architectural Design

The 18-storey/55 meters AGL proposal includes approximately 1,500 square meters gross area of flexible commercial tenancies at street level fronting a 300 square meters covered courtyard plaza overlooking McMinn Street and a further 250 square meters of "open space" to the rear of the site, which overlooks the greenbelt public open space of the former railway corridor.

The project design has varied the ratio of glass balustrade to solid balustrade which increases up the tower in a pyramidal form. To help signify the developments gateway location into and out of the city, the upper three storeys of the central 2 bedroom apartments extend the balcony out to form an arch or "gateway" in the façade allowing the design to be emblematic with the developments namesake of "Gateway Plaza". Shading of the openings and external walls has been incorporated to almost 100% of the external envelope and has been achieved through continuous wrap-around balconies allowing all habitable rooms accessible to natural light and breezes as well as promoting airflow through apartments. In addition, each balcony acts as a sun-hood to the apartment below shading walls and windows/sliding doors. This improves the energy efficiency of each unit and helps lower the reliance on using air-conditioners all year round.

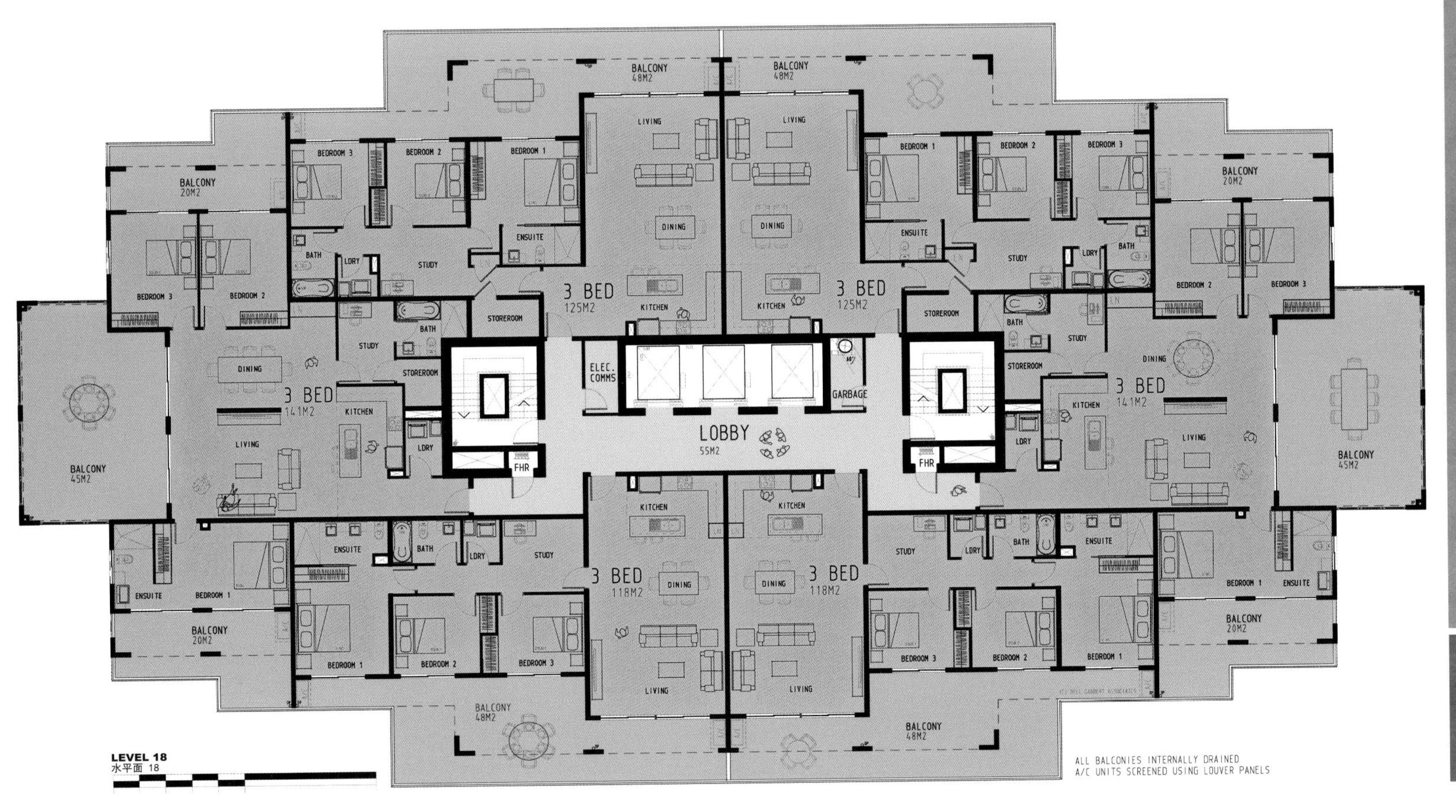

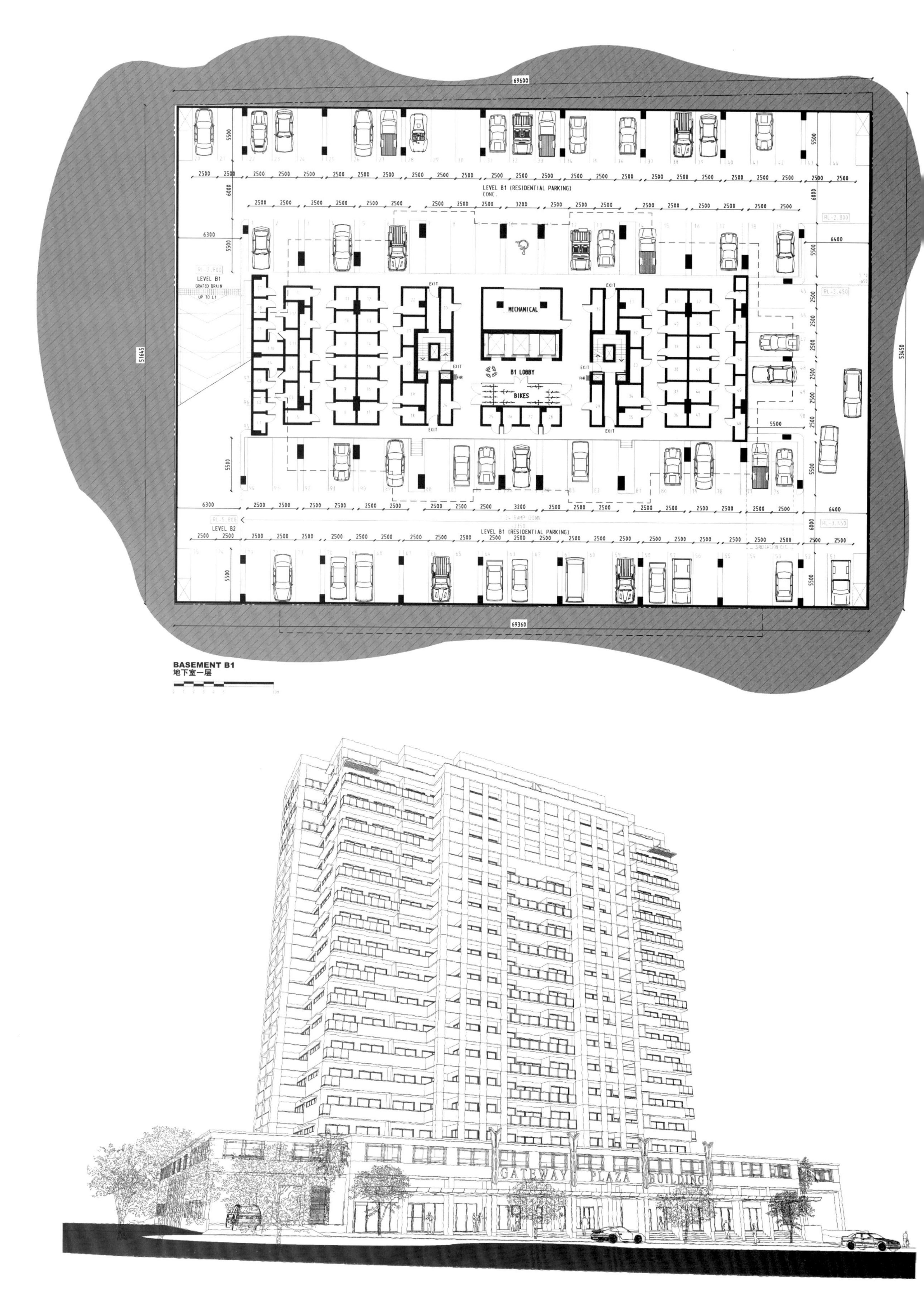

BASEMENT B1
地下室一层

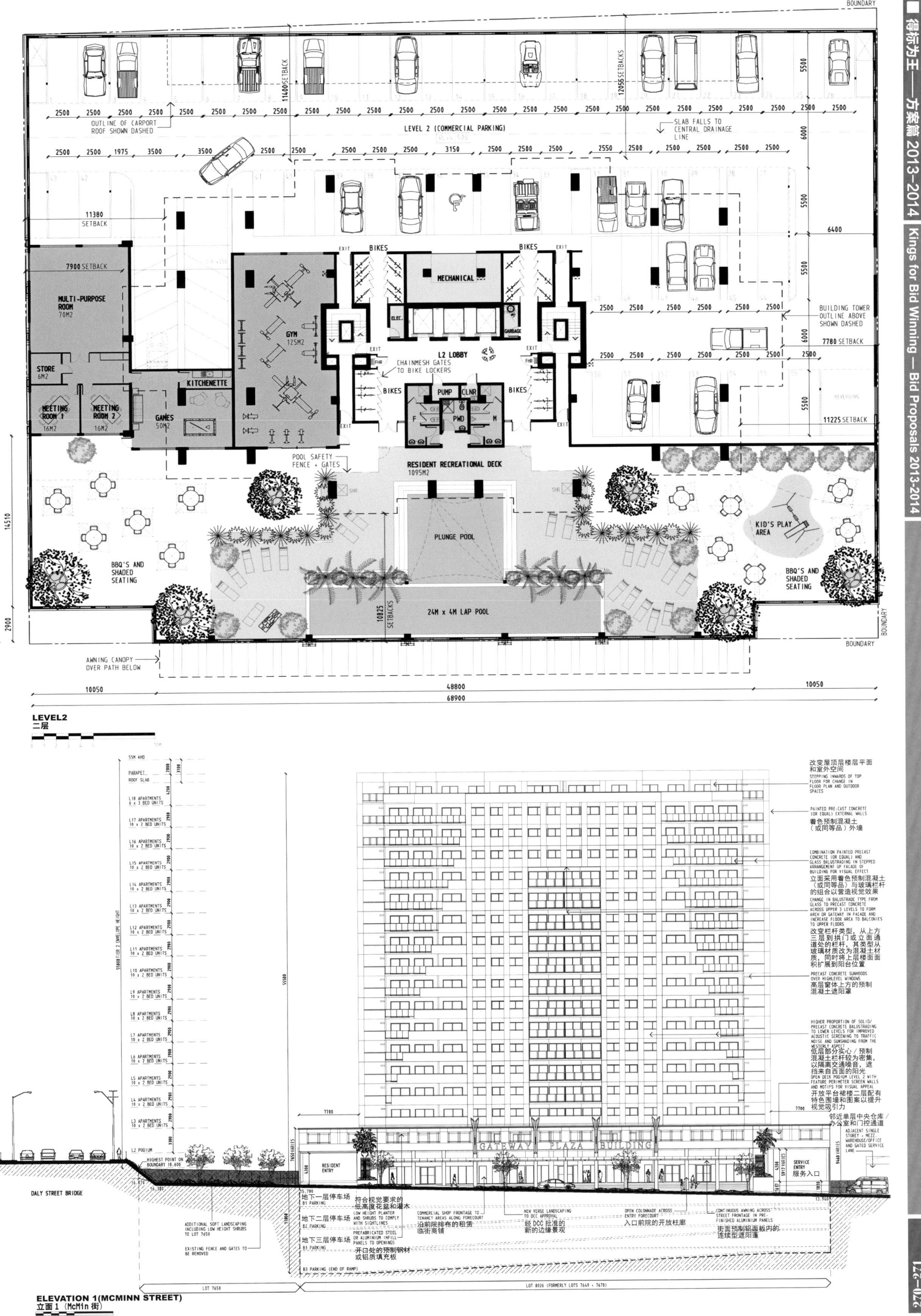
LEVEL 2 (COMMERCIAL PARKING)
MECHANICAL
GYM
MULTI-PURPOSE ROOM
L2 LOBBY
RESIDENT RECREATIONAL DECK
PLUNGE POOL
24M x 4M LAP POOL
KID'S PLAY AREA
BBQ'S AND SHADED SEATING
KITCHENETTE
GAMES
STORE
MEETING ROOM 1
MEETING ROOM 2
LEVEL2
二层
GATEWAY PLAZA BUILDING
改变屋顶层楼层平面和室外空间
着色预制混凝土（或同等品）外墙
立面采用着色预制混凝土（或同等品）与玻璃栏杆的组合以营造视觉效果
高层窗体上方的预制混凝土遮阳罩
低层部分实心／预制混凝土栏杆较为密集，以隔离交通噪音，遮挡来自西面的阳光
开放平台裙楼二层配有特色围墙和图案以提升视觉吸引力
邻近单层中央仓库／办公室和门控通道
服务入口
地下一层停车场
地下二层停车场
地下三层停车场
符合视觉要求的低高度花盆和灌木
开口处的预制钢材或铝质填充板
沿前院排布的租赁临街商铺
经DCC批准的新的边缘景观
入口前院的开放柱廊
街面预制铝面板内的连续型遮阳蓬
DALY STREET BRIDGE
ELEVATION 1(MCMINN STREET)
立面1（McMin街）

澳大利亚达尔文市海滨公寓

Oceanfront Apartments

设计单位：Bell Gabbert Associates

项目地址：澳大利亚达尔文市

摄影：Mark Bell

Designed by:* *Bell Gabbert Associates

Location: Darwin, Australia

Photography: Mark Bell

项目概况

这个 2 030 平方米的场地位于达尔文滨海大道区的显要位置，在此可俯瞰达尔文港和双世纪公园。项目旨在在这个宽度仅为 31.8 米的场地上构建一栋双朝向的精品住宅公寓大楼。

建筑设计

这栋精品公寓大楼每层设有 36 × 3 个公寓单元，每一套公寓都设有一个私人休息室。大楼容纳了 72 个有顶棚的全天候居民停车分隔间以及 12 个办公租户停车分隔间。此外，大楼还配备了一个抬高的水池平台区和位于街道层面的休闲裙楼平台，这一休闲裙楼平台将包括一个 25 米长的小型健身游泳池、网球场、野外烧烤设施和多功能游戏室等休闲空间。

双层高度的弧形玻璃车道入口大门凸显了建筑的入口，既加强了建筑的视觉效果，也进一步柔化了大楼立面棱角分明的外观形象。在选材上，建筑外立面的预制墙体采用有质感的混合型涂漆，门、窗体和栏杆则用粉末喷漆进行装饰，玻璃窗则采用有色玻璃。

生态设计

项目采用了被动能源设计，利用“太阳帽”和大进深的阳台控制入射阳光，起到遮阳的效果。建筑前方大型的开敞式窗体经由前后方连续的阳台的防护，为住户提供了良好的遮蔽。建筑前方和后方的全高滑动玻璃门以及侧墙的百叶窗则为建筑提供了横向的通风条件。

Profile

The 2,030 square meters site occupies prime frontage along Darwin's prestigious Esplanade precinct overlooking Darwin Harbor and Bicentennial Park. The narrow width of the 31.8 meters wide site required a design that accommodated dual-facing apartments in a proposal that develops a boutique residential apartment tower.

Architectural Design

The residential apartment tower with 36 × 3 bedroom units positioned each floor allowing a private lobby for each apartment. The building will include 72 secure, all-weather covered on-site resident car bays plus an additional 12 office tenant car bays. In addition, an elevated pool deck and a street level entertainment podium deck will provide on-site recreational space with a 25 meters lap pool, tennis court, BBQ facilities as well as a multi-purpose games room.

Visual interest is added to the building through the inclusion of a curved two-storey high glazed Porte Cochere which defines the entrance to the building and softens the angular look used further up the tower façade. The proposed finishes to the exterior facades include a mixture of textured paint finish to the structural precast walls, powder coat finish to all door, window and balustrades and tinted glass to all glazing.

Ecological Design

The design adopts proven principles of passive energy design including the control of incident sun by the use of sun hoods and deep balconies for sun protection. The building elevation will feature extensive fenestration to the front of the building protected by deep continuous balconies across the front and rear, which will provide good shading. Cross-flow ventilation is provided by the location of front and rear full-height sliding glass doors supplemented by shaded sliders and louvers to the side walls.

THE ESPLANADE ELEVATION
海滨大道立面

ELEVATION FACING LOT 627
立面面向地块 627

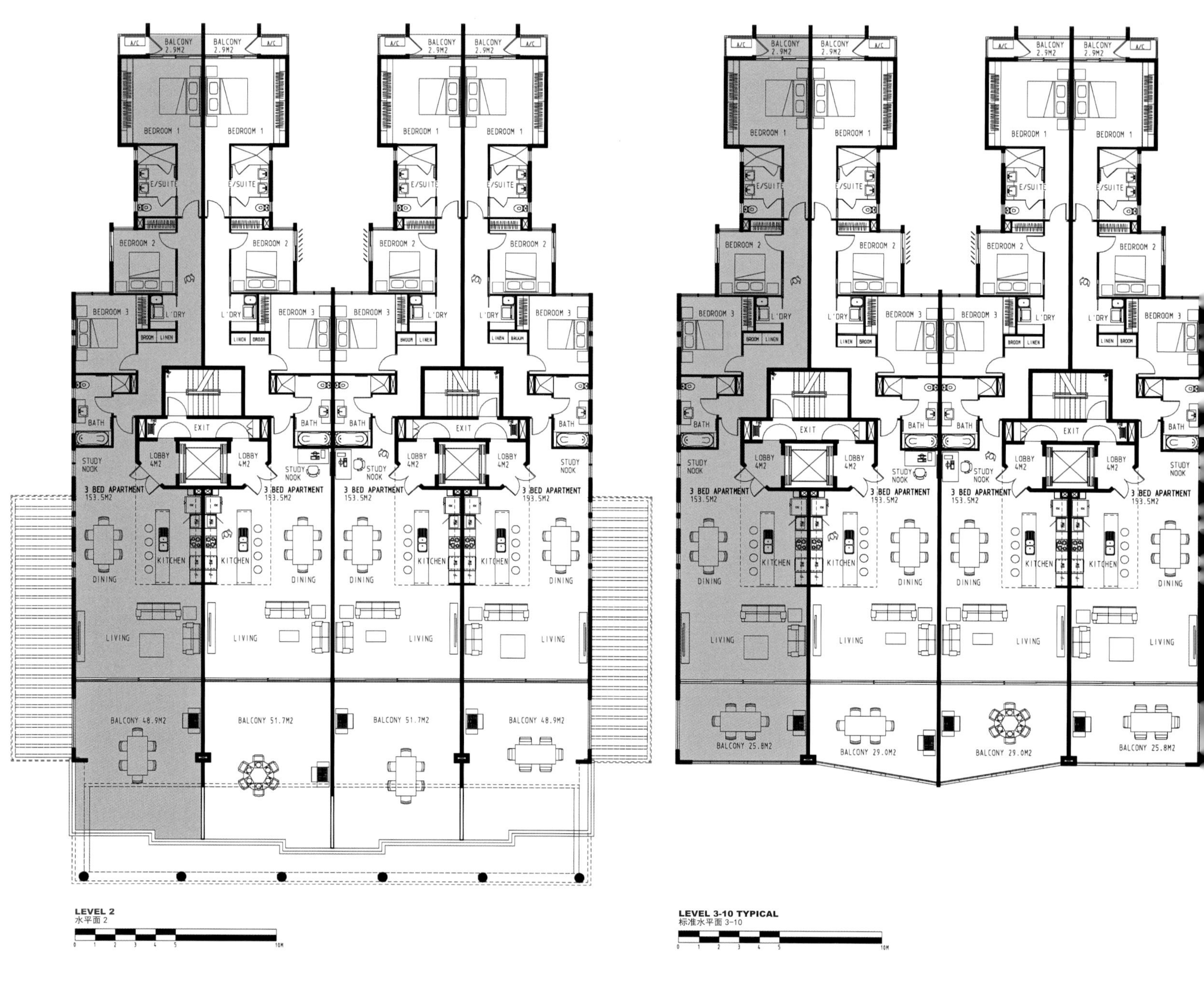

LEVEL 2
水平面 2

LEVEL 3-10 TYPICAL
标准水平面 3-10

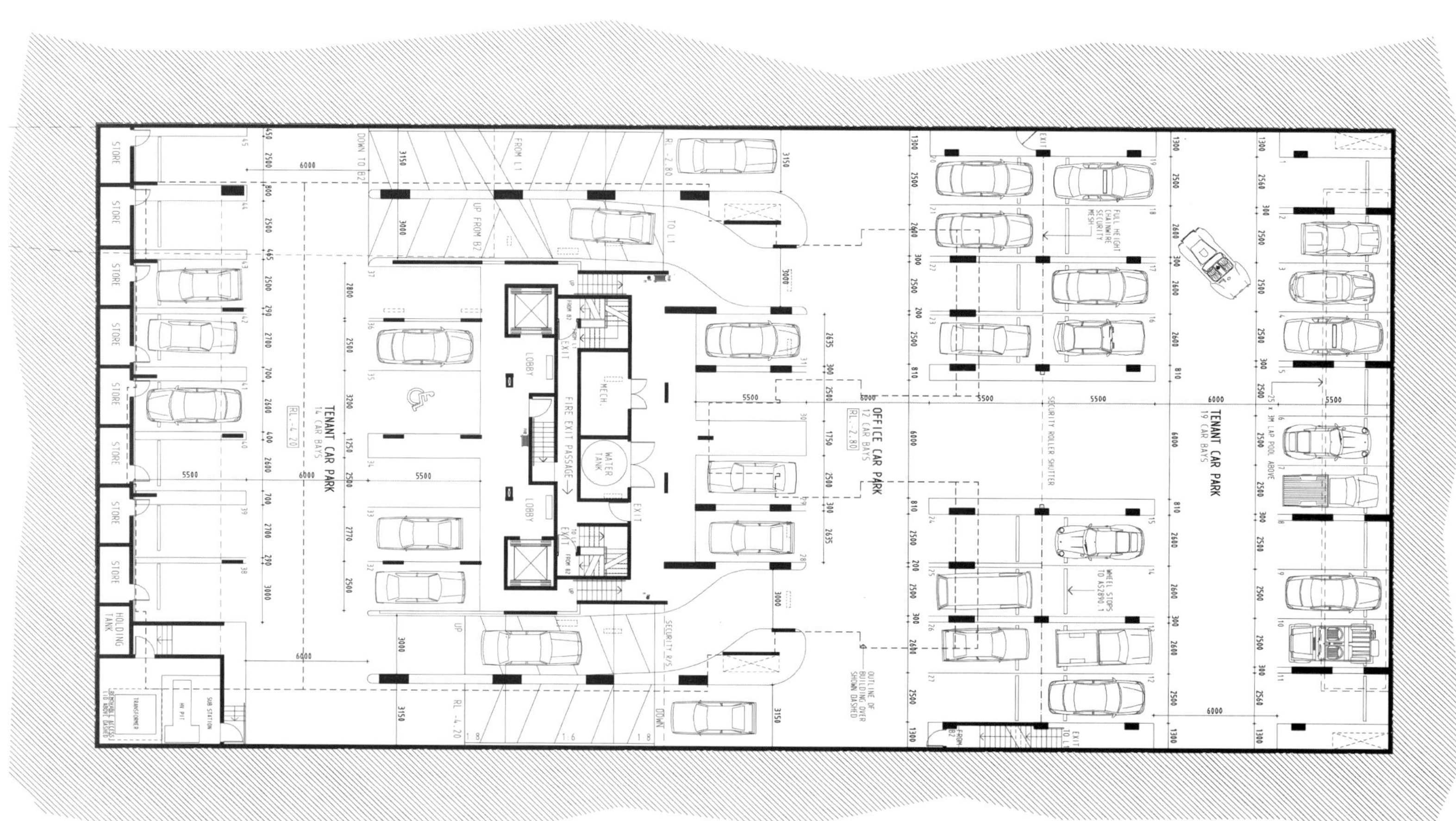

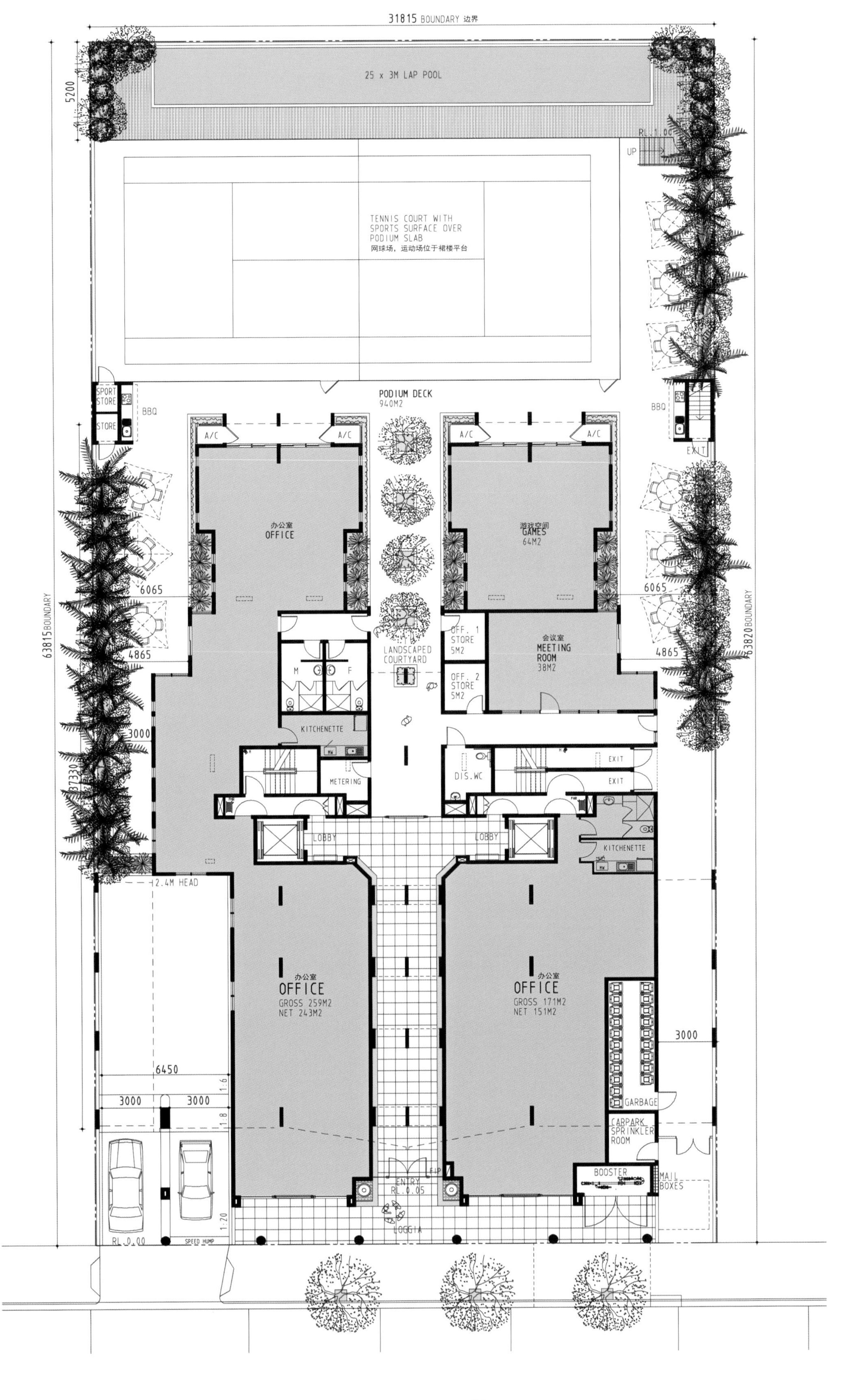
31815 BOUNDARY 边界
25 x 3M LAP POOL
5200
RL.1.00
UP
TENNIS COURT WITH
SPORTS SURFACE OVER
PODIUM SLAB
网球场，运动场位于裙楼平台
PODIUM DECK
940M2
SPORT STORE
STORE
BBQ
A/C
A/C
A/C
A/C
EXIT
办公室
OFFICE
游戏空间
GAMES
64M2
6065
6065
63815BOUNDARY
63820BOUNDARY
4865
4865
OFF. 1
STORE
5M2
OFF. 2
STORE
5M2
会议室
MEETING
ROOM
38M2
LANDSCAPED
COURTYARD
M
F
KITCHENETTE
3000
37330
METERING
DIS.WC
EXIT
EXIT
LOBBY
LOBBY
KITCHENETTE
2.4M HEAD
办公室
OFFICE
GROSS 259M2
NET 243M2
办公室
OFFICE
GROSS 171M2
NET 151M2
3000
6450
3000
3000
1.6
1.8
GARBAGE
CARPARK
SPRINKLER
ROOM
BOOSTER
MAIL
BOXES
ENTRY
RL.0.05
LOGGIA
1.20
RL.0.00
SPEED HUMP
31815 BOUNDARY 31815 边界
THE ESPLANADE 滨海大道

澳大利亚达尔文市日出大厦

Sunrise Towers

设计单位：Bell Gabbert Associates

项目地址：澳大利亚达尔文市

摄影：Mark Bell

Designed by: Bell Gabbert Associates

Location: Darwin, Australia

Photography: Mark Bell

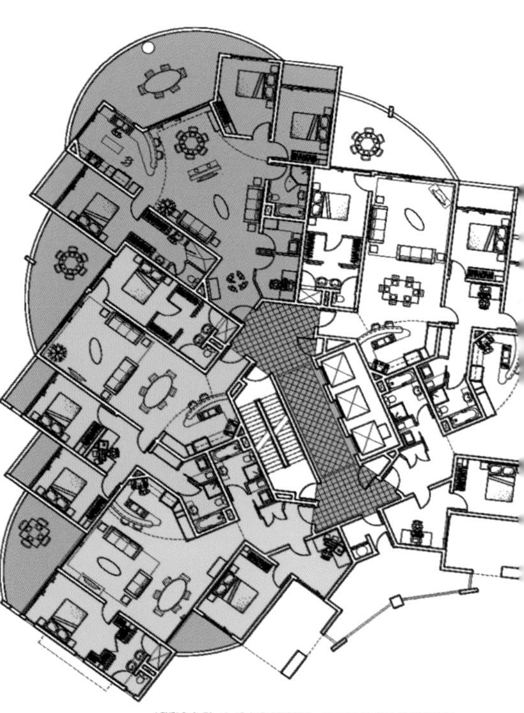

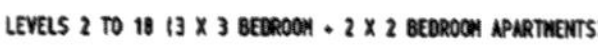

LEVELS 2 TO 18 (3 X 3 BEDROOM + 2 X 2 BEDROOM APARTMENTS)

项目概况

项目位于达尔文市中心商务区的边缘地带，这两栋独具特色的哨兵式大楼呈现出鲜明的现代主义风格，将成为进入达尔文市中心商务区的标志。

设计特色

设计利用曲线环绕式的玻璃栏杆突出展现了一个未来主义建筑，这一特色通过建筑上层向外延伸的阳台得到加强，赋予两栋大楼标志性的建筑外观。两栋大楼的位置经过了精心的计算，在最大限度内确保了公寓的私密性，同时也有效地扩展了公寓的视线范围，特别是使较低的楼层拥有了良好的视野。

本着"尽可能地提高居住者的舒适度"的设计理念，设计师在裙楼平台为居住者提供了近 3 120 平方米的特色空间，这些空间包括一个围合的网球场、两个高尔夫练习场和果岭、一个围合的儿童游乐区、一个游泳池，这些休闲区间通过景观道路和草坪连接起来。

裙楼平面设计旨在最大限度地实现自然通风和自然采光，建筑外围约 3 米的景观缓冲区为实现这一目标提供了条件。设计将停车场下降约 1.2 米，同时将高架桥平面抬升高出街道平面 1.5 米，场地已有的落差则形成低层停车场，这使包括停车场在内的北面空间都能实现自然采光和自然通风。

Profile

The project is located on the fringe of the CBD of Darwin City. The two sentinel-like towers will become an entry statement on the approach into the CBD representing a modernist architectural style.

Design Feature

The design of the precinct has focused on presenting a futuristic type of architecture by using curvilinear wrap-around glazed balustrading to give the two buildings their signature appearance. This is further enhanced towards the top of each tower with the splaying outwards of upper balconies, incorporating a v-split to give it further visual interest up the facade. The sitting of each tower maximizes privacy by reducing the overlooking of apartments, increases view sight lines from all apartments, and especially those lower down.

As part of the overall concept to introduce added amenity for occupants, the podium deck provides approximately 3,120 square meters of area accommodating a variety of on-site features for occupants. These include a fenced tennis court, two golf practice nets and putting green, a fenced children's play area, a shaded toddler/wading pool and swimming pool with ablutions plus a BBQ area all linked by landscaped paths and lawn verges.

The podium deck has been designed to maximize the cross ventilation and introduce as much natural light as possible. This is achieved by a perimeter landscaped buffer of 3 meters or more, a set-down of the first car park level of about 1.2 meters and an elevated deck of approximately 1.5 meters above street level. The existing fall of the site is then used to accommodate the lower parking level resulting in the entire northerly aspect of both parking levels being open to light and breezes.

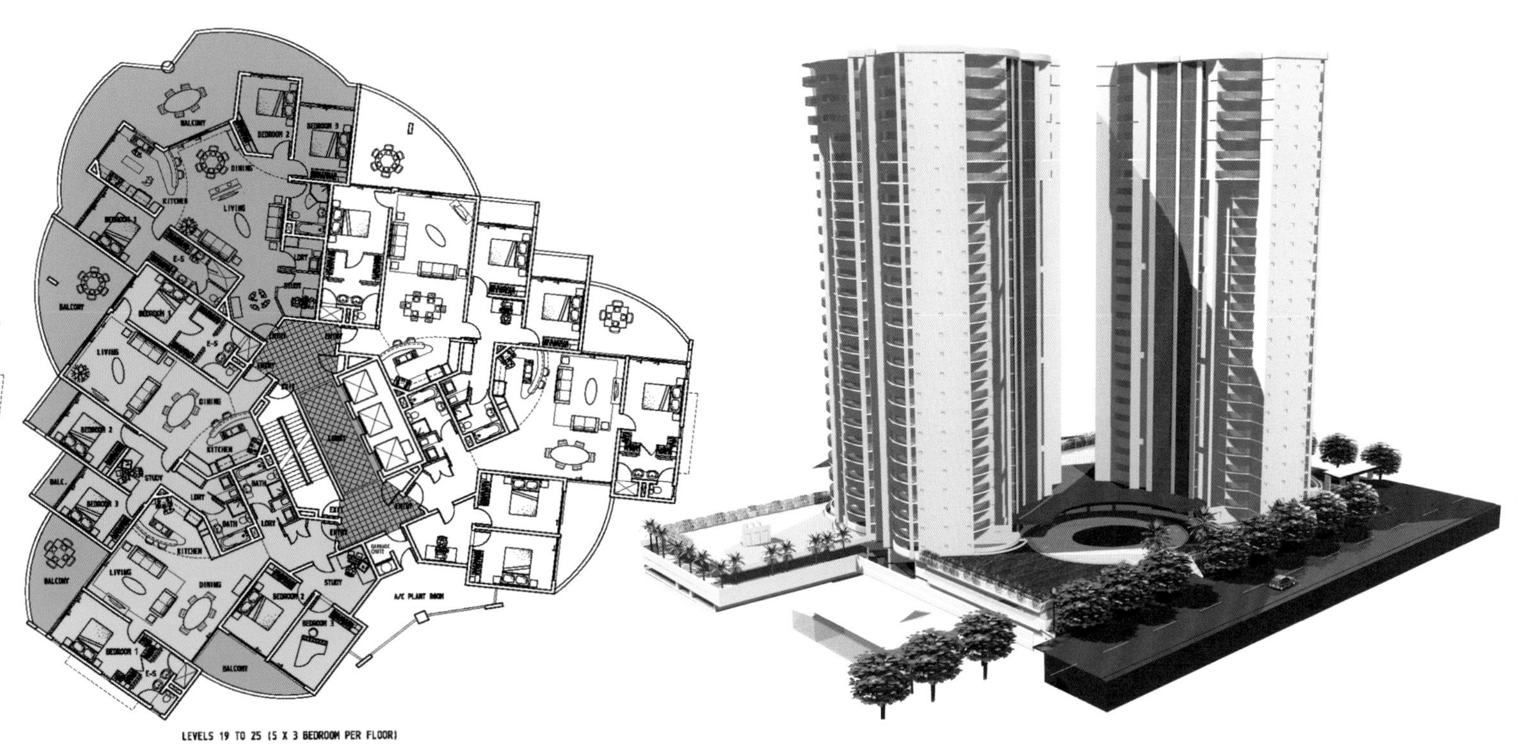

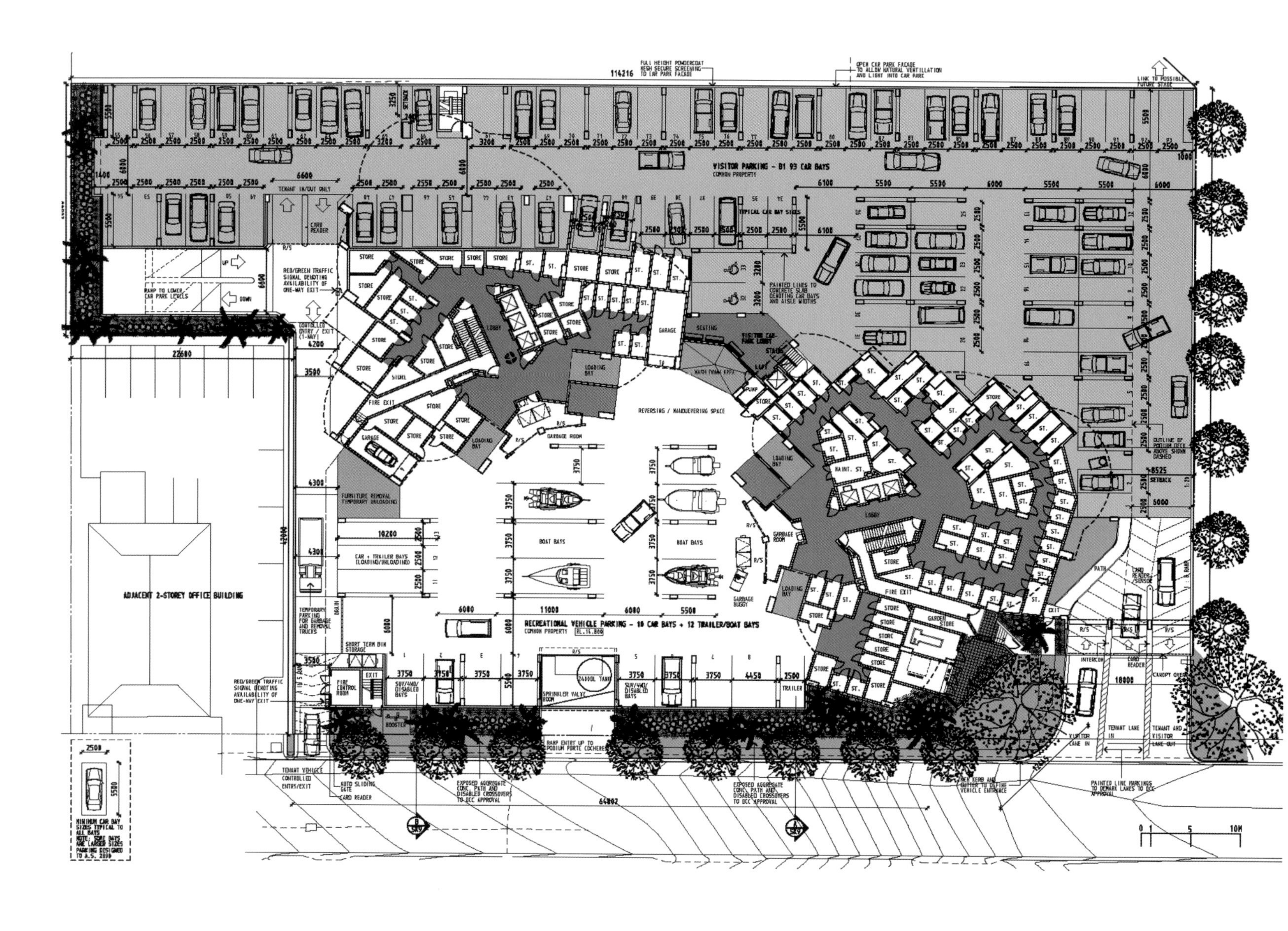

VISITOR PARKING - B1 93 CAR BAYS
COMMON PROPERTY
REVERSING / MANOEUVERING SPACE
BOAT BAYS
RECREATIONAL VEHICLE PARKING - 10 CAR BAYS + 12 TRAILER/BOAT BAYS
ADJACENT 2-STOREY OFFICE BUILDING
LOBBY
FIRE EXIT
GARBAGE ROOM
LOADING BAY
SPRINKLER VALVE ROOM
TYPICAL CAR BAY SIZES

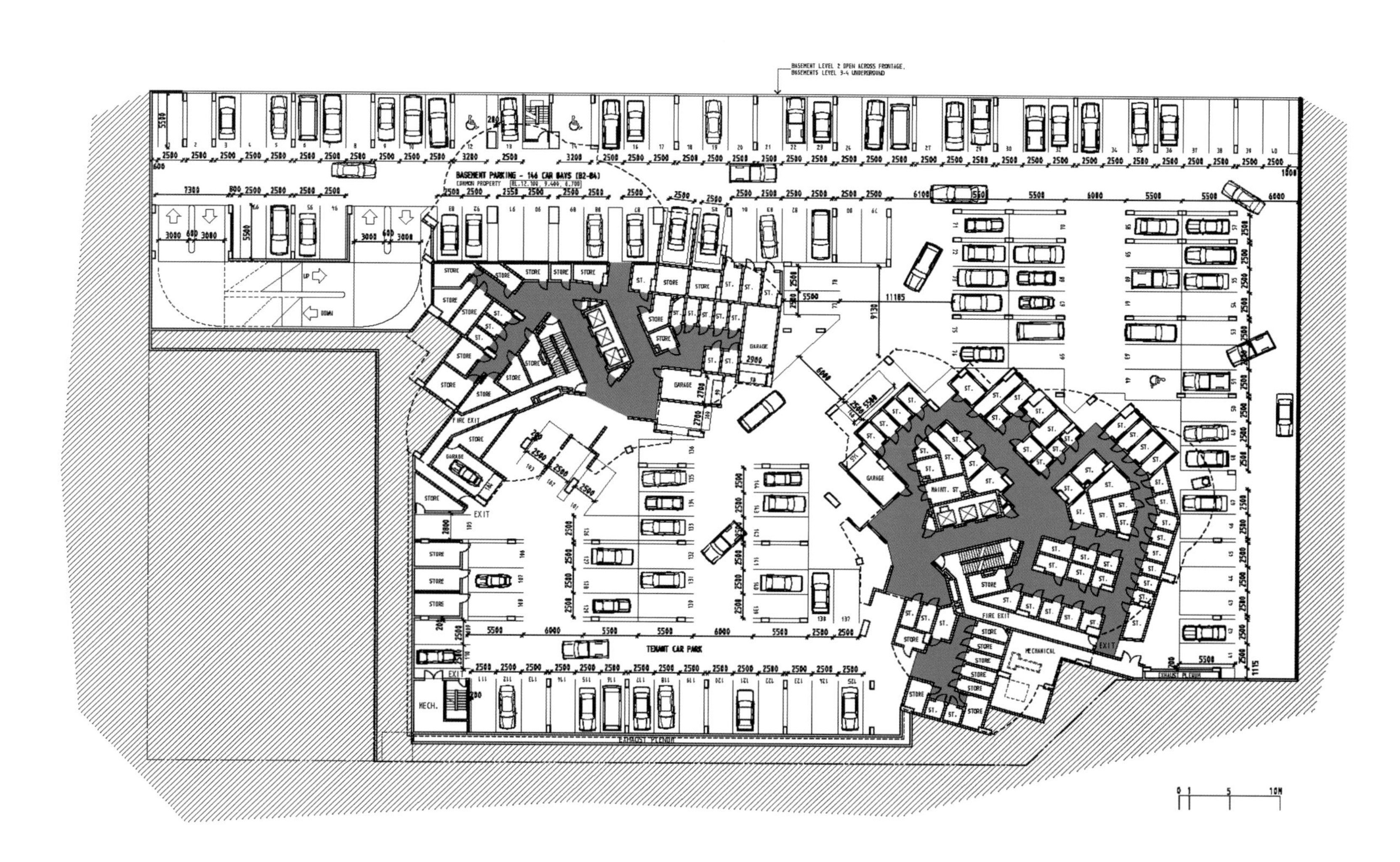

BASEMENT PARKING - 146 CAR BAYS (B2-B4)
COMMON PROPERTY
TENANT CAR PARK
MECHANICAL
FIRE EXIT
GARAGE
EXIT
MECH.

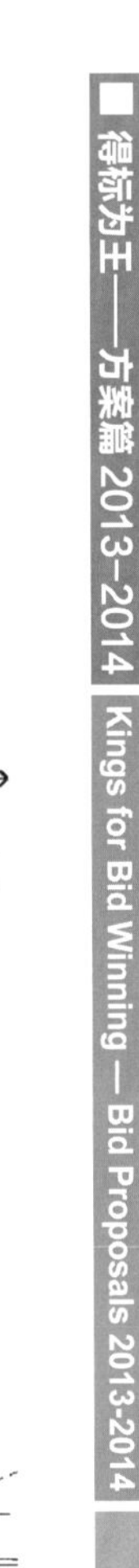

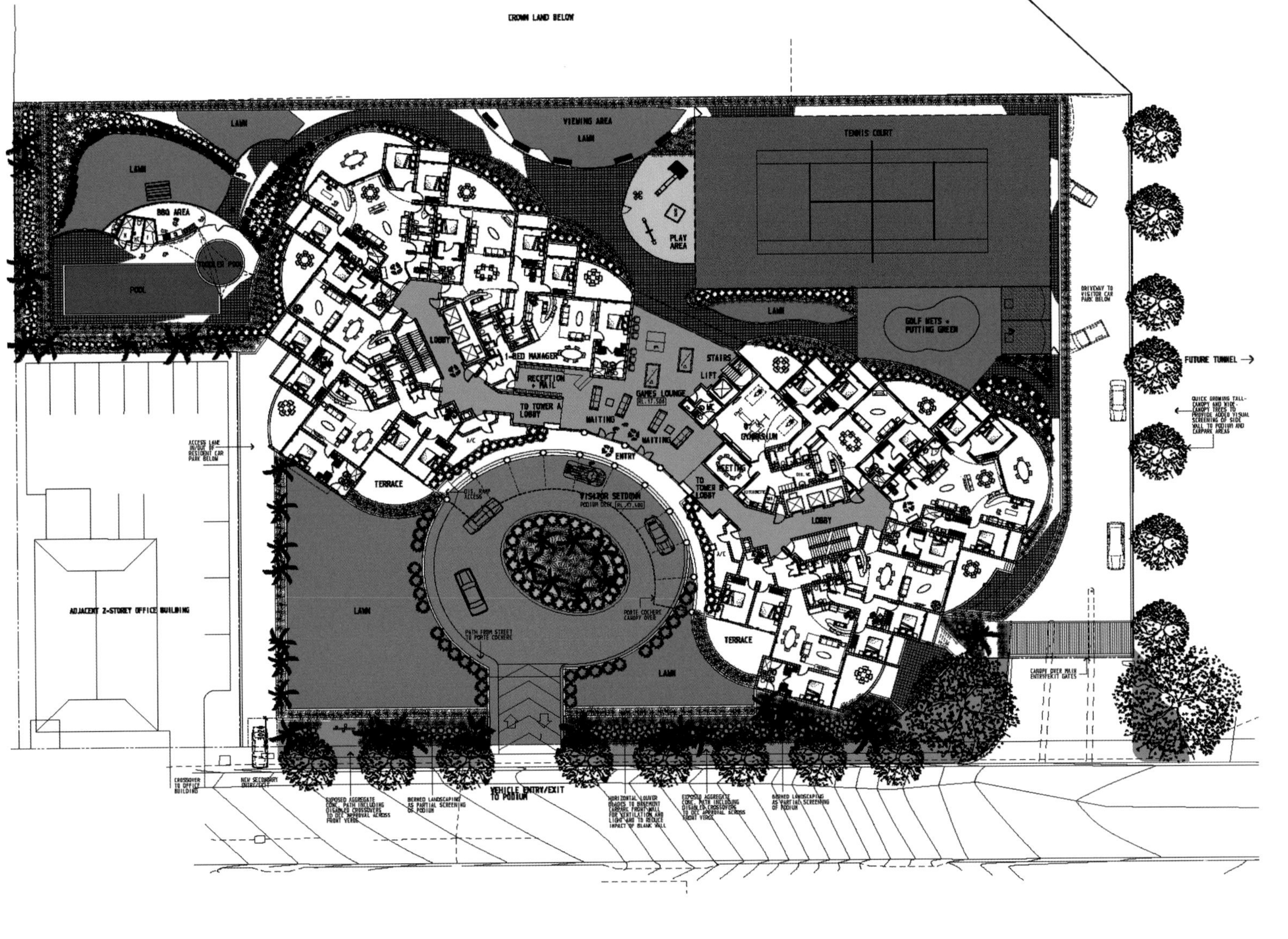
CROWN LAND BELOW
TENNIS COURT
VIEWING AREA
LAWN
BBQ AREA
POOL
TODDLER POOL
PLAY AREA
GOLF NETS + PUTTING GREEN
LOBBY
1-BED MANAGER
RECEPTION
GAMES LOUNGE
STAIRS
LIFT
WAITING
ENTRY
VISITOR SETDOWN
TO TOWER A LOBBY
TO TOWER B LOBBY
TERRACE
FUTURE TUNNEL
ADJACENT 2-STOREY OFFICE BUILDING
VEHICLE ENTRY/EXIT TO PODIUM

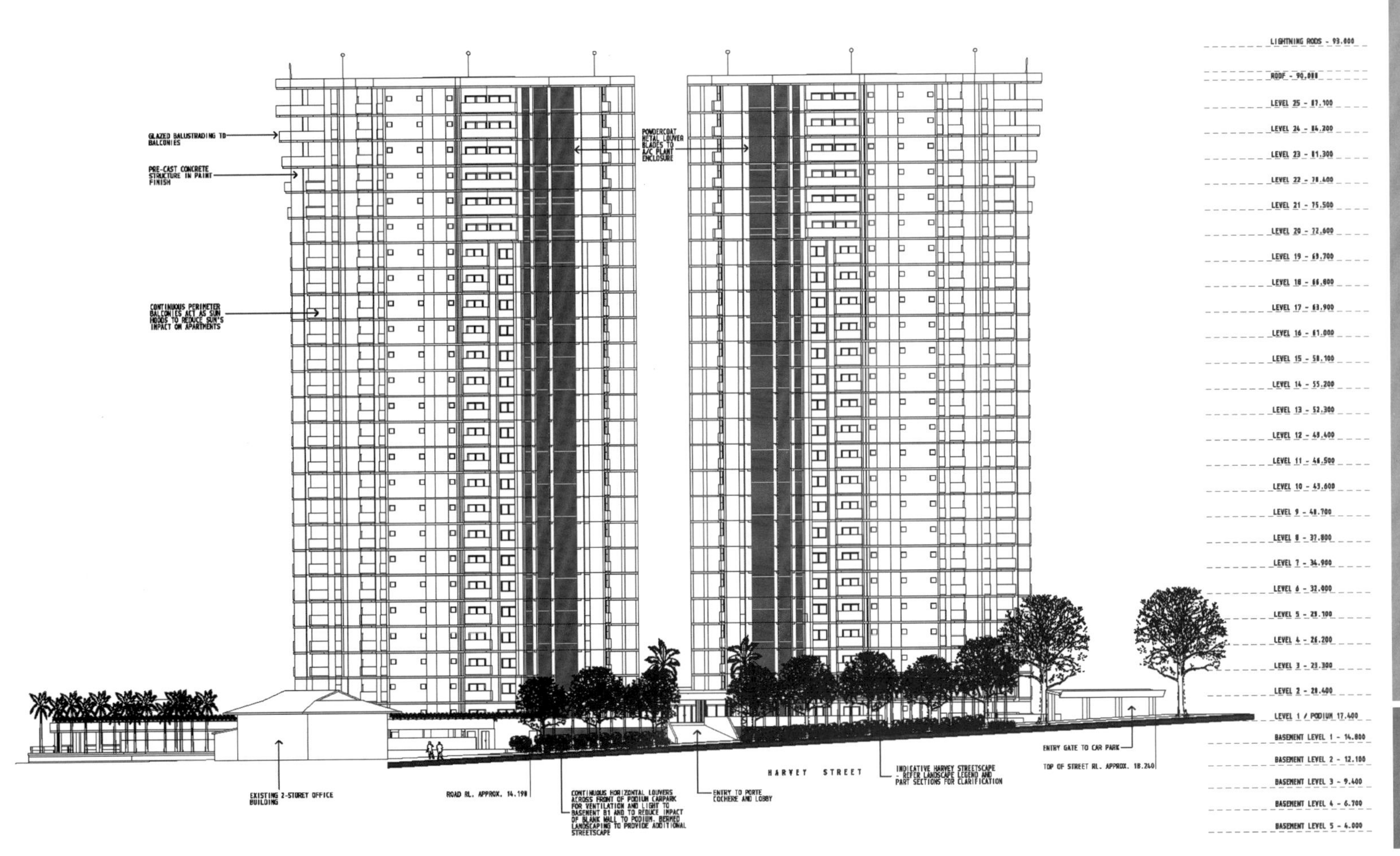
LIGHTNING RODS - 93.000
ROOF - 90.000
LEVEL 25 - 87.100
LEVEL 24 - 84.200
LEVEL 23 - 81.300
LEVEL 22 - 78.400
LEVEL 21 - 75.500
LEVEL 20 - 72.600
LEVEL 19 - 69.700
LEVEL 18 - 66.800
LEVEL 17 - 63.900
LEVEL 16 - 61.000
LEVEL 15 - 58.100
LEVEL 14 - 55.200
LEVEL 13 - 52.300
LEVEL 12 - 49.400
LEVEL 11 - 46.500
LEVEL 10 - 43.600
LEVEL 9 - 40.700
LEVEL 8 - 37.800
LEVEL 7 - 34.900
LEVEL 6 - 32.000
LEVEL 5 - 29.100
LEVEL 4 - 26.200
LEVEL 3 - 23.300
LEVEL 2 - 20.400
LEVEL 1 / PODIUM 17.400
BASEMENT LEVEL 1 - 14.800
BASEMENT LEVEL 2 - 12.100
BASEMENT LEVEL 3 - 9.400
BASEMENT LEVEL 4 - 6.700
BASEMENT LEVEL 5 - 4.000
GLAZED BALUSTRADING TO BALCONIES
PRE-CAST CONCRETE STRUCTURE IN PAINT FINISH
CONTINUOUS PERIPHERAL BALCONIES ACT AS SUN HOODS TO REDUCE SUN'S IMPACT ON APARTMENTS
POWDERCOAT METAL LOUVER BLADES TO A/C PLANT ENCLOSURE
HARVEY STREET
EXISTING 2-STOREY OFFICE BUILDING
ROAD RL. APPROX. 14.190
ENTRY TO PORTE COCHERE AND LOBBY
ENTRY GATE TO CAR PARK
TOP OF STREET RL. APPROX. 18.240

北京温都水城国际公寓

Beijing Hot Spring Leisure City International Apartment

设计单位：澳大利亚 SDG 设计集团
开发商：北京宏福集团
项目地址：中国北京市
占地面积：36 000 ㎡
建筑面积：135 980 ㎡
建筑密度：45%
容积率：2.6
设计团队：聂建鑫　胡乃昆　张春红
　　　　　翟　莹　钱　捷　左凡杰

Designed by: Shine Design Group Pty. Ltd.
Developer: Beijing Hongfu Group
Location: Beijing, China
Site Area: 36,000 m^2
Floor Area: 135,980 m^2
Building Density: 45%
Plot Ratio: 2.6
Design Team: Nie Jianxin, Hu Naikun, Zhang Chunhong, Di Ying, Qian Jie, Zuo Fanjie

项目概况

项目位于北京昌平区温都水城，占地面积为 3.6 公顷，总建筑面积达 135 980 平方米。设计在充分利用土地和周边配套设施的前提下，试图构建一个高效、绿色、人性化的养老宜居场所。

设计特色

在最大限度地提高土地利用率的前提下，设计师通过将规划中的五栋大楼的塔楼平面扭转 45 度，不仅使每个居室都有开阔的视野、良好的通风条件，也使整个楼体几乎没有阳光死角，户户都能享有充足的阳光。

设计将基底的裙房与塔楼连成一体，通过裙房内完善的配套设施、屋顶花园、露台、绿色共享空间、屋顶休闲运动、水疗、酒店式居室等非传统老年居住方式，提倡适合老年心理的全新老年居住形式。

空中连廊将项目与医院、酒店以及东侧的商业娱乐广场连接起来，既独立又安静，为老年人提供了生活上的便利。

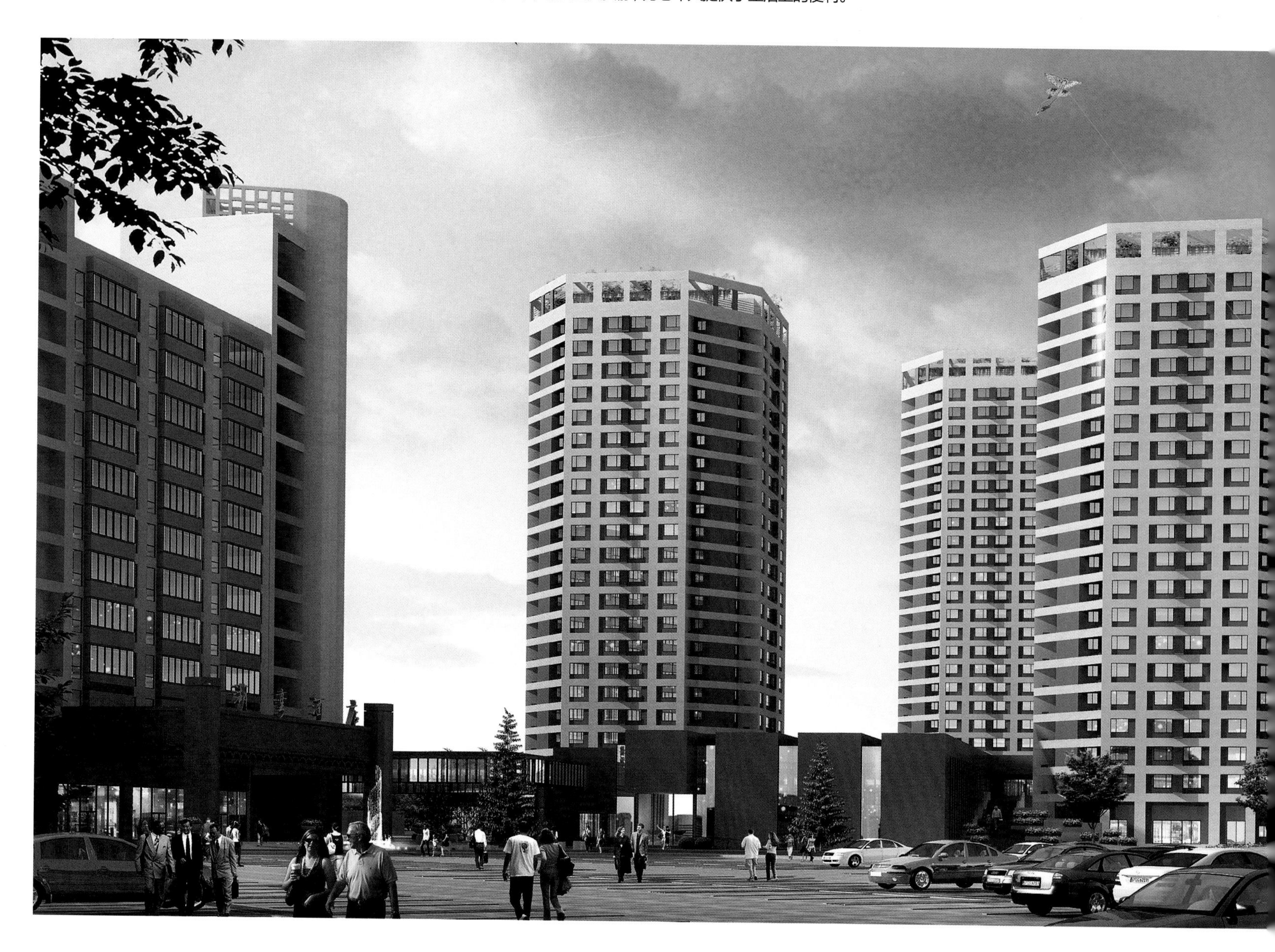

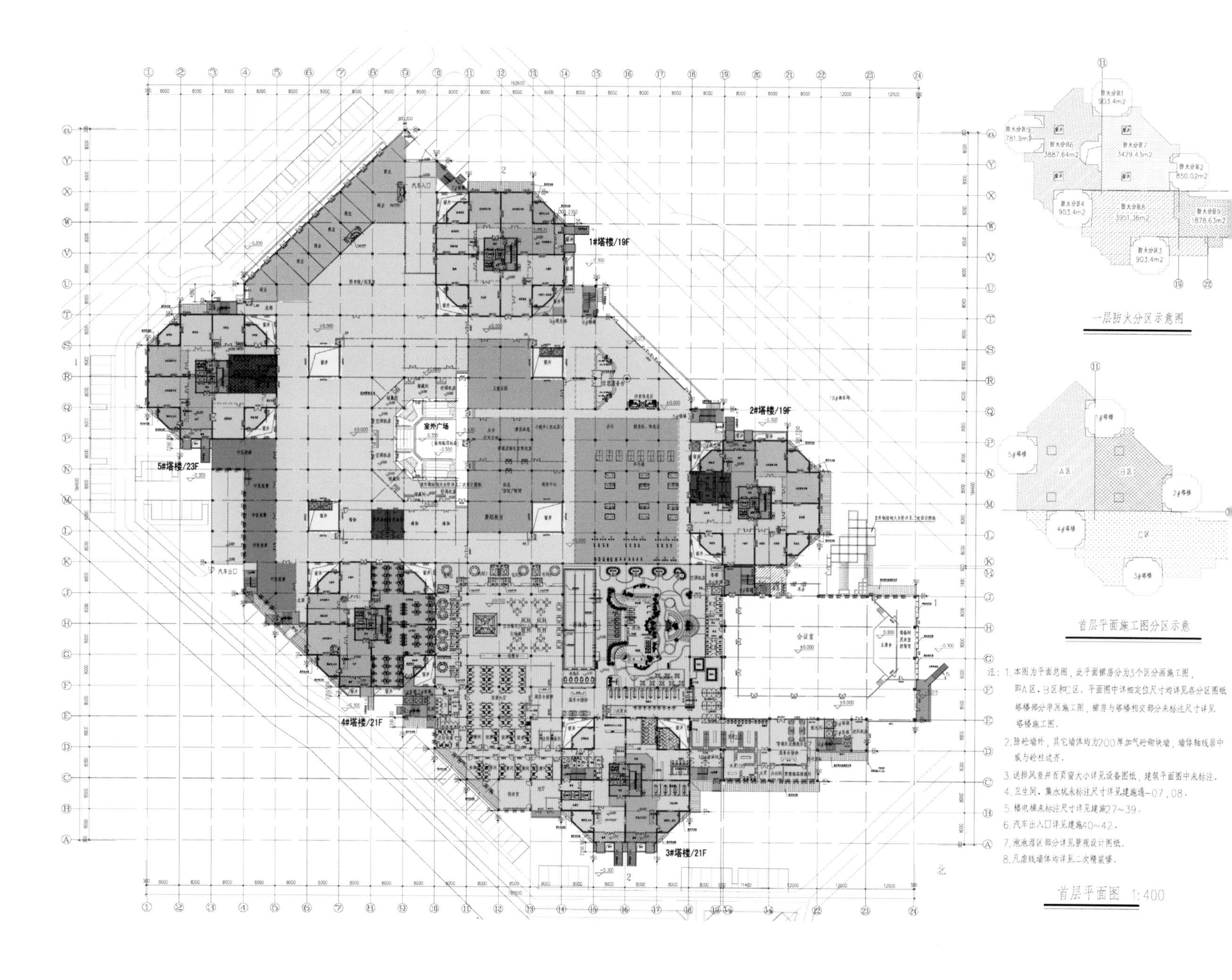

首层平面图 1:400

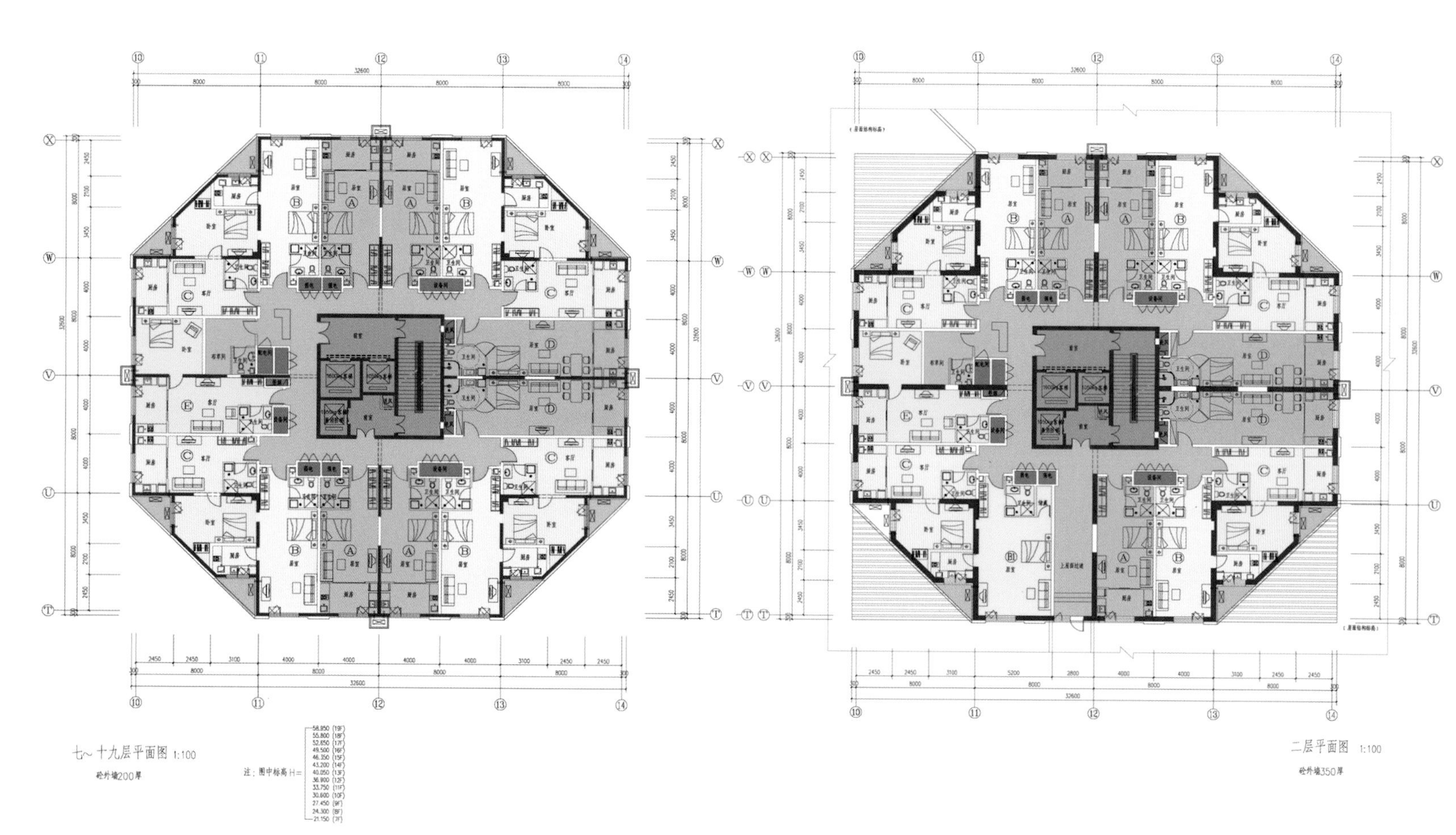

七~十九层平面图 1:100

砼外墙200厚

二层平面图 1:100

砼外墙350厚

Profile

The project is located in Hot Spring Leisure City of Beijing Changping District with a site area of 3.6 hectares and GFA of 135,980 square meters. Based on full utilization of the site and supporting facilities around, the project tries to build an efficient, green and people-oriented livable place.

Design Feature

On the premise of maximally improving land use capability, the floors of the five towers under planning are rotated for 45°, which ensures every unit to enjoy broad view and favorable ventilation. The whole building almost has no shady dead corner; all units are free to enjoy ample sunlight.

The podium and the tower are connected. Unconventional living mode for the aged, comprising complete supporting facilities, roof garden, terrace, green public space, roof leisure sports, SPA and hotel-style room, encourages a brand new living style suitable for the mental health of the aged.

Space corridor connects the project with hospital, hotel and the commercial entertainment plaza on the east side. It is independent and tranquil, providing the aged with convenient transportation.

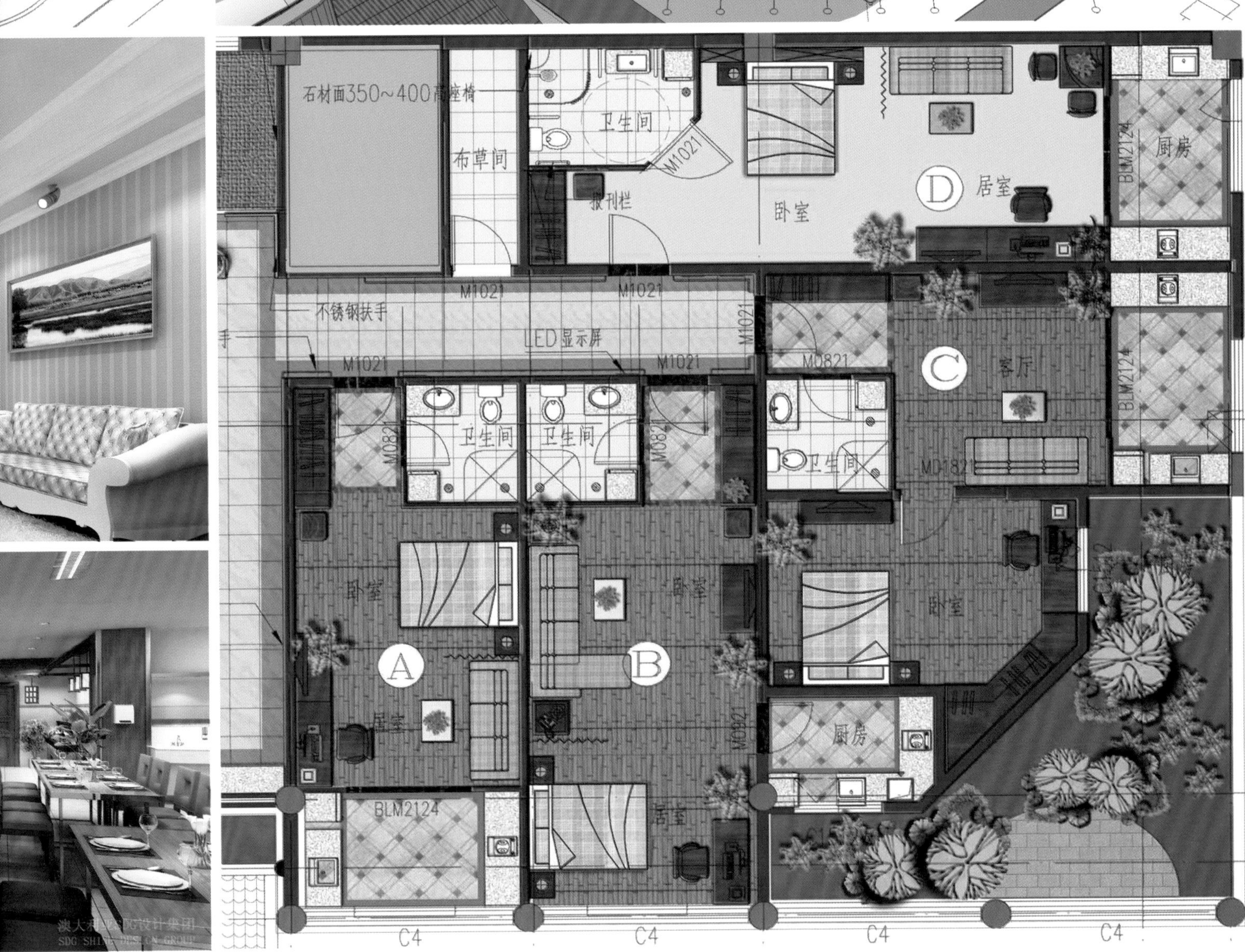
石材面350~400高座椅
布草间
卫生间
M1021
报刊栏
卧室
D
居室
厨房
BLM2124
不锈钢扶手
LED显示屏
M0821
C
客厅
卫生间
MD1821
卧室
A
居室
B
卧室
居室
厨房
BLM2124
C4
C4
C4
C4

印度班加罗尔住宅

Houses at Bangalore

设计单位：Alessio Patalocco

开发商：MAARK VISION

项目地址：印度班加罗尔

总规划面积：42 000 ㎡

建筑密度：60%

绿化率：10%

Designed by: Alessio Patalocco

Client: MAARK VISION

Location: Bangalore, India

Total Planning Area: 42,000 m^2

Building Density: 60%

Greening Ratio: 10%

设计理念

设计师认为，个性化对每一栋住宅来说都是至关重要的，只有这样，才能赋予这些建筑独有的形象和特征。同时，在追求建筑独特性的同时，也要遵循该住宅区的整体形象。

建筑设计

在表现设计上，设计师采用了智能表面的设计手法，综合考虑了美学、热物理学以及主动能源系统等因素，使建筑既美观，又可实现能源的自给自足。在空间布局方面，设计师将卧室设置在建筑的东侧，这既便于空间灵活布局，合理规划功能区间，也有利于营造舒适的内部生活环境。

设计特色

建筑整体布局紧凑，建筑间的间距也显得较为狭窄，因此，设计师借助树叶和花草，设计了一道道“绿色墙壁”，不仅可以遮阳，提供荫蔽，而且赋予了建筑赏心悦目的外观。建筑有一个大型的屋顶花园，屋顶花园复制了场地地表原有的面貌，遵循了可持续建筑的原则。同时，屋顶花园也成为了一个可以聚会、晒日光浴的私人花园。

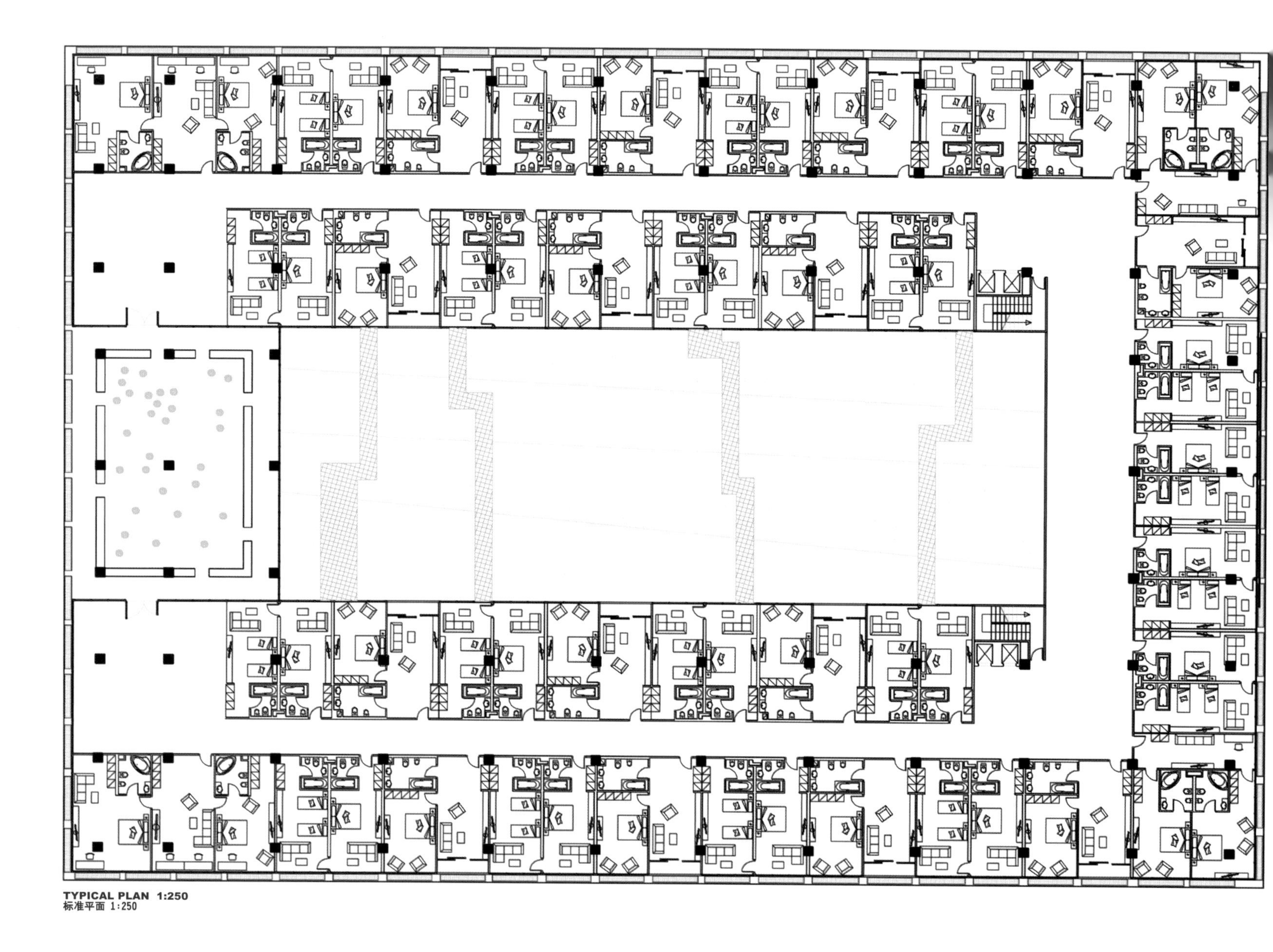

TYPICAL PLAN 1:250
标准平面 1:250

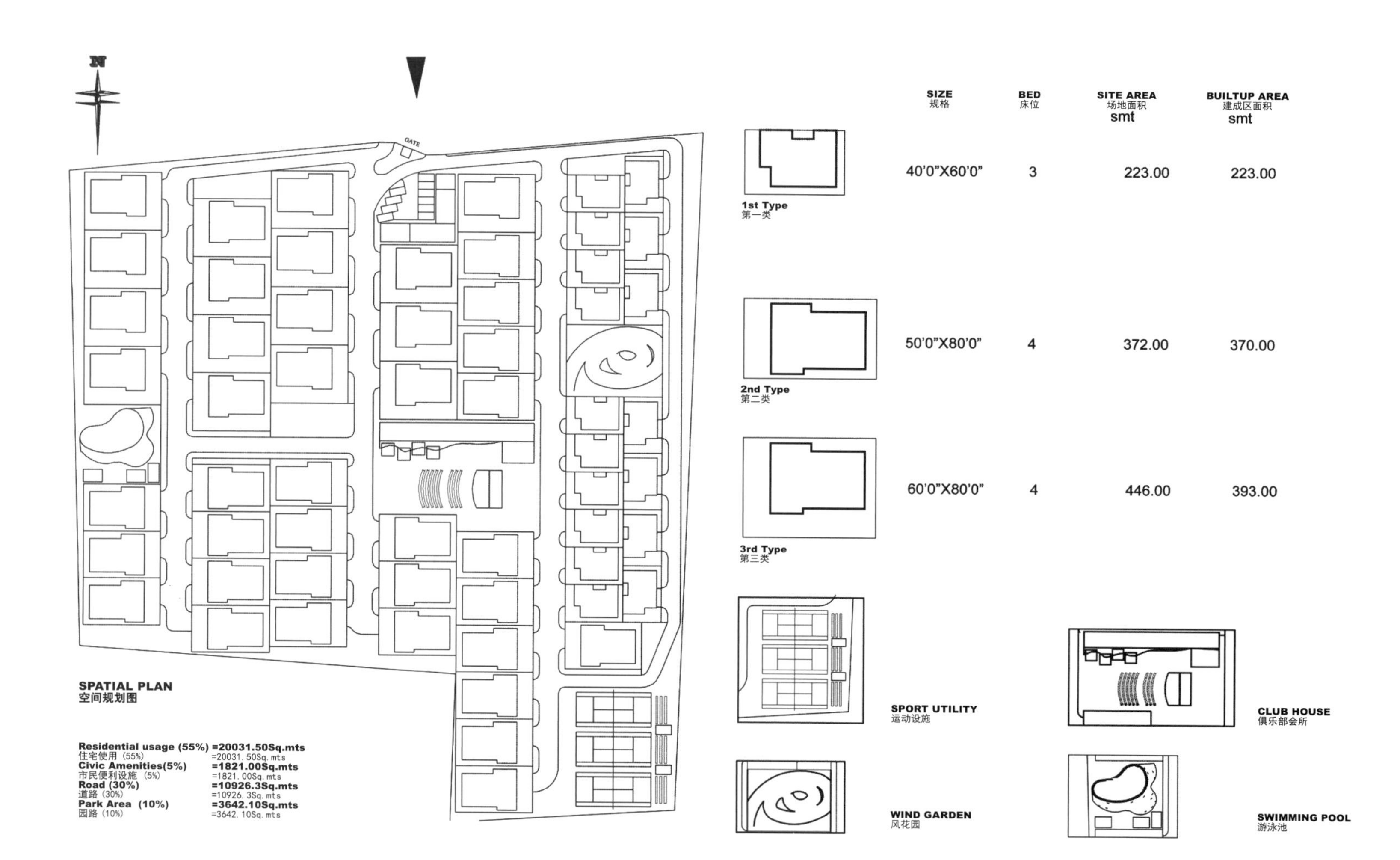
N
GATE
SIZE
规格
BED
床位
SITE AREA
场地面积
smt
BUILTUP AREA
建成区面积
smt
1st Type
第一类
40'0"X60'0"
3
223.00
223.00
2nd Type
第二类
50'0"X80'0"
4
372.00
370.00
3rd Type
第三类
60'0"X80'0"
4
446.00
393.00
Residential usage (55%) =20031.50Sq.mts
住宅使用 (55%) =20031.50Sq.mts
Civic Amenities(5%) =1821.00Sq.mts
市民便利设施 (5%) =1821.00Sq.mts
Road (30%) =10926.3Sq.mts
道路 (30%) =10926.3Sq.mts
Park Area (10%) =3642.10Sq.mts
园路 (10%) =3642.10Sq.mts
SPORT UTILITY
运动设施
CLUB HOUSE
俱乐部会所
WIND GARDEN
风花园
SWIMMING POOL
游泳池

SPATIAL PLAN
空间规划图

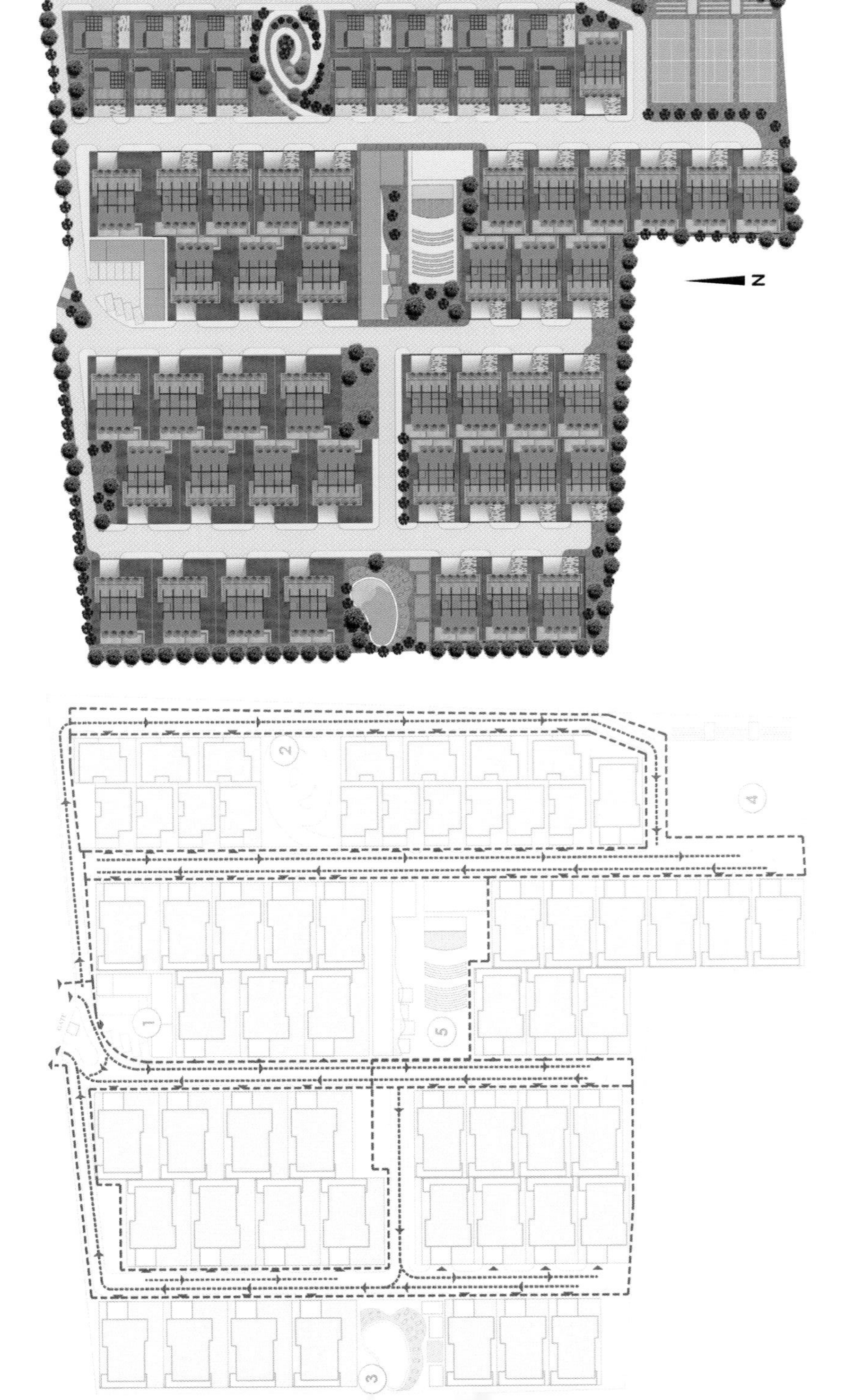

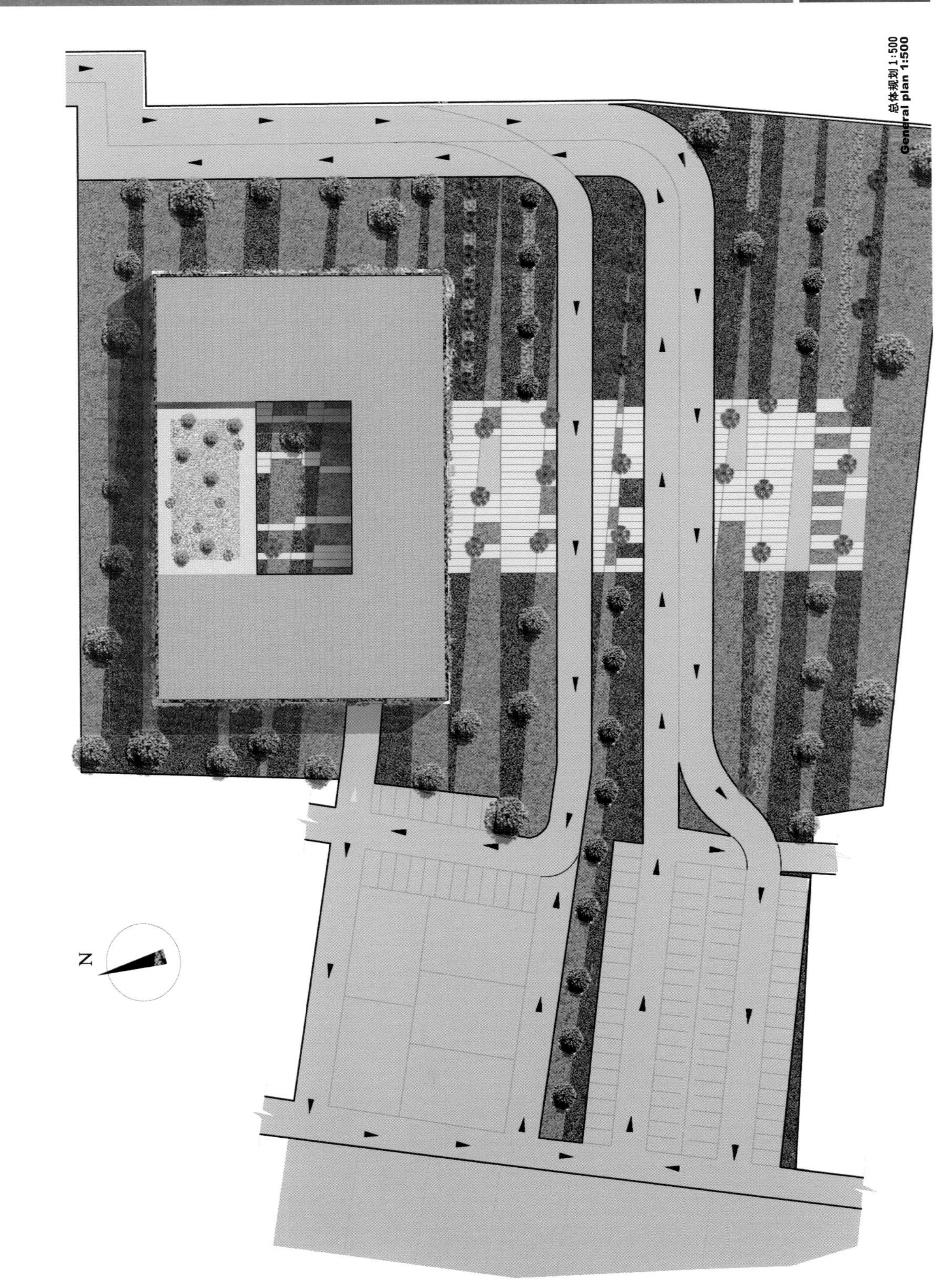

总体规划 1:500
General plan 1:500

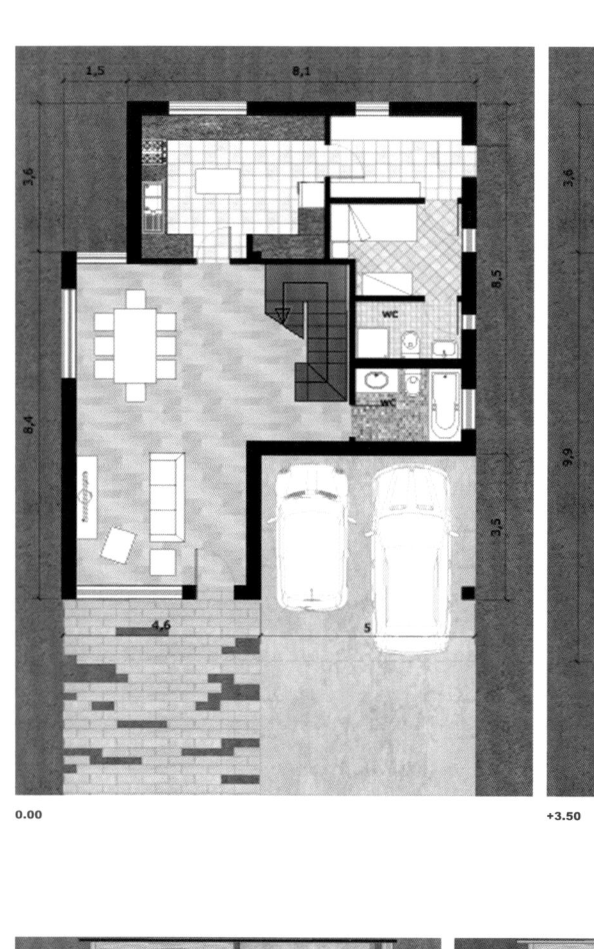

0.00

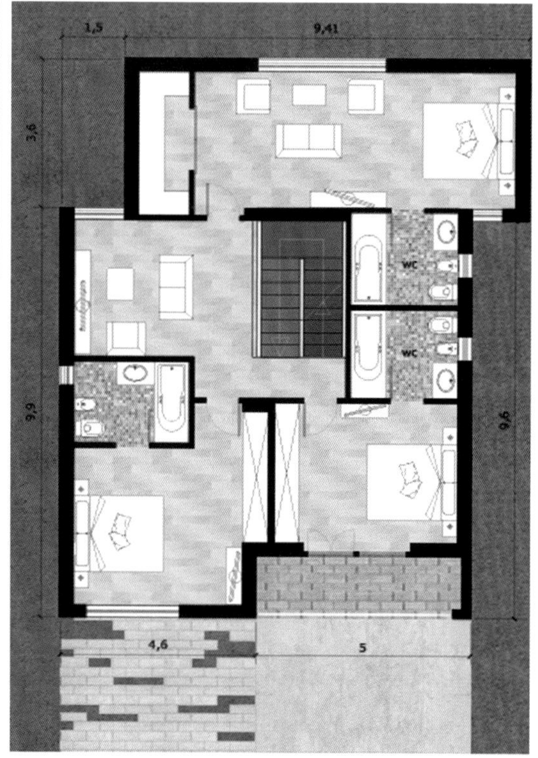

+3.50

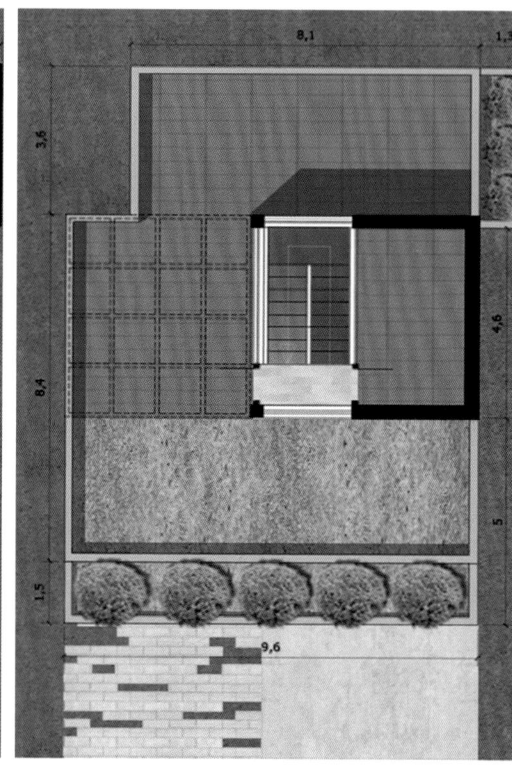

+7.00

1st type (40×60ft)
第一类 (40×60ft)

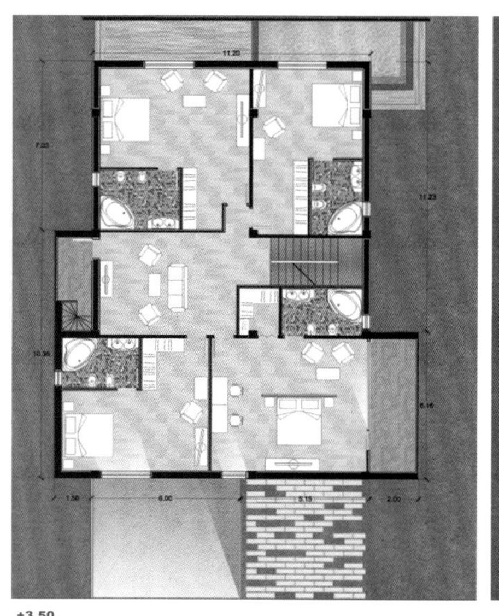
+3.50

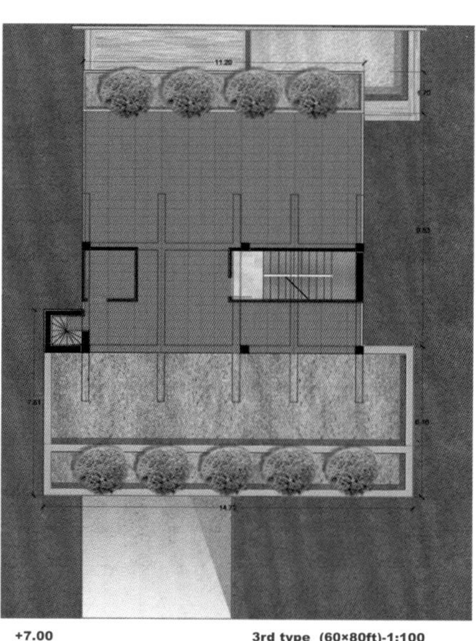
+7.00

3rd type (60×80ft)-1:100
第三类 (50×80ft)-1:100

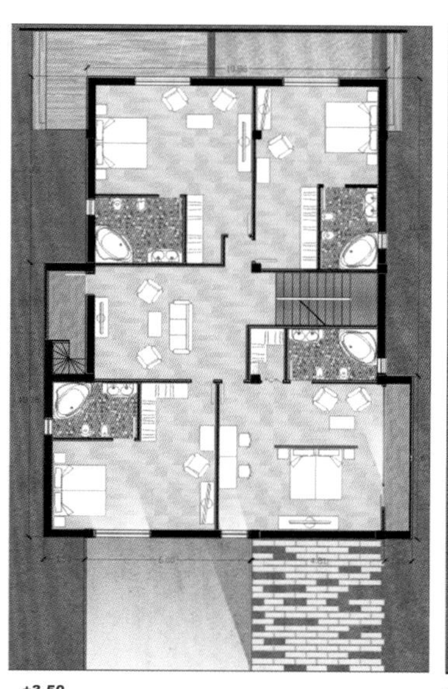
+3.50

+7.00

2nd type (50×80ft)-1:100
第二类 (50×80ft)-1:100

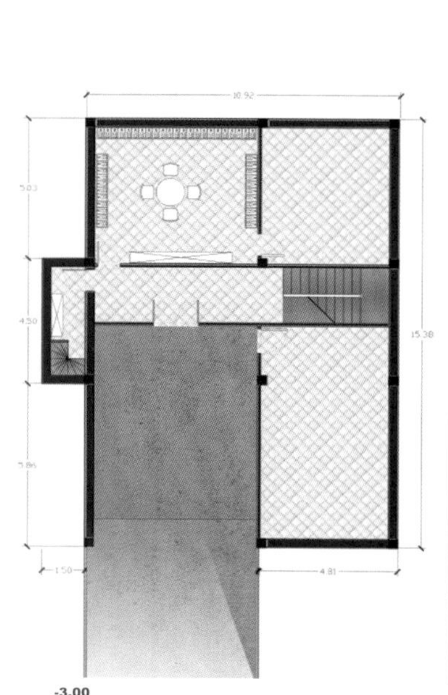
-3.00

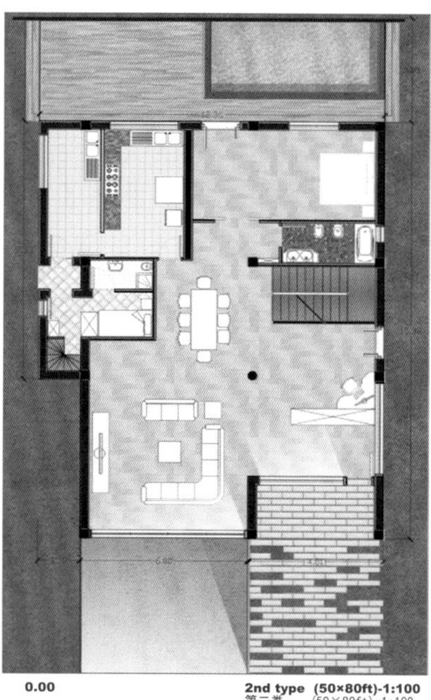
0.00

2nd type (50×80ft)-1:100
第二类 (50×80ft)-1:100

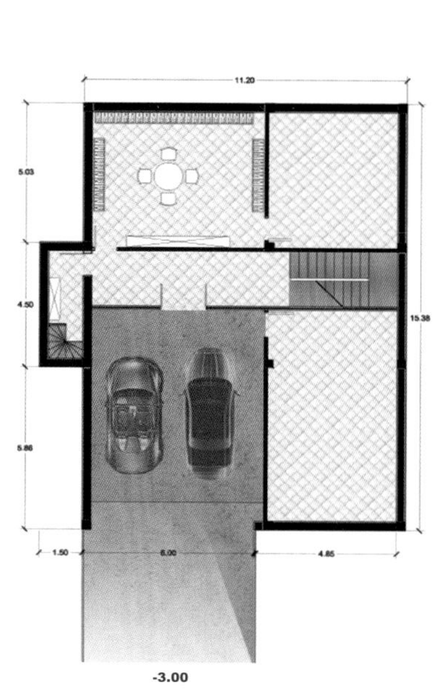
-3.00

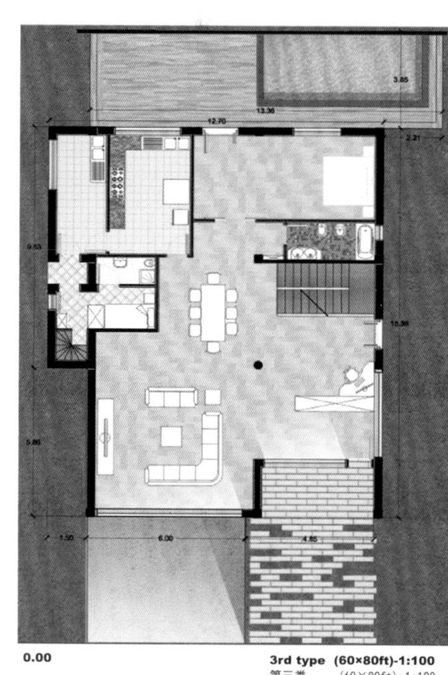
0.00

3rd type (60×80ft)-1:100
第三类 (60×80ft)-1:100

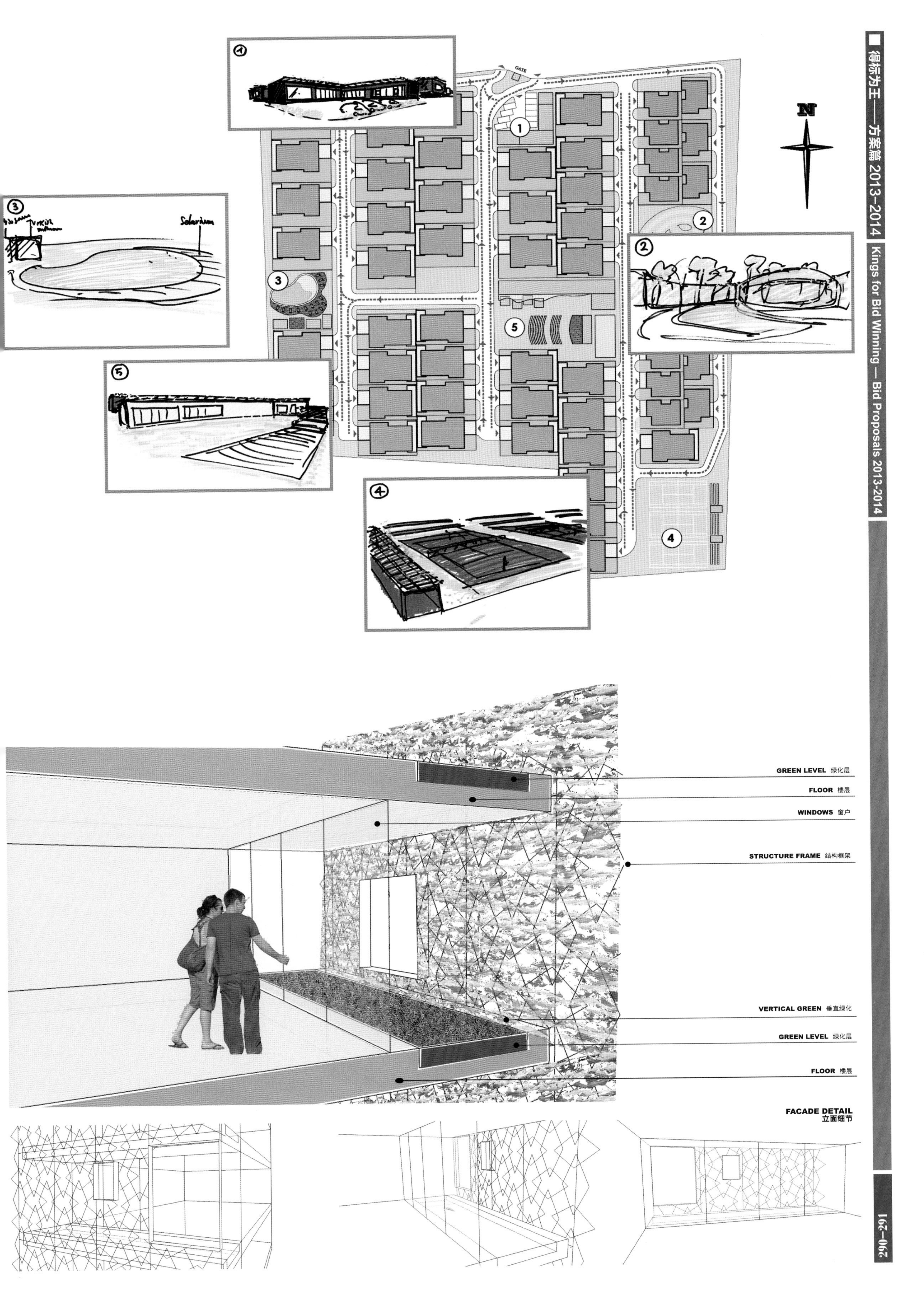

FACADE DETAIL
立面细节

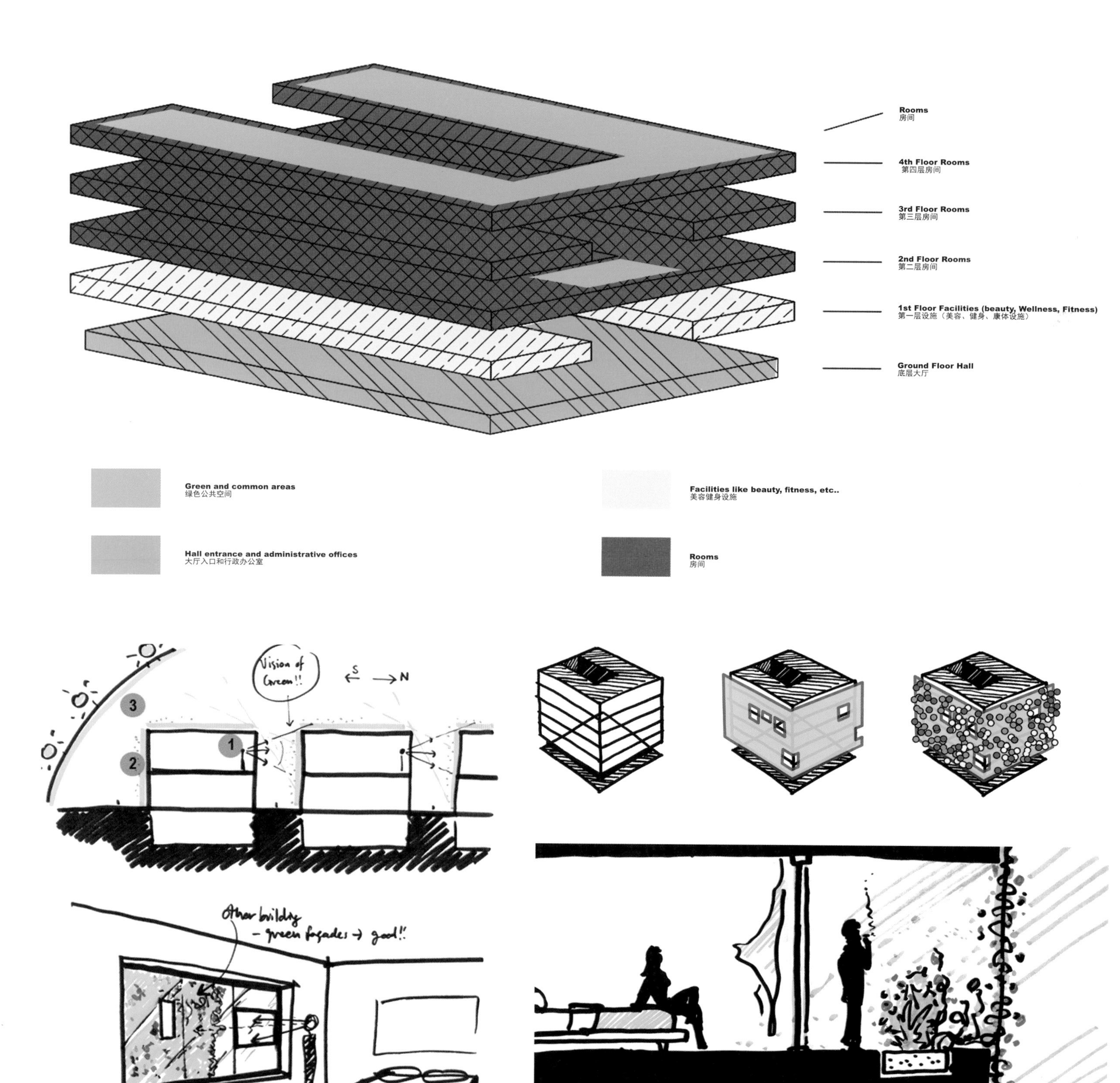
Rooms
房间
4th Floor Rooms
第四层房间
3rd Floor Rooms
第三层房间
2nd Floor Rooms
第二层房间
1st Floor Facilities (beauty, Wellness, Fitness)
第一层设施（美容、健身、康体设施）
Ground Floor Hall
底层大厅
Green and common areas
绿色公共空间
Facilities like beauty, fitness, etc..
美容健身设施
Hall entrance and administrative offices
大厅入口和行政办公室
Rooms
房间
Vision of Green!!
S
N
3
2
1

Design Concept

It's important to customize/personalize each home to have different images of these villas. Each house will be different from the nearby, while respecting the overall image.

Architectural Design

Smart surfaces are used in this design. Aesthetical, thermic benefits and active systems offer gorgeous building appearance and self-sufficient energy supply. Setting bedrooms on the east side allows for a flexible spatial layout and well-planned functional area as well as comfortable internal living environment.

Design Feature

The overall layout of this project is compact leaving relatively narrow space between buildings. Leaves and flowers are used to create a "green wall" for sunshading and shadowing. A large roof garden of the building restitutes the original ground surface, which follows sustainable criteria and, at the same time, makes it possible to have a private garden into the top, for parties or solarium.

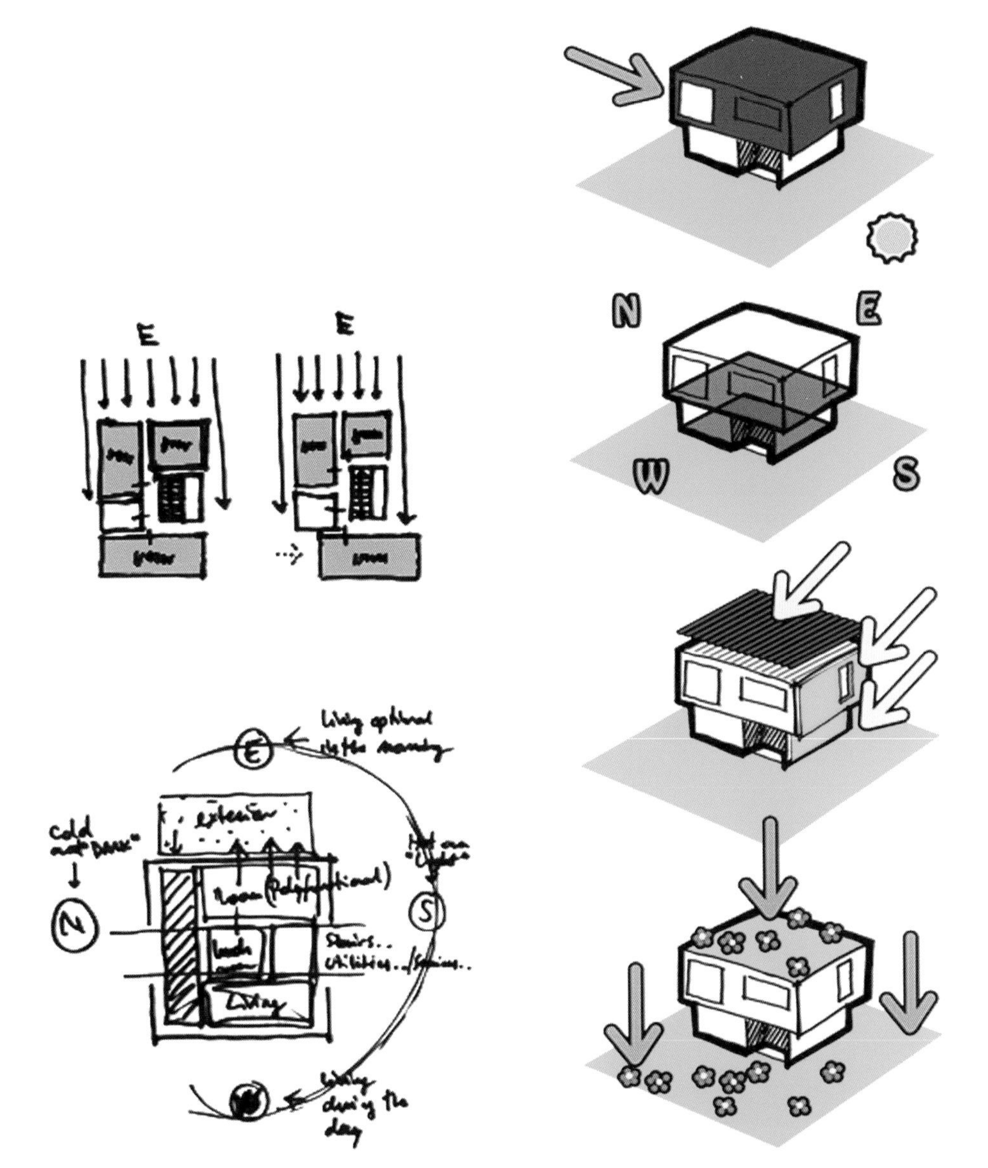

意大利卡蒂尼亚诺 Della Volpe 住宅

Della Volpe House

设计单位：Lorenzo Martella Architetto

项目地址：意大利佩斯卡拉省卡蒂尼亚诺

总规划面积：1 200 ㎡

Designed by: Lorenzo Martella Architetto

Location: Catignano, Pescara, Italy

Total Planning Area: 1,200 m²

项目概况

项目位于意大利佩斯卡拉省卡蒂尼亚诺，是该区域一个住宅建筑群的一部分。

建筑设计

设计界定了周边区域，定义了建筑体量，也规划了交通流线。建筑分为三个层次，且包括了一个地下室。其主要空间位于首层，设计师在不同的水平区域内规划了生活区和休息区。环形的楼梯可将人们带到屋顶的开放空间，该空间位于建筑的第三层，可用作日光浴室。设计师还在建筑外围的边界上框定了一片区域，作为研究植物的自由空间。

面向道路的前立面是建筑最显眼的外观，这个立面由三个简单的矩形结构构成，分别采用混凝土面板、大理石和木材等不同的材料对表面进行装饰。建筑的内部前立面设置了更大尺度的开口，调节了起居室和厨房之间的小环境。同时，来自屋顶的自然光线可以照亮住宅内部最核心的区域。

Profile

The project is located in Catignano, Pescara, Italy. It is part of a group of lots for the construction of residences.

Architectural Design

The Master Plan defines the volumes, areas as well as access. The project of interest is divided into three levels, including a basement. The main space is on the ground floor with a plan based on two different levels for the living area and the sleeping area. The circular staircase leads to the open space on the roof (third floor) used as a solarium. The structural frame is positioned on the edge of the perimeter and this has allowed complete freedom in the study of plant.

The front on the road is also the most visible face. This front is based on a simple composition of three rectangles with different surface finishes including concrete facing, marble and wood. The internal fronts provide openings with larger dimensions, which illuminate part of the environment of the living room and kitchen. The light from the roof gets to light the most central areas of the house.

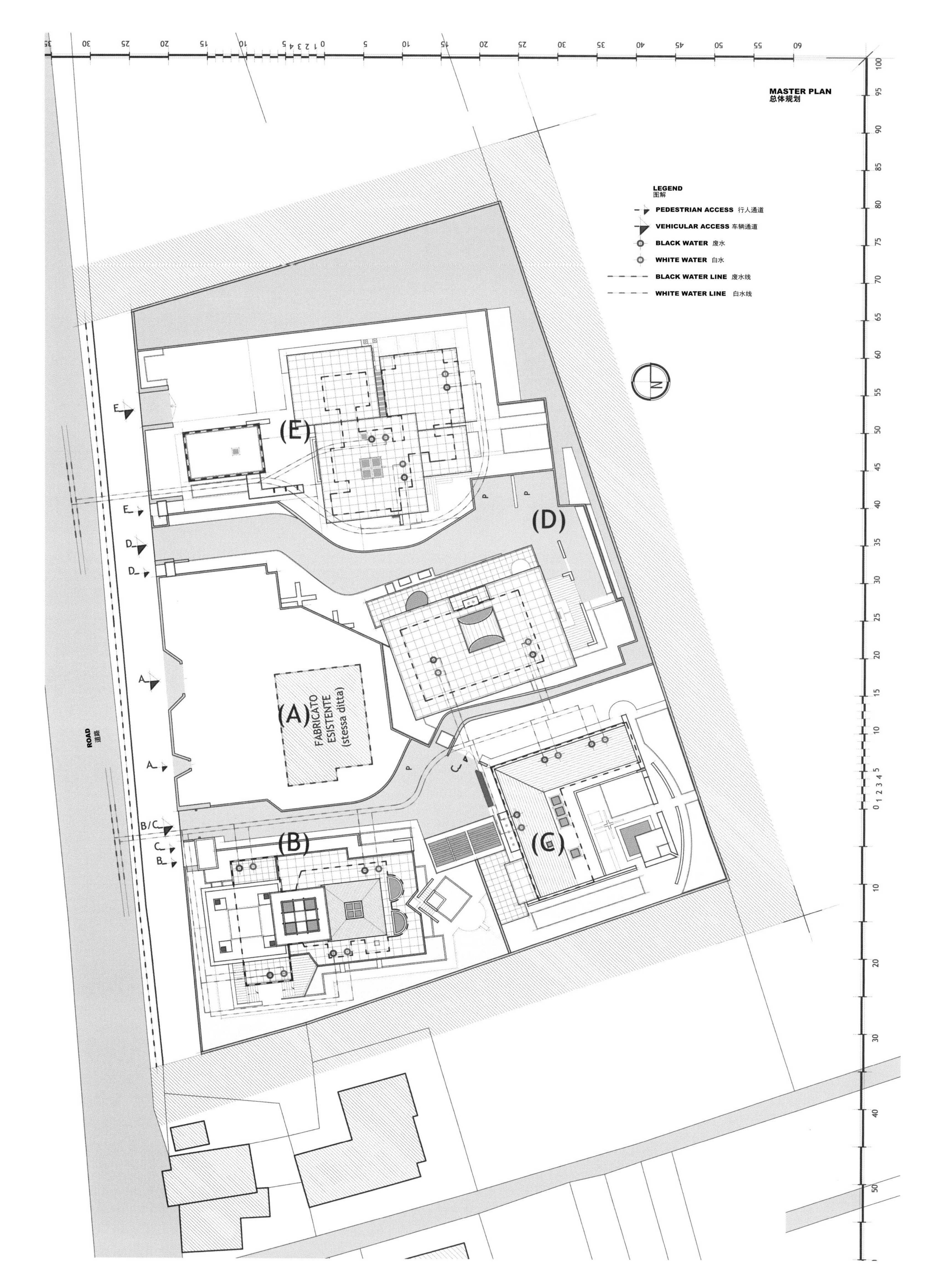
MASTER PLAN
总体规划
LEGEND
图解
PEDESTRIAN ACCESS 行人通道
VEHICULAR ACCESS 车辆通道
BLACK WATER 废水
WHITE WATER 白水
BLACK WATER LINE 废水线
WHITE WATER LINE 白水线
(A)
FABRICATO ESISTENTE (stessa ditta)
(B)
(C)
(D)
(E)
ROAD
道路

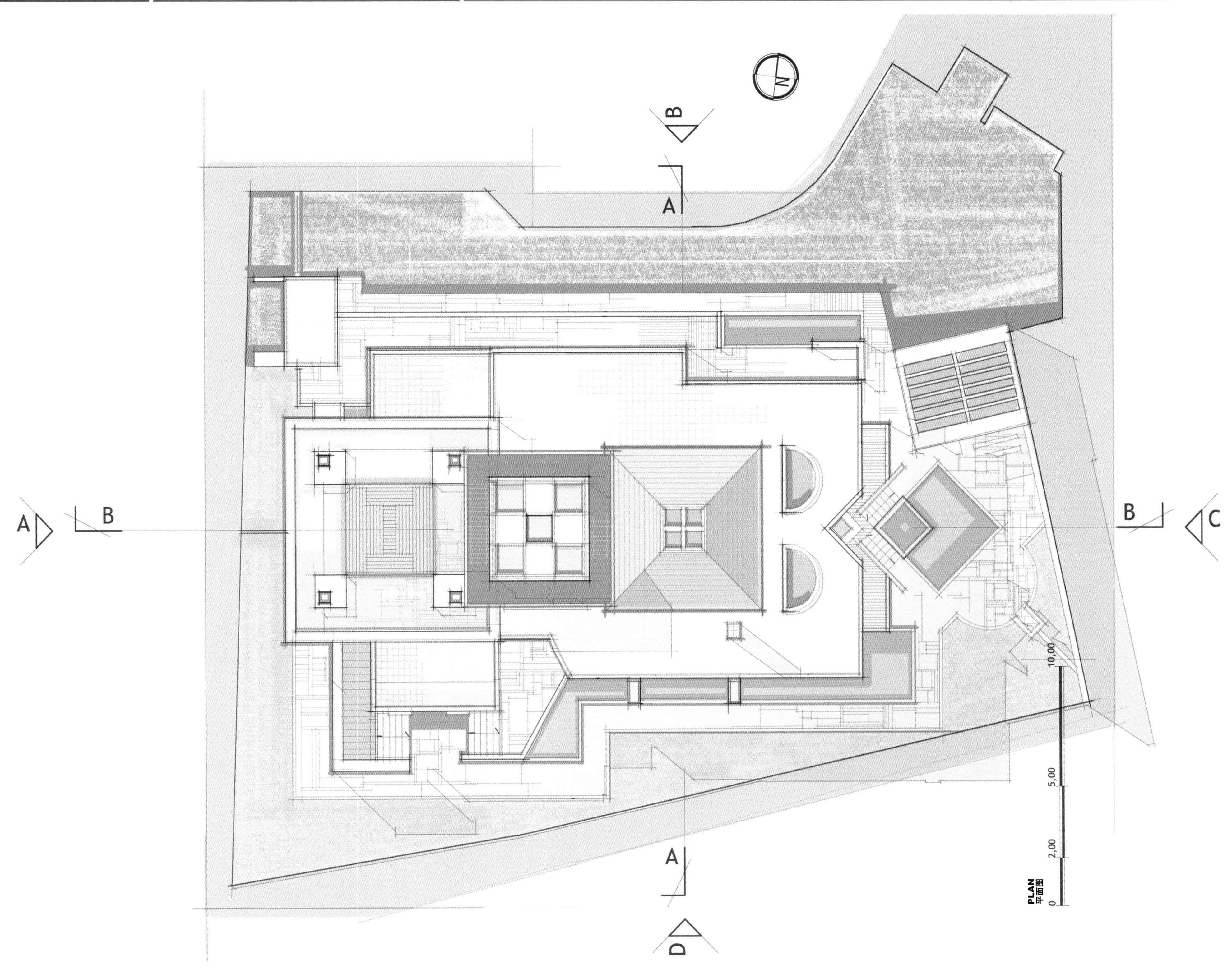
A
B
A
B
A
B
C
A
D
N
PLAN
平面图
0
2,00
5,00
10,00

ROAD
道路

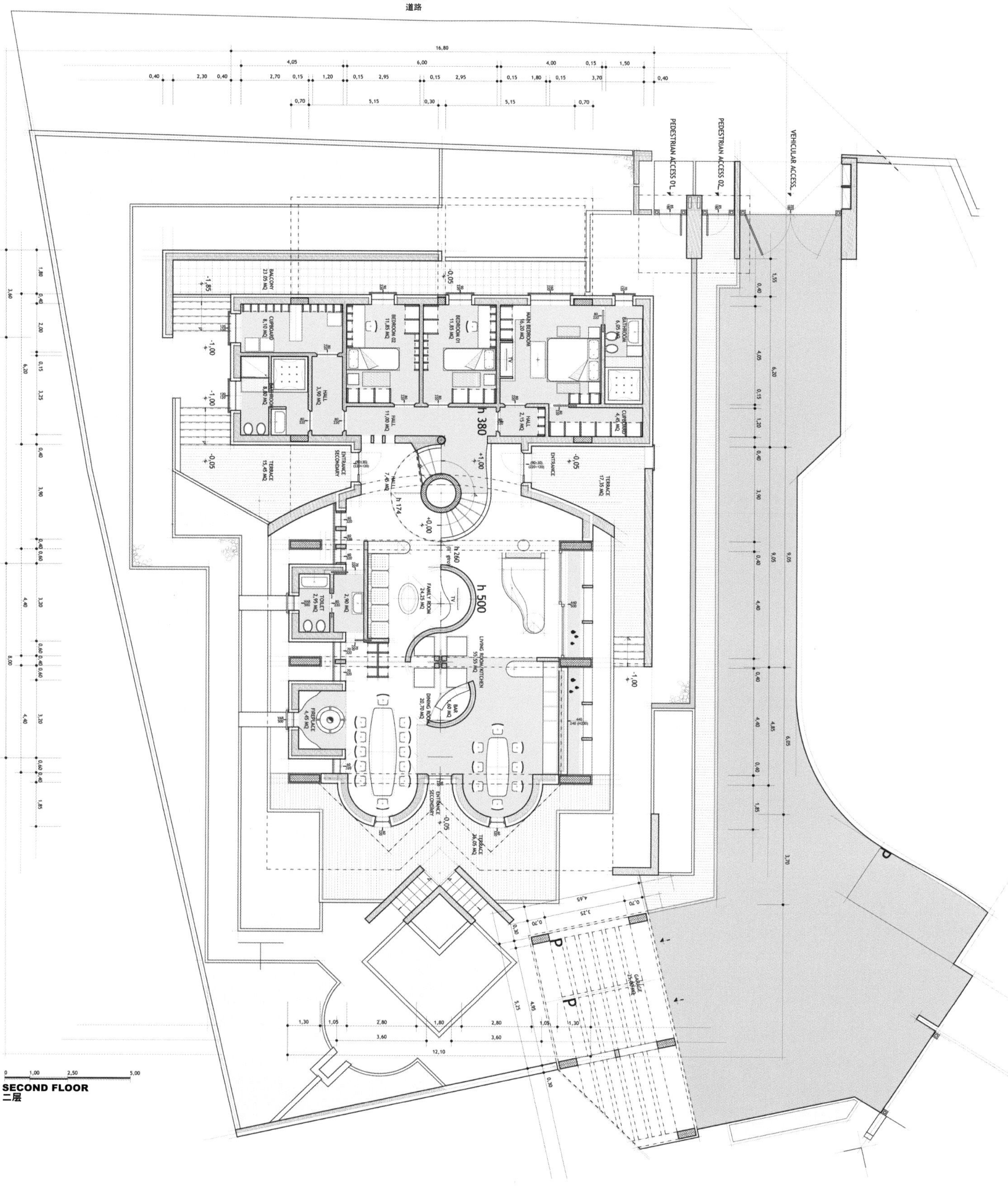

SECOND FLOOR
二层

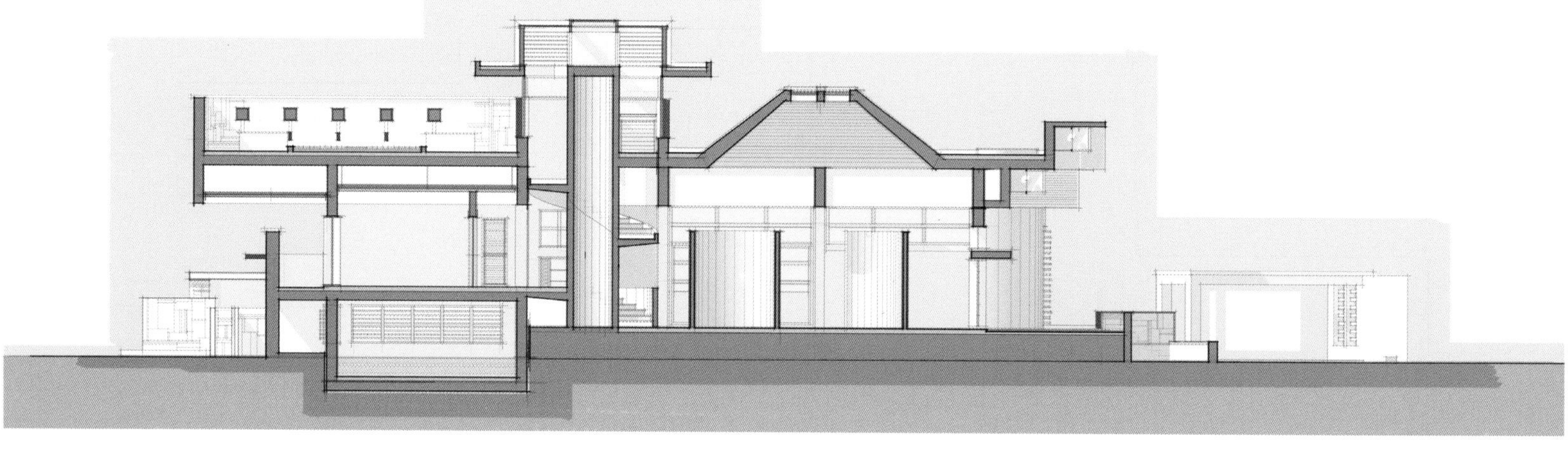

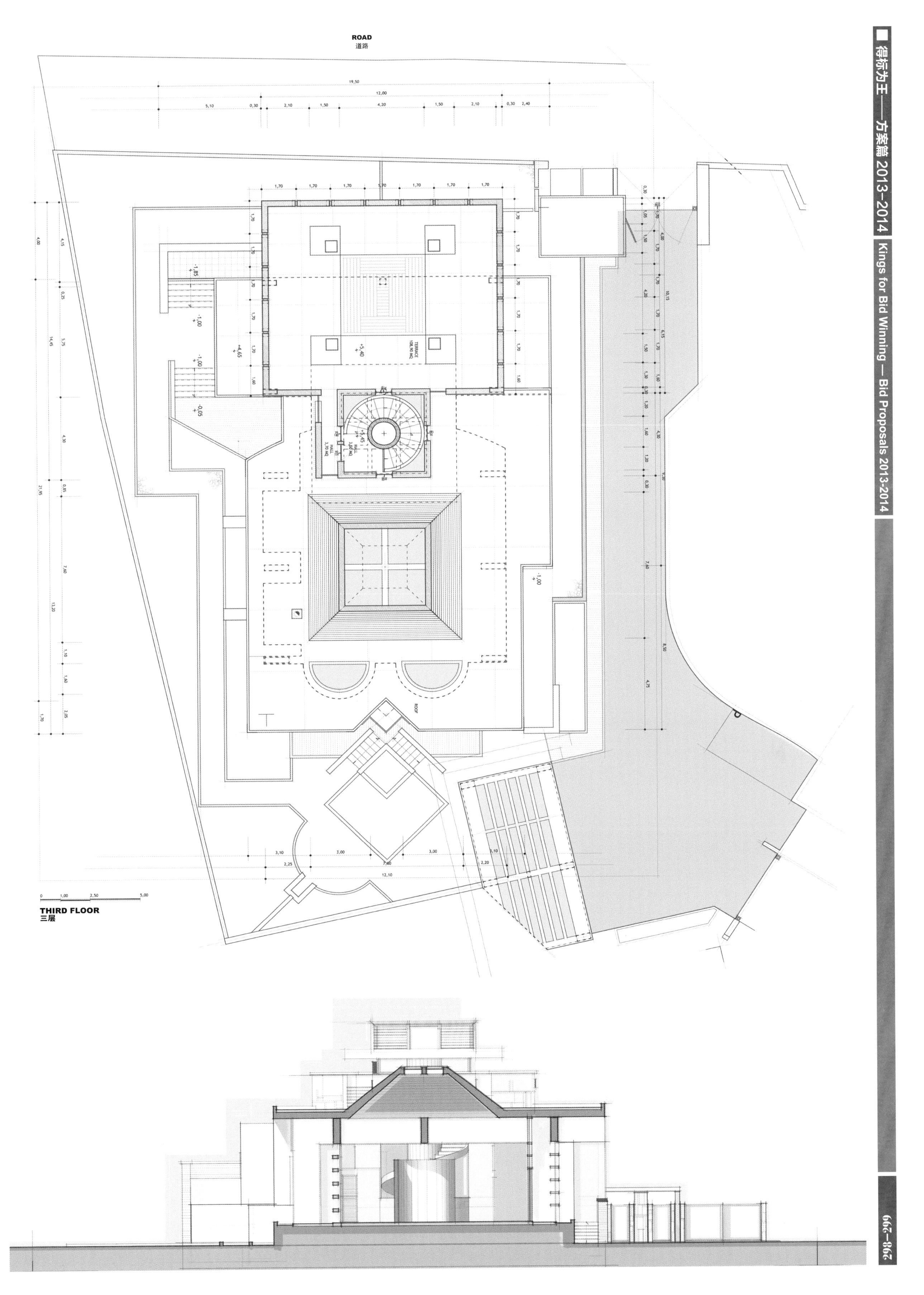
ROAD
道路
ROOF
THIRD FLOOR
三层

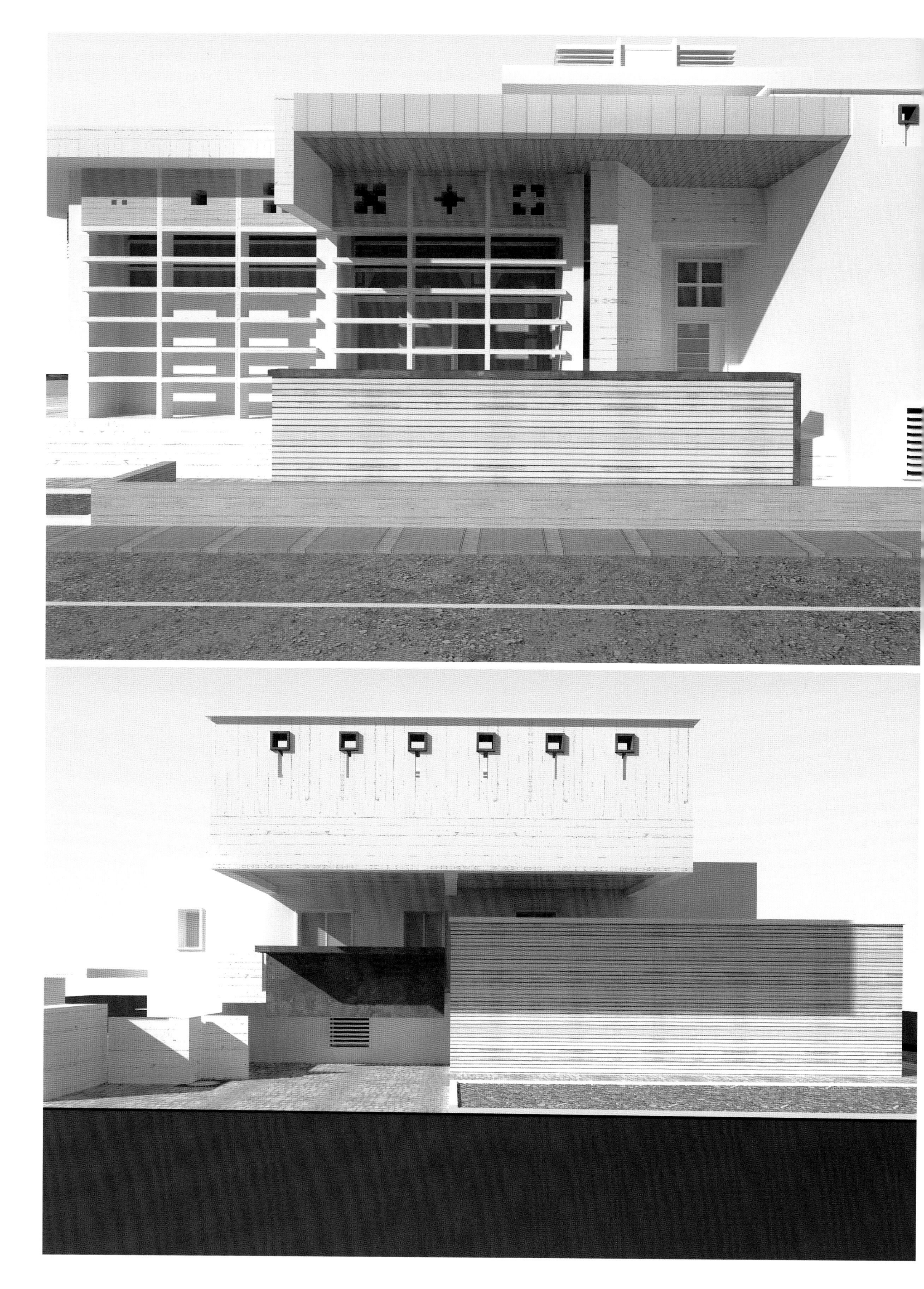

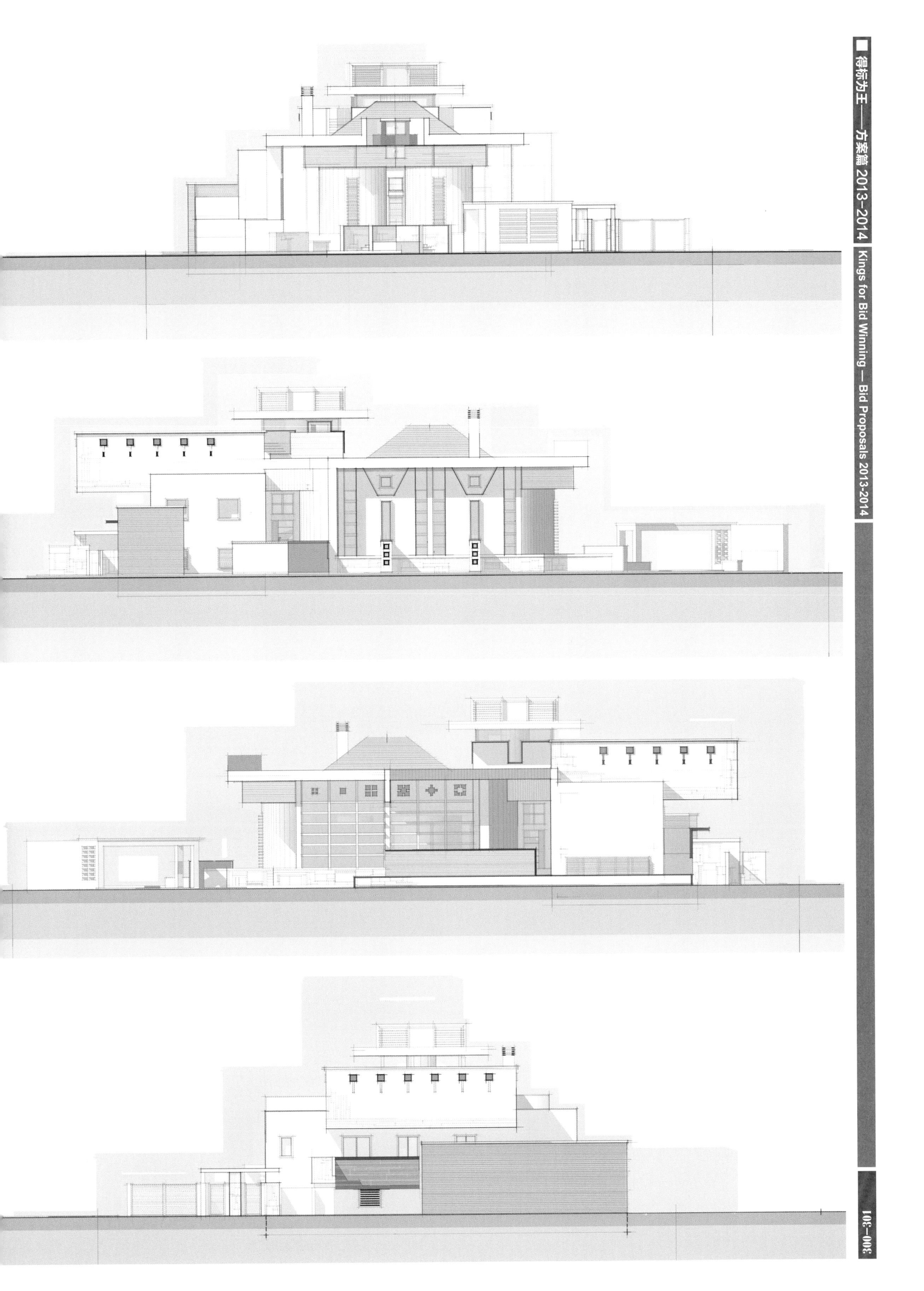

澳大利亚达尔文港 Aspire 公寓

Aspire Apartments

设计单位：Bell Gabbert Associates
项目地址：澳大利亚达尔文港
摄影：Mark Bell

Designed by: Bell Gabbert Associates
Location: Darwin, Australia
Photography: Mark Bell

项目概况

Aspire 豪华公寓大楼将会成为达尔文港地区一栋极具纪念性意义的标志性建筑，建成后的建筑将在达尔文港新兴的中央商务区的住宅建筑设计领域设立新的标准。

建筑设计

建筑呈现出鲜明的未来主义建筑风格和特征，将之与周围的建筑区分开来。为了与现代时尚的建筑外观保持一致，建筑的边角处将被切割磨平。一个连续的曲线形外观覆盖了主要的生活区，为每个住宅单元提供了宽敞的户外阳台空间，这个紧挨着室内的大型开敞式空间将成为连接室外空间与室内空间的过渡空间。

为了减少建筑吸收来自北面的热量，同时营造一个舒适的生活环境，承担了最大热负荷的外立面透过原有的建筑外墙，将阳台延展出来。阳台与光滑的栏杆一起作为“太阳帽”，既为建筑提供了荫蔽、最小化外墙的吸热量，同时也为居者提供了户外空间。这一结构也赋予了建筑独特的外观。

建筑南立面的太阳能摄取量较少，设计师将之设计成一个富有趣味性的、视觉效果良好的外观。延伸了的阳台仿佛建筑表面安装的“太阳帽”，并使建筑外观呈现出曲线形的流畅形态。

Profile

Aspire will be a signature development delivering a landmark building to the city of Darwin. When completed, the Aspire Luxury Apartment Tower will set a new standard in architectural design for residential living in Darwin's emerging CBD.

Architectural Design

Aspire's distinctive futuristic style architecture will set it apart from other buildings. In keeping with the building's modern appearance will be cutting edge electronics, building infrastructure and energy efficiency. When designing the building, an almost continuous curvilinear façade was placed across the majority of living spaces. This provides residences with generous outdoo balcony areas annexed to large open-plan interior creating an indoor-outdoor space.

To achieve a workable level of energy efficiency reducing incident heat gain across the northerly aspec as well as providing a pleasing appearance, the facade that have the greatest heat loads have extende their balconies out past the external building walls Combined with glazed balustrading the balconies act a sunhoods providing shade and usable outdoor spac for occupants whilst minimizing the amount of incider sun on exterior walls. The result is a curved, dynami façade, which gives the tower its unique appearance.

On the southern facades where the sun's impact i less, an attempt has been made to provide interes and a façade with visual appeal. The use of extende balconies negates the need to provide surface mounte sunhoods and in doing so, provides a façade that i streamlined.

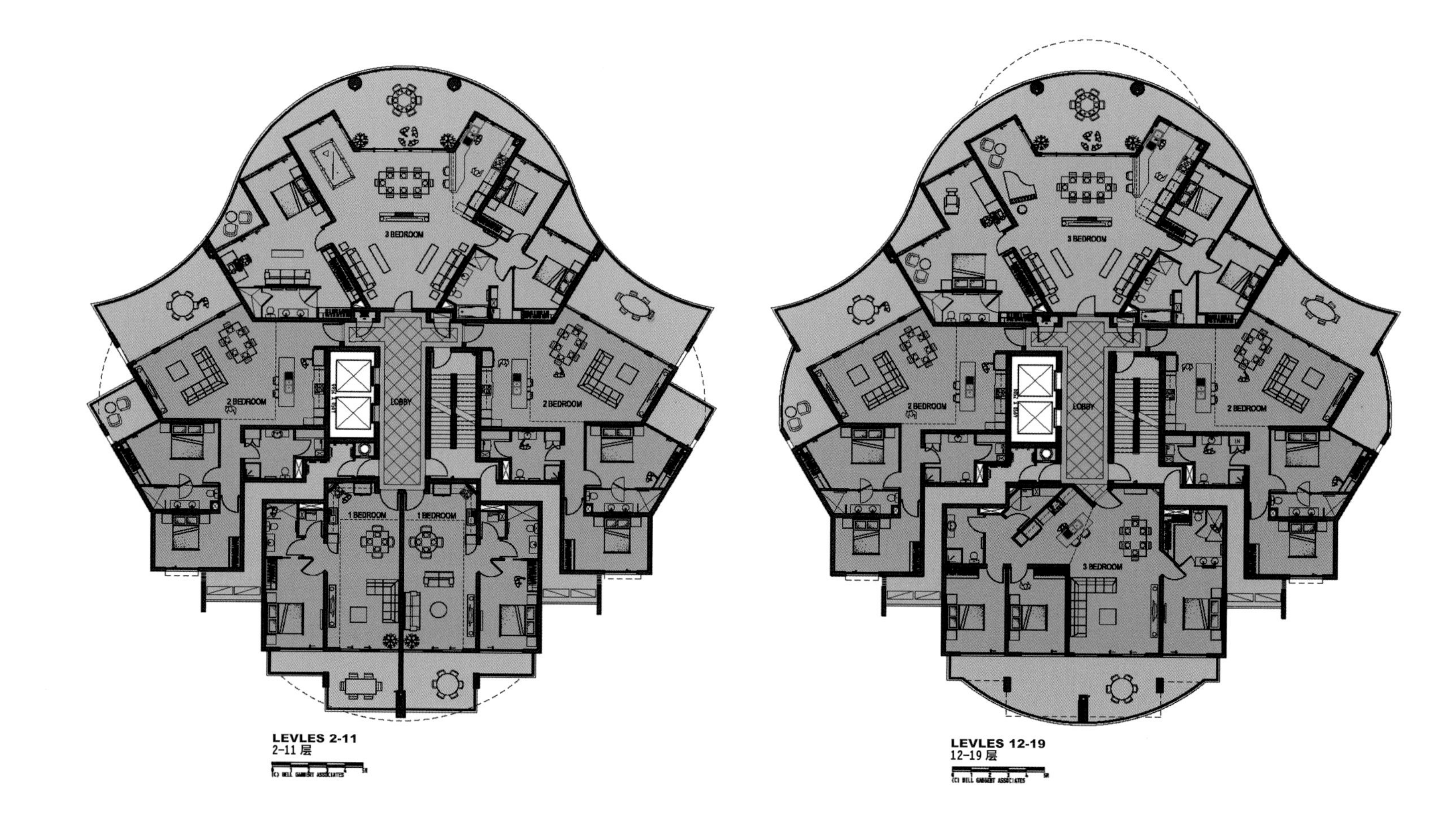

LEVLES 2-11
2-11 层

LEVLES 12-19
12-19 层

DRY SEASON BREEZES 干季微风

DRY SEASON BREEZES 干季微风

TIGER BRENNAN DRIVE

DRY SEASON BREEZES 干季微风

LOT 5953

单层仓库
SINGLE STOREY SHED

SINGLE STOREY OFFICE
单层办公室

WET SEASON BREEZES

SHED
仓库

SHED
仓库

SHED
仓库

SINGLE STOREY OFFICE
单层办公室

CAREY STREET

MCMINN STREET

WET SEASON BREEZES
湿季微风

WET SEASON BREEZES
湿季微风

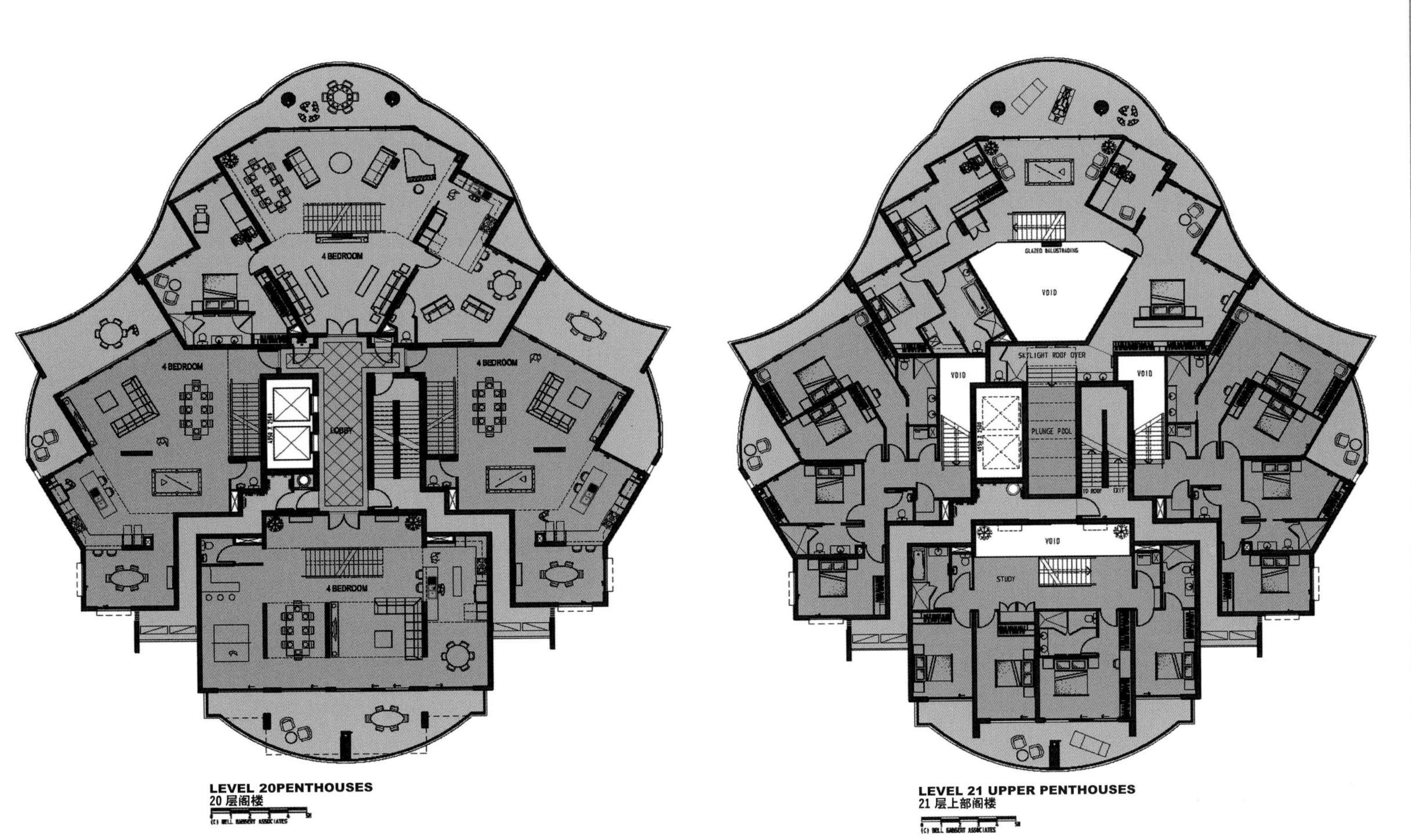

LEVEL 20PENTHOUSES
20 层阁楼

LEVEL 21 UPPER PENTHOUSES
21 层上部阁楼

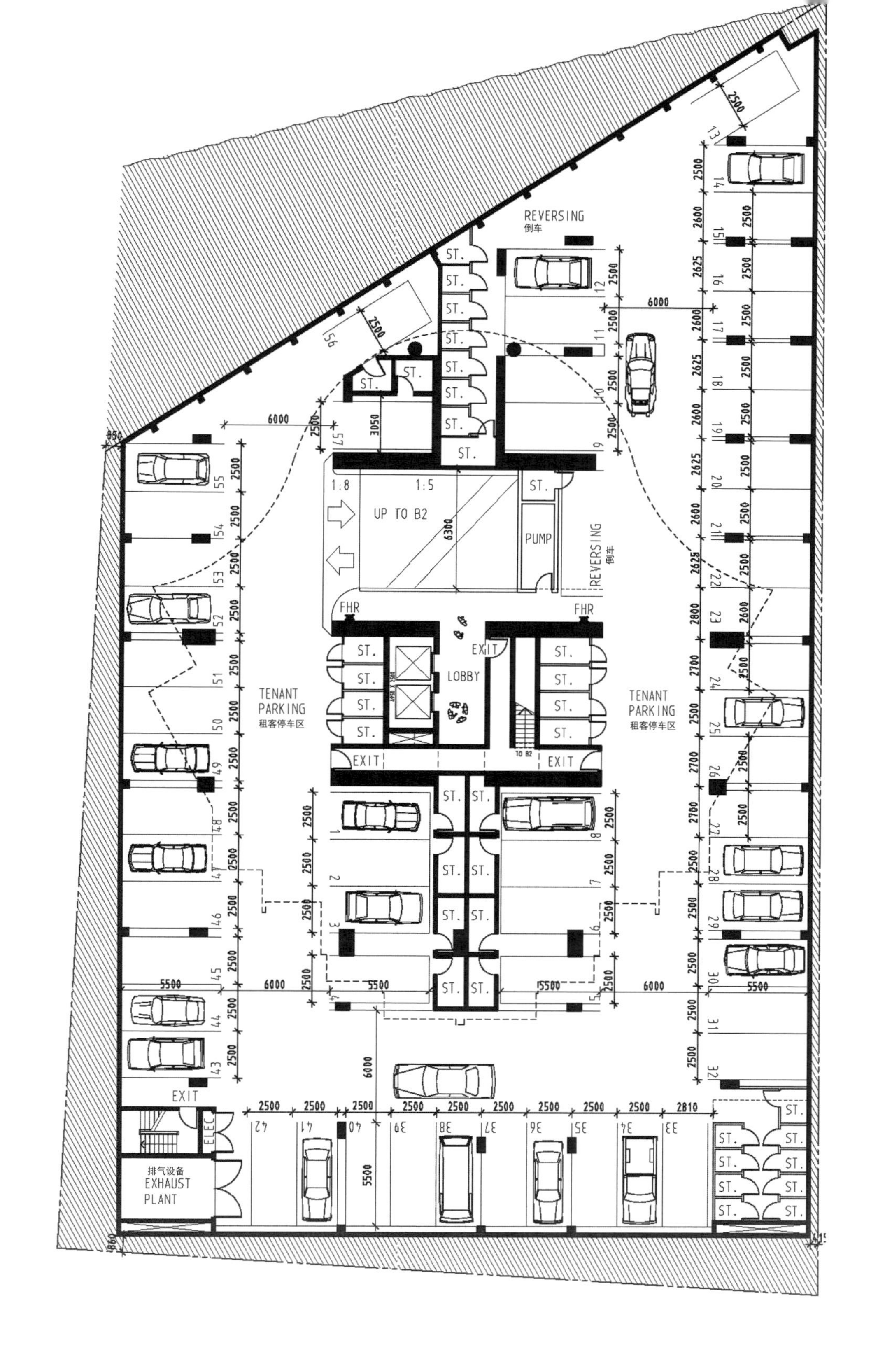

REVERSING
倒车
UP TO B2
PUMP
FHR
LOBBY
EXIT
TENANT PARKING
租客停车区
ELEC
排气设备
EXHAUST PLANT

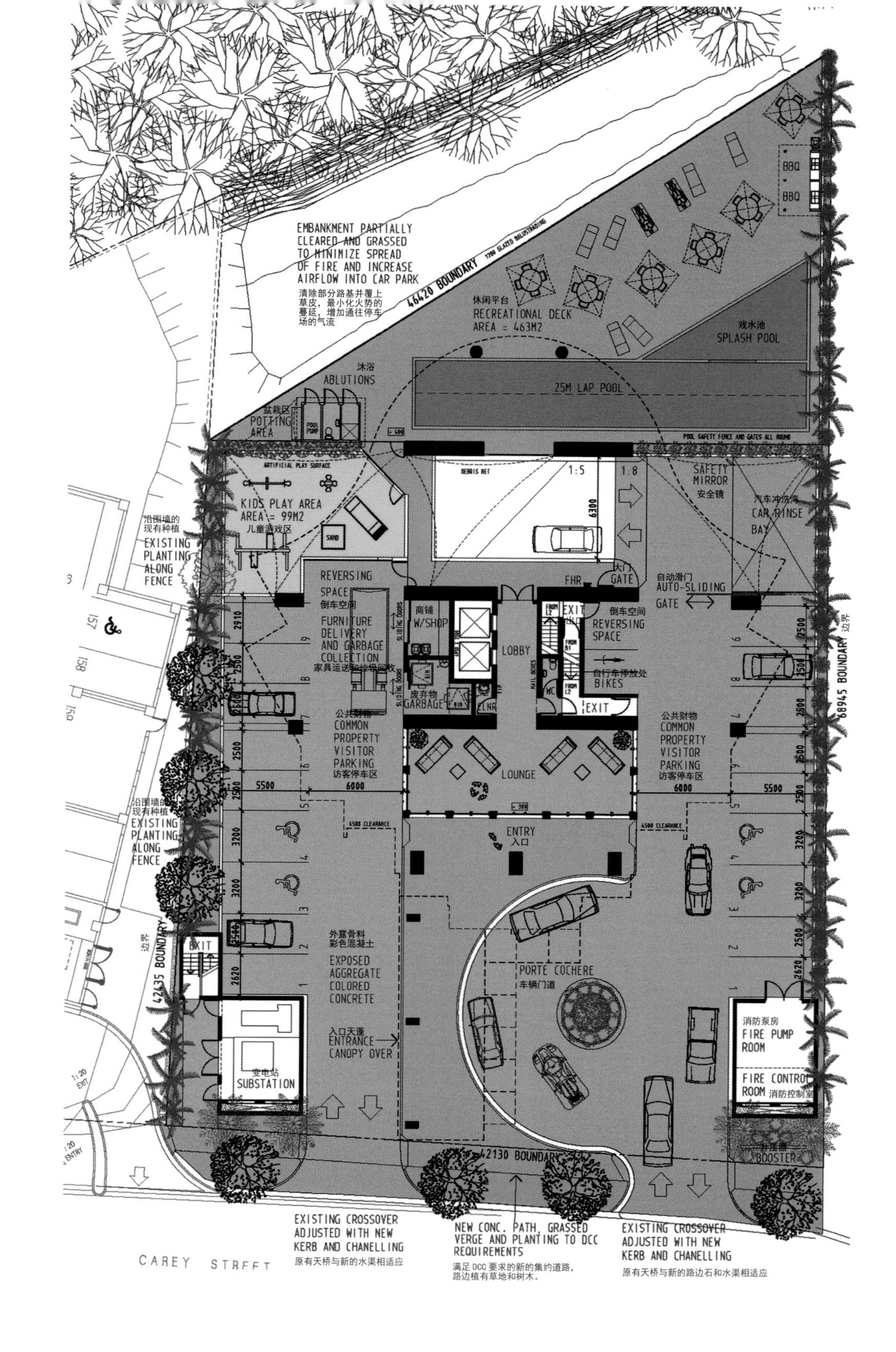

EMBANKMENT PARTIALLY CLEARED AND GRASSED TO MINIMIZE SPREAD OF FIRE AND INCREASE AIRFLOW INTO CAR PARK
清除部分路基并覆上草皮，最小化火势的蔓延，增加通往停车场的气流
46420 BOUNDARY
休闲平台
RECREATIONAL DECK
AREA = 463M2
戏水池
SPLASH POOL
25M LAP POOL
沐浴
ABLUTIONS
POTTING AREA
KIDS PLAY AREA
AREA = 99M2
沿围墙的现有种植
EXISTING PLANTING ALONG FENCE
SAFETY MIRROR
安全镜
CAR RINSE BAY
REVERSING SPACE
倒车空间
FURNITURE DELIVERY AND GARBAGE COLLECTION
W/SHOP
LOBBY
GARBAGE
GATE
自动滑门
AUTO-SLIDING GATE
自行车停放处
BIKES
公共财物
COMMON PROPERTY VISITOR PARKING
访客停车区
LOUNGE
ENTRY
入口
68945 BOUNDARY
外露骨料彩色混凝土
EXPOSED AGGREGATE COLORED CONCRETE
PORTE COCHERE
车辆门道
入口天篷
ENTRANCE CANOPY OVER
变电站
SUBSTATION
消防泵房
FIRE PUMP ROOM
FIRE CONTROL ROOM 消防控制室
BOOSTER
42635 BOUNDARY
42130 BOUNDARY
EXISTING CROSSOVER ADJUSTED WITH NEW KERB AND CHANELLING
NEW CONC. PATH, GRASSED VERGE AND PLANTING TO DCC REQUIREMENTS
满足 DCC 要求的新的集约道路，路边植有草地和树木。
原有天桥与新的路边石和水渠相适应

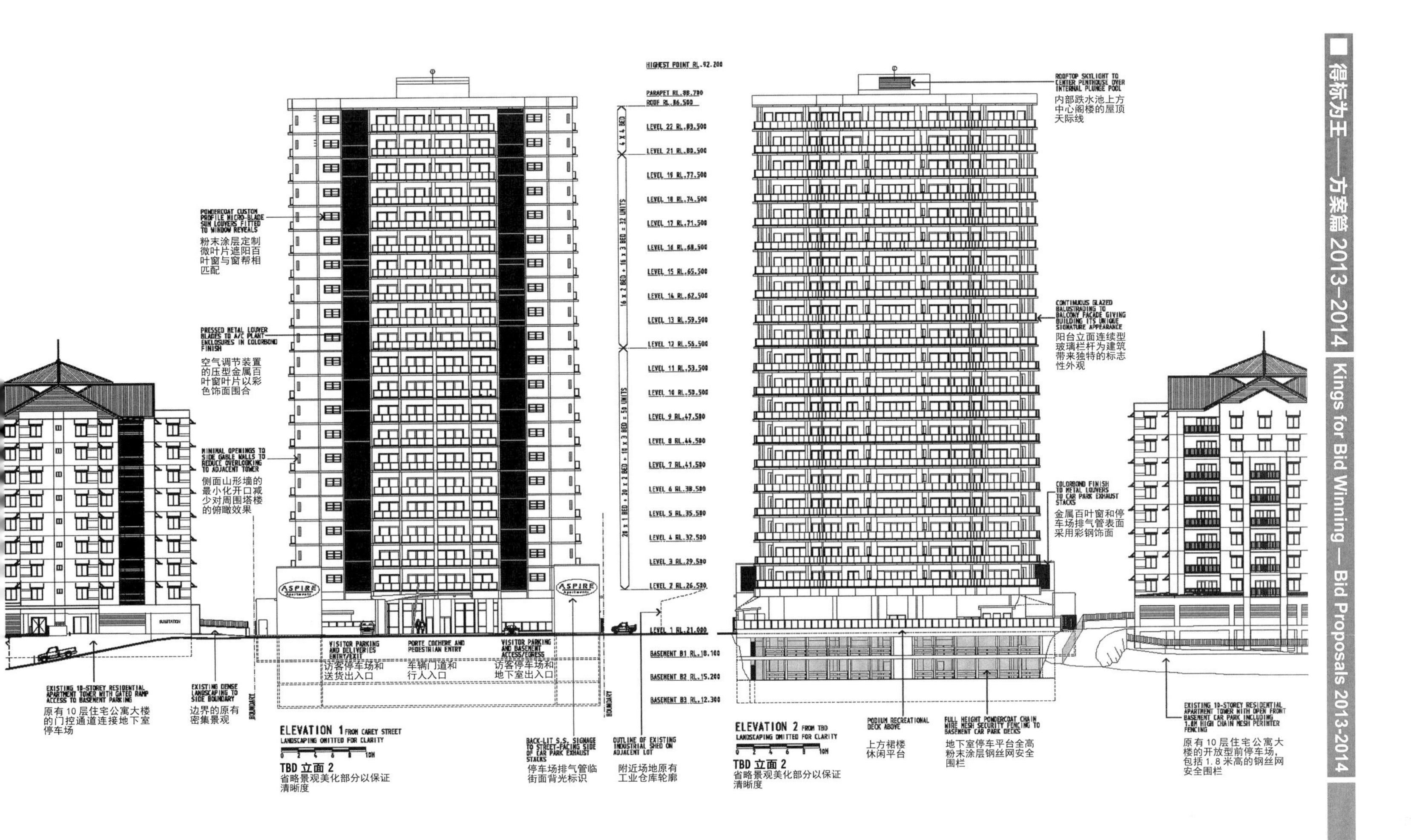
HIGHEST POINT RL.92.200
PARAPET RL.88.780
ROOF RL.86.500
LEVEL 22 RL.83.500
LEVEL 21 RL.80.500
LEVEL 19 RL.77.500
LEVEL 18 RL.74.500
LEVEL 17 RL.71.500
LEVEL 16 RL.68.500
LEVEL 15 RL.65.500
LEVEL 14 RL.62.500
LEVEL 13 RL.59.500
LEVEL 12 RL.56.500
LEVEL 11 RL.53.500
LEVEL 10 RL.50.500
LEVEL 9 RL.47.500
LEVEL 8 RL.44.500
LEVEL 7 RL.41.500
LEVEL 6 RL.38.500
LEVEL 5 RL.35.500
LEVEL 4 RL.32.500
LEVEL 3 RL.29.500
LEVEL 2 RL.26.500
LEVEL 1 RL.21.000
BASEMENT B1 RL.18.100
BASEMENT B2 RL.15.200
BASEMENT B3 RL.12.300
POWDERCOAT CUSTOM PROFILE MICRO-BLADE SUN LOUVERS FITTED TO WINDOW REVEALS
粉末涂层定制微叶片遮阳百叶窗与窗帮相匹配
PRESSED METAL LOUVER BLADES TO A/C PLANT ENCLOSURES IN COLORBOND FINISH
空气调节装置的压型金属百叶窗叶片以彩色饰面围合
MINIMAL OPENINGS TO SIDE GABLE WALLS TO REDUCE OVERLOOKING TO ADJACENT TOWER
侧面山形墙的最小化开口减少对周围塔楼的俯瞰效果
EXISTING 10-STOREY RESIDENTIAL APARTMENT TOWER WITH GATED RAMP ACCESS TO BASEMENT PARKING
原有10层住宅公寓大楼的门控通道连接地下室停车场
EXISTING DENSE LANDSCAPING TO SIDE BOUNDARY
边界的原有密集景观
VISITOR PARKING AND DELIVERIES ENTRY/EXIT
访客停车场和送货出入口
PORTE COCHERE AND PEDESTRIAN ENTRY
车辆门道和行人入口
VISITOR PARKING AND BASEMENT ACCESS/EGRESS
访客停车场和地下室出入口
BACK-LIT S.S. SIGNAGE TO STREET-FACING SIDE OF CAR PARK EXHAUST STACKS
停车场排气管临街面背光标识
OUTLINE OF EXISTING INDUSTRIAL SHED ON ADJACENT LOT
附近场地原有工业仓库轮廓
ELEVATION 1 FROM CAREY STREET
LANDSCAPING OMITTED FOR CLARITY
TBD 立面 2
省略景观美化部分以保证清晰度
ROOFTOP SKYLIGHT TO CENTER PENTHOUSE OVER INTERNAL PLUNGE POOL
内部跌水池上方中心阁楼的屋顶天际线
CONTINUOUS GLAZED BALUSTRADING TO BALCONY FACADE GIVING BUILDING ITS UNIQUE SIGNATURE APPEARANCE
阳台立面连续型玻璃栏杆为建筑带来独特的标志性外观
COLORBOND FINISH TO METAL LOUVERS TO CAR PARK EXHAUST STACKS
金属百叶窗和停车场排气管表面采用彩钢饰面
ELEVATION 2 FROM TBD
LANDSCAPING OMITTED FOR CLARITY
TBD 立面 2
省略景观美化部分以保证清晰度
PODIUM RECREATIONAL DECK ABOVE
上方裙楼休闲平台
FULL HEIGHT POWDERCOAT CHAIN WIRE MESH SECURITY FENCING TO BASEMENT CAR PARK DECKS
地下室停车平台全高粉末涂层钢丝网安全围栏
EXISTING 10-STOREY RESIDENTIAL APARTMENT TOWER WITH OPEN FRONT BASEMENT CAR PARK INCLUDING 1.8M HIGH CHAIN MESH PERIMETER FENCING
原有10层住宅公寓大楼的开放型前停车场，包括1.8米高的钢丝网安全围栏

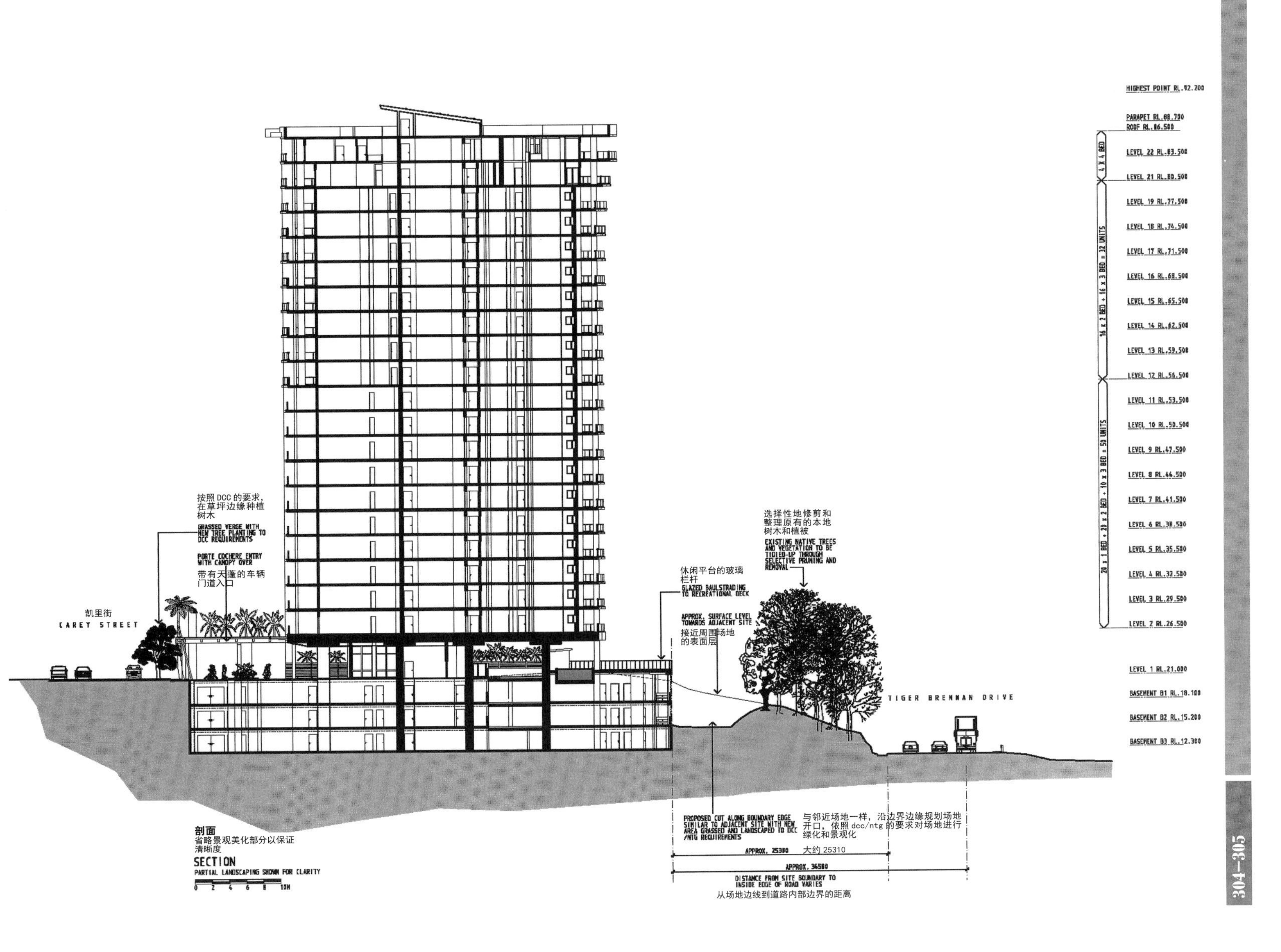
HIGHEST POINT RL.92.200
PARAPET RL.88.780
ROOF RL.86.500
LEVEL 22 RL.83.500
LEVEL 21 RL.80.500
LEVEL 19 RL.77.500
LEVEL 18 RL.74.500
LEVEL 17 RL.71.500
LEVEL 16 RL.68.500
LEVEL 15 RL.65.500
LEVEL 14 RL.62.500
LEVEL 13 RL.59.500
LEVEL 12 RL.56.500
LEVEL 11 RL.53.500
LEVEL 10 RL.50.500
LEVEL 9 RL.47.500
LEVEL 8 RL.44.500
LEVEL 7 RL.41.500
LEVEL 6 RL.38.500
LEVEL 5 RL.35.500
LEVEL 4 RL.32.500
LEVEL 3 RL.29.500
LEVEL 2 RL.26.500
LEVEL 1 RL.21.000
BASEMENT B1 RL.18.100
BASEMENT B2 RL.15.200
BASEMENT B3 RL.12.300
按照DCC的要求，在草坪边缘种植树木
GRASSED VERGE WITH NEW TREE PLANTING TO DCC REQUIREMENTS
PORTE COCHERE ENTRY WITH CANOPY OVER
带有天篷的车辆门道入口
凯里街
CAREY STREET
休闲平台的玻璃栏杆
GLAZED BAULSTRADING TO RECREATIONAL DECK
APPROX. SURFACE LEVEL TOWARDS ADJACENT SITE
接近周围场地的表面层
选择性地修剪和整理原有的本地树木和植被
EXISTING NATIVE TREES AND VEGETATION TO BE TIDIED-UP THROUGH SELECTIVE PRUNING AND REMOVAL
TIGER BRENNAN DRIVE
PROPOSED CUT ALONG BOUNDARY EDGE SIMILAR TO ADJACENT SITE WITH NEW AREA GRASSED AND LANDSCAPED TO DCC /NTG REQUIREMENTS
与邻近场地一样，沿边界边缘规划场地开口，依照dcc/ntg的要求对场地进行绿化和景观化
APPROX. 25300
大约25310
APPROX. 34580
DISTANCE FROM SITE BOUNDARY TO INSIDE EDGE OF ROAD VARIES
从场地边线到道路内部边界的距离
剖面
省略景观美化部分以保证清晰度
SECTION
PARTIAL LANDSCAPING SHOWN FOR CLARITY

澳大利亚悉尼 The Eliza 豪华公寓

The Eliza

设计单位：Tony Owen Partners
项目地址：澳大利亚悉尼
设计团队：Paul Demaine　Esan Rahmani
Angela Selke　Diana Quintero Saul
Andres Caceres　Michael Civovic
Bryan Ahn

Designed by: Tony Owen Partners
Location: Sydney, Australia
Design Team: Paul Demaine, Esan Rahmani,
Angela Selke, Diana Quintero Saul,
Andres Caceres, Michael Civovic, Bryan Ahn

项目概况

The Eliza 位于悉尼海德公园中心商务区的伊丽莎白街上，这栋 17 层楼的建筑项目仅容纳了 19 套公寓，将为澳大利亚现代、豪华的住宅设计确立新的标准。

设计特色

项目设计参照了 20 世纪的经典设计，以数码设计原则提出一份独特的、环保的设计方案，来构建一栋真正独一无二的世界级建筑。

大楼的立面采用了独特的参数化设计，建筑每一层的立面形态都不尽相同，又在整体上呼应了不断改变的日光以及周边的景色，同时也满足了规划的要求。

设计以配备有 2 室或 4 室可以欣赏到公园全景和海景的公寓为特色，3 层的阁楼也是设计的另一特色。阁楼因屋顶游泳池、私人电梯、豪华楼梯的设置以及层高的设定，将成为世界级的休闲空间。

Profile

The Eliza is located on Elizabeth Street in Sydney's CBD on prestigious Hyde Park. With only 19 apartments, this exclusive 17 storey development will create a new standard in luxury and contemporary design for Australia.

Design Feature

The project design is inspired by the 20th century classic design. With digital design practices to create unique and exciting environmental solutions in a new approach to urban design, the project is to create a world class building for a truly unique site.

The units have a unique parametrically designed façade. It responds to the changing solar, view and planning conditions on each level such that the façade on every level is different.

The design features 2 and 4 bedroom units with panoramic park and harbor views as well as a 3 storey penthouse. The penthouse has been conceived as a unique world class address with roof top pool, private elevator and grand staircase and ceiling heights.

阿特金斯 ATKINS

阿特金斯是世界领先的设计顾问公司之一，专业知识的广度与深度使之能够应对具有技术挑战性和时间紧迫性的项目。阿特金斯是国际化的多专业工程和建设顾问公司，能为各类开发建设项目提供一流专业服务，从摩天大楼设计到城市规划、铁路网络改造以及防洪模型的编制，都能提供规划、设计、实施的全程解决方案。

作为阿特金斯集团远东区的全资子公司，阿特金斯中国于1994年正式进入中国市场。借助集团总部的强大支持，阿特金斯中国以提供多专业、多学科的"一站式"全方位服务的核心优势区别于竞争对手，并通过国际经验和本地知识的有机结合在中国近年来迅猛推进的城市化进程中取得了骄人的项目业绩。

其主要作品和奖项有：中央景城一期荣获2010年中国土木工程詹天佑奖住宅小区优秀规划奖；伊顿小镇荣获2010年中国土木工程詹天佑奖住宅小区金奖；上海办事处荣获LEED绿色建筑商业内部装修金奖等多个奖项。

Atkins

Atkins is one of the world's leading design consultancies. It has the breadth and depth of expertise to respond to the most technically challenging and time-critical projects and to facilitate the urgent transition to a low carbon economy. Atkins' vision is to be the world's best design consultant.

Whether it's the architectural concept for a new supertall tower, the upgrade of a rail network, master planning a new city or the improvement of a management process, designers plan, design and enable solutions.

With 75 years of history, 17,700 employees and over 200 offices worldwide, Atkins is the world's 13th largest global design firm (ENR 2011), the largest global architecture firm, the largest multidisciplinary consultancy in Europe and UK's largest engineering consultancy for the last 14 years. Atkins is listed on the London Stock Exchange and is a constituent of the FTSE 250 Index.

In 1994 Atkins established its first Asian office in Hong Kong followed by Singapore in 1996. Today Atkins also has offices in Hong Kong, Beijing, Shanghai, Chengdu, Chongqing and Ho Chi Minh and Sydney, all part of an integrated network that delivers innovative multidisciplinary projects and employs approximately 1,000 staff across the region from China to South East Asia and Australia.

UNStudio

UNStudio 由本·范·伯克尔和卡洛琳·博斯于1988年组建，是一家专门从事建筑设计、城市开发和基础工程建设的建筑设计事务所。公司名"UNStudio"代表的是United Network Studio，强调团体的协作与配合。事务所致力于在设计、技术、专业知识和管理等方面不断提升自身质量，以在建筑领域做出应有的贡献。

2009年亚洲UNStudio建立，其第一个办事处设在中国上海，该办事处由最开始致力于杭州来福士广场项目的设计，逐渐扩展成为一个全方位服务的设计公司，有着全面、专业的跨国建筑师团队。

UNStudio 设计的作品是环境可持续发展、市场需求与客户意愿的完美结合，其主要作品有：杭州来福士广场、Karbouw、Remu发电站、Villa Wilbrink、伊拉斯谟斯大桥、摩比斯宅邸、Het Valkhof博物馆、Prince Claus大桥、梅赛德斯奔驰博物馆、Arnhem中央火车站等。

UNStudio

UNStudio, founded in 1988 by Ben van Berkel and Caroline Bos, is a Dutch architectural design studio specializing in architecture, urban development and infrastructural projects. The name, UNStudio, stands for United Network Studio, referring to the collaborative nature of the practice. In 2009 UNStudio Asia was established, with its first office located in Shanghai, China. UNStudio Asia is a full daughter of UNStudio and is intricately connected to UNStudio Amsterdam. Initially serving to facilitate the design process for the Raffles City project in Hangzhou, UNStudio Asia has expanded into a full-service design office with a multinational team of all-round and specialist architects.

斯蒂文·霍尔建筑师事务所

斯蒂文·霍尔建筑师事务所是一所由斯蒂文·霍尔(Steven Holl)于1976年创立的建筑设计事务所，业务范围包括建筑设计与城市规划。事务所擅长有关艺术和高等教育类型的建筑设计，作品包括赫尔辛基当代美术馆、纽约普拉特学院设计学院楼、爱荷华大学艺术与艺术史学院楼、西雅图圣伊格内修斯小教堂。

斯蒂文·霍尔建筑师事务所在国际享有盛誉，以其高质量的设计多次获奖，其作品多次获得出版或展览，所获奖项有：北京当代MOMA荣获由国际高层建筑与城市住宅协会(CTBUH)所颁发的"2009年世界最佳高层建筑"大奖、2008年美国建筑师协会纽约分会可持续设计大奖、2009年西班牙对外银行(BBVA)基金会知识前沿奖、尼尔森·阿特金斯美术馆获得2008年美国建筑师协会纽约分会建筑荣誉大奖。斯蒂文·霍尔本人被授予"美国建筑师协会金奖"，这个奖项旨在认可对建筑界做出持久贡献的个人。

Steven Holl Architects

Steven Holl Architects is a 40-person innovative architecture and urban design office working globally as one office from two locations; New York City and Beijing. Steven Holl leads the office with senior partner Chris McVoy and junior partner Noah Yaffe. Steven Holl Architects is internationally-honored with architecture's most prestigious awards, publications and exhibitions for excellence in design. Steven Holl Architects has realized architectural works nationally and overseas, with extensive experience in the arts (including museum, gallery, and exhibition design), campus and educational facilities, and residential work. Other projects include retail design, office design, public utilities, and master planning.

Steven Holl Architects has been recognized with architecture's most prestigious awards and prizes. Most recently, Steven Holl Architects' Cite de l'Ocean et du Surf received a 2011 Emirates Glass LEAF Award, and the Horizontal Skyscraper won a 2011 AIA National Honor Award. The Knut Hamsun Center received a 2010 AIA NY Honor Award, and the Herning Museum of Contemporary Art received a 2010 RIBA International award. Linked Hybrid was named Best Tall Building Overall 2009 by the CTBUH, and received the AIA NY 2008 Honor Award. Steven Holl Architects was also awarded the AIA 2008 Institute Honor Award and a Leaf New Built Award 2007 for The Nelson-Atkins Museum of Art (Kansas City).

RTA-Office 建筑事务所

RTA-Office建筑事务所是Santiago Parramon建筑事务所在上海设立的一个分支机构。Santiago Parramon建筑事务所于1991年在西班牙巴塞罗那成立，并于1999年启动了它的全球计划。

自成立以来，RTA在建筑设计行业中稳步发展。其设计宗旨是将现代元素与中国传统文化相结合，全力为社会打造人与自然和谐的空间。其独特的艺术性设计在深圳华侨城欢乐海岸总体规划、深圳华侨城海景街改造、深圳华侨城Loft创意产业园区规划设计、天津市南河镇镇区规划、金华市月亮湾公园等多个国内项目中得到了充分体现，成为国内规模、技术实力、业绩等方面都处于行业最前沿的专业建筑设计公司。

RTA-Office

RTA-Office is a branch of Santiago Parramon architectural office, which since 1991 has its headquarters in Barcelona. Since the development of their work recognized at the local level, the architectural office begins around the year 1999 to show its global claim.

Since its establishment, RTA develops with a steady pace in architectural design industry. It aims to combine modern elements with Chinese traditional culture, creating spaces that man and nature coexist harmoniously. The unique artistic design of RTA enjoys full expression in Shenzhen OCT Harbor Master Plan, Shenzhen OCT Seaview Street Renovation, Shenzhen OCT Loft Creative Industrial Park Planning & Design, Tianjin Nanhe Town Planning and Jinhua Moon Bay Park. RTA has become a cutting-edge professional architectural design company from aspects of scale, technology and performance.

Alonso Balaguer y Arquitectos Asociados

Alonso Balaguer y Arquitectos Asociados 由 Luis Alonso Sergio Balaguer 建立于1978年，总部设在巴塞罗那，在马德里、利马和纽约设有办事处。事务所的业务范围涵盖建筑设计、城市规划、室内设计、工业设计和平面设计。其作品包括摩天大厦、体育中心、酒店、医院、商业和休闲中心、住宅项目、多功能建筑等。该事务所认为"时尚终究会变得过时"，故而，他们更注重作品的功能性和可持续性以及对环境的尊重与保护。

事务所的主要作品包括：巴塞罗那的西班牙大厦、O2健康中心、巴塞罗那国际高级医疗中心、佩尼亚菲尔新的普罗多思酒庄、巴塞罗那拉斯阿雷纳斯斗牛场改造项目、奥斯皮塔莱特欧洲广场内的摩天大厦、加泰罗尼亚住宅大楼、阿尔及尔住宅大楼、马德里伊比德罗拉商务培训校区、哈萨克斯坦阿拉木图体育馆等。

Alonso Balaguer y Arquitectos Asociados

Established in 1978 by Luis Alonso and Sergio Balaguer, their head office is presently in Barcelona and they also have offices in Madrid, Lima and New York. Their 50-strong team specializes in architecture, urban planning, interior design, industrial design and graphic design. It is precisely that determined and unswerving international gamble which underlines and defines their present standing with more than 700 projects and completed commissions illustrating the geographically diverse extent of their accomplishments (Chile, Colombia, Peru, Kazakhstan, Ukraine, Poland, Morocco, Algeria, China, etc.). Work which includes skyscrapers, sports centers, hotels, hospitals, commercial and leisure centers, housing projects, multifunctional buildings, etc.

The office has never gambled on solutions where snobbery or attractiveness predominate over functionality, respect for the environment or the necessary sustainability and energy savings of their proposals.

Their most representative projects include the Hesperia Tower skyscraper in Barcelona, the O2 Wellness Centre and the International Advanced Medicine Centre (CIMA) in Barcelona, the new Protos winery in Peñafiel, the former Las Arenas bullring reconverted into a Recreational and Shopping Centre in Barcelona, the four new skyscrapers in the Europa Square in l'Hospitalet and the two I'Illa del Celtowers in Diagonal Mar.

amphibianArc

amphibianArc 由王弄极创建于1992年，其总部设在洛杉矶，并在上海设有办事处。该公司认为，设计既是艺术的表现形式，也是解决问题的技巧。而在实践过程中将创作元素与现实元素完美融合，则是该公司设计贯彻的信条。自公司成立以来，amphibianArc一直关注于"液体建筑"概念，这一概念不仅表现为充斥于近代建筑舞台的曲线形体，更包含了华夏文明中表意文字的逻辑。

amphibianArc 在美国和中国设计了各类型的建筑项目，包括屡获殊荣的北京天文馆、佛山东平新城交通枢纽中心以及红星美凯龙世界家具展示中心等。其作品也被各地杂志竞相刊登，包括《建筑实录》、《世界建筑》以及《洛杉矶建筑师》等。

amphibianArc

amphibianArc, founded by Nonchi Wang in 1992, is an architecture design firm headquartered in Los Angeles, with branch office in Shanghai, China. The practice endeavors to synthesize between artistic expression and problem solving technique. Through a shared disclosure called Liquid Architecture, its work encompasses not only the curvilinear forms prevalent in contemporary architectural scene, but also ideographic methodology which is the foundation of Chinese word making.

Since its founding, amphibianArc has designed a wide range of projects in both the U.S. and China, including the award winning Beijing Planetarium, Foshan Dongping New City Mass Transit Center, and Hongxing Macalline Furniture Beijing Flagship Store. The practice's work has also been featured in publications including Architectural Record, World Architecture, and LA Architect.

C.F.Møller Architects

C.F.Møller Architects 是斯堪的纳维亚半岛历史最悠久、规模最大的建筑机构之一，业务范围涵盖方案分析、城市规划、总体规划、景观设计、建筑工程设计及其他诸多领域。事务所于 1924 年创立，以简约、明快、朴素的理念指导各项实践，并根据每个项目的基址特点，结合国际发展趋势和地域差异对理念进行重新解读和诠释。

事务所以改革和创新为发展理念，力图打造独具吸引力和发展前景的工作环境，使每位员工都能接受高要求设计项目的挑战。多年来，事务所屡获国内外设计大奖，其作品多次在国内外的展会上展出。其主要作品有：法尔斯特岛新封闭式州立监狱、奥尔胡斯 Incuba 科学公园、奥尔胡斯大学礼堂、奥尔胡斯艺术大楼扩展项目、奥尔胡斯低能耗办公大楼、国家海事博物馆扩建、巴里考古博物馆等。

C.F. Møller

C.F. Møller is one of Scandinavia's oldest and largest architectural practices. Its award-winning work involves a wide range of expertise that covers all architectural services, landscape architecture, product design, healthcare planning and management advice on user consultation, change management, space planning, logistics, client consultancy and organisational development.

Simplicity, clarity and unpretentiousness, the ideals that have guided its work since the practice was established in 1924, are continually re-interpreted to suit individual projects, always site-specific and based on international trends and regional characteristics.

C.F. Møller regards environmental concerns, resource-consciousness, healthy project finances, social responsibility and good craftsmanship as essential elements in its work, and this holistic view is fundamental to all its projects, all the way from master plans to the design.

Today C.F. Møller has about 320 employees. The head office is in Aarhus, Denmark and it has branches in Copenhagen, Aalborg, Oslo, Stockholm and London.

LAVA

LAVA 是由 Chris Bosse、Tobias Wallisser 和 Alexander Rieck 于 2007 年创立的一家跨国公司，它是创造性研究和设计思维的交汇，在悉尼、上海、斯图加特和阿布扎比设有办事处。其业务包括了城市中心总体规划、家具、酒店、住宅以及机场设计。

LAVA 热衷于从雪花、蜘蛛网、肥皂泡沫等自然形态中探索新的建筑类型和结构，这种源自自然的几何体实现了功能与美观的统一。LAVA 擅长将智能系统和建筑外皮融合，以此对空气压力、温度、湿气、太阳辐射、污染等外界因素作出反应。

LAVA 获得的主要奖项有：澳大利亚室内设计奖、多伦多"零脚印公司"颁发的国际建筑表面改造奖、国际设计杂志年度设计回顾奖和 IDEA 奖、英国伦敦建筑设计大奖、迪拜都市景观建筑奖。该公司还曾获得 Iakov Chernikhov 国际大奖、2011 年 Dedalo Minosse 国际建筑奖和科技创新奖提名。

LAVA

Chris Bosse, Tobias Wallisser and Alexander Rieck founded multinational firm, Laboratory for Visionary Architecture (LAVA) in 2007 as a network of creative minds with a research and design focus and with offices in Sydney, Shanghai, Stuttgart and Abu Dhabi. The business scope of LAVA covers urban center master plan, furniture, hotel, houses, airports, etc.

The potential for naturally evolving systems such as snowflakes, spider webs and soap bubbles for new building typologies and structures has continued to fascinate LAVA – the geometries in nature create both efficiency and beauty. LAVA's projects incorporate intelligent systems and skins that can react to external influences such as air pressure, temperature, humidity, solar radiation and pollution.

Awards include the Australian Interior Design Awards, UN partnered ZEROprize Re-Skinning Award, I. D. Annual Design Review, IDEA Awards, AAFAB AA London, Cityscape Dubai Award Sustainability; commendations include Well Tech Award and Dedalo Minosse International Prize in 2011; and nominations for the Iakov Chernikhov International Prize and the Index Award.

蓝天组

1968 年由沃尔夫·德·普瑞克斯和海默特·斯维茨斯基在奥地利维也纳设立了建筑设计事务所，取名"蓝天组"，其业务范围涉及建筑设计、城市规划、平面和艺术设计多个服务领域。经过 20 年的奋斗，他们在 1988 年于美国洛杉矶成立了境外第一家分支事务所。

蓝天组的主要项目有：法国默伦塞纳城市总体规划、荷兰 Groninger 博物馆东馆、瑞士 2002 年世博会比尔市艺术长廊会议中心、1998 年竣工的德国德莱斯顿 UFA 影院中心、德国慕尼黑宝马中心、洛杉矶视觉与表演艺术高中、法国里昂汇聚博物馆、丹麦奥尔堡音乐之屋、德国法兰克福欧洲央行新总部和韩国釜山电影中心等。

在过去的三十年间，蓝天组也获得了许多国际竞赛大奖：1982 年柏林建筑艺术促进大奖；1988 年维也纳建筑奖；1992 年 Erich Schelling 建筑大奖；1989-1991 年连续三年 P.A. 大奖；1999 年奥地利城市建筑大奖；2001 年欧洲钢结构设计大奖；2005 年凭借阿克隆艺术馆设计方案赢得美国建筑奖；2008 年凭借 BMW 展厅与阿克隆艺术馆赢得 RIBA 大奖等。

COOP HIMMELB(L)AU

COOP HIMMELB(L)AU was founded by Wolf D. Prix, Helmut Swiczinsky and Michael Holzer in Vienna, Austria in 1968, and is active in architecture, urban planning, design and art. In 1988, a second studio was opened in Los Angeles, USA. Further project offices are located in Frankfurt, Germany and Paris, France. COOP HIMMELB(L)AU employs currently 150 team members from nineteen nations.

COOP HIMMELB(L)AU's most well-known projects include the Rooftop Remodeling Falkestraße; the master plan for the City of Melun-Sénart; the Groninger Museum, East Pavilion in Groningen; the design for the EXPO.02—Forum Arteplage in Biel; the multifunctional UFA Cinema Center in Dresden; the Academy of Fine Arts and the BMW Welt in Munich; the Akron Art Museum in Ohio; the Pavilion 21 MINI Opera Space in Munich; the Martin Luther Church in Hainburg; the Busan Cinema Center in Busan and the Dalian International Conference Center in China.

COOP HIMMELB(L)AU has received numerous international awards. In 2005, for the design of the Akron Art Museum, the studio received the American Architecture Award. The studio was awarded for the 2007 International Architecture Award for four projects. In 2008 COOP HIMMELB(L)AU received the RIBA International Award for the Akron Art Museum. In the same year the RIBA European Award and the World Architecture Festival Award Production were given for the project BMW Welt. In 2010, COOP HIMMELB(L)AU won the MIPIM Architectural Review Future Projects Award in the category sustainability for their project Town Town Erdberg. In 2011 the office received the Wallpaper Design Award 2011 (Category: "Best Building Sites") for the project Dalian International Conference Center, the Dedalo Minosse International Prize for the project BMW Welt as well as the Red Dot Award: product design (Category: "Architecture") for the Central Los Angeles Area High School #9 of Visual and Performing Arts.

Barbosa & Guimarães

Barbosa & Guimarães 可从事建筑领域各方面的设计项目，包括办公楼、商业设施、教育大楼、卫生中心、体育设施、文化设施、酒店装备、私人住宅楼等建筑类型的设计和城市设计。事务所认为处理项目的能力在特定的历史文化背景中与客户是息息相关的，充分发挥专业技能，甚至出人意料地采用一些新技术、新方法，将对建筑的最终建成起决定性作用。事务所擅长于通过逻辑性的思维方式寻找与每个项目相适应的设计方法，这种设计方式使之能够很好地把握当代国际建筑设计的趋势。

Barbosa & Guimarães

Barbosa & Guimarães develops projects in various fields of architecture, ranging from office buildings, commercial facilities, educational buildings, health centers, sports facilities, cultural facilities, hotel equipment as well as the private residential buildings. The firm has the capacity to approach a program, inseparable from a client in a specific physical context, history and culture, embracing the expertise, new techniques and innovations, always trying to introduce an artistic intention, which will prove decisive in the final shape of the object. The projects are the result of logic and a proper way intended to be consistent, always aware of trends in contemporary international architecture.

3LHD

3LHD 是一家专业从事建筑设计、城市规划的建筑师事务所。该公司不断探索建筑、社会和个人之间新的互动，以现代的设计方法完成了耶卡纪念桥、日本 2005 年世博会、伊斯特里亚巴勒体育馆、里耶卡扎梅特中心、萨格勒布舞蹈中心、萨拉戈萨 2008 年世博会克罗地亚馆、斯普利特滨水景观、罗维尼罗恩酒店等多个设计项目。

3LHD 的设计作品荣获了多项克罗地亚以及国际大奖，其中包括：2011 年国际体育建筑奖银奖、英国 AR 新兴建筑奖、美国 ID 杂志奖；2009 年国际体育建筑奖铜奖；2008 年第一届世界建筑节体育类最佳建筑奖等多项卓越奖项。

3LHD 还参加了国内外的众多展览，包括：2010 年代表克罗地亚参加威尼斯建筑双年展、第 12 界国际建筑展；2008 年"地中海文明"鹿特丹第二届国际建筑双年展；以及在波士顿哈佛大学的一个集体展——"新的轨迹：克罗地亚和斯洛文尼亚的当代建筑"。

3LHD

3LHD is an architectural practice, focused on integrating various disciplines – architecture, urban planning, design and art.

3LHD architects constantly explore new possibilities of interaction between architecture, society and individuals. With contemporary approach, the team of architects resolves all projects in cooperation with many experts from various disciplines.

Projects, such as Memorial Bridge in Rijeka, Croatian Pavilion in EXPO 2005 in Japan and EXPO 2008 in Zaragoza, Riva Waterfront in Split, Sports Hall Bale in Istria, Centre Zamet in Rijeka, Zagreb Dance Centre in Zagreb and Hotel Lone in Rovinj are some of the important highlights.

They represented Croatia at the Venice Biennale 2010, 12th International architecture exhibition with the group of authors, and took part in the 2008 "Mare Nostrum" exhibition on the second International Architecture Biennale in Rotterdam, and in a group exhibition in Boston at Harvard University: "New trajectories: Contemporary Architecture in Croatia and Slovenia".

The work of 3LHD has received important Croatian and international awards, including the award for best building in Sport category on first World Architecture Festival WAF 2008, IOC/IAKS Bronze Medal Award 2009 and IOC/IAKS Silver Medal Award 2011 for best architectural achievement of facilities intended for sports and recreation, AR Emerging Architecture Award (UK), the ID Magazine Award (USA); and Croatian professional awards Drago Galić (2008), Bernando Bernardi (2009; 2005), Viktor Kovačić (2011; 2001), and Vladimir Nazor (2009; 1999). In 2012 Croatian President Ivo Josipović awarded 3LHD with the Charter of the Republic of Croatia for exceptional and successful promotion of contemporary architecture in Croatia and abroad.

捷得国际建筑师事务所

捷得国际建筑师事务所（The Jerde Partnership）是一家创造视觉体验的建筑设计公司。自 1977 年以来，公司在世界各地都拥有自己的设计项目，并曾荣获美国建筑学会、*Progressive Architecture* 杂志、美国规划协会、芝加哥历史博物馆等授予的多项荣誉。

作为国际知名的城市设计和建筑公司，美国捷得国际建筑师事务所一直坚持设计作品中要体现社区意识的理念，致力于城市翻新、城市再创造以及建造具有长期社会经济影响的项目。

捷得从事的并非是单纯的建筑设计，其作品深刻地体现了各地区独特的文化特色，在捕捉当地传统精髓的同时，赋予其现代的笔触。事务所的主要作品有：运河城华盛顿酒店、百年运动中心、马尼拉世纪城总体规划、昌原市 7 商城、新濠天地、大邱彩色广场体育场购物中心、福冈君悦酒店、横琴岛旅游开发区、霍顿广场、吉隆坡生态城、圣莫尼卡广场、厦门双子塔等。

The Jerde Partnership

The Jerde Partnership is a visionary architecture and

urban design firm that creates dynamic places that deliver memorable experiences and attract over 1 billion people annually. Founded in 1977, the firm has pioneered "placemaking" throughout the world with projects that provide lasting social, cultural and economic value and promote further investment and revitalization. Based in a design studio in Los Angeles with project offices in Shanghai, Hong Kong, and Seoul, Jerde takes a signature, co-creative approach to design and collaborates with private developers, city officials, specialty designers and local executive architects to realize the vision of each project. The firm has received critical acclaim from the American Institute of Architects, Progressive Architecture, American Planning Association, International Council of Shopping Centers, and Urban Land Institute. To date, over 110 Jerde Places have opened in diverse cities, including Atlanta, Budapest, Hong Kong, Istanbul, Las Vegas, Los Angeles, Osaka, Rotterdam, Seoul, Shanghai, Tokyo and Warsaw.

Ingarden & Ewý 建筑师事务所

Ingarden & Ewý 建筑师事务所由克拉科夫建筑师 Krzysztof Ingarden 和 Jacek Ewý 成立于 1998 年。Ingarden & Ewý 涉及的工作领域主要是创新型公共建筑设计，其主要项目有 2012 年小波兰艺术花园、2010 年宗座大学图书馆、2008 年隆多商业园、2007 年维斯皮安斯基 2000 展览和信息馆、2005 年爱知世博会波兰馆等多个项目。除了公共建筑设计外，Ingarden & Ewý 还承担了众多非正式的住宅、办公、商业和工业项目的设计，受到建筑师、投资者和使用者的一致好评。

事务所主要负责人之一 Krzysztof Ingarden1987 年毕业于克拉科夫技术大学，曾在东京的 Arata Isozaki 工作室和纽约的 J.S.Polshek & Partners 工作室工作。事务所另一主要负责人 Jacek Ewý 于 1983 年毕业于克拉科夫技术大学，于 1998 年与 Krzysztof Ingarden 建立了自己的事务所。

Ingarden & Ewy

Ingarden & Ewý Architects – architecture atelier established in 1998 by Kraków architects Krzysztof Ingarden and Jacek Ewý collaborating for over 20 years, first in JET Atelier, founded in 1991.

The domain of Ingarden & Ewý are innovative public building and site designs, notably Małopolska Garden of Arts (2012), Manggha Museum of Japanese Art and Technology in Kraków conducted in collaboration with Japanese architect Arata Isozaki (1994), the Polish Embassy in Tokyo (2001), the Polish Pavilion at EXPO 2005 in Aichi (2005), and plenty of sites in Kraków, including Wyspiański 2000 exhibition and information pavilion (2007), Rondo Business Park office centre (2008), Garden of Experience in the Polish Aviators Park (2008), and Library of the Pontifical University (2010). Being constructed currently in Kraków is the ICE Congress Centre, a prestigious investment of the Municipal Office of Kraków designed by Ingarden & Ewý.

Besides public architecture, Ingarden & Ewý Atelier developed plenty of unorthodox residential, office, commercial, and industrial designs applauded by architects, investors, and users.

JSA–Jensen & Skodvin Arkitektkontor

JSA–Jensen & Skodvin Arkitektkontor 由 Jensen 和 Skodvin Borre 于 1995 年在挪威建立的一家建筑公司。Jensen 和 Skodvin Borre 曾就读于奥斯陆建筑和设计学院，并曾在 NSB Architects 就职。Jensen & Skodvin 认为建筑的最终形态都是自然、选择和特定条件产生的结果，他们也将这一理念和原则贯彻到项目设计中。该公司已设计了许多广受赞誉的建筑项目，并获得众多奖项。

JSA-Jensen & Skodvin Arkitektkontor

JSA-Jensen & Skodvin Arkitektkontor is a Norwegian architectural firm established in 1995 by Jan Olav Jensen and Skodvin Borre. Both partners were educated at the Oslo School of Architecture and Design. Meanwhile they have previously worked at NSB Architects. Jensen & Skodvin follows a design philosophy: the final shape is a result of natural, selected and specified terms. The philosophy is implemented in their project design. The firm has designed many acclaimed buildings, and received many awards.

Oppenheim Architecture+Design

Oppenheim Architecture+Design（OAD）是一家提供建筑设计、室内设计和城市规划等全方位服务的公司。其总部位于美国佛罗里达州迈阿密市，在洛杉矶、瑞士、巴塞尔设有办事处。公司由 Chad Oppenheim 创立，专注于为富有挑战性的复杂项目提供强有力的实用方案，拥有设计世界级医院、住宅和多用途建筑的丰富经验。

该公司从环境和相关规划中汲取精华，以创造一种戏剧性的、强有力的体验，同时也赋予建筑舒适感。其设计首先着眼于对客户需求的全方位分析，从而构建人性化的建筑和环境。该公司的主要作品有：Dellis Cay 别墅群、拉斯维加斯的 Hard Rock、索尼斯达毕士倾岛酒店、马可岛万豪酒店、哥伦比亚特区和亚特兰大 1 酒店等。

Oppenheim Architecture + Design

Oppenheim Architecture + Design (OAD) is a full service architecture, interior design and urban planning firm located in Miami, Florida with offices in Los Angeles and Basel, Switzerland. The firm, founded by Chad Oppenheim, specializes in creating powerful and pragmatic solutions to complex project challenges and has extensive experience in world class hospitality, residential and mixed-use design.

The firm's design strategy is to extract the essence from each context and relative program— creating an experience that is dramatic and powerful, yet simultaneously sensual and comfortable. The firm's approach begins by a comprehensive analysis of a client's vision in relation to the projects typology, context, zoning parameters and financial realities.

Since its inception, OAD has accumulated over $100 billion in project work and has won over 45 awards and honors for its unique design sensibility and ability to pioneer innovative concepts, optimize challenging sites and revitalize blighted urban areas. Projects designed by the firm cover a broad spectrum of programmatic requirements, budgets and building types. This award-winning work is based on both a physical and spiritual contextual sensitivity, supported by evocative and economic design solutions. More specifically, the firm's pragmatic yet poetic architectural solutions in unproven and undeveloped areas have served to revitalize many neighborhoods throughout the world.

欧博迈亚工程咨询（北京）有限公司

欧博迈亚工程咨询（北京）有限公司由德国著名土木工程师莱昂哈特·欧博迈亚博士（Leonhard Obermeyer）于 1958 年成立于德国慕尼黑，是德国现今历史最悠久的独立工程咨询公司之一。50 多年来，该公司秉承欧博迈亚博士提出的一项基本原则，即：为钻研最先进的工程技术和独特的设计理念而不断奋斗，并将钻研成果、最高水准的工作方法和在相关领域积累的项目经验有机结合起来，为客户提供质量最优的服务。

公司的创立者欧博迈亚博士因其在工程领域以及规划和建筑设计领域的卓越贡献，而获得德意志联邦共和国颁发的一等十字勋章。他认为，规划设计、建筑设计和工程建设三者之间存在着紧密的内在联系和拓扑型逻辑关系。该公司的服务宗旨也由此得以奠定，即：为客户的工程建设项目提供从前期咨询到项目管理、从规划设计到建设质量监控的全方位服务。

OBERMEYER Engineering Consulting Co., Ltd.

Founded in Munich in 1958, the OBERMEYER Corporate Group now ranks as one of Germany's largest independent engineering consultancies.

After 20 years of successful design and planning experience in China, OBERMEYER incorporated OBERMEYER Engineering Consulting Co., Ltd. in 2005.

With more than 1,200 employees in its numerous subsidiary and associated companies throughout Germany and in many countries abroad the OBERMEYER Group offers integrated planning for all fields of construction planning: buildings, transport and environment.

Among the permanent staff are experts from the disciplines architecture, civil engineering, traffic planning, structural planning, supply engineering, electrical engineering and surveying as well as natural scientists, IT specialists, economists and business management experts.

The OBERMEYER quality management system has been certified in accordance with DIN EN ISO 9001 since 1997.

Cervera & Pioz Arquitectos

Cervera & Pioz Arquitectos 成立于 1979 年，由设计师 Javier Pioz 和 Rosa Cervera 共同创立。该公司致力于构建独特而又具有标志性意义的建筑，其设计作品获得多项荣誉和奖项。

自公司成立后，Cervera 和 Pioz 在对逻辑原则的灵活性与适应性以及能源的高效利用等方面有了全面理解的基础上，为提出新的建筑设计和城市规划方案而奋斗。

1992 年，Cervera 和 Pioz 开始了将生物工程应用到现代结构工程和建筑设计中的研究，并由此提出一种工程结构和设计的新概念——生物结构。既而，他们将这一概念运用到高层空间中，发展可持续的垂直生物结构工程，这是一种可持续的、以人为本的城市发展新模型。这一理论强调对自然规律的尊重和利用，以实现自然与技术之间的平衡，在国际上具有重大的影响力。

Cervera & Pioz

Founded by Javier Pioz and Rosa Cervera in 1979, the Firm has, since then, devoted itself to the creation of unique and emblematic buildings. Their work has won many distinguished awards and prizes, including:

Administrative Centre of the Spanish town in Shanghai (1); High School Complex (1); Xixi Wetland Museum in Hangzhou (2); P.R. of China Embassy in Madrid (1); Citibank-Spain Competition in Cadiz (1); Islamic Cultural Centre in Madrid (3); the National Festival of Video-Art in Cadiz (2); Official College of Architects of Madrid Award; Antonio Maura Award from the City Council of Madrid; Diploma from the Architecture Biennial in Sofia, Bulgaria; and Honor Award by the "Foundation For Architectural & Environmental Awareness" and "ArchiDesign Perspective", New Delhi, India.

NL Architects

NL Architects 总部位于阿姆斯特丹，由 Pieter Bannenberg、Walter van Dijk、Kamiel Klaasse 和 Mark Linnemann 于 1997 年建立。在 NL 看来，当今的建筑与郊区存在的问题及发展策略是密不可分的，这也是他们构建创新、前瞻性项目的关键。NL 认为建筑是以一种实验性形式展现的活动，融合了经济、文化、技术和环境等因素，城市则是一个平衡了经济发展与环境保护的生态系统。

事务所的精选项目包括格罗宁根广场、阿姆斯特丹菲英 K 街区、韩国环形住宅、赞斯塔德 A8ernA、乌特勒支篮球吧、乌特勒支 WOS8。其获得的奖项包括：2006 年城市公共空间欧洲地区大奖、2005 年密斯·凡·德罗大奖、2004 年 N.A.I 大奖、2003 年里特维德奖以及 2001 年鹿特丹设计奖。

NL Architects

Amsterdam-based NL Architects was founded in 1997 by Pieter Bannenberg (1959), Walter van Dijk (1962), Kamiel Klaasse (1967) and Mark Linnemann (1963).

In the view of NL, architecture today cannot be separated from suburban issues and strategies, which is their key to develop innovative and forward-looking devices and arrangements. Architecture is presented as a field of experimental activity, at the crossroads of economic, programmatic, technical and environmental thinking. The city is perceived as an ecosystem, an environment where logical systems of urban growth and natural factors, consumption and production, flux and stasis are balanced; and where recycling is set up as a method of stable and sustainable functioning, whether it has to do with waste, energy, materials or even architecture.

Its selected works include Groninger Forum (Groningen, 2006–2011), Funen Blok K (Amsterdam, 2010), Loop House (Korea, 2006), A8ernA (Zaanstad, 2006), Basket Bar (Utrecht, 2003), WOS 8 (Utrecht, 1997).

NL Architects participated in many exhibitions around the world: Out There, Architecture Biennale Venice (Venice, 2008), New Trends of Architecture in Europe and Asia-Pacific (Shanghai, 2007), Sign as Surface (traveling Exhibition USA, 2003), Fresh Facts, Biennale Venice (Venice, 2002), NL Lounge, Dutch Pavilion Biennale Venice (Venice, 2000).

Among the numerous awards there are the European Price for Urban Public Space (2006), Mies van der Rohe award (2005), N.A.I award (2004), Rietveld Price (2003), and the Rotterdam design price (2001).

Jaspers-Eyers Architects

Jaspers-Eyers Architects 是一家国际性的建筑师事务所，在布鲁塞尔、鲁汶和哈塞尔特等地都设有工作室。事务所由 John Eyers 和 Jean-Michel Jaspers 领导，可为公共领域和私人领域的客户提供建筑设计、建筑方案制作、城市规划、项目整体规划、图形设计以及室内设计等全方位服务。

事务所的设计风格简洁明快，其设计关注生态环保，注重建筑和人之间的交流，以构建对人类和环境无危害的建筑。凭借多年的行业经验以及多样化的服务特色，事务所已完成了多个成功的项目，包括 IBM 总部、Alcatel-Lucent 总部、Ageas 金融中心、KBC 银行以及各种酒店、机场、体育馆项目。当前，Jaspers-Eyers Architects 已成为欧洲和比利时最著名的建筑设计事务所之一。

Jaspers-Eyers Architects

Jaspers-Eyers Architects is an international practice with studios located in Belgium in Brussels, Leuven and Hasselt and a growing presence in Eastern Europe. The practice led by John Eyers and Jean-Michel Jaspers provides full architectural design, programming, urban design, master planning, graphic and interior design services for clients in both the public and the private sector.

More than 30 years ago the firm designed the first atrium building in Belgium equipped with a series of glass-enclosed elevators; an atrium building which was probably the first of its kind in Europe. With numerous completed atrium buildings the firm helped popularize this building type since it brings natural light to office spaces and communication between building occupants is greatly enhanced. Over the years the firm has specialized itself in a wide variety of building types ranging from headquarters for major global companies such as IBM, Alcatel-Lucent, Ericsson, HP, Sony and Xerox, financial headquarters such as Ageas, Axa, Belfius and KBC Bank, to shopping malls, educational and cultural buildings as well as high-rise projects to name a few.

奥雷·舍人事务所

奥雷·舍人事务所由奥雷·舍人建立于 2010 年，在北京、香港和伦敦设有办公室。奥雷·舍人（Ole Scheeren）是国际知名的德国籍建筑师，现为奥雷·舍人事务所的设计主持和香港大学客座教授。奥雷·舍人曾在瑞士洛桑大学和德国卡尔斯鲁厄大学求学，后毕业于伦敦建筑协会学院，并获得英国皇家建筑师协会银奖。

奥雷·舍人设计的主要作品有北京中央电视台和电视文化中心、曼谷 MahaNakhon 综合大楼、翠城新景（The Interlace）大型住宅综合体、新加坡高端公寓 The Scotts 大厦深圳新市中心规划以及台北艺术中心等。

Buro Ole Scheeren

Buro Ole Scheeren was founded by Ole Scheeren in 2010 with offices set in Beijing, Hong Kong and London. Ole Scheeren is an internationally renowned German architect, the chief designer of Buro Ole Scheeren, and the visiting professor of the University of Hong Kong. Educated at the universities of Karlsruhe and Lausanne, he graduated from the Architectural Association in London and was awarded the RIBA Silver Medal.

The masterpieces of Buro Ole Scheeren include China Central Television Station CCTV and the Television Cultural Centre TVCC in Beijing, MahaNakhon mixed-use tower, the Interlace large-scale complex, the Scotts Tower high-end apartments in Singapore, Shenzhen's new city center, the Taipei Performing Arts Center in Taiwan and etc.

Moore Ruble Yudell Architects & Planners

30 多年前，Charles Moore、John Ruble、Buzz Yudell 创立了 Moore Ruble Yudell Architects & Planners。他们三人热衷于探索基于场所与当地居民之间的对话而衍生的新建筑，关注人文环境，鼓励社区群体间的交流和联系，这些价值观也成为了该公司在建筑规划与设计时的核心理念。

作为一家拥有近 60 名成员的建筑事务所，该公司强调成员间的合作与交流，同时不断引入各类型的资深设计师和工程顾问，以灵活、高效地为不同难度的各类型项目提供解决方案。

Moore Ruble Yudell Architects & Planners

The founding partners – Charles Moore, John Ruble, and Buzz Yudell established Moore Ruble Yudell Architects & Planners over thirty years ago. They share a passion for an original architecture that grows out of an intense dialogue with places and people, celebrates human activity, enhances and nurtures community. These values continue to guide their process, providing the core principles for a wide-ranging exploration of planning and architecture.

With an office of some sixty people, Moore Ruble Yudell has been able to meet the challenges of complex programs and contemporary project delivery, while maintaining a close involvement by the partners. As the practice has expanded – both programmatically and geographically – it has grown in its technical capability and its skill in leveraging the multiple talents of the firm, often bringing in increasingly specialized design and engineering consultants. Moore Ruble Yudell has formed successful national and international alliances, gaining flexibility to effectively address the needs of large and small-scale projects with a global reach.

Henning Larsen 建筑事务所

Henning Larsen 建筑事务所于 1959 年由建筑师 Henning Larsen 创立，是一家根植于斯堪的纳维亚文化的国际建筑公司。Henning Larsen 建筑事务所由首席执行官 Mette Kynne Frandsen 以及设计总监 Louis Becker 和 Peer Teglgaard Jeppesen 负责管理，已在哥本哈根、慕尼黑、贝鲁特以及利雅得建立了办事处。

Henning Larsen 建筑事务所在环境友好型建筑和综合节能设计等方面卓有建树，以创造怡人、可持续的项目为目标，给当地的人们、社会及文化创造持久的价值。设计师对项目负有高度的社会责任感，注重与客户、业主以及合作者的交流，积累了从建议书草案到细节设计、监督和施工管理等各方面的建筑知识和理论。

其主要作品有：在都柏林设计的"钻石"大厦、哥本哈根商业学校、瓦埃勒"波浪住宅"、格鲁吉亚巴统水族馆、瑞典于默奥 (Umea) 大学建筑学院楼等。

Henning Larsen Architects

Henning Larsen Architects is an international architecture firm based in Copenhagen, Denmark. Founded in 1959 by noted Danish architect and namesake Henning Larsen, it has around 200 employees. In 2011 the company worked on projects in more than 20 countries.

In 2008 Henning Larsen Architects opened an office in Riyadh, Saudi Arabia named Henning Larsen Middle East and in 2011 an office in Munich, Germany was inaugurated. Most recently Henning Larsen Architects opened two offices, one in Oslo, Norway and one in Istanbul, Turkey.

Henning Larsen Architects is known for their cultural and educational projects. Last year Harpa Concert Hall and Conference Centre in Reykjavik was selected as one of the ten best concert halls in the world by the British magazine Gramophone. Henning Larsen Architects also designed the Copenhagen Opera.

Current projects include a new headquarter for Siemens in Munich, Germany and a 1.6 mill m² masterplan for the King Abdullah Financial District in Riyadh, Saudi Arabia.

New Wave Architecture

New Wave Architecture 建立于 2006 年，位于伊朗德黑兰，公司共有 70 名员工，是一家创新型的建筑设计公司。该公司致力于探索全球性的建筑语言，在新兴理念、高标准的创新空间美学、人性化设计等方面探索新的途径，以实现创新又富有挑战的现代设计。公司凭借在校园和教育类建筑的卓越表现，以及在医疗和保健设施类建筑的专业设计，获得了国内外多项建筑大奖和荣誉。

事务所的主要作品有：Broojen 健康管理中心、Chelgerd 健康管理中心、扎黑丹大学展览厅、中心礼堂和图书馆、艺术和建筑系大楼、人类科学系大楼、Hirmand 文化综合体、Dalgan 文化综合体、德黑兰 Shareiati 医院、德黑兰内分泌和新陈代谢中心、伊朗 Abali Onlooker 住宅、伊朗纳米科技研究中心等。

New Wave Architecture

New Wave Architecture (Lida Almassian/Shahin Heidari) founded in 2006 is a 70-person innovative architecture design firm in Tehran, Iran. It has been nationally and internationally honored with architecture's prestigious awards, publications, competitions and citation for design excellence with extensive experience in the campus and educational faculties, specialized hospitals and health care facilities. Other projects include retail design, residential works and recreational facilities. Over 150 projects have been designed, accomplished or due to be completed.

New Wave Architecture seeks for global language of architecture to approach an innovative and challenging contemporary movement. It explores the new ways of emerging ideas, demanding and distinctive spaces regarding the aesthetic aspects, humanity and global communication.

OAD 欧安地建筑设计事务所

OAD 欧安地建筑设计事务所是成立于美国新泽西州的国际化建筑师事务所，主要合伙人由美国注册建筑师及美籍华人建筑师组成。OAD 一直致力于向业主提供最优秀的设计、技术、管理和服务。在对多种建筑类型深入了解的基础上，事务所对设计中的创新孜孜以求，努力创造独特而成功的空间环境。

OAD 具有国际化设计理念，强调从城市和环境的角度理解建筑，对规划、建筑、室内、景观全方位整合，将人性化空间、绿色生态等观念融入设计中，以创新作为设计的灵魂，积极面对每项工程的特定条件及要求的挑战，提供创造性的解决方案。OAD 多次在国内外设计竞赛中获奖，其中北京富力丹麦小镇、成都花水湾名人度假酒店项目获得 2009 年度联合国人居奖；方恒国际商业中心于 2006 年 7 月获第二届中国国际建筑艺术双年展建筑设计优秀奖。

OAD Office for Architecture & Design

OAD (Office for Architecture & Design) was founded by a group of American architects and oversea Chinese architects who were educated and trained in both China and the United States. The company is based in New Jersey, USA. OAD had designed and completed a wide range of project types of varying scope and complexity. The work of OAD includes commercial and institutional buildings, housing, urban landscape, hotel and office interior design.

OAD's effort is to deliver quality design products and intelligent services that enable clients to achieve their goals and enrich their building environment.

OAD's overall philosophy is that each project is unique. A high degree of energy and enthusiasm is devoted to every undertaking project in order to reach a natural solution to each design challenge. Designers begin with an assessment of each project's physical, social and economical condition, in a process of inquiry, research and analysis frame experiments in material and construction. Drawings, models, and research are the tools that shape each design. Designers keep close involvement with the construction process and collaboration with consultants, engineers, artists, and contractors to assure the finished quality of the project. This is an art that weaves together both context and contents.

Efficiency, intelligence, and beauty are the means and ends of each project. Since the company's founding, it has developed interactive collaboration with clients.

深圳市天方建筑设计有限公司

深圳市天方建筑设计有限公司 (TAF) 成立于 2005 年，是一家在英国伦敦注册的专业建筑设计顾问公司。TAF 极具创新精神，并关注当代建筑与文化问题，通过超前的理念为未来的城市提供独特、有效的设计方案。TAF 以东方的自然体验和现代科技为基础，使建筑成为与自然和谐相处的人造复合体，让生活在其中的人们感受建筑所具有的生命力。

该公司将国际先进的建筑理论与实践经验相结合，同时结合本土文化，以国际性的视野解决地域性问题，为设计项目注入新鲜的思想与活力，为业主提供高端优质的设计服务。其设计作品涵盖了各种类型的公共建筑、住宅规划建筑、城市规划等，多类型的设计实践使之积累了丰富的设计理念与实践经验。

Shenzhen TAF Architectural Design Co., Ltd.

Shenzhen TAF Architectural Design Corporation funded in 2005 is a professional architectural design consulting company registered in London, UK. With spirit of innovation, TAF concerns issues of modern architecture and culture. It devotes to provide unique effective design proposals for future cities by advanced concepts. Based on natural experience and modern technology of the Orient, architecture becomes man-made complex that coexists harmoniously with nature. TAF's clients are able to feel the vitality of architecture.

The firm, combining international advanced architectural theory with practical experience, and referring to local culture, devotes to settle regional issues with international vision, brings new ideas and vitality to projects, and provides high quality design service for owners. The design projects of TAF cover public buildings, residential planning buildings, urban planning and so on. It has accumulated rich design concepts and practical experience from multi-type design projects.

OFIS 建筑事务所

OFIS 建筑事务所是由 Rok Oman 和 SpelaVidecnik 共同创建的一家位于斯洛文尼亚卢布尔雅那的事务所。OFIS 参与设计活动可以追溯到 19 世纪 90 年代，其工作范围包括建筑设计、城市设计、艺术设计和舞台设计。其项目类型包括了新建筑设计和旧建筑改造重建；其设计的项目包括社会住房、豪华别墅、体育设施、文化建筑和办公楼等类别。其代表项目包括位于斯洛文尼亚的蜂巢公寓、卢布尔雅那市立博物馆扩建、四季帐篷塔楼——梅赛德斯奔驰酒店设计方案等。

OFIS 赢得了多项国内和国际比赛，其多个项目曾得到密斯·凡·德罗奖提名。2009 年 OFIS 的足球场项目获得了国际奥林匹克委员会国际体育与休闲建筑协会银奖；2005 年的别墅扩展项目获得迈阿密双年展荣誉奖；2004 年的城市博物馆整修和扩建项目获得了《英国建筑评论》的高度好评；2000 年，OFIS 获得英国伦敦“年度青年建筑师”荣誉。

OFIS Arhitekti

OFIS Arhitekti is a firm of architects established in 1996 by Rok Oman and Špela Videčnik, both graduates of the Ljubljana School of Architecture and the London Architectural Association. Upon graduation they had already won several prominent competitions, such as Football Stadium Maribor and the Ljubljana City Museum extension and renovation. In 2001 they were awarded with the UK and Ireland's "Young architect of the year award".

The company is based in Ljubljana, Slovenia, but works internationally. They won a large business complex in Venice Marghera, Italy and a residential complex in Graz, Austria. However, it was by winning 180 apartments in Petit Ponts, Paris, their first large scale development abroad, which led them to open a branch office in France, 2007. This has been followed by a second large scale development with the construction of a football stadium for FC BATE in Borisov, Belarus, due for completion in 2012. They also have partner firm agreements in London, Paris and Moscow.

de Architekten Cie.

de Architekten Cie. 是一家全球性的建筑设计公司，具有 30 多年的建设设计和规划经验。其主要业务范围包括总体规划、城市规划、建筑设计和室内设计。

公司的主要创始人 Pi de Bruijn1967 年毕业于代尔夫特理工大学建筑学院，其后分别在伦敦萨瑟克区的伦敦市委员会建筑系和阿姆斯特丹市政房屋署工作，并于 1978 年成为 Oyevaar Van Gool De Bruijn 建筑事务所 BNA 办事处的合伙人。1988 年，Pi de Bruijn 和 Frits van Dongen、Carel Weeber、Jan Dirk Peereboom Voller 共同建立了 de Architekten Cie.。

de Architekten Cie.

Branimir Medić & Pero Puljiz, de Architekten Cie. is a global architectural design company with over 30 years of architectural design and planning experience. The business scope covers overall planning, urban planning, architectural design and interior design. Pi de Bruijn, the chief founder of the company, completed his studies at the Faculty of Architecture at Delft University of Technology in 1967. He then left to work at the Architects Department of the London City Council in Southwark, London. On his return to Amsterdam, he worked for the Municipal Housing Department, until he established himself as an independent architect in 1978, as a partner in the Oyevaar Van Gool De Bruijn Architecten BNA bureau. In 1988 he founded de Architekten Cie. together with Frits van Dongen, Carel Weeber and Jan Dirk Peereboom Voller, and has been a partner ever since.

日本 M.A.O. 一级建筑士事务所

日本 M.A.O. 一级建筑士事务所是一家国际化的建筑事务所，一个拥有创作理想的主流事务所，在城市综合体、旅游综合体、商业、办公、高端住宅、个性化开发项目等领域业绩显著，并逐渐成为此领域开发商的首选设计公司。M.A.O. 以其强大的创造力与全新的设计理念得到众多国内知名大型房地产企业及建筑业同行的一致认可，在业界享有很高的美誉度，同时也为业主赢得了广泛的社会效益和经济效益。

该事务所获得的奖项主要包括：时代楼盘上海主角荣获 2011“年度最佳写字楼”二等奖；罗浮·天赋荣登中国十大超级豪宅；荣获 Di·2011 年度最佳商业地产品牌设计机构奖；被评为 2005 年度 CIHAF 中国建筑十大品牌影响力规划建筑设计公司；被评为 2004 年度 CIHAF 中国建筑十大品牌影响力景观设计公司。

M.A.O.

M.A.O. is a Japan-based level architect firm. Established in Tokyo in 1994, M.A.O. was a professional firm led by Mr. KOTOKU MO, a famous architect in Japan and engaged in urban planning, architectural design and landscape design. It is a design unit with qualification of a legal person in Japan. M.A.O. owns a batch of high-level professional designers who boast excellent design strength and abundant overseas design experience. As of the beginning of 2010, the company had about 100 employees. The designers are divided into three parts among which about 10 designers are from the original architecture design team in Japan, about 60 designers form the architecture & planning team and about 30 designers form the landscape design team.

Relying on tremendous creativity and brand new design concepts, M.A.O. has obtained consistent recognition from many famous large domestic real estate enterprises and those of the same occupation in architecture industry and enjoyed high reputation in the industry. Meanwhile, M.A.O. has also created extensive social and economic benefits for the owners.

Studio Nicoletti 建筑事务所

Studio Nicoletti 建筑事务所建立于 1957 年，经过多年的发展，该事务所已成为意大利最古老、规模最大的建筑设计公司之一。作为可持续性和先进技术理念的先导，这个多样化的建筑事务所以世界级的专业设计和管理服务而著称，其业务范围包括建筑设计、高端项目建设和管理等综合建筑服务。

该事务所的主要作品有：普特拉贾亚滨水住宅楼、2012 韩国丽水世界博会大蓝鲸馆、哈萨克斯坦中央音乐厅、吉隆坡城市发展展览馆、马来西亚 Petaling Jaya 商业城、台北疾病控制中心、台北艺术表演中心、松兹瓦尔艺术中心、拉梅齐亚泰尔梅机场、马哈奇卡拉卫星城、阿雷佐法院、坎波巴索法院、阿布贾 Millennium 公园等项目。

Studio Nicoletti Associati

Studio Nicoletti Associati, founded in 1957, delivers world-class professional design and management services. Studio Nicoletti Associati offers award winning architectural design, highly ranked project and construction management and is known for his quality and professionalism.

As a pioneer in Sustainability and Advanced Technologies, this diversified practice offers comprehensive architectural services with a wide range of related disciplines, from development of master plans to furniture detail.

At over 30 professionals, it becomes one of the oldest and largest firms in Italy. It can deliver total services to a project and form flexible work teams that can integrate into a larger team in partnering and sub-consulting roles. S.N.A. practice's expanded through all major aspects of urban and building design in Italy, Europe, Africa, Asia, USA, and Middle East.

ECDM Architectes

ECDM Architectes 是一家通过建筑来定义居住环境的建筑设计事务所。ECDM 以社会的演变和转型为基础来实现其设计目标，并通过对当地自然环境和人文背景的分析来强化建筑的场所感和动态感。ECDM 将环境、景观、功能、技术以及当地居民的生活方式纳入考量范围， 通过严谨的逻辑思维，构建简洁、稳重、舒适的空间环境。

ECDM 的主要作品有：巴黎皇宫 VIII、de Ville-Bezons 酒店、法国 Zac Seguin 办公大楼、巴黎大众运输公司公交中心、法国残疾人运动场、图卢兹住宅等多类型项目。ECDM1993 年获得 Albums de la Jeune 建筑奖，并于 1996 年获得 Villa Médicis Hors les Murs 奖。

ECDM Architectes

Founded after being the recipient of the Albums de la Jeune Architecture award in 1993 and the Villa Médicis Hors les Murs – scholarship in 1996, Emmanuel Combarel and Dominique Marrec Architects (ECDM) is leading a work focused working on defining a living environment through an architectural project. The dynamics of the project emerges from a confrontation with the context, and a hierarchization of the problematics induced by the program and the site. Environmental quality, landscape, uses, ways of life, and technical choices are all structuring elements of the office's projects. The architecture that materializes the firm's approach is underpinned by the evolutions and mutations of society. It tends to be a simple, sober architecture following a rigorous logic, with no preconceptions, nostalgia or stylistic preoccupations.

ZNA 泽碧克建筑设计事务所

ZNA 泽碧克建筑设计事务所总部位于美国波士顿，在规划与建筑设计领域积累了 40 年的实践经验。事务所集优秀的设计创意及卓越的项目管理水平于一身，是一个在国际享有盛誉的知名设计公司。ZNA 泽碧克在各类规划与设计国际竞赛中屡获佳绩，项目类型跨越了从大型城市公共空间、混合功能综合体、超高层建筑综合体到度假居住、办公展览、教育科研等相关的规划、建筑、景观及室内设计。事务所曾为美国联邦政府、沙特阿拉伯王国政府、阿联酋王室、哈佛大学及其他知名商业、政府机构提供了各类重要项目的设计服务，所实施的项目遍及全球五大洲数十个国家。

ZEYBEKOGLU NAYMAN ASSOCIATES. INC

ZNA is a Boston based firm of Architecture, Urban Design, Master Planning and Interior Design. Responding to the need of excellence in both design and project management, with a proven design creativity and professional competence, ZNA presents credentials that reflect an exceptional and rare array of experience. With executed projects valued at over $10 billion, ZNA is a firm with national and international credits. ZNA projects span campus master plans, urban design, residential developments, and educational, commercial and institutional facilities. Clients have included the Federal Government, the Commonwealth of Massachusetts, Kingdom of Saudi Arabia, Harvard University, University of Sarajevo, and a roster of major institutions, commercial and industrial concerns.

澳大利亚 SDG 设计集团

澳大利亚 SDG 设计集团是澳大利亚建筑设计和景观规划行业非常专业化的设计公司，其在大型综合开发项目、旅游度假项目、居住环境项目、市政项目及公园与娱乐项目等方面所表现出的极富个性的创造力，得到了广泛的认可。

公司的优势体现在一体化综合设计能力上，从项目规划、建筑设计到环境设计、室内设计以及现场督导，一体化服务展现的活力有目共睹，创造了巨大的商业价值。 2002 年，公司在中国成立了一支由澳大利亚总部资深设计师与国内优秀设计人员组成的多元化设计团队，使总体设计充满了活力与创意，适应了不同地区、不同类型、不同文化背景的项目需求。

Shine Design Group Pty. Ltd.

Shine Design Group is a specialized Australian design company in industries of architectural design and landscape planning. The profound creativity that SDG shows in large comprehensive developments, resorts, residences, civic facilities, parks and entertainment projects has recieved extensive recognition.

The advantages of the firm are reflected on its integrated comprehensive design ability. From project planning, architectural design to environment design, interior design and then to field supervision, its integrated services show tremendous vitality and great commercial value. In 2002, SDG organized a diverse design team in China composed of senior designers from the Headquater and excellent designers from China. They work together to fill master plans with vitality and creativity, catering for different demands of projects of different regions, types and backgrounds.

SAMYN and PARTNERS, Architects & Engineers

SAMYN and PARTNERS，Architects & Engineers 是一家由 Ir Philippe SAMYN 博士领导的私人公司。随着其附属公司 Ingenieursbureau Jan MEIJER、FTI、DAE、AirSR 的相继建立，该公司也在建筑设计和建筑工程各领域表现得极为活跃。其业务范围涵盖了规划设计、城市规划、景观设计、建筑设计、室内设计、建筑物理、MEP 和工程结构、工程建设管理、成本规划与控制、工程造价管理等多方面。

SAMYN and PARTNERS，Architects & Engineers 的设计方案建立在"质疑"的基础上，可用"为什么"理念来概括。该公司尝试着接手各种类型的项目，并悉心听取客户的意见和需求。

SAMYN and PARTNERS, Architects & Engineers

SAMYN and PARTNERS, Architects & Engineers is a private company owned by its partners and lead by its design partner Dr Ir Philippe Samyn. With the establishment of its affiliated companies Ingenieursbureau Jan MEIJER, FTI, DAE, and AirSR, it is active in all fields of architecture and building engineering. The firm's client services include Planning and Programming, Urban Planning, Landscaping and Architectural Design, Interior Design, Building Physics, MEP and Structural Engineering, Project and Construction Management, Cost and Planning Control, Quantity Surveying, Safety and Health Coordination.

Philippe Samyn's architectural and engineering design approach is based on questioning, which can be summarized as a "why" methodology. The firm approaches projects openly to all sorts of possibilities whilst listening closely to its clients' demands.

Synthesis Design + Architecture

Synthesis Design + Architecture 是由 Alvin Huang 创建的一家新兴的现代设计公司，在建筑设计、基础设施建设、室内设计、装置设计、展览设计、家具设计等领域积累了多年的专业实践经验，其卓越的设计工作已获得国际认可。该团队由多学科的专业设计人员组成，包括注册设计师和建筑设计者，以及在美国、英国、丹麦、葡萄牙、中国台湾接受了专业性教育与培训的计算机专家。

Synthesis Design + Architecture 也是一家极具前瞻性的国际性设计公司，其设计统筹了性能、技术和工艺之间的关系，平衡了现实与想象之间的差距，以实际、实用的手法实现超凡的设计。

Synthesis Design + Architecture

Synthesis is an emerging contemporary design practice with collective professional experience in the fields of architecture, infrastructure, interiors, installations, exhibitions, furniture, and product design. The firm's work has already begun to achieve international recognition for its design excellence. Its diverse team of multidisciplinary design professionals includes registered architects, architectural designers and computational specialists educated, trained, and raised in the USA, UK, Denmark, Portugal, and Taiwan. This diverse cultural and disciplinary background has supported our expanding portfolio of international projects in the USA, UK, Russia, Thailand, and China.

北京世纪安泰建筑工程设计有限公司

北京世纪安泰建筑工程设计有限公司（SJAT）成立于 1994 年，隶属中国建筑学会，持有国家颁发的建筑工程甲级设计资质证书。公司的主要业务包括：建筑设计、城市规划设计、环境艺术设计；居住区规划设计、古建园林设计；室内装饰及设计；建设项目可行性研究、建筑工程开发咨询服务等。

世纪安泰对建筑设计与城市规划保持着深度思考，其主要作品包括：太原长丰商务区、太原高铁南站站前广场、唐山凤凰新城、北亚国际中心、青岛市北区综合培训中心、亚胜撒丁湾假日酒店及度假公寓、伯朗峰国际中心、首都图书馆二期、上海"欢乐谷"少儿活动馆等。其建筑设计作品屡获国际大奖，并受到委托方一致好评及业界的高度认可，荣获"中国最具影响力品牌设计机构"奖、"北京领袖建筑设计机构"等多项荣誉。

Beijing SJAT Architecture & Engineering Design Company

Beijing SJAT Architecture & Engineering Design Company is founded in 1994, subordinate to Architectural Society of China and holding the national Class A Construction Engineering Certificate. SJAT keeps in-depth thinking to architectural design and urban planning. Its architectural design works have won consistent praises from clients and high recognition in the filed of architecture. The Company has won "the Most Influential Design Organization in China", "the Leading Architectural Design Institution in Beijing" and so on.

Tony Owen Partners

2004 年，在商业建筑和住宅项目方面有着丰富经验的 NDM 与有设计天赋的 Tony Owen 合作，建立了 Tony Owen Partners。这是一家新兴的建筑事务所，其设计师将先进的设计与可持续性原则和商业价值结合起来，以构建挑战常规的、可实践的建筑。事务所采用最新的 2D CAD 技术，并致力于推进 3D 建模和可视化软件的发展，在建筑设计、城市规划和室内设计等领域获得发展。

其主要作品有：悉尼多佛海茨区的莫比斯住宅、Eliza 豪华公寓、比尔私人住宅、波士顿大学学生宿舍、哈雷戴维森总部、Paramount 酒店、Fractal 咖啡厅、奥斯陆歌剧院、阿布扎比女士俱乐部、NSW 教师联盟、波浪住宅、坎特伯雷城市中心等。

Tony Owen Partners

Tony Owen Partners was formed in 2004 when NDM, with 10 years of commercial and residential experience combined with the awarded design talents of Tony Owen. Since that time Tony Owen Partners has grown rapidly to be an emerging mid-sized practice focusing on an idea based approach to commercial projects. This combination combines a genuinely progressive approach to design with a firm with a strong track record in proven documentation, management and deliverability.

Tony Owen Partners offer Architectural, Interiors and Urban Planning services. Tony Owen Partners currently has a full time staff of 20 including 15 architectural staff, 3 interior designers and in house 3-D rendering facilities. Tony Owen Partners utilizes the latest 2D CAD technology and continues to push the boundaries in 3D Modelling and visualization software. Tony Owen Partners's core capabilities include: progressive design that sets projects apart; design based on sound commercial and sustainable principles; a proven track record in deliverability; proven results with authorities.

ZPLUS 普瑞思国际

ZPLUS 普瑞思国际注册于法国巴黎和中国天津两地，由欧洲资深建筑师、规划师和有留学背景的中国设计师组成。普瑞思以建筑设计专业为主干，可提供从项目策划到规划建筑设计再到工程实施监督的全程服务。其业务主要集中在高层办公、公寓都市综合体、中高档住宅区、大型专业社区、五星级酒店等高端产品上。目前，该公司已在高层建筑、科技园区、酒店、金属幕墙设计等领域积累了独特而宝贵的经验。

普瑞思在设计中旨在寻求商业与艺术精神的完美结合，创造价值、打造品牌、彰显魅力。频繁的国际合作与交流，使 ZPLUS 在引进国外先进理念、把握国际时尚元素和潮流方面，拥有得天独厚的优势。其主要作品有：ICTC、海河之子、赛顿中心、丽晶大厦、百合春天、水岸江南、华门名筑、天齐国际广场、五矿商务大厦、滨海浙商大厦、大岛商业街及办公楼等。

ZPLUS

ZPLUS, registered in Paris (France) and Tianjin (China), is composed of senior architects, planners and landscapers both from Europe and China, with international working background. Specialized in architectural design, ZPLUS also provides full service from project planning to architectural design and construction supervision. The business scope involves high-rise office building, urban apartment complex, middle & high grade residential area, large community and five-star hotels. Currently, ZPLUS has accumulated experience on the design of high-rise, science park, hotel and metal curtain wall.

ZPLUS aims to achieve a perfect combination of business and artistic spirit, creating value, building brand and exerting charm. Frequent international cooperation and communication enable ZPLUS to import advanced concept from abroad, capture international fashion elements and trend, and possess inherent advantages. The masterpieces of ZPLUS includes: ICTC, Son of Sea & River, Saidun Center, Lijing Tower, Lily Spring, Jiangnan Waterfront, Huamenmingzhu, Tianqi International Plaza, Minmetals Commercial Building, Waterfront Zhejiang Commercial Building, Dadao Commercial Street & Office Building.

三磊设计

三磊设计是一家拥有建筑工程甲级设计资质、城乡规划乙级资质、国家高新技术企业认证、ISO-9001 质量体系认证的综合性设计咨询服务机构。三磊设计注重团队合作，具备扎实的工程技术经验，可在公共建筑、都市与住区、结构设计与咨询、机电设计与咨询等领域为客户提供创造性和专业性的服务。

三磊善于对各个项目的功能、成本、市场、营销、文化、文脉、运营、可持续发展等因素进行整合思考，凭借专业能力，在业主利益、使用者利益与公共利益之间建立沟通桥梁，努力寻求完美的解决方案。为实现这一目标，三磊在所有项目的规划设计中建立了整合资源、以人为本、锐意创新和可持续发展的四项基本原则。

Sunlay Design

Sunlay Design is a comprehensive design consulting firm possessing Class A design qualification of construction engineering, Class B qualification of town & country planning, national qualification of high and new tech enterprise, ISO-9001 Quality System Certification. Sunlay Design emphasizes on team cooperation. With solid engineering technology, it can offer creative professional services for clients on public buildings, urban & residential area, structural design & consultation, electromechanical design & consultation and etc. Sunlay Design is good at integrating different factors at each project: cost, market, marketing, culture, context, operation, sustainable development and reaching comprehensive balance. By virtue of professional competence, it aims to build the communication bridge between the interests of the owners, users and public interests, and to seek the perfect solution. To achieve this goal, Sunlay Design establishes the four cardinal principles in the planning and design of all projects. They are integration of resources, to be people-oriented, innovation and sustainable development.

广州市纬纶建筑设计公司

广州市纬纶建筑设计公司成立于 1998 年，至今，该公司已形成了高素质、实力雄厚的设计团队。纬纶专注于房地产开发建筑设计，对房地产项目总体规划、市场状况、小区空间环境及住宅户型有专业的研究，可为客户提供商业、零售中心、公建、酒店、住宅、规划等多个领域的建筑设计服务。

纬纶以"设计无限可能"为核心理念，以精益求精的态度完成每一项设计作品，以精品的设计和优质的服务在客户群中树立了极佳的口碑。其主要作品有：合景泰富 · 科汇金谷大型商住项目、万科 · 天景花园、万科 · 红郡、美林湖畔别墅、美林海岸住宅小区、天伦 · 林和村城中村改造、天伦 · 湘潭隆平论坛、苏宁 · 天润城、佛山奥园超大型住宅小区、佛奥 · 星光广场商住项目、佛山宾馆等。

Win-land Architecture Design Co., Ltd.

Win-land Architecture Design Company was founded in 1998. It has established a high-quality, powerful design team which concentrates on real estate development and architectural design. They are specialized in real estate overall planning, research on market conditions, community space environment and housing type. The firm can offer architectural design services for commercial buildings, retail, public building, hotel, residence, planning and etc.

"Design Infinite Possibilities" as core concept, Win-land strives to perfectly complete each design work, striving to achieve good reputation by exquisite design and high-quality service. The masterpieces of its works include: KWG Property · International Creative Valley, Vanke · Tianjing Garden, Vanke · Sunshine City, Meilin Lake Villa, Meilin Waterfront Residential Community, Tianlun · Linhe Village Urban Village Rennovation, Tianlun · Xiangtan Longping Forum, Suning · Tianrun Town, Foshan Aoyuan ultra large residential community, Foao · Star Plaza residential & commercial project, Foshan Hotel and etc.

加拿大 CPC 建筑设计顾问有限公司

加拿大 CPC 建筑设计顾问有限公司于 1994 年在加拿大温哥华成立，1995 年公司开始进入中国市场，为政府及房地产开发商提供城市规划、高端住宅设计、商业项目策划及设计、公共建筑设计、景观设计、室内设计等全方位的设计服务。随着工程项目的日益增加，2001 年 CPC 公司将其中国办事处设置于上海，以便更好地为业主提供即时的服务。

其主要作品有：深圳国际会展中心、杭州文化广场、苏州金鸡湖酒店、大连大学城、北京万柳购物中心、外滩 15#、上海嘉定新城、上海浦江智谷工业园区规划及建筑设计、上海宝山顾村规划等。

其获得的主要奖项和荣誉包括：2012 年梅陇镇西地块配套商品房荣获"房型设计奖"； 2010 年都江堰友爱学校重建项目荣获年度工程勘察设计"四优"一等奖；2005 年"古北瑞仕花园"项目荣获全国人居经典建筑规划设计方案竞赛"规划、建筑双金奖"。

Coast Palisade Consulting Group

The Coast Palisade Consulting Group (C.P.C. Group) is an architectural design firm operating in Vancouver and Shanghai. The firm specializes in architectural design, planning, urban design and interior design. Most members of the firm have more than one university degree in architecture or related fields such as planning and urban design.

The C.P.C. Group offers clients creative design solutions that are market and user responsive and feasible to construct from both a technical and budget perspective. The company consists of a core group of long-term staff and a large pool of designers, planners, computer technologists, model makers, etc.

10 DESIGN（拾稼设计）

10 DESIGN（拾稼设计）是一家国际性的建筑设计事务所，其设计中心分别设于中国香港、上海、英国爱丁堡及美国丹佛。事务所的业务范围包括总体规划、建筑设计、建筑可持续性的研究、景观园林设计及室内设计。其作品涵盖大规模的城市规划、综合性功能开发、公共文化建筑、度假及酒店设施、企业总部、科研及办公设计及高端住宅设计等类型。

艾高登（Gordon Affleck）是拾稼设计的创始人之一，拥有超过 18 年在亚洲、英国、美国和中东等地带领国际设计团队的工作经验。他深信绿色低碳的可持续性技术与建筑的强力结合将成为建筑设计及与客户建立紧密合作关系的核心原则，这也是确保每个项目成功的关键。艾高登已建成的地标性作品有：北京 2008 奥运会多用途场馆、关联场馆设计（现为中国国家会议中心）、南京华为研发总部园区等。

10 DESIGN

10 DESIGN is a leading International partnership of Architects, Urbanists, Landscape Designers and Animators. It works at all scales and sectors, including corporate, cultural, hospitality, retail, education and residential. 10 DESIGN's Architecture evolves from the multicultural nature of clients, and is driven by their respective social, economic and ecological conditions. 10 DESIGN has offices in Hong Kong, Edinburgh, Shanghai, Istanbul and Dubai. The current projects include the Central Business District in Chongqing, China; the Fujian Photonic University in Xiamen, China; Badshahpur Business Park in Gurgaon, India; Mavişehir Residential Development in Izmir; and the Route De Jussy 22 Thônex Project in Geneva, Switzerland.

Projects under construction include Chengtou New Jiangwan Business Park in Shanghai, China; Bohai Bank Headquarters in Tianjin, China; and the Fujian Photonic University in Xiamen, China.

美国 KDG 建筑设计咨询有限公司

美国 KDG 建筑设计咨询有限公司是集建筑设计、景观设计、城市规划、地产咨询等业务为一体的综合性设计机构。自 1993 年在美国洛杉机地区成立以来，其作品受到中美地产界和地方政府机构的普遍赞誉。KDG 信奉"设计是创造价值的重要手段"，领先的设计理念和严谨的设计作风是其核心竞争力。

Kalarch Design Group

Kalarch Design Group is a multi-disciplinary design firm offering services in four core areas of practice: Architectural Design, Landscape Architecture, Urban Design & Planning, Development Consultation. Founded in 1993 in Los Angeles, California, KDG has developed into an experienced and respected practice among clients in China and US. KDG believes that "Design Creates Value". Leading design concept and rigorous design style are the core competitiveness of KDG.

M.S.B Arquitectura e Planeamento

M.S.B Arquitectura e Planeamento 是一家年轻的建筑设计事务所，于 2004 年由 Miguel Mallaguerra、Susana Jesus 和 Bruno Martins 建立于丰沙尔。MSB 的设计师在项目设计上有丰富的经验，在建筑设计、城市规划和室内设计等领域颇有研究。近几年该事务所在住宅、酒店、设施、城市开发领域设计的作品展现了同时期建筑的魅力，在国内外比赛中受到认可。

事务所主张向世界开放的哲学，这一思想不仅表现在它们的沟通策略上，同时，也通过其与葡萄牙本国及国外伙伴的合作中彰显出来。在设计工作中，事务所十分注重多学科之间的团队合作，并将此理念视为成功的关键。

M.S.B Arquitectura e Planeamento

M.S.B Arquitectura e Planeamento, as a young workgroup, was established in Funchal in 2004 by architects Miguel Mallaguerra, Susana Jesus and Bruno Martins. The MSBs are experienced in producing projects and studies directed to the areas of architecture, urban planning and interior design. The works produced inside this young workgroup reveal the fascination by contemporaneous architecture accompanying its time, having developed over recent years a number of projects in the area of housing, hospitality, equipment and urban development, noted for its quality recognized by the victories achieved in national and international competitions. The office maintains an open philosophy to the world, both in its communication policy, but also in how it works with its partners inside and outside of Portugal, privileging the work in multidisciplinary teams, and promoting ideas as the crucial factor in their projects.

Richard Kirk Architect

Richard Kirk Architect 以设计尊重环境和满足人类需求的不朽建筑为目标。该事务所侧重于设计不同类型、不同规模的标志性项目，这些项目极大地考验了事务所的设计能力。在设计过程中，其设计团队会综合考虑建筑的定位、选址、功能、建筑与环境的关系等多项因素。同时，事务所也十分注重细节设计，以对当地人的需求作出回应。事务所以"合作"为企业文化，将团队凝为一个活跃的、极具创造力的整体。

Richard Kirk Architect

Richard Kirk Architect aims to design buildings of enduring beauty that are confident in their environment and respectful to the needs of the communities they serve. The focus is to design sensitive and landmark projects of great diversity and scale that demand a challenging response which only comes from a great design capability. Richard Kirk Architect thinks deeply about the role that a building is to play before starts imagining its future, carefully considering where it sits, how it functions and its relationships within its context. Also, the firm pays attention to detail design, responding to demands of the locals. "Cooperation" as essence of corporate culture devotes to turn the design team into an active creative whole.